CEREALS FOR FOOD AND BEVERAGES

RECENT PROGRESS IN CEREAL CHEMISTRY AND TECHNOLOGY

ACADEMIC PRESS RAPID MANUSCRIPT REPRODUCTION

Proceedings of an International Conference on Cereals for Food And Beverages, Held in Copenhagen, Denmark, August 13–17, 1979

CEREALS FOR FOOD AND BEVERAGES

RECENT PROGRESS IN CEREAL CHEMISTRY AND TECHNOLOGY

edited by

GEORGE E. INGLETT
U.S. Department of Agriculture
Science and Educational Administration
Northern Regional Center
Peoria, Illinois

LARS MUNCK
Department of Biotechnology
Carlsberg Research Center
Copenhagen, Denmark

ACADEMIC PRESS
A Subsidiary of Harcourt Brace Jovanovich, Publishers
New York London Sydney Toronto San Francisco 1980

ACADEMIC PRESS, INC.
111 Fifth Avenue, New York, New York 10003

Library of Congress Cataloging in Publication Data

Conference on Cereals for Food and Beverages, Carlsberg
Reference Center, 1979.
Cereals for Food and Beverages.

Includes index.
1; Cereals as food--Congresses. I. Inglett, G. E.,
Date II. Munck, Lars. III. Title.
TX557.C66 1979 641.3'31 80-10981
ISBN 0-12-370960-1

PRINTED IN THE UNITED STATES OF AMERICA

80 81 82 83 9 8 7 6 5 4 3 2 1

Contents

Contributors

Ahrenst-Larsen, Bent (365), Department of Brewing Chemistry, Carlsberg Research Center, DK-2500 Valby, Copenhagen, Denmark

Anderson, James W. (447), Veterans Administration Hospital, Lexington, Kentucky 40507

Berndorfer-Kraszner, Éva (429), Department of Biochemistry and Food Technology, Technical University of Budapest, Budapest, Hungary

Bizot, Hervé (97), Laboratoire de Biophysique des Aliments, Institut National de la Recherche Agronomique, 44072 Nantes Cedex, France

Bos, Cornelis (75), DDS Krøyer A/S, DK-2100, Copenhagen, Denmark

Bosveld, Pieter (171), Institute for Cereals, Flour and Bread TNO, Wageningen, The Netherlands

Brandt, Anders (339), Department of Physiology, Carlsberg Research Center, DK-2500 Valby, Copenhagen, Denmark

Cameron-Mills, Verena (339), Department of Physiology, Carlsberg Research Center, DK-2500 Valby, Copenhagen, Denmark

Daley, Laura H. (275), Biochemistry Department, Miles Laboratories, Inc., Elkhart, Indiana 46514

Earp, C. F. (513), Cereal Quality Laboratory, Texas A & M University, College Station, Texas 77843

Erdal, Kenneth (365), Department of Brewing Chemistry, Carlsberg Research Center, DK-2500 Valby, Copenhagen, Denmark

Eriksen, Svend A. (395), Bioteknisk Institut, DK-6000 Kolding, Denmark

Fagerson, Irving S. (395), Department of Food Science and Nutrition, University of Massachusetts, Amherst, Massachusetts 01002

Feil, Carol (27), Department of Biotechnology, Carlsberg Research Center, DK-2500 Valby, Copenhagen, Denmark

Feillet, Pierre (183), Laboratoire de Technologie des Céréales, Institut National de la Recherche Agronomique, 9 Place Viala, 34060 Montpellier Cedex, France

Fujita, Akio (381), Central Research Laboratory, Nisshin Flour Milling Co., Ltd., Saitama, Japan

Fulcher, R. G. (1), Canadian Department of Agriculture, Ottawa Research Station, Ottawa, K1A 0C6, Canada

Gibbons, Gregory C. (27), Department of Biotechnology, Carlsberg Research Laboratory, DK-2500 Valby, Copenhagen, Denmark

Gould, Max R. (447), Quaker Oats Company, Barrington, Illinois 60010

Graveland, Aris (171), Institute for Cereals, Flour and Bread TNO, Wageningen, The Netherlands

Hemmingsen, Steen Helde (61), Novo Research Institute, DK-2880 Bagsvaerd, Denmark

Hood, Lamartine F. (41), Department and Institute of Food Science, Cornell University, Ithaca, New York 14853

Hoseney, R. C. (461), Department of Grain Science and Industry, Kansas State University, Manhattan, Kansas 66506

Hug, H. (287), Swiss Breweries Research Station, 8059 Zurich, Switzerland

Huszár, Magda (429), Department of Biochemistry and Food Technology, Technical University of Budapest, Budapest, Hungary

Ingversen, John (339), Department of Physiology, Carlsberg Research Center, DK-2500 Valby, Copenhagen, Denmark

Jende-Strid, B. (365), Department of Physiology, Carlsberg Research Center, DK-2500 Valby, Copenhagen, Denmark

Juliano, Bienvenido O. (409), Chemistry Department, The International Rice Research Institute, Los Baños, Laguna, Philippines

Khan, M. N. (513), Cereal Quality Laboratory, Texas A & M University, College Station, Texas 77843

Landry, Jacques (255), Laboratoire d'Etude des Proteines, Institut National de la Recherche Agronomique, Versailles 78000, France

Larsson, Kåre (121), Department of Food Technology, University of Lund, Lund, Sweden

Lásztity, Radomir (429), Department of Biochemistry and Food Technology, Technical University of Budapest, Budapest, Hungary

Lichtendonk, Willem J. (171), Institute for Cereals, Flour and Bread TNO, Wageningen, The Netherlands

Mossé, Jacques (255), Laboratoire d'Etude des Proteines, Institut National de la Recherche Agronomique, Versailles 78000, France

Multon, Jean-Louis (97), Laboratorie de Biophysique des Aliments, Institut National de la Recherche Agronomique, 44072 Nantes Cedex, France

Munck, Lars (27), Department of Biotechnology, Carlsberg Research Center, DK-2500 Valby, Copenhagen, Denmark

Norman, Barrie E. (61), Novo Research Institute, DK-2800 Bagsvaerd, Denmark

Ogawa, Gengo (381), Central Research Laboratory, Nisshin Flour Milling Co., Ltd., Saitama, Japan

O'Mahony, Sean (447), Nabisco Company, Fairlawn, New Jersey 07410

Palmer, G. H. (301), Department of Brewing and Biological Sciences, Heriot–Watt University, Edinburgh, Scotland

Pfenninger, H. (287), Swiss Breweries Research Station, Zurich, Switzerland

Phillips, R. D. (275), Department of Food Science, University of Georiga College of Agriculture, Athens, Georgia 30602

Pomeranz, Y. (201), U.S. Grain Marketing Research Laboratory, Science and Education Administration, Agricultural Research, U.S. Department of Agriculture, Manhattan, Kansas 66502

Rooney, L. W. (513), Cereal Quality Laboratory, Texas A & M University, College Station, Texas 77843

Sarkki, Marja-Leena (155), Raision Tehtaat, SF-21200 Raisio, Finland

Saunders, R. M. (137), Western Regional Research Center, Science and Education Administration, Agricultural Research, U.S. Department of Agriculture, Albany, California 94710

Savet, Bernadette (97), Foss–Electric, 75009 Paris, France

Sternberg, M. (275), Biochemistry Department, Miles Laboratories, Inc., Elkhart, Indiana 46514

Tsen, Cho C. (245), Department of Grain Science and Industry, Kansas State University, Manhattan, Kansas 66506

Varriano-Marston, E. (461), Department of Grain Science and Industry, Kansas State University, Manhattan, Kansas 66506

Weipert, D. (495), Bundesforschungsanstalt für Getreide- und Kartoffelverarbeitung, Institut für Müllereitechnologie, Postfach 23, D-4930 Detmold, West Germany

Wingfield, John (233), Department of Grain Science and Industry, Kansas State University, Manhattan, Kansas 66506

Wong, S. I. (1), Canadian Department of Agriculture, Ottawa Research Station, Ottawa, K1A 0C6, Canada

Zwingelberg, H. (495), Bundesforschungsanstalt für Getreide- und Kartoffelverarbeitung, Institut für Müllereitechnologie, Postfach 23, D-4930 Detmold, West Germany

Preface

Cereals are grown and consumed worldwide in various foods and beverages. The chemistry and technology of the major cereals related to their usage in foods and beverages are covered in this book, which consists of the proceedings of an international conference on the subject held in Copenhagen, Denmark, August 13–17, 1979. Recent progress in cereal chemistry and technology is extensively reviewed.

Cereals in this volume include wheat, maize, barley, oats, rye, sorghum, rice, and millet. The chemistry of wheat is reviewed as related to water activity, particle analysis, dietary fiber, proteins, and properties in breadmaking. The technology of wheat is covered in a milling chapter on bread and soft wheats. Maize science includes a new protein concentrate, starch and protein chemistry. Maize technology covers the recent progress in sugar production by enzymes from starch, germ products in baked foods, and utilization in brewing.

Barley studies include its morphology and physiology in malting, proanthrocyanidin-free barley in beer, and the basic science of hordein. Chemistry and technology of oats are covered in two chapters, followed by chapters on sorghum, rice, millet, soy sauce production, and hydrolyzed vegetable proteins.

This book should be a useful reference for students, scientists, technologists, and executives who are involved in any facet of food and beverage production.

INSIDE CEREALS - A FLUORESCENCE MICROCHEMICAL VIEW

R.G. Fulcher
S.I. Wong

Ottawa Research Station
Agriculture Canada
Ottawa, Ontario
Canada

I. INTRODUCTION

Mature cereal grains are complex biological systems. They contain an awesome array of biochemical constituents, from the major reserves of structural and storage carbohydrates, proteins and lipids, through numerous other amines, lignins, waxes, sterols, phytin, nucleic acids and enzymes. Moreover, these constituents are synthesized, packaged, and stored in different tissues and vary considerably in concentration or chemical and morphological form depending on their genetic background and the environmental conditions in which the plants were grown. In short, the cereal grain is differentiated *morphologically* into compartments which also display considerable *chemical* variation. As these differences ultimately determine the physiological, nutritional and processing characteristics of cereals, it is important to establish the extent of this variation within seeds in order to provide a sound basis for further improvement.

Unfortunately, mature grains are among the most difficult of biological materials to examine microscopically. Indeed, until little more than a decade ago, it was considered impossible to obtain high resolution micrographs of mature grain tissues (McLeod *et al.* 1964). This problem has been largely overcome by the introduction of improved fixation methods and low viscosity embedding resins for both light

ISBN: 0-12-370960-1

(Feder and O'Brien 1968) and transmission electron microscopy (Spurr 1969, Pease 1973). Scanning electron microscopy is also an invaluable tool for high resolution examination of internal seed morphology.

Despite high resolution preparation methods, there is a paucity of specific microscopic markers for many grain constituents. Scanning electron microscopy provides superb morphological information (e.g. surface organization, starch and protein matrices, etc.), but provides limited chemical data. Likewise, transmission electron microscopes allow maximum resolution of component relationships but the requirement for electron-dense stains has limited development of specific cellular tags. Even bright-field (light) microscopy is limited to a few chemically specific reagents of sufficient optical density for visible contrast and high resolution in thin sections. Nonetheless, all of these microscopic techniques will continue to assist in unravelling the internal complexities of cereal grains.

Fluorescence is the property of many organic compounds by which high intensity (short wavelength) incident illumination is absorbed by the molecules and re-emitted as fluorescent (lower intensity, longer wavelength) light. Using this principle, the fluorescence microscope combines many of the advantages of other microscopic techniques, with few of the disadvantages. It is a simple modification of the bright-field microscope, yet the fluorescence mode permits greater chemical specificity and sensitivity than conventional systems (as little as 10^{-18} moles of fluorescent substance can be detected by microspectrofluorometry -- von Sengbusch and Thaer 1973). In view of its sensitivity, it is surprising that the instrument has been used only occasionally for cereal studies, usually for special applications such as immunofluorescence (Jacobsen and Knox 1971, Barlow *et al.* 1973a,b) and autofluorescence and microspectrofluorometry (Fulcher *et al.* 1972a). Fluorescence microscopy also allows high resolution, rapid sample preparation, and may be used in combination with conventional bright-field techniques (phase-contrast, polarizing optics) for added flexibility. These advantages imply an expanded role for fluorescence microscopy in cereal analysis.

Recently we have established a microchemical facility for the purpose of defining more precisely for breeding programs (and for industry) some of the physiological and structural interactions within cereal grains which influence grain quality and performance. Occasionally we employ conventional light microscope staining methods for assessing variation within and between cultivars, but mostly we rely on a fluorescence microscope equipped with a high intensity mercury arc lamp and epi-illuminating system. The instrument is used in conjunction with simple sectioning techniques and a range of fluorescence methods for visualizing major grain reserves. Examples of several useful techniques are described briefly, and their advantages over conventional microscopy are outlined. Special topics such as immunofluorescence and microspectrofluorometry are beyond the scope of this paper, but must be considered as important extensions of fluorescence microchemical analysis.

II. MATERIALS AND METHODS

A. The Fluorescence Microscope

This microscope is one of the most sensitive chemical instruments available for cereal analysis, and it is also one of the simplest. A suitable instrument consists of a bright-field microscope with a high intensity illuminator and filter systems for (a) modifying the wavelength of excitation (exciter filters) and (b) eliminating unwanted illumination from the fluorescent image (barrier filters). Exciter filters are inserted between the illuminator and the specimen to maximize the excitation of the fluorescent compounds under investigation and transmit in the ultraviolet, blue or green regions of the spectrum. Barrier filters are inserted between the specimen and the detector to remove wavelengths (including excitation) shorter than those of the induced fluorescence. Thus the fluorescent image is viewed on a dark or black background and the high contrast provides considerable sensitivity. A wide range of filter systems is available for diverse applications.

It is preferable to equip a fluorescence microscope with an epi- (or incident-) illuminating system. The epi-illuminator collects the excitation beam and reflects it (by dichroic mirrors) through the microscope objective to the *upper* surface of the specimen (conventional sub-stage condensers illuminate the lower surface of the specimen). It is this modification which has dramatically increased the efficiency of fluorescence microscopes in recent years, for several reasons. First, because the epi-illuminator excites the top surface of the specimen, earlier problems with thick specimens (undesirable diffusion and absorption within the specimen and hence loss of intensity and resolution) have been minimized. Second, excitation occurs only in the area of the specimen being examined, and fading of some fluorochromes due to prolonged exposure to the excitation source is restricted to the area in view. (Sub-stage condensers illuminate large areas of the sample.) Third, epi-illuminators dramatically increase the excitation intensity at the specimen surface with each increment in objective power. Therefore, fluorescence intensity (and hence sensitivity) increases as resolving power increases. This "intensity factor" of epi-illuminators has permitted streamlining of fluorescence microscopes with lower intensity (50 Watt), cooler, lighter, cheaper illuminators which are mounted closer to the specimen and provide the same intensity as larger (200 Watt) sources.

Several high quality fluorescence microscopes are available commercially and include a range of modifications to suit diverse purposes. We employ a Zeiss Universal Research Microscope equipped with a IIIRS epi-illuminating condenser, a 200 Watt mercury-arc illuminator, and a sub-stage brightfield illuminator and condenser for additional flexibility. The instrument is simple to use and provides adequate intensity for detection of most fluorochromes.

The most important element in the fluorescence microscope is the exciter/barrier filter system, many combinations of which are available. Excitation filters are selected to approximate the excitation wavelengths of the fluorescing substances being examined. Their transmission curves (maximum between 360 and 560 nm) are generally broad and each may be used for several fluorochromes. Barrier filters also exhibit broad transmission curves (with steep cut-offs) and the lower limits of transmission may range from 330 to 590 nm. Monochromators are available for more precise fluorescence analysis. For special applications such as fluorescence immunochemistry or detection of a limited number of components, a single filter combination (with appropriate dichroic mirror) may be adequate. For our purpose of devising fluorogenic reactions for several cereal components, we employ three standard filter sets (*Table I*). They provide excitation in the ultraviolet, blue, and green regions of the spectrum and are referred to as FC I, II and III in the text.

Other factors affect the quality of the fluorescence image. For example, fluorite objectives (e.g. Zeiss Neofluor) transmit a high percentage of incident illumination and are preferred for routine work (apochromatic and achromatic objectives are color-corrected and absorb some of the excitation wavelengths). Many commercial mounting media degrade the image by adding non-specific fluorescence to the preparation -- a non-fluorescent immersion oil is essential.

Consistent with improvements in fluorescence microscope design (e.g. epi-illuminators), several standard photographic films are suitable for fluorescence recording. High speed

TABLE I. Spectral Characteristics of Fluorescence Filter Combinations

Combination	*Exciter Filter*	*Barrier Filter*
(FC)	*maximum transmission (nm)*	*maximum transmission (nm)*
I	*365*	$\geq$ *418*
II	*450-490*	$\geq$ *520*
III	*546*	$\geq$ *590*

(ASA approximately 400) color, and black and white, films are adequate for most applications in 35 mm format or larger. Suitable exposures are obtained in less than 60 seconds (compared to several minutes with slower films).

The instrument which we have described briefly is designed for routine applications. Microspectrofluorometric equipment is also available for quantitative work and rapid growth in other areas of fluorescence chemistry (laser excitation, infra-red fluorescence) may soon allow further modifications for fluorescence microscopy.

B. *Preparation of Samples for Fluorescence Microscopy*

Methods for sample preparation may be as varied as the morphological and chemical entities under investigation. Mature cereal grains, like most biological materials, must be sectioned prior to microscopic examination and careful consideration should be given to selecting methods for retaining, *in situ*, whatever molecular species are of interest. In some instances (e.g. detection of lipids) certain solvent systems must be avoided to prevent extraction of reserves if tissues are to be plastic-embedded and thinly sectioned for high resolution microscopy. In other cases (e.g. detection of aromatic amines), it is desirable to avoid *all* embedding procedures and to conduct fluorogenic reactions on unprocessed sections in the vapour phase to ensure retention of the deposits.

In view of these requirements, we have adopted a limited range of preparative techniques for routine use such that most major cereal components may be localized in mature grains after a minimum of manipulation. The methods are complimentary and involve two simple techniques: (a) hand-sectioning for low resolution and rapid scanning, and (b) glycol methacrylate (hydroxyethyl methacrylate) embedding and sectioning for high resolution examination. Other preparative methods may be equally useful (e.g. epoxy resin embedding, cryostat sectioning), but in our experience hand sections and methacrylate-embedded sections are jointly very compatible with resolution and fluorescent staining requirements.

1. Hand Sectioning for Rapid Scanning. With the widespread availability of epi-illuminating fluorescence condensers, improved cytochemical resolution can be achieved with relatively thick (10-50 µm) sections (in contrast to substage illuminators which yield low intensity and low resolution images with all but the thinnest sections). Hand sections are obtained by simply cutting the thinnest possible sections from intact seeds with a sharp, cleaned (acetone or xylene) double-edged razor blade. Occasionally, it may be desirable to slightly hydrate particularly hard grains before sectioning, but a minimum of practise can result in routine production of 10-20 µm thick sections. After cutting, the sections are viewed in a drop of immersion oil under a cover glass on a microscope slide to detect primary fluorescence, or they may be manipulated through a wide range of non-destructive staining procedures. Hand sections impart the obvious advantage of providing samples which have not been modified or extracted by dehydrating and plastic embedding chemicals. Furthermore, many fluorescent reagents can be applied directly to slide mounted sections without washing or additional manipulation. It is frequently possible to observe the distribution of particular compounds within 30-60 seconds, and a large number of samples can be analyzed in a few hours. Hand sections are also useful for comparison with plastic-embedded sections to determine whether the embedding procedures result in significant losses of compounds (e.g. standard glycol methacrylate embedment removes 60-70% of nicotinic acid from wheat bran).

2. Plastic (Glycol Methacrylate) Embedding for High Resolution Microscopy. For high resolution work, it is desirable to examine sections which are 0.1-2.0 µm thick. This is less than the minimum dimension of many cereal grain structures (e.g. endosperm protein bodies, aleurone grain inclusions, lipid droplets, starch grains) and their internal morphology or spatial relationships may be analyzed with considerable precision.

Glycol methacrylate (GMA) is admirably suited to producing

sections of the requisite thickness: (a) its low viscosity in the unpolymerized state permits complete penetration into most cereal grain structures; (b) polymerized blocks of GMA-infiltrated tissue are easily sectioned on glass knives; (c) the sections are hydrophilic and permeable to most histochemical dyes, reagents, and enzymes (features not shared by epoxy resins), permitting a wide range of chemical reactions to occur in the sections; (d) the sections are permanent. Typically, we process tissues through two GMA-based media:

a. *Standard GMA embedment*. This is essentially the procedure described by Feder and O'Brien (1968). Briefly, 1-2 µm thick slices of cereal grains are placed in aldehyde fixative for 24-48 hr at 0-4°C, dehydrated 12-24 hr each in a methyl cellosolve, ethanol, *n*-propanol, *n*-butanol series at 0-4°C and infiltrated with GMA monomer mixture at room temperature for 3-5 days. Tissues and fresh GMA are then placed in gelatine capsules or other suitable air-tight transparent containers and polymerized to hardness in an oven at 60°C or under UV light (366 µm) at room temperature. Polymerized tissue blocks may be stored indefinitely or sections are cut on an ultramicrotome using glass knives. Sections floated on drops of water on glass slides are dried down gently, and are then ready for cytochemical treatment.

Because of the dense structure of mature grains, it is important that tissues remain in the fixative, dehydration, and GMA monomer solutions for a sufficient period to permit complete infiltration. This will vary with different tissues and may be reduced dramatically for developing or germinating grains, or for vegetative tissues, but the times we have outlined are acceptable for most fractions.

Similarly, the composition of the initial fixative solution for preserving cellular integrity may be varied considerably. Glutaraldehyde (3 to 6% in 0.025 to 0.1 M phosphate buffer pH 6.8-7.2) yields excellent preservation of cellular detail and may be improved by the addition of acrolein (1-2%) and/or formaldehyde (2-4%). Aldehyde fixation reduces protein and amine staining somewhat, but this is seldom a problem in view of the extreme sensitivity of fluorescence assay. Oxidizing fixatives such as osmium tetroxide and potassium permanganate seriously reduce primary fluorescence or interfere with fluorogenic reactions and should be avoided.

b. *Modified GMA embedment for retention of lipids*. While standard GMA embedment is satisfactory for retaining major grain components *in situ*, most lipids (with the exception of those in the seed cuticle) are lost during processing.

However, hexane-soluble lipids may be preserved by employing the glycol methacrylate-glutaraldehyde-urea resin mixture described by Pease (1973). Initially, tissues are fixed in 4% glutaraldehyde in 0.025 M phosphate buffer for 4-24 hr (pH 6.8-7.2), taken stepwise through increasing concentrations of neutralized glutaraldehyde and infiltrated with 50% glutaraldehyde before final infiltration in the modified GMA mixture. Pease's (1973) protocol may be followed closely although mature grains require prolonged infiltration times in each solution (up to 24 hr). Adjusting the water content in the final resin mixture to 35% before polymerization produces blocks which are more easily sectioned.

The processes outlined above (hand-sectioning, GMA, and GMA-glutaraldehyde-urea embedding) are relatively simple techniques which can be applied routinely for grain microscopy. Sections prepared by these methods are amenable to a wide range of microchemical procedures, the resolution of components is generally high, and embedded sections provide a permanent record for continuing comparisons. With little modification, the procedures are also applicable to other types of samples including vegetative tissues and manufactured products. Flour samples require little preparation and may be examined directly on microscope slides after applying one or more of the following fluorescent staining procedures.

C. *Applications*

A few cereal constituents exhibit *primary* (or auto-) fluorescence of sufficient intensity to be detected microscopically without further enhancement. Most, however, are relatively non-fluorescent and *secondary* fluorescence must be induced using fluorogenic reactions or by specifically labelling with fluorescent molecules. Often, both approaches may be combined to considerable advantage. The following is a brief description of several simple fluorescence methods which may be used to localize most major grain constituents

FIGURE 1. Diagram of the anatomy of an Avena sativa L. grain showing the tissue relationships in longitudinal (left) and cross section (lower right). Enlargements show cellular relationships in the bran (A), central endosperm (B) and germ-endosperm interface (C) and are representative of the regions shown in subsequent micrographs. The morphology of protein and starch deposits in the endosperm varies considerably in wheat, oats and barley; otherwise the tissue relationships are similar.

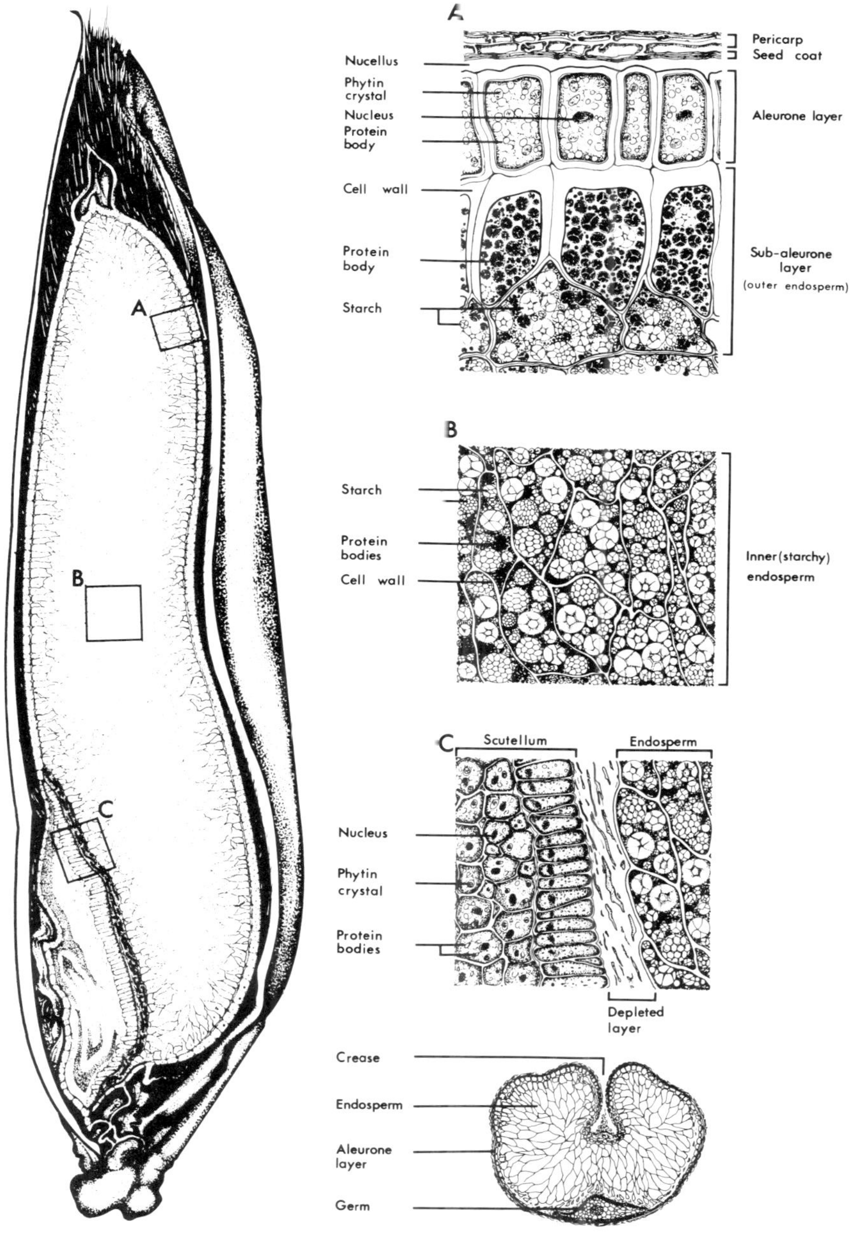
A
B
C
A
Pericarp
Seed coat
Nucellus
Phytin crystal
Nucleus
Protein body
Aleurone layer
Cell wall
Protein body
Starch
Sub-aleurone layer
(outer endosperm)
B
Starch
Protein bodies
Cell wall
Inner (starchy) endosperm
C
Scutellum
Endosperm
Nucleus
Phytin crystal
Protein bodies
Depleted layer
Crease
Endosperm
Aleurone layer
Germ

1.

in situ or in flours and processed material. Examples include tissues derived from wheat (*Triticum aestivum* L. cv Fredrick), barley (*Hordeum vulgare* L. cv Betzes, Himalaya, and Vanier), oats (*Avena sativa* L. cv Hinoat, and low protein or high oil breeding lines from Ottawa Research Station), and sorghum (*S.vulgare* Pers.). Illustrations include both hand sections and GMA sections (see previous section). *Figure 1* is a view of the internal organization of an oat grain and is included for reference. Wheat and barley grains are similar in general organization and terminology is used interchangably. However, all cereals show sub-cellular and chemical differences which are elaborated in the remaining figures.

1. *Phenolic Compounds.* Many phenolic compounds autofluoresce strongly in the blue region of the spectrum. Therefore, cereal structures which contain phenols often can be examined directly without added fluorochromes.

a. *Lignin.* Lignin deposits are characteristic of most outer surfaces of higher plant organs. Both leaf (*Figure 2*) and seed surfaces (*Figure 2b*) of cereals are extensively lignified and guard cells, trichomes and epidermal or pericarp cells are fluorescent under short wavelength excitation (FC I). Specimens are mounted in a few drops of oil under a cover glass or fluorescence may be enhanced by mounting in an alkaline medium (e.g. 0.07 M K_3PO_4, pH 10).

b. *Phenolic acids.* The intense blue autofluorescence of the aleurone cell walls of wheat (*Figure 3*) is due primarily to high concentrations of ferulic acid (Fulcher *et al* 1972a). This characteristic fluorescence also occurs in the aleurone walls of other cereals (*Figure 4*) and in the scutellum (Fulcher *et al* 1972a). Considerable use has also been made of ferulate fluorescence in following cell wall formation in developing wheat endosperm (Morrison *et al* 1975) and in other grass structures (Harris and Hartley 1976). The fluorescence is readily distinguishable from other emissions (e.g. lignin) using FC I and undergoes characteristic pH shifts in alkaline conditions (Fulcher *et al* 1972a). Microscopic analysis of autofluorescence is particularly useful in detecting pericarp, aleurone and scutellar contaminants in flours and for assessing the degree of disruption of these tissues during processing.

c. *Phenolic amines.* *Ortho*-aminophenol and *o*-aminophenyl glucose have been identified as constituents of wheat bran by Mason *et al* (1973) and Mason and Kodicek (1973a,b) but

FIGURES 2 - 37 are fluorescence micrographs. Excepting Figures 2a, 2b, 34 and 35, all illustrations are of transverse GMA, modified GMA or hand-cut (HS) sections of mature wheat, oats or barley kernels. Scale bar numbers indicate µm. Abbreviations: aleurone layer (al); cell wall (cw); cuticle (c); scutellum (sc); starch (s); starchy endosperm (end); stomata (st); trichomes (t).

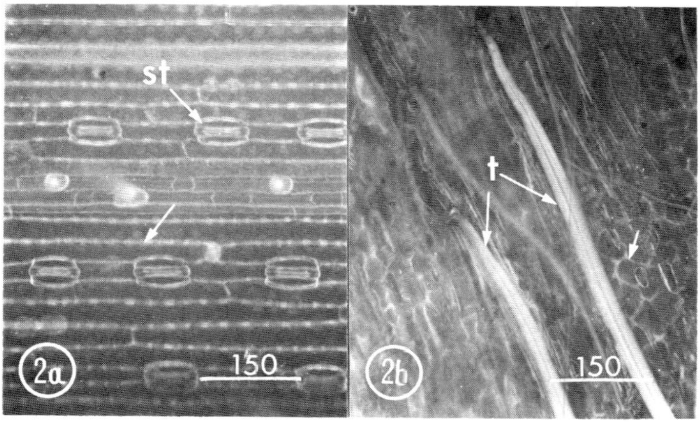

FIGURE 2a,b. Primary fluorescence patterns on the surface of (a) barley leaf showing stomata and epidermal cell outlines (arrow) and (b) wheat grain with trichormes and underlying aleurone layer (arrow).

the sub-cellular location of these compounds has not been determined. Using Ehrlich's reagent (0.5% 2,4-dimethylaminobenzaldehyde in ethanol containing 1% conc. HCl) as a fluorochrome for primary aromatic amines, fluorescent deposits are detectable in the aleurone layer of wheat, barley and oats (*Figures 5-8*) using FC II. The compounds are visualized by adding a few drops of Ehrlich's reagent to hand sections on a microscope slide. The reagent is evaporated to dryness at 50-60°C and replaced with a few drops of immersion oil under a cover glass. Alternatively, to ensure that the compounds are not mobilized during staining, sections may be suspended over the reagent solution at 60°C for 1 hr on a slide.

Ehrlich-positive structures are found only in aleurone cells but are common to all cereals which we have examined. That they represent the aromatic amines described by Mason and Kodicek (1973a,b) is likely -- they are confined to the bran, we have isolated substances which co-chromatograph with *o*-aminophenol from hydrolysed wheat, oats and barley bran fractions, and the fluorescence color is similar to that of

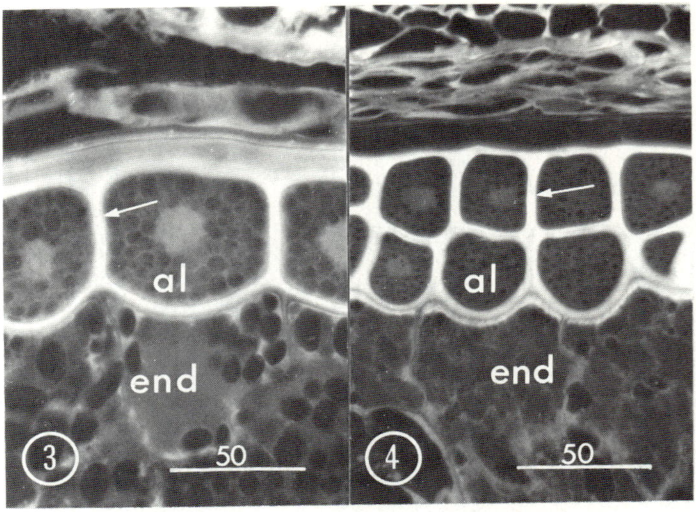

FIGURES 3 and 4. Primary fluorescence of ferulic acid in the aleurone layer cell walls (arrows) of wheat (Figure 3) and Vanier barley (Figure 4). GMA sections.

authentic *o*-aminophenol after reaction. Control treatments with acidified ethanol induce no fluorescence in the structures.

d. Flavonoid compounds. Little attention has been paid to localizing flavonoid compounds cytochemically in cereals since Chaze remarked on their occurrence in 1933. Recent work suggests that diphenyl borinic acid in 80% methanol is a sensitive marker for flavonoids and related compounds on thin-layer chromatograms (F.W. Collins, personal communication). Application of this compound to sections results in intense fluorescence in the sub-aleurone tissues of oats (*Figure 9*) while control treatments show only limited autofluorescence (*Figure 10*). Experiments are in progress to determine the chemistry of the fluorescing substance(s) and details will be published elsewhere.

2. Nicotinic Acid. Wheat bran is one of the richest natural sources of nicotinic acid. Much of the vitamin is nutritionally unavailable and the nature of its association with other bran components has been the subject of considerable investigation for several decades (e.g. Kodicek 1940; Christianson *et al* 1968; Mason and Kodicek 1973a,b). However, the vitamin has not been located cytochemically although such information might be expected to shed some light

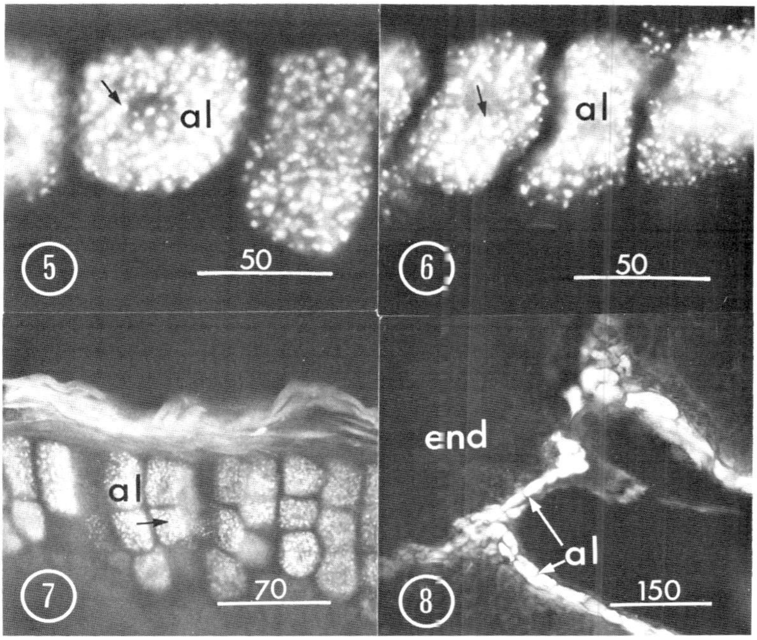

FIGURES 5 - 8. Ehrlich-positive residues (arrows) in GMA sections of the aleurone layers of wheat (Figure 5), Hinoat (Figure 6), and Betzes barley (Figure 7). Low magnification micrograph of the crease in hand sectioned Hinoat (Figure 8) shows no fluorescence in the endosperm.

on its association with other cellular compounds and consequently the reasons for its low nutritional value. Therefore, we have developed the following method for fluorescence detection of nicotinic acid *in situ*. The test is rapid and specific for nicotinic acid and pyridine-containing residues (Feigl 1975).

Hand-cut or GMA sections are mounted on microscope slides and suspended over a freshly-prepared solution of cyanogen bromide (slowly add 10% potassium cyanide dropwise to 5-10 ml of saturated bromine water on ice until the solution is just decolorized by one drop of KCN. This solution is very toxic.) The reaction is complete in 5-10 minutes and the free aldehyde groups induced in the pyridine rings are reacted with *p*-aminobenzoic acid (2 g in 75 ml of 0.75N HCl + 25 ml ethanol) by immersing the sections in the reagent for 5-10 minutes. An intense yellow-orange color (FC II) is generated in specific inclusions of the aleurone

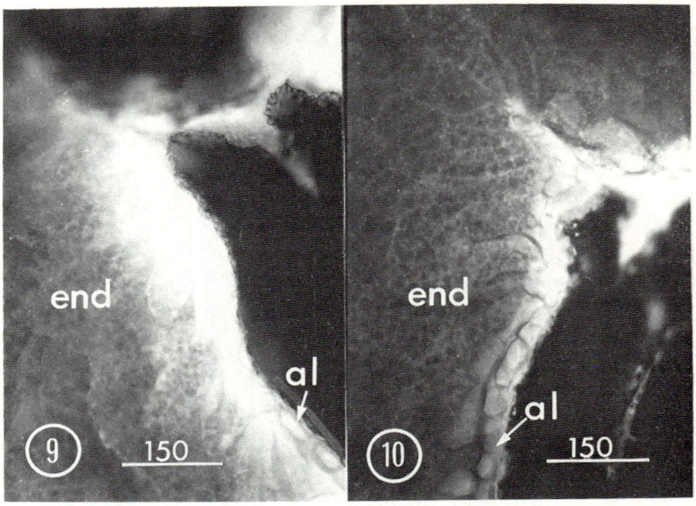

FIGURES 9 and 10. Hand sections of Hinoat in the crease region showing fluorescence differences in the endosperm with (Figure 9) and without (Figure 10) diphenylborinic acid treatment. The reaction may indicate flavonoid residues.

protein bodies of all cereals. Alternatively, orange-red fluorescence (FC III) is induced by substituting saturated barbituric acid solution (in 3% KH_2PO_4) for *p*-aminobenzoic acid (Feigl 1975). Sections are rinsed briefly in water, air dried, and mounted in oil for examination. Wheat, barley, oats and sorghum (*Figure 11-15*) all contain significant concentrations of the reaction product in the Type II aleurone inclusions described by Fulcher (1972) and Morrison *et al* (1975). The cytological location of the nicotinic acid residues (within dense, membrane-bound protein bodies which are in turn enclosed by a thick, and somewhat impermeable ferulic acid-rich cell wall) demands consideration as at least a partial explanation for the low nutritional availability of the vitamin. Other chemical constraints also apply however (Mason and Kodicek 1973a,b). Because the pyridine residues are found only in the aleurone layer, fluorescence tests similar to those outlined may prove useful for identifying aleurone-derived structures in processed material.

FIGURES 11 - 15. Nicotinic acid deposits (arrows) in the aleurone layer of wheat (Figures 11 and 12), Betzes barley (Figure 13), Hinoat (Figure 14) and sorghum (Figure 15). The deposits are discrete structures within each aleurone protein body (Figure 11). GMA sections.

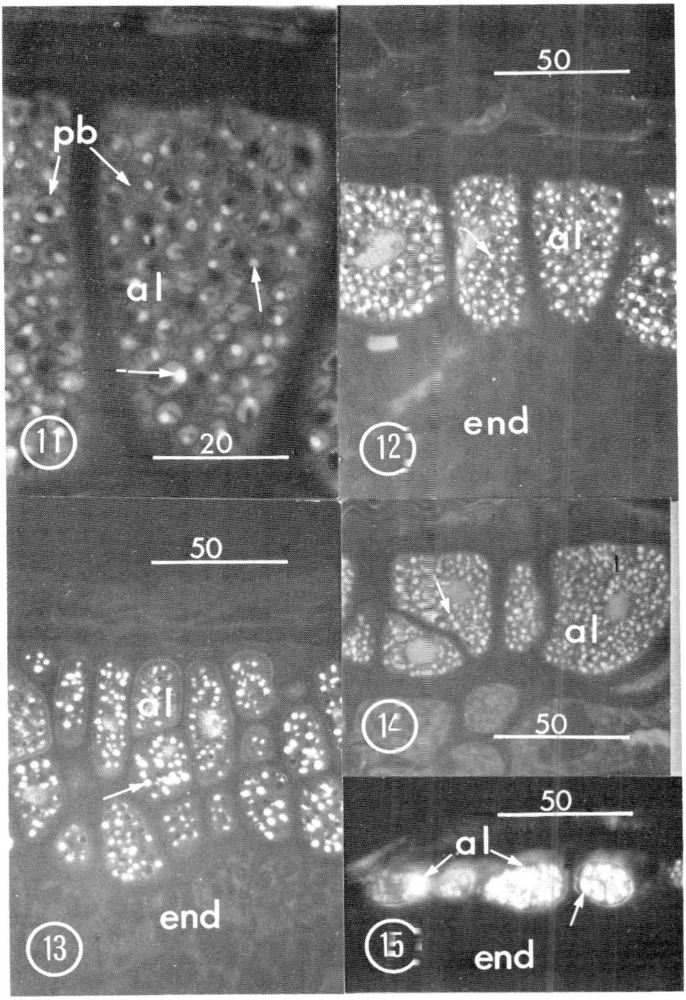

3. *Carbohydrates*. Polysaccharides are the largest single class of compounds in mature cereal grains, contributing 80% or more of the dry weight of the grain primarily as starch, or structural (cell-wall) carbohydrate. These substances vary considerably in morphology and distribution in different cultivars and the following methods can be used routinely to assess variation.

a. *Cell walls*. Calcofluor White M2R New is one of the most intensely fluorescent compounds available for microscopic use and is well known as a general stain for plant cell walls (Hughes and McCully 1975). Its affinity for endosperm walls of barley and oats is spectacular and recent studies have shown a specific precipitation reaction between Calcofluor and the mixed-linkage β-glucan of barley and oats *in vitro* (Wood and Fulcher 1978). A similar dye, Congo Red, undergoes an equivalent reaction with β-glucan (Wood and Fulcher 1978). *In vitro* reactions may be extrapolated to

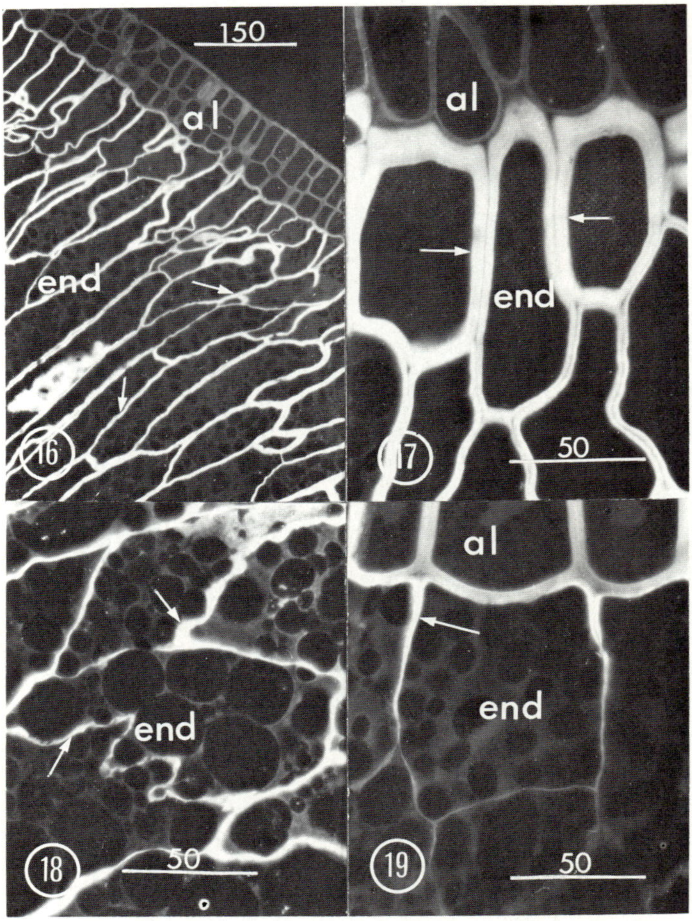

FIGURES 16 - 19. Congo Red-stained endosperm cell walls (arrows) of Betzes barley (Figure 16), Hinoat (Figures 17 and 18) and wheat (Figure 19). Hinoat outer endosperm walls (Figure 17) are much thicker than those of the inner endosperm (Figure 18). Wheat endosperm walls are thinner and less intensely stained (Figure 19).

reactions in tissue sections only with caution, but the two fluorochromes possess excellent structural specificity for endosperm cell walls in barley and oats, and to a lesser extent, wheat (*Figures 16-19*). They have limited affinity for cell walls containing high concentrations of ferulic acid or related autofluorescent compounds.

Sections are rapidly stained by immersion in Calcofluor or Congo Red (0.01-0.001% in water) for 30-60 seconds. After brief washing in water (2 changes, 1-2 minutes each) they are air-dried, mounted in immersion oil, and examined. Some protein staining may occur in unfixed hand sections but this can be minimized by adjusting the dye solutions to pH 8 with phosphate or carbonate buffers. Calcofluor-stained walls exhibit blue fluorescence with FC I; Congo Red is intensely fluorescent with all three filter systems. The red fluorescence of the latter is preferred for most analyses as it cannot be confused with the blue ferulate-related autofluorescence.

Periodate-sensitive cell wall structures can also be demonstrated by the periodate-Schiff's reaction (see Starch, below and *Figure 23*).

b. Starch. The periodic acid/Schiff's (PAS) reaction has been used extensively as a bright field microscopic stain to detect carbohydrates containing adjacent hydroxyl groups (see Feder and O'Brien 1968). Starch, and certain cell wall components, are well stained by the reaction and are very fluorescent under green excitation (filter combination III).

Sections are first oxidized in 1% aqueous periodic acid for 10 minutes followed by a 5-10 minute water rinse. The sections are stained by immersion in fresh Schiff's reagent for 1-2 minutes (longer treatment results in over-staining and quenching of fluorescence) and washed to remove excess reagent. For critical assessment, it is desirable to eliminate native or fixative-induced Schiff's-positive aldehydes prior to oxidation (see Feder and O'Brien 1968 for suitable methods).

For high resolution work, fluorescence analysis of PAS-positive residues is far superior to bright-field examination. The PAS reaction provides sufficient contrast for most bright-field work, but in thin GMA sections a positive reaction is often questionable in structures of small dimension or with limited periodate sensitivity. The fluorescence approach dramatically increases the sensitivity of the reaction and starch and cell wall structures are easily demonstrated (*Figures 20-23*). Flour samples may be analyzed after staining in suspension using the method of Dahlqvist *et al* (1965). Additional fluorescence tests for periodate-sensitive substances are also available (Stoward 1967).

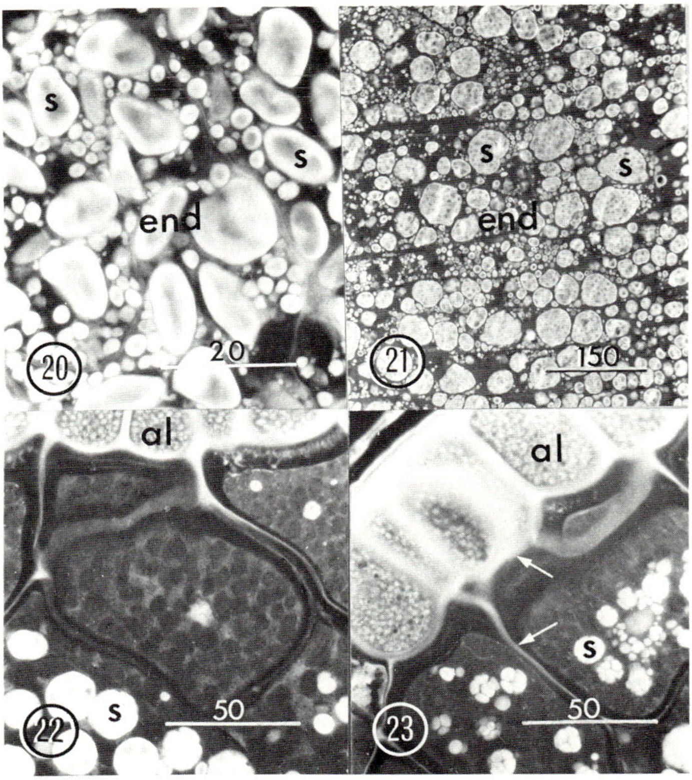

FIGURES 20 - 23. PAS-stained wheat (Figure 20) and low protein oat (Figure 21) endosperm sections (GMA) showing morphological differences in starch deposits. Differences in starch accumulation are sometimes evident in the outer endosperm (sub-aleurone cells) of oat varieties, such as Hinoat (which contains little starch, Figure 22) and high oil types (which show considerable accumulation in the same region, Figure 23). Cell walls are also fluorescent but the stained regions (arrows) do not correspond to Congo Red-positive material (cf Figure 17).

c. β-1,3-glucans. Aniline blue is a useful fluorescent indicator of certain cell wall-associated deposits in endosperm tissues. Empirical evidence suggests the dye has some specificity for β-1,3-glucans (Kessler 1958; Nakanishi *et al* 1974) but recent work has questioned such specificity (Smith and McCully 1978). Regardless of the chemical nature of the interaction, aniline blue has pronounced structural specificity for discrete deposits in most cereal endosperms, including wheat (Morrison and O'Brien 1976) and barley (Fulcher *et al* 1977). The structures may represent remnants of more common substances formed during cellular differentiation (Fulcher *et al* 1976; Morrison and O'Brien 1976). More importantly, the aniline blue-fluorescent deposits in barley sub-aleurone cells show some morphological variation in different cultivars (*Figures 24-26*) and the method may have value as a varietal marker. The central endosperm contains numerous aniline blue-positive structures associated with the cell wall (*Figure 27*).

Sections are stained rapidly (in 10-60 seconds) by placing a drop of water-soluble Aniline Blue (0.001% in 0.01 M phosphate buffer pH 7) on each section and adding a cover glass. The structures are bright yellow using FC I or II and fluorescence is enhanced by increasing the pH of the dye solution to pH 8-11.

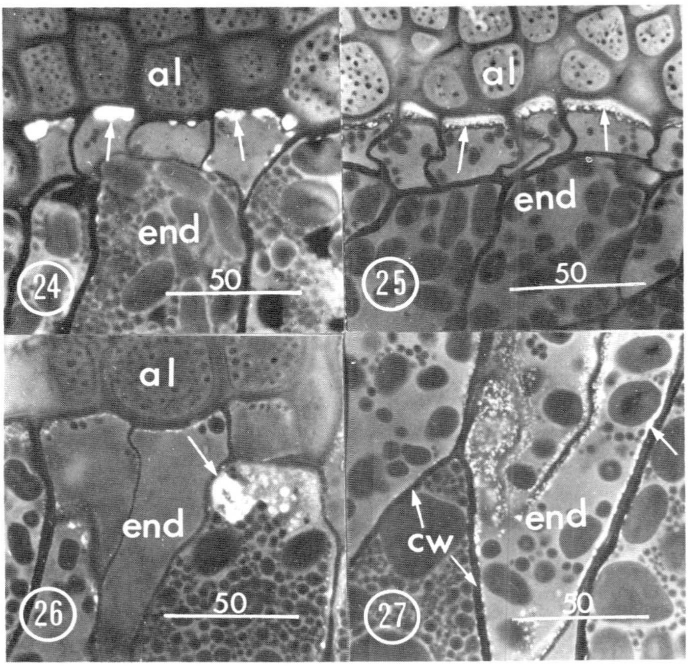

FIGURES 24 - 27. Aniline Blue-stained GMA sections of Himalaya (Figure 24), Vanier (Figure 25) and Betzes barley (Figure 26) showing different morphology of the fluorescent deposits (arrows) near the aleurone layer in the three varieties. Numerous deposits are also associated with the cell walls of the central endosperm of Betzes (Figure 27).

4. Storage Lipids. Until recently it has been difficult to visualize lipid deposits in cereal grains except by using fat-soluble substances such as the Sudan dyes. These dyes often produce non-specific stain precipitates in sections, they must be carefully differentiated to remove excess dye, and they frequently yield low contrast images in thin sections. In contrast, Nile Blue A induces intense fluorescence in hexane-soluble structures and is ideally suited to lipid detection. The distribution of fluorescence after staining correlates with the known distribution of neutral lipids in cereals (high concentrations in the aleurone layer and germ, lower levels in the starchy endosperm) and with the exception of the seed cuticle, no fluorescence is observed after hexane extraction for 1-2 minutes prior to staining. As an aqueous dye, Nile Blue A has the added advantage of minimizing lipid mobilization and extraction during staining.

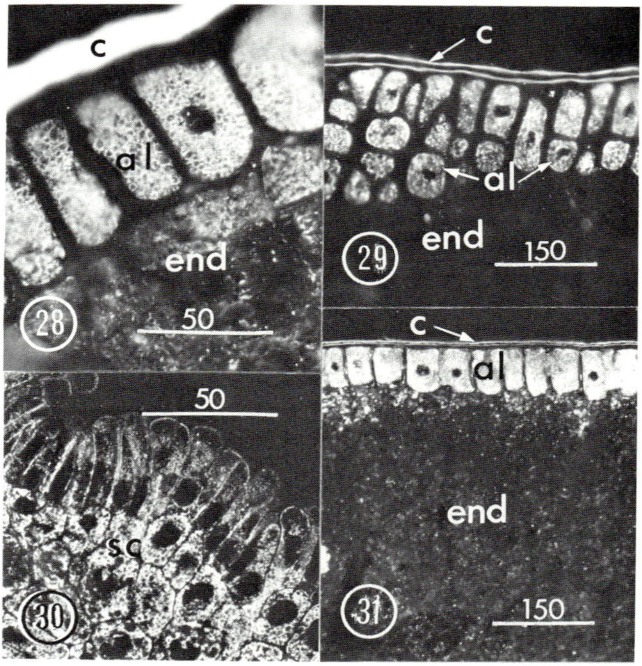

FIGURES 28 - 31. Nile Blue-stained modified GMA sections showing intense fluorescence in the cuticle and aleurone layers of Hinoat (Figure 28) and Betzes barley (Figure 29), and in the scutellum (Figure 30) and aleurone layer (Figure 31) of wheat.

The fluorescence chemistry of the interaction between Nile Blue A and lipid deposits is not defined although Gurr (1960 p 299) noted a minor fluorescent component in Nile Blue A preparations, and the dye has been used to induce fluorescence in a variety of other biological systems, including muscle (Bezanilla and Horowitz 1975) and tumour cells (Bastos and Marques 1972). Nile Blue A has a distinct affinity for cereal triglycerides isolated on thin layer chromatograms (unpublished observations) but further studies will be necessary to define the nature of the interaction. It is likely that the fluorescent component of Nile Blue A is simply absorbed selectively into neutral lipid fractions.

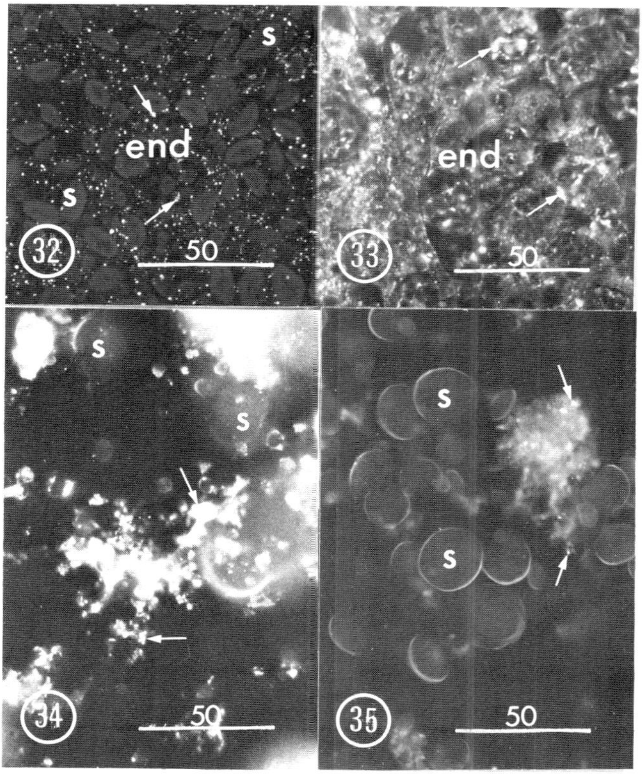

FIGURES 32 and 33. Modified GMA sections of the central endosperm of wheat (Figure 32) and Hinoat (Figure 33) illustrating differences in concentration of Nile Blue A-stained deposits (arrows).

FIGURES 34 and 35. Flour particles stained with Nile Blue A showing fluorescent lipid deposits (arrows) before (Figure 34) and after (Figure 35) hexane extraction.

Tissues are stained by placing a few drops of Nile Blue A solution (0.001% aqueous) on the sections under a cover glass. Fluorescence appears within a few seconds and is most intense using FC II and III. The method is useful for high resolution analysis (provided that the non-extractive GMA-glutaraldehyde-urea resin is used), and for rapid screening of hand sections. Examples of Nile Blue A fluorescence are shown in *Figures 28-33*. Hexane-soluble lipids in flour fractions are easily detected by suspending flour particles in a drop of dye solution on a microscope slide (*Figures 34, 35*).

5. *Proteins*. Several reagents are effective as fluorescent protein markers *in situ*. One of these, Acid Fuchsin, is an acid dye with marked affinity for endosperm proteins at neutral or acid pH. It has been used as a bright-field stain in many areas of cytochemistry for almost a century (see Gurr 1960 p 202), but it is much more sensitive as a fluorochrome under blue or green excitation (FC II and III). Acid Fuchsin is used on both hand-cut and GMA sections by immersing the sections for 1-2 minutes in a 0.01% aqueous solution of the dye (adding glacial acetic acid to 1% concentration provides more intense staining). The sections are rinsed briefly and mounted in water, or air-dried and examined under immersion oil. *Figures 36* and *37* show typical fluorescence patterns in barley and oats. The method is especially useful in emphasizing differences in protein distribution in different cultivars.

Orange G, another acid dye which forms the basis of the Udy dye-binding technique (Udy 1956) is also fluorescent (FC II and III) and can be used in the same manner as Acid Fuchsin. These acid dyes provide some differentiation of proteins in wheat and barley, and we assume that the more intensely stained structures correspond to the basic amino acid-enriched deposits described previously (Fulcher *et al* 1972b).

Two additional fluorochromes, ANS (1-anilino-8-naphthalene sulfonic acid) and Fluorescamine, are more sensitive markers for proteins than the acid dyes and add flexibility in labelling cereal tissues. ANS is unique in fluorescing in aqueous solutions only when absorbed to proteins (Lawrence 1952, Weber and Lawrence 1954) which eliminates the necessity of washing excess fluorochrome from samples. Therefore, it is particularly useful for observing proteinaceous structures in flour samples. ANS is employed by applying 0.001% aqueous solution to the specimens under a cover glass and viewing with FC I or II.

In contrast, Fluorescamine undergoes rapid hydrolysis

in aqueous solution (Udenfriend *et al* 1972) and is used in acetone or similar solvents. Thus the fluorochrome provides an intense and very specific non-aqueous stain for circumstances in which aqueous systems extract or disrupt specimens (e.g. oat β-glucans swell in water and cause severe distortion of unfixed hand sections). Specimens are stained by applying a few drops of 0.001% Fluorescamine in acetone directly to the sample. The acetone is allowed to air-dry completely, and sections are mounted in immersion oil for viewing with FC I.

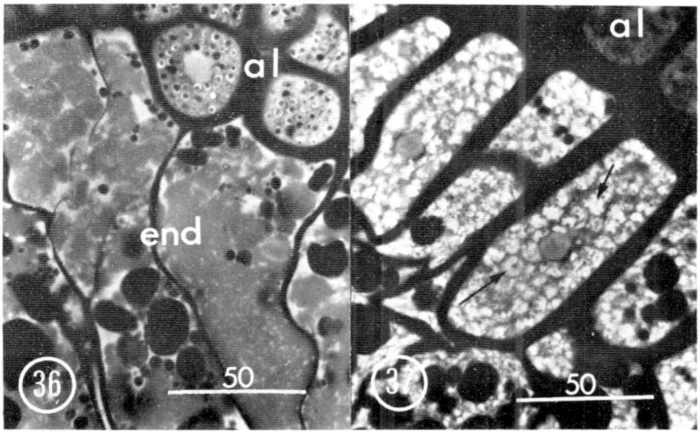

FIGURES 36 and 37. Acid Fuchsin-stained GMA sections of Betzes barley (Figure 36) and Hinoat (Figure 37) sub-aleurone tissues. Storage protein bodies are very fluorescent in oats (arrows) while barley proteins vary in their fluorescence intensity.

6. *Other Applications.* Other fluorescent stains are also applicable to cereals, including the standard Feulgen or Acriflavine-Feulgen (Levinson *et al* 1977), and ethiduim bromide (Hsung *et al* 1976) nuclear stains. These and several of the preceding assays can be applied sequentially to single grain sections for simultaneous demonstration of two or more endosperm constituents and many of the methods are suitable for detecting grain-associated microorganisms.

D. *Sources of Histochemical Reagents*

Embedding and fluorogenic chemicals may be obtained from the following sources: Calcofluor White M2R New - American Cyanamid, Bound Brook, N.J.; Nile Blue A (C.I.

51180) - J.T. Baker Co., Phillipsburg, N.J.; Acid Fuchsin (C.I. 42685); Congo Red (C.I. 22120) - Fisher Scientific Co., Fair Lawn, N.J.; Aniline Blue (C.I. 42755), glutaraldehyde, glycol methacrylate; Orange G (C.I. 16230) - Polysciences Inc., Warrington Park, Pa.; ANS (1-anilino-8-napthalene sulforic acid), Fluorescamine (4-phenylspirofuran-2(3H),1-phthalan-3,3'-dione) - Sigma Chemical Co., St. Louis, Mo.

III. SUMMARY

The methods outlined in this chapter represent but a partial list of potentially useful fluorogenic reactions for detecting cereal compounds. Some are chemically quite specific, while others are empirical markers for chemically undefined structures. Many other fluorochromes are also available for cell analysis and the rising interest in fluorescence spectroscopy as a standard analytical tool is spawning an increasing number of highly specific (and extremely sensitive!) fluorescent reagents. As they become available, many of these compounds may prove to be adaptable to microscopic work.

For cereal analysis the advantages offered by fluorescence microchemistry are many. First, and most important, is the superior sensitivity of fluorescence assay in comparison with older bright-field staining methods, a characteristic which is well known to fluorescence spectroscopists. Small structures which were previously difficult to detect by conventional microscopic methods are now dramatically emphasized by fluorescence characteristics. In addition, the chemical specificity of fluorescent reagents invariably surpasses that of most bright-field stains. Both the excitation and fluorescence spectra are diagnostic of individual compounds and this fact may prompt the increased use of microspectrofluorometers for identification and quantitation of substances *in situ*. Similarly, the fluorescence spectrum of a particular substance often shifts in varying circumstances (different composition or concentration of substrate, different pH, etc.) and this feature of fluorescence has been exploited only minimally in evaluating cereal components (e.g. Fulcher *et al* 1972a; Wood and Fulcher 1978). The potential for quantitative and qualitative refinements is considerable.

Other advantages of fluorescence microscopy may not be so obvious. For example, the high contrast afforded by fluorescence renders the technique very suitable for quantitative image analysis and stereoscopy. It may also be combined with other forms of microscopy (e.g. phase contrast) for greater flexibility. Finally, the speed and ease with which fluorescence images can be obtained is a distinct benefit -- most of the fluorescent reactions which we have described are complete within seconds. And, in addition to providing increased sensitivity, epi-illuminators allow rapid scanning of the surface of even the crudest hand sections with improved clarity.

We have confined our remarks to cereal grain components, but it should be emphasized that these techniques are useful in a wide range of circumstances. Indeed, several of the methods are adapted from other diverse areas of investigation. We have applied similar assay techniques to vegetative tissues, pathogen morphology and infection processes (Holland and Fulcher 1971), immunochemistry (Fulcher and Holland 1971), and a wide variety of materials derived from industrial processes. In short, we are of the opinion that fluorescence microscopy, when appropriate fluorochromes are available, is preferred to most other microscopic techniques for morphological and microchemical evaluation of cereals.

ACKNOWLEDGMENTS

We wish to thank Drs. V. Burrows, G. Fedak and P.J. Wood for providing samples for analysis, and Dr. F.W. Collins for providing diphenylborinic acid.

REFERENCES

Barlow, K.K., Buttrose, M.S., Simmonds, D.H. and Vesk, M. (1973a). *Cereal Chem. 50,* 443.

Barlow, K.K., Simmonds, D.H., and Kenrick, K.G. (1973b). *Experientia 29,* 229.

Bastos, A.L. and Marques, D. (1972). *Z. Naturforsh. 27,* 1395.

Bezanilla, F. and Horowitz, P. (1975). *J. Physiol. 246,* 709.

Chaze, M.J. (1933). *Compt. Rend. Acad. Sci. Paris. 196,* 952.

Christianson, D.D., Wall, J.S., Dimler, R.J. and Booth, A.N. (1966). *J. Agric. Fd. Chem. 16,* 100.

Dahlqvist, A., Olsson, I., and Norden, A. (1965). *J. Histochem. Cytochem. 13,* 423.

Feder, N. and O'Brien, T.P. (1968). *Amer. J. Bot. 55,* 123.

Feigl, F. (1975). "Spot Tests in Organic Analysis" (Seventh Edn.). p. 384. Elsevier Scientific Publishing Co., Amsterdam.

Fulcher, R.G. (1972). Ph.D. dissert. Monash University, Clayton, Victoria, Australia.
Fulcher, R.G. and Holland, A.A. (1971). *Arch. Mikrobiol. 75,* 281.
Fulcher, R.G., O'Brien, T.P. and Lee, J.W. (1972a). *Aust. J. Biol. Sci. 25,* 23.
Fulcher, R.G., O'Brien, T.P., and Simmonds, D.H. (1972b). *Aust. J. Biol. Sci. 25,* 487.
Fulcher, R.G., McCully, M.E., Setterfield, G. and Sutherland, J. (1976). *Can. J. Bot. 54,* 539.
Fulcher, R.G., Setterfield, G., McCully, M.E. and Wood, P.J. (1977). *Aust. J. Plant Physiol. 4,* 917.
Gurr, E. (1960). "Encyclopaedia of Microscopic Stains" Leonard Hill (Books) Ltd., London.
Harris, P.J. and Hartley, R.D. (1976). *Nature 259,* 508.
Holland, A.A. and Fulcher, R.G. (1971). *Aust. J. Biol. Sci. 24,* 819.
Hsung, H., Lown, W. and Johnson, D. (1976). *Can. J. Biochem. 54,* 1047.
Hughes, J. and McCully, M.E. (1975). *Stain Technol. 50,* 319.
Jacobsen, J.V. and Knox, R.B. (1971). *In* "Plant Growth Substances" (D.J. Carr, Ed.) Angus and Robertson, Sydney.
Kessler, G. (1958). *Ber. Schweiz, Bot. Ges. 68,* 5.
Kodicek, E. (1940). *Biochem. J. 34,* 724.
Lawrence, D.S.R. (1952). *Biochem. J. 51,* 168.
Levinson, J. Retzel, S. and McCormick, J. (1977). *J. Histochem. Cytochem. 25,* 355.
Mason, J.B. and Kodicek, E. (1973a). *Cereal Chem. 50,* 637.
Mason, J.B. and Kodicek, E. (1973b). *Cereal Chem. 50,* 646.
Mason, J.B., Gibson, N., and Kodicek, E. (1973). *Br. J. Nutr. 30,* 297.
McLeod, A.M., Duffus, J.H. and Horsfall, D.J.L. (1964). *J. Inst. Brew. 70,* 303.
Morrison, I.N., Kuo, J. and O'Brien, T.P. (1975). *Planta. 123,* 105.
Morrison, I.N. and O'Brien, T.P. (1976). *Planta. 130,* 57.
Nakanishi, I., Kimura, K. Kusui, S. and Yamazaki, E. *Carbohydr. Res. 32,* 47.
Pease, D. (1973). *J. Ultrastruct. Res. 45,* 124.
Smith, M.M. and McCully, M.E. (1978). *Protoplasma. 95,* 229.
Spurr, A.R. (1969). *J. Ultrastruct. Res. 26,* 31.
Stoward, P.J. (1967). *J. Roy. Microsc. Soc. 87,* 237.
Udenfriend, S., Stein, S., Bohlen, P., Dairman, W., Teimbruber, W., and Weigole, M. (1972). *Science 178,* 871.
Udy, D.C. (1956). *Cereal Chem. 33,* 190.
von Sengbusch, G. and Thaer, A.A. (1973). In "Fluorescence Techniques in Cell Biology" (A.A. Thaer and M. Sernetz, Ed.), p. 31. Springer-Verlag (New York).
Weber, G. and Lawrence, D.S.R. (1954). *Biochem. J. 56,* xxxi.
Wood, P.J. and Fulcher, R.G. (1978). *Cereal Chem. 55,* 952.

ANALYSIS OF BOTANICAL COMPONENTS IN CEREALS AND CEREAL PRODUCTS - A NEW WAY OF UNDERSTANDING CEREAL PROCESSING

L.Munck, C.Feil
and G. Gibbons

Carlsberg Research Laboratory
Copenhagen

I. INTRODUCTION

Ideally one should consider the processing of cereals as a partition and separation of the discrete botanical entities of grain. If we could develop suitable techniques for quantification of the characteristic botanically-defined cereal grain tissues, we should be able to study cereal processing in an entirely new way, both in terms of precision, and of the functional properties of the various fractions. Such sophisticated analyses represent a significant advance from the traditional ash and flour-colour measurements used at present. A clear advantage of this new type of approach is measurement of the cereal aleurone layer. Dissection of the outer integuments of the cereal grains reveals that the aleurone layer, freed from the adhering testa and pericarp, is essentially colourless. It is well established that the aleurone layer of wheat contains up to 60 percent of the total minerals of the seed as well as substantial lipid, nicotinic acid, and phosphorus reserves. The direct result of these observations is that flour colour cannot give a reliable indication of the ash content. If we can accurately measure the aleurone content of a flour, then bakers could use such defined aleurone-enriched flours as sources of both vitamins and minerals, as well as the dietarily important arabinoxylans of the aleurone cell walls. An added advantage of such botanical analysis of aleurone content in cereal products is that we indirectly can estimate the level of phytic acid, allowing the cereal processer to plan his

ISBN: 0-12-370960-1

processing so as to include measures for the removal of the phytic acid of the aleurone phosphate reserves, as it is well known that phytate can bind calcium and zinc.

The results and perspectives presented here can hopefully lead to entirely new methods for the analysis of cereals and their products. The work described in this paper is primarily based upon the intrinsic autofluorescence of the different botanical components of the cereal seed. The future of the use of other fluorescence techniques in the analysis of cereals has clearly been shown by Fulcher (1979). The practical utilization of the options available is the major concern of our group. By studying a cereal product in the microscope, we want to estimate the distribution of the different botanical components - testa, aleurone, scutellum, embryo and endosperm. Based upon chemical analysis of these discrete anatomical entities we have then the basis for describing the distribution and availability of cereal enzymes, vitamins, minerals, starch and cell wall components in cereal seeds and products.

II. MICROSCOPIC OBSERVATIONS

Fluorescence microscopy of cereal seed sections using ultraviolet excitation revealed a strong bluish-white autofluorescence of the aleurone cell walls and a weaker bluish-white autofluorescence in the endosperm cell walls. Using the same microscope filter combination, the outer integuments of the seed showed a yellowish-green autofluorescence of lower intensity than that of the aleurone layer. Ferulic acid is responsible for the bluish-white autofluorescence of both aleurone and endosperm cell walls (Fulcher 1972). The chemical basis for the light yellowish-green autofluorescence has however, not as yet been established. Because the aleurone cell walls are much thicker than the endosperm cell walls the observed autofluorescence of the former is greatly accentuated as seen in a black and white photograph of a sectioned wheat kernel (Fig. 1). Examination of a wheat flour using the same excitation and barrier filters shows how easily one can recognize aleurone cell walls in a cereal product using fluorescence microscopy (Fig 2). Another example of the possibility of identifying botanical components by fluorescence microscopy is the yellow-green autofluorescence observed in the pericarp and testa when sections of cereal seeds are excited by blue light (Fig. 3). Examination of cereal products using blue-light exitation allows in the same way as for the aleurone immediate identification of testa plus pericarp.

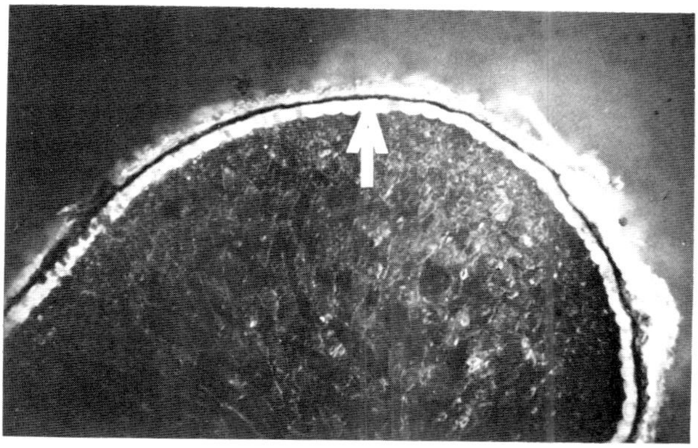

FIGURE 1. Micrograph of polished section of a wheat kernel seen under UV excitation. The bright layer at the arrow is blue fluorescent aleurone cells.

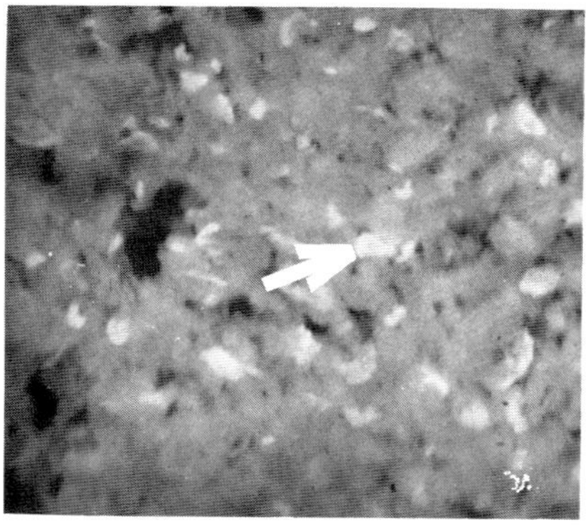

FIGURE 2. Micrograph of surface of a compact sample of wheat flour. The arrow points at one of the bright particles which is a blue fluorescent aleurone particle.

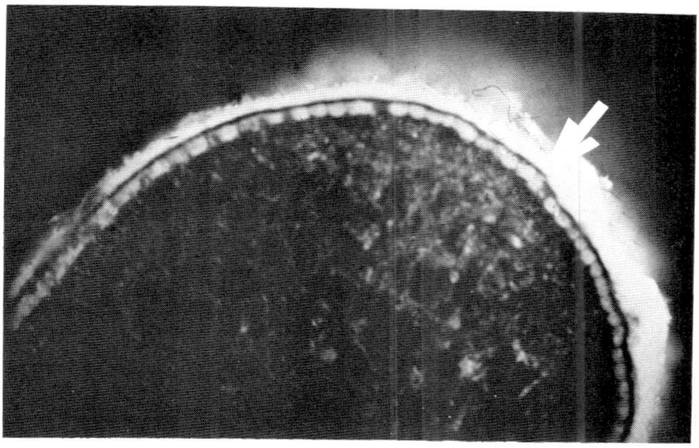

FIGURE 3. Micrograph of polished section of a wheat kernel seen under blue excitation. The outer bright layer at the arrow is yellow fluorescent pericarp and testa.

Inspired by the fluorescence observations we decided to approach the problem of identification and quantification of botanical components of cereal products along two lines:

1. Physical characterisation of the autofluorescence of the different components by spectrofluorimetry.
2. Individual particle analysis of cereal products using fluorescence microscopy with optimized exciter/barrier filter combinations.

III. SPECTROFLUORIMETRIC ANALYSIS OF CEREAL SEED COMPONENTS

As starting material for the spectrofluorimetric studies we have used both hand-dissected pericarp, aleurone and endosperm material, as well as milling fractions highly enriched in these components. The results for the enriched endosperm fraction are shown in figure 4. This fraction is characterized by a very strong autofluorescence at 330 nm which is maximally excited by light of about 280 nm (Fig. 7). The observed emission spectrum is typical of a mixture of tyrosine and tryptophane indicating the presence of these amino acids in the endosperm proteins.

The results for the aleurone (Fig. 5) were entirely different from those of the endosperm and revealed the presence of two major peaks at 420 and 470 nm optimally excited by light between 350-360 nm (Fig. 7). The presence of tyrosin and tryptophane can also be seen in this aleuron fraction by the emission peak at 330 nm excited by 275 nm light. The peaks observed at 420 and 470 nm are characteristic for ferulic acid (Fulcher 1972). Turning to the pericarp and testa (Fig. 6) we find the main fluorescence component is now an emission peak at 520 nm excited by light of 415,450 and maximally 487 nm (Fig. 7). The presence of a small amount of ferulic acid can also be seen in this fraction as peaks at 420 and 470 nm partly because of the slight contamination with aleurone material and partly due to the occurence of slight amounts of bluish-white fluorescent material in the pericarp. On the other hand, in this fraction the peak at 320 nm excited at 275 nm is virtually absent in spite of the fact that both tryptophane and tyrosin should be present in the pericarp-testa tissues. The reason for this lack of response is discussed in the following.

Following the initial experiments with highly enriched fractions, a series of defined mixtures of the various botanical components were made using the enriched fractions as raw material. Emission spectra of such mixtures showed that with excitation at the optimum wavelength for any one botani-

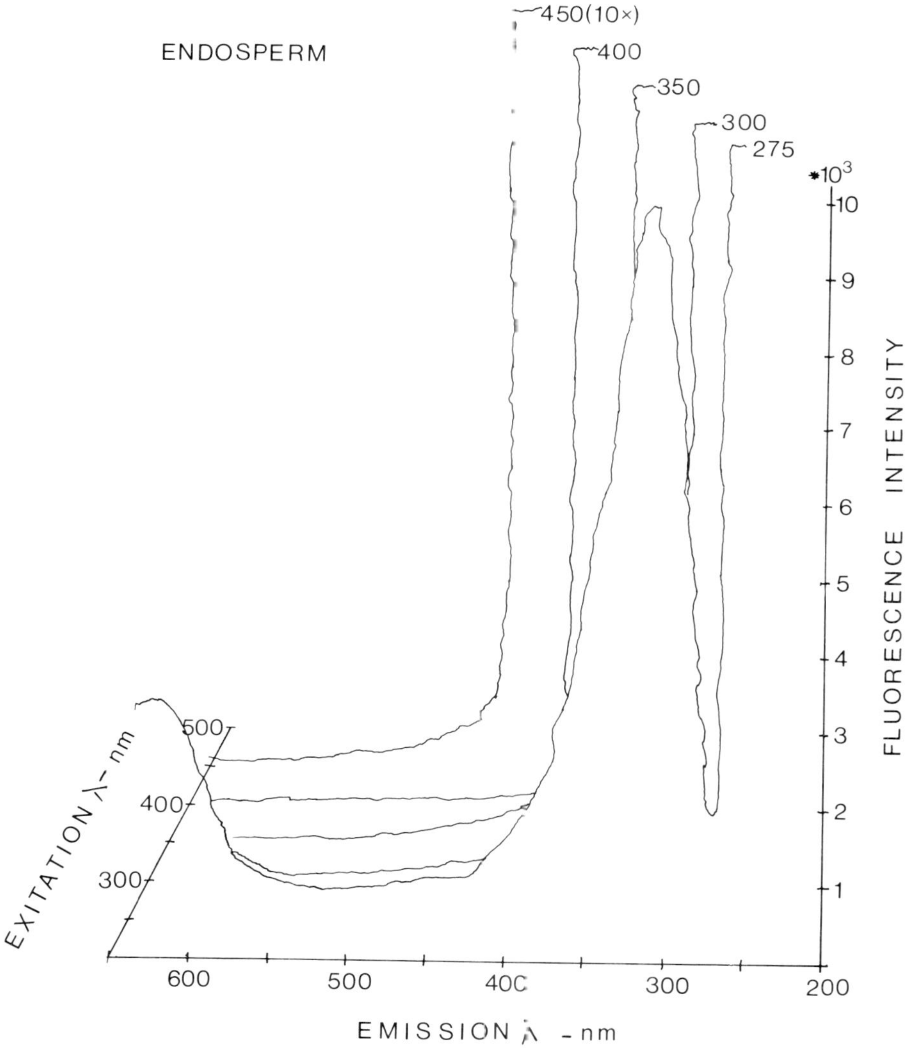

FIGURE 4. Autofluorescence characteristics of endosperm. Endosperm was purified by dehulling and milling in the Carlsberg Research Mill. The endosperm contained 86.9% starch, 0.233% fiber and 0.46% ash (on a dry weight basis). 50mg of endosperm were suspended in 3ml fluorescent free glycerine and the emission spectrum recorded in a Jasco FP-550 spectrofluorimeter at the indicated excitation wavelengths. The 450nm excited spectrum were amplified 10 X to show detail.

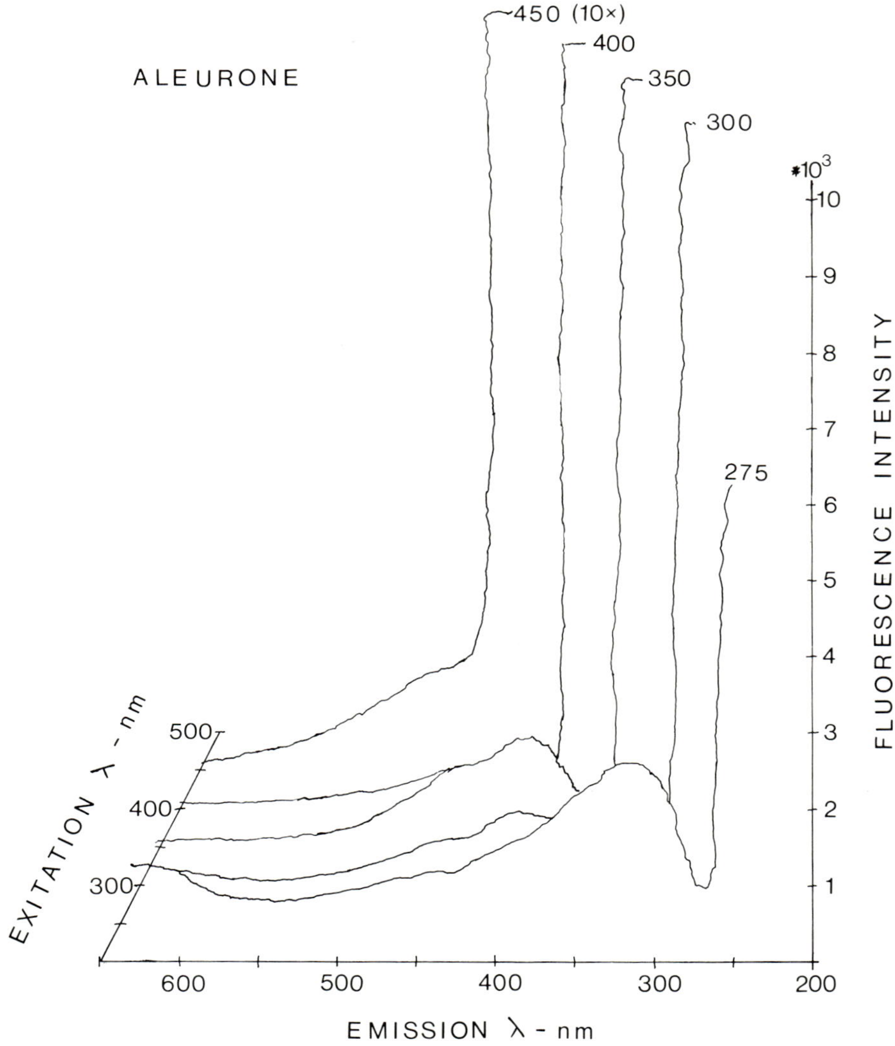

FIGURE 5. Autofluorescence characteristics of aleurone. Aleurone was isolated by dehulling and debranning in the Carlsberg Research Mill. The aleurone contained 53.6% starch, 3.9% fiber and 2.9% ash (on a dry weight basis). 50mg of aleurone were measured as described in Fig. 4.

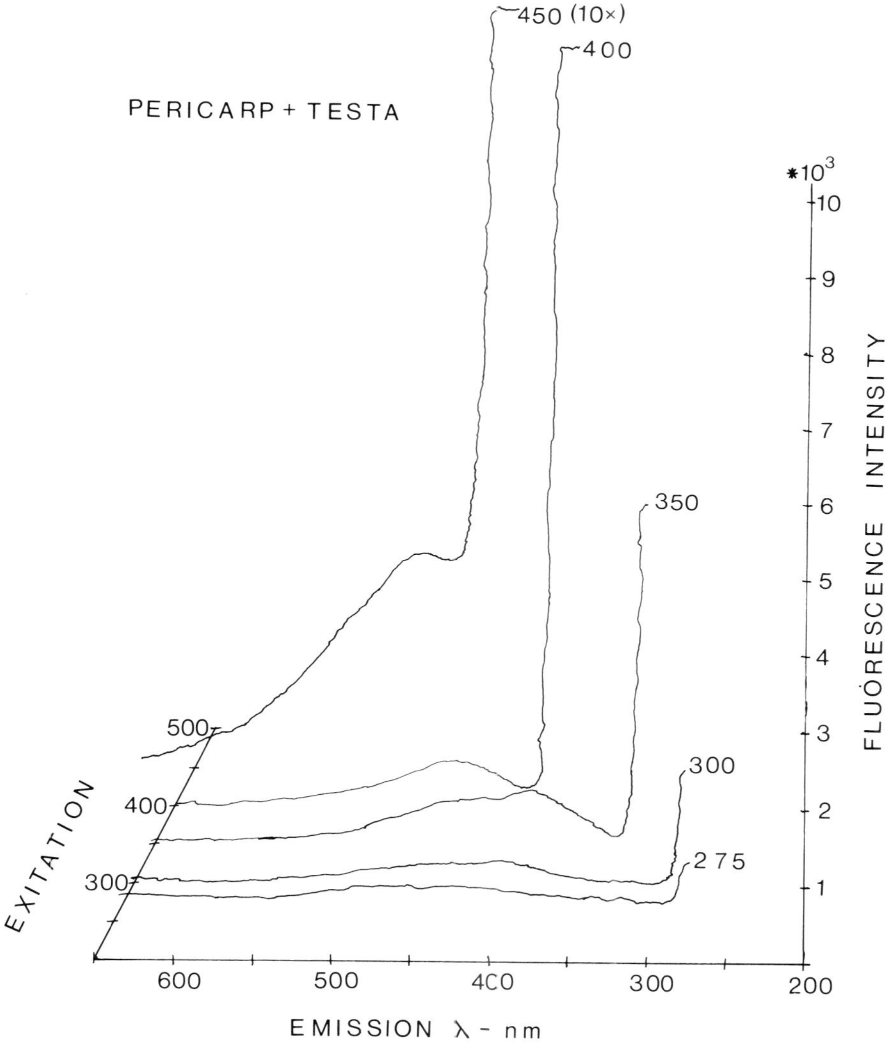

FIGURE 6. Autofluorescence characteristics of pericarp and testa. Pericarp and testa were isolated by dehulling and milling in the Carlsberg Research Mill. The pericarp and testa contained 19.3% starch, 14.2% fiber and 3.96% ash (on a dry weight basis). 50mg of pericarp and testa were measured as described in Fig. 4.

cal component the emission yield of that component was highly affected by the presence of other components of the mixture. For example when a 1:1 mixture of aleurone and endosperm was examined, excitation at 280 nm resulted in a lower emission intensity at 330 nm and a higher emission yield at 420 and 470 nm than expected. This fact can be explained by looking at the excitation spectrum for the 420 and 470 nm peaks of the aleurone (Fig. 7). It can be seen that there is considerable excitation of these peaks at 330 nm indicating that in the mixture of the two botanical components energy transfer is occuring, from the 320 emission peak to the aleurone peaks. Similar results were seen in mixtures of aleurone plus pericarp and testa as well as in mixtures of endosperm plus aleurone plus pericarp and testa, where excitation for aleurone respectively endosperm resulted in a higher 520 nm emission from the pericarp and testa than expected.

IV. IMAGE ANALYSIS OF CEREAL FLOURS

Due to the observed energy transfer phenomena it is not an easy task to perform quantitative measurements of the cereal seed components in a flour by spectrofluorimetry. But if we return to the visual image of the flour as seen in the flourescence microscope, it is in principle quite easy to obtain a quantitative measurement by planimetry of the various fluorescent particles. To do this manually in the microscope by comparison with calibrated grids or on photographs with a planimeter or with a computerized digitizer is a very tedious and time-consuming job, as one usually has to measure thousands of particles to get a reliable value.

For this reason we turned to automatic image analysis using TV scanning techniques, which can speed up the measuring process dramatically. This technique also offered additional advantages. By spectrofluorometric measurement only the concentration of a certain component is obtained, while with image analysis information on particle sizes and shapes is available, as well as the size and shape distribution for all particle types <u>and</u> for individual components such as aleurone and hull.

The image analyser used in this work was a Quantimet 720 from Cambridge Instruments Ltd., and fig. 8 shows a simplified block diagram of the system. The principle of its operation can be described as follows:

1) The optical image, in this case from a microscope, is scanned by a TV-camera, giving a TV-image composed by a great number of picture points, each having a specific "grey" value.

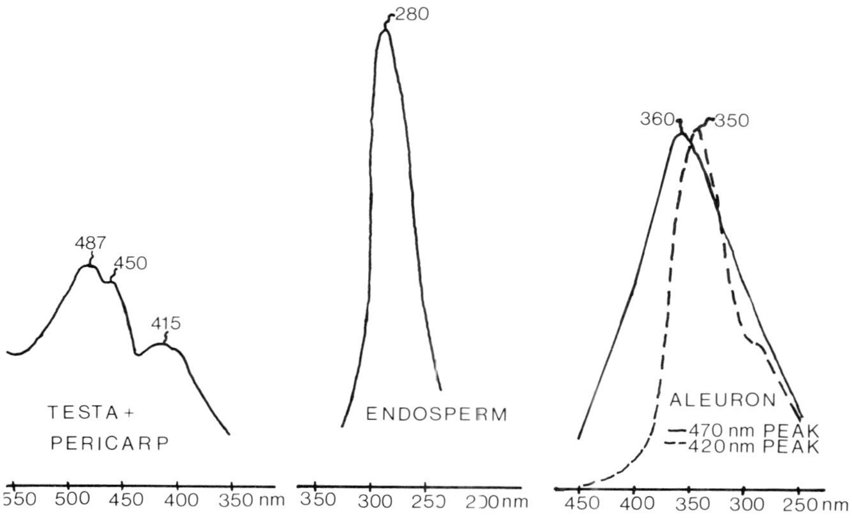

FIGURE 7. Excitation spectrum from pericarp and testa starchy endosperm and aleurone.

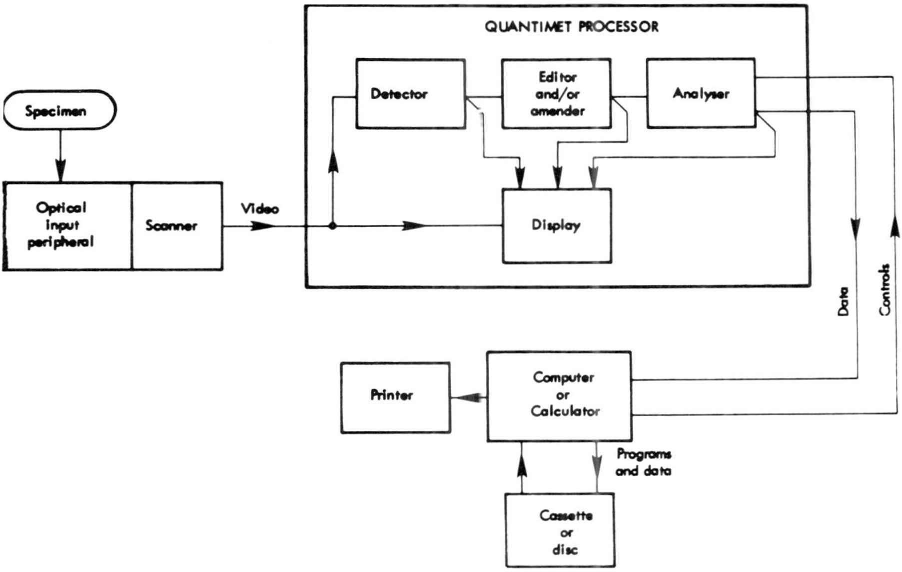

FIGURE 8. Quantimet 720, simplified block diagram.

2) The particles or phase we want to measure are detected by setting an upper and/or lower grey level threshold.
3) The detected image may be "corrected" in various ways either automatically (e.g. by erosion, dilation or sizing) or manually by means of a light pen.
4) On the detected and eventually corrected image the desired parameters (e.g. area, perimeter, Feret-diameters, etc.) are now measured electronically. This can be done per field or individually per particle.
5) The data are transferred to a calculator in which they are treated and stored according to the software in use. The calculator controls the whole system including the stepping stage on the microscope.

Fig. 9 shows the Reichert Polyvar microscope used in our laboratory as well as the plumbicon TV-camera, and fig. 10 gives a view of the entire system.

The most critical process in the above measurement chain is the detection.

Our first problem in detecting the autofluorescent aleurone and hull fragments was the fact that the spectral characteristics of the TV-tube were quite different from that of the eye. Therefore to get the optimal selection of exactly the desired components in terms of grey values on the TV-camera we had to design new filter systems for the microscope. Based upon the spectra obtained of the purified components by spectrofluorimetry, the spectral characteristics of the TV-tube, and the emission spectrum of the mercury lamp used on the polyvar microscope, a series of exciter filters, dichroic mirrors and barrier filters were designed. These filters and dichroic mirrors were specially made as interference filters and mirrors by Optisk Laboratorium, Lyngby, Denmark, and optimized for sharp cutoff and a high transmission of 80-90% by computer control of layer thickness during coating. This was of special value for image analysis of fine flours where the fluorescence of minute fragments is very weak.

For the measurement of aleurone, UV-excitation with the 365 nm peak of the mercury lamp spectrum was used with the beam splitter dichroic mirror edge at 400 nm. The measurement of pericarp and testa was made using blue excitation of 455-500 nm and a beam splitter edge at 520 nm. It is very essential to cut off excitation below 455 nm in order to avoid the aleurone autofluorescence.

The barrier filters were optimized with regard to the spectral characteristics of the plumbicon TV-tube and to the eye to facilitate identification of the particles by the operator when setting the detection threshold. A magenta filter which cut off the green and yellow colours (475-590 nm) was

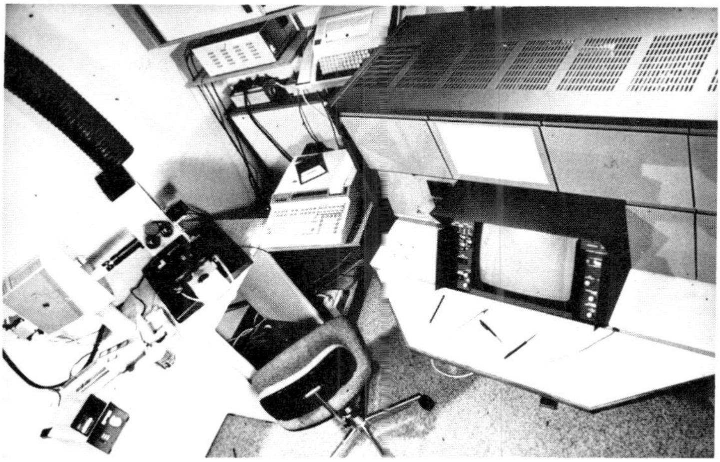

FIGURE 9. Reichert Polyvar microscope with Quantimet plumbicon camera.

FIGURE 10. View of total image analyser system.

found best for the aleurone measurement. For pericarp and testa fragments a barrier filter with a cut-off at 520 nm is used. As an example, Fig. 11 is a photomicrograph of a barley flour using blue excitation to identify the pericarp and testa components. Fig. 12 shows that it is possible on the image analyser to set the detection so that only the strongly fluorescent pericarp and testa components will be measured.

Once the selective detection of features of the image is satisfactory, the remaining part of running the image analysis system to its full potential is a software problem. We have now a very flexible multiple purpose programme running, as well as special programmes for specific problems.

The sample preparation techniques are also of importance to the software development. If it is impossible to separate the individual flour particles on the slide, erosion routines are required to separate them electronically. The better the sample preparation in terms of separation of the particles, and the better the adjustment of the optical system to give good contrast, the easier and more accurate will be the image analysis. We are still in the exploration phase regarding the adaption of image analysis for the quantitative analysis of botanical components. The main effort at present is optimization of the sample preparation techniques to the software routines, and critical measurement of the precision by comparing photographs from the microscope with photographs from the detected images on the TV-screen.

V. CONCLUSION

The results discussed in the present communication represent, we believe, the basis of a new, highly sensitive approach to the quantification and qualification of the distribution of the discrete botanical entities of the grain in cereal seed products. As these discrete botanical entities possess, in the majority of cases, discrete functional properties we believe that measurement of these must result in an invaluable addition to the information obtained from the traditional chemical analyses. The combination of both "botanical" and chemical analysis of cereal seed products will allow a greater understanding of the various steps of milling, baking etc, and as such, provide a strong starting point for future cereal processing development.

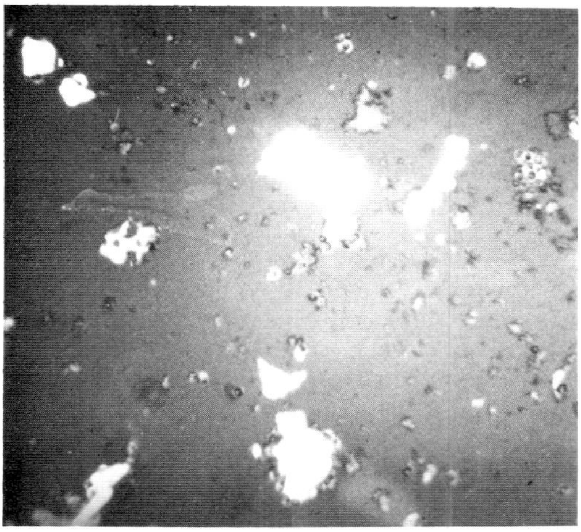

FIGURE 11. Barley flour, photo from microscope showing autofluorescence of pericarp blue light excitation.

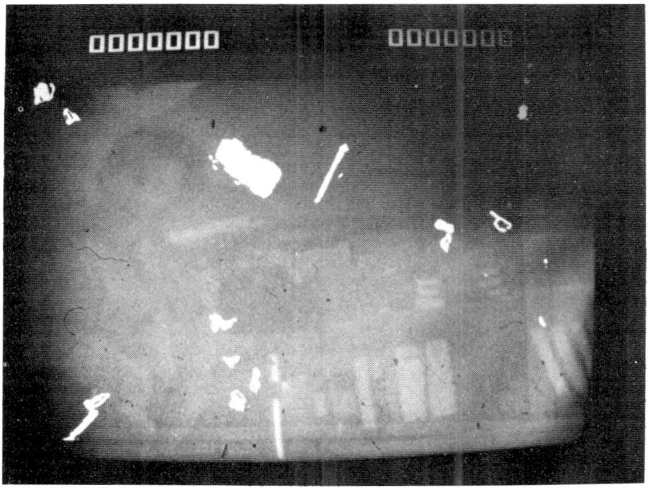

FIGURE 12. Image analyser display of the field from Fig. 11 with fluorescent pericarp particles detected.

In order to fully utilize the potential contained within fluorescence analysis of cereal seed products a series of problems remain. These include optimization of sample preparation and further developement of software to characterize the complicated fluorescence energy transfer systems observed in cereal product fractions. The ultimate goal is to obtain quantitative estimates of the various botanical components in cereal seeds and cereal seed products.

ACKNOWLEDGMENTS

The authors wish to thank Ms. Jane Tinning, Ms. Inge Hallberg and Ms. Lise Tang Petersen for excellent technical assistance as well as Ms. Britha Jensen and Ms. Eva Christiansen for typing the manuscript. Discussions with Dr. Gary Fulcher, Ms. Tine Olesen, Mr. Werner Olsen and Dr. Klaus Schindl are gratefully acknowledged.

REFERENCE

Fulcher (1979) Cereals - a microfluorescence view. These proceedings.

ADVANCES IN MAIZE CARBOHYDRATES

Lamartine F. Hood

Department and Institute of Food Science
Cornell University
Ithaca, New York

I. INTRODUCTION

In the United States, the bulk of corn produced annually is utilized in animal feeds. In 1977-78, 6.2 billion bushels disappeared as feed, food and through export markets. Only 548 million bushels (8.8% of total) were utilized as food. Harness (1978) has reported that in 1977, 375 million bushels were utilized by the corn wet-milling industry. This was an increase from 250 million bushels in 1970, largely due to the increased usage of corn for corn syrup production. Starch is the component recovered in highest yield (Table I).

Sales of corn-derived ingredients are shown in Table II. From these data, it is evident that only about 27% of the corn transformed into food ingredients was marketed as starch *per se*. The balance was converted to sweetners and dextrins.

TABLE I. Yields from Corn Wet-Milling

	Percent
Starch	61
Gluten and hull	19
Oil	4
Water	16

ISBN: 0-12-370960-1

TABLE II. Sales of Corn-Derived Ingredients in U.S., 1977

	Billion Lbs.
Unmodified starch	2.0
Modified starches	1.5
Conventional corn syrups and maltodextrins	5.0[a]
High fructose corn syrups	3.5[a]
Dextrose	1.0
Corn syrup solids	.15

[a]Wet weight basis
From Harness, 1978.

Maize has been a basic component of man's diet for centuries; indeed, numerous references are made to it in the Bible. In this century, plant breeders have developed two strains of maize: waxy maize in 1909 and amylomaize in 1952. Waxy maize starch came into its own as a food ingredient during World War II as a replacement for tapioca starch when supplies of tapioca from southeast Asia dwindled.

The various maize genotypes vary in starch and amylose content (Table III).

TABLE III. Starch and Amylose Composition of Maize Genotypes

Genotype	Starch Content (% of dry weight)	Amylose Content (%)
Corn	69	25
Waxy (wx)	69	0
Amylomaize (ae)	63	50-80

From Shannon and Creech, 1973.

The mature kernel is comprised of four principal parts: tip cap, pericarp (i.e. hull or bran), germ and endosperm. The corn wet-milling process results in essentially complete separation of the endosperm from the other components. Not only is the endosperm the largest component, percentage wise, but it also has the highest percent starch of any of the components (Table IV). According to Watson (1965), the endosperm contains 98.5% of the starch found in the whole kernel.

TABLE IV. Composition of Mature Dent Corn Kernels

Part	% of total[a]	Carbohydrate (% of the part)	
		Starch	Sugars
Whole kernel	100	72.4	1.9
Endosperm	82	86.6	0.6
Germ	12	8.3	11.0
Tip cap	1	5.3	1.5
Pericarp	5	7.3	0.3

[a]Dry weight basis
From Earle et al., 1946

The purity and the properties of the starch recovered during the wet-milling process are influenced by the post harvest handling treatments applied to the corn. Drying temperatures above 80-100°C result in a reduction in starch yield (Lasseran, 1973; Peplinski et al., 1975; Vojnovich et al., 1975). The magnitude of this effect depends upon the moisture of the corn at harvesting, the rate of drying, and the design of the dryer. Paste viscosities of starches recovered from dried corn are affected by the drying temperature (Vojnovich et al.,1975. Starches dried at low temperatures (50°C) have viscosity characteristics similar to corn dried at ambient temperature. There is a gradual decrease in peak viscosity as drying temperatures are increased above 50°C. The stability of the starch paste beyond the peak viscosity is the same for all starches regardless of harvest moisture or drying temperature.

II. SUGARS

A number of sugars have been identified in the endosperm of the corn kernel (Gentinetta et al., 1979). Sucrose, glucose, maltose, fructose, galactose, cellobiose, ribose, mannose and xylose are all present. Raffinose was detected by Cerning (1970), but not by Gentinetta (1979). Although starch content increases during endosperm development, sucrose, glucose, fructose, and maltose concentrations decrease.

III. STARCH

The three types of maize starch commonly used in the food industry, corn, waxy maize, and amylomaize, have different physicochemical properties that can in part be explained by differences in granular and molecular structure. Corn starch with about 25% amylose, has granules of 5-25 μm. These granules are multisided (polygonal) (Fig. 1). Although it contains essentially no amylose, waxy maize granules are about the same size and shape as corn granules (Fig. 2). Amylomaize or high amylose starches have some long, irregularly shaped granules whose shape and quantity varies with amylose content (Figs. 3 and 4).

A. Granule Structure

The internal structure of the maize starch granule has been more difficult to study. Granules in their ungelatinized form are insoluble and highly crystalline, and therefore extremely difficult to penetrate with fixatives, and dehydrating and embedding agents. As a result, thin sections taken for transmission electron microscopy often reveal unusual morphological characteristics which have been attributed to artifacts produced by the preparation techniques. Several methods have been used to overcome this problem of penetration into the granule. Mussulman and Wagoner (1968) utilized acid treatment (7% HCl, room temperature, 35 days) to show that corn and waxy maize starch granules have similar well defined lamellar structures with alternating concentric electron-dense and electron-transparent rings. Amylomaize starch granules have a heterogenous internal structure and in most cases do not have a lamellar structure. Many amylomaize granules have bud-like protuberances that have various sizes, shapes and orientations relative to the main part of the granule. The percentage of the anomalous granules increases as the amylose content increases. Lamellae are never observed in the protuberances. Based on the results of acid treated starch, Mussulman and Wagoner (1968) suggested that amylomaize granules be classified as those which have 1) alternating rings of comparatively high and low degrees of crystallinity, the difference in degree being less than in corn and waxy maize starches, 2) a high degree of crystallinity distributed evenly throughout the granule, 3) a high degree of crystallinity randomly distributed into irregularly shaped areas about 0.1 to 1.0 μm, across, and 4) a low degree of crystallinity distributed evenly throughout the granule. Based on their work and that of Wolf et al. (1964), they suggested that there is significant

FIGURES 1 and 2. Scanning electron micrographs of uncooked corn (Fig. 1) and waxy maize (Fig. 2) starch granules.

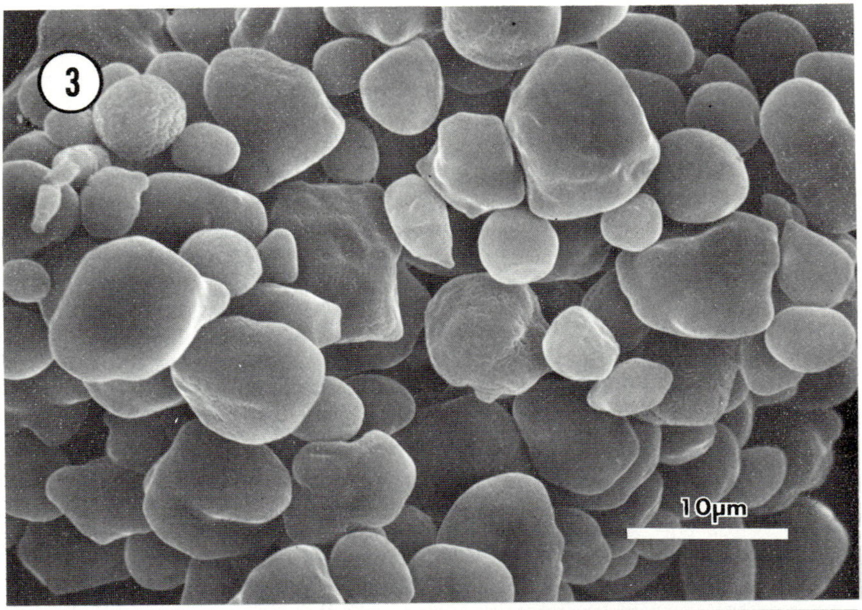

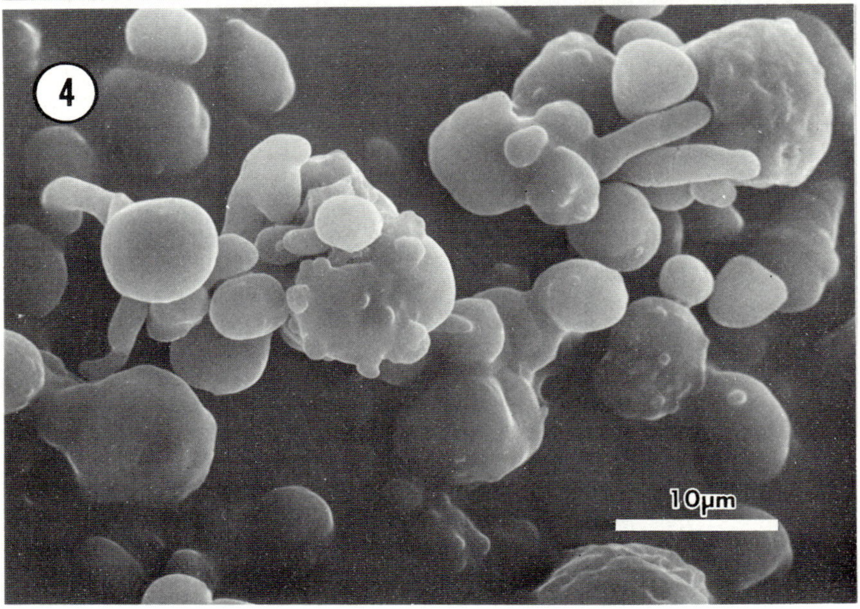

FIGURES 3 and 4. Scanning electron micrographs of uncooked amylomaize starch granules. Fig. 3: 55% amylose; Fig. 4: 70% amylose.

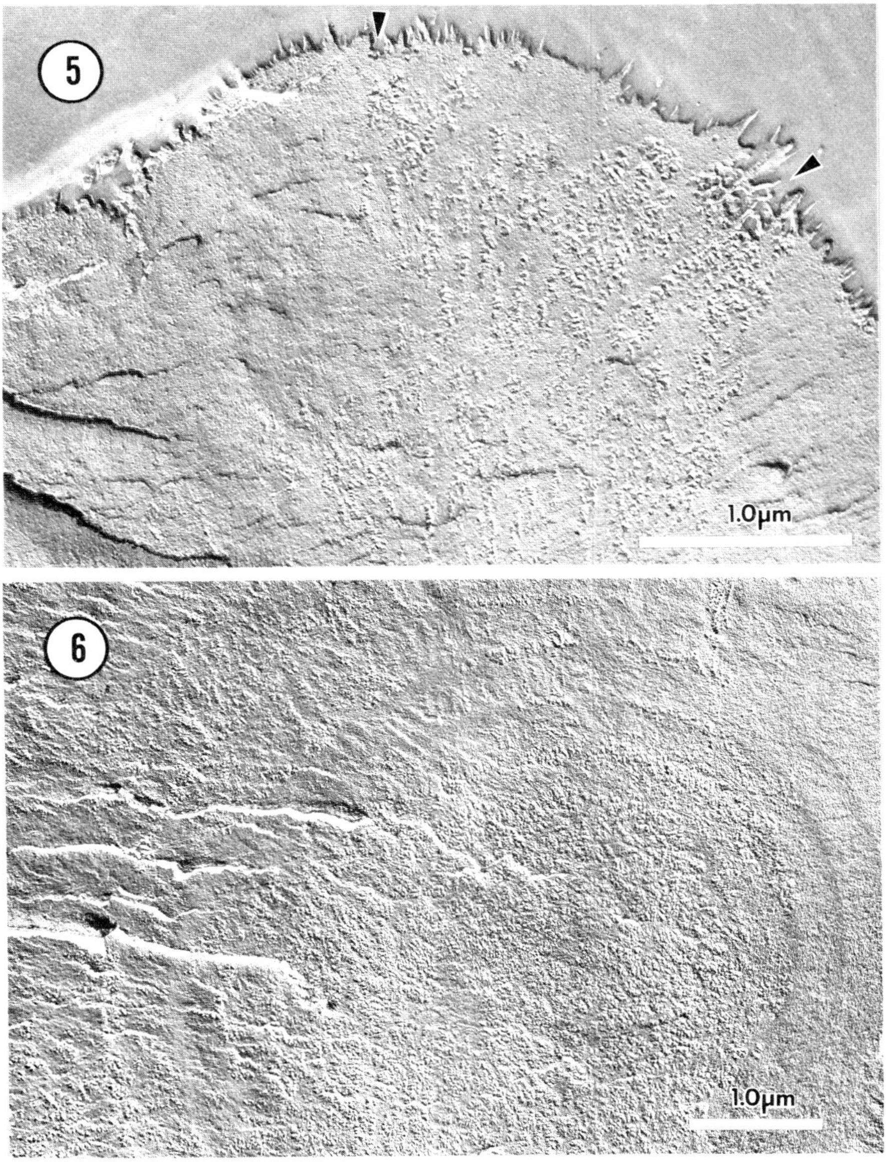

FIGURES 5 and 6. Cross-fracture of acid-treated waxy maize starch granules. Note regions of acid attack on granule surface (Fig. 5, arrows) and lamellar ring structure (Fig. 6). (Chabot et al., 1978).

variation in amylose content among amylomaize granules.

Recently, Chabot et al. (1978) have used the tedious freeze-etching technique to study the ultrastructure of waxy maize and acid-treated waxy maize starches. The freeze-etching method does not entail the use of chemical reagents and thus reduces the potential for artifact generation during preparation of the material for transmission electron microscopy. They showed that the outer surfaces of waxy maize granules are relatively smooth. After acid hydrolysis, extensive surface pitting is evident (arrows, Fig. 5). Acid hydrolysis starts at a few locations on the granule surface and proceeds radially rather than laterally into the interior of the granule. The concentric lamellar structure, as observed by Mussulman and Wagoner (1968) for corn and waxy maize starches, was confirmed (Fig. 6)

Similar morphological patterns are observed after alpha amylase action on starch granules. Gallant et al. (1973) have shown that corn, waxy maize and amylomaize granules have an endocorrosion pattern when treated with alpha amylase. Enzyme hydrolysis proceeds slowly at the periphery of the granule but rapidly in the interior. Pores are formed in the surface of the granule which ultimately are enlarged to large cavernous areas in the interior of the granule. There are differences in the patterns of enzyme penetration among the maize starches. From these enzymatic studies and the freeze-etching studies described, it is evident that the surface of all types of maize granules differs from the interior. It is unknown whether this difference occurs when the granule is present _in situ_ in the plant cell or is as a result of case hardening induced by the drying step in starch isolation.

Electron microscopy also has been a useful technique in evaluating changes in maize starch granule structure during gelatinization. At 60°C, amylomaize (55% amylose) granules show no sign of change. Hill and Dronzak (1973) reported that no morphological changes were evident until about 80°C, although changes in birefringence were apparent at temperatures as low as 69°C. At 85°C (Figs. 7 & 8), granules begin to swell and to take on the typical donut-shape, characteristic of partially gelatinized amylomaize and corn starch granules. The protuberances elongate and expand. In the case of the 70% amylose amylomaize, the protuberances seem to be interconnected. Some extragranular material is evident, especially in the 55% amylose starch.

Corn starch gelatinizes at a lower temperature than amylomaize. At 65°C, extragranular fibrillar material is evident even though little granule swelling is apparent (Fig. 9). Changes in granule shape are obvious with some granules taking on the characteristic donut-shape.

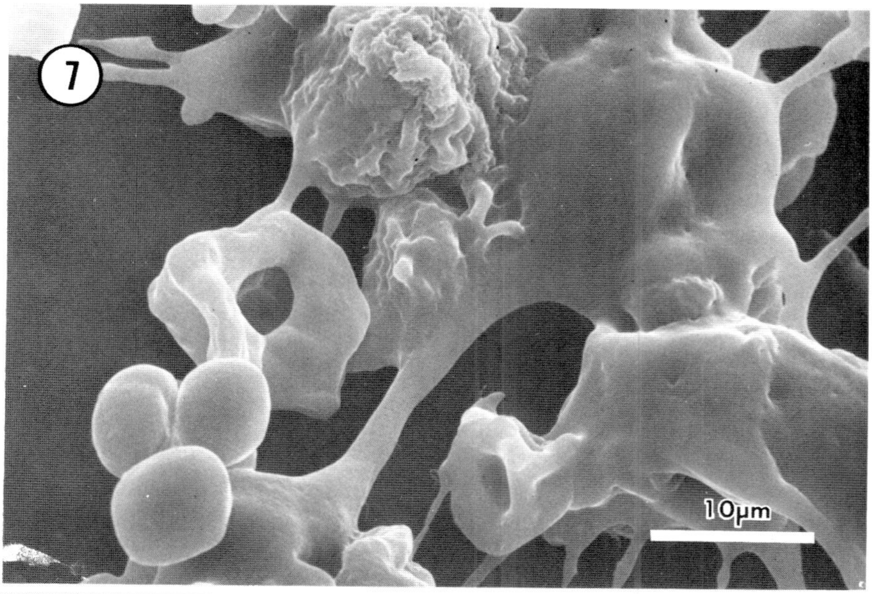

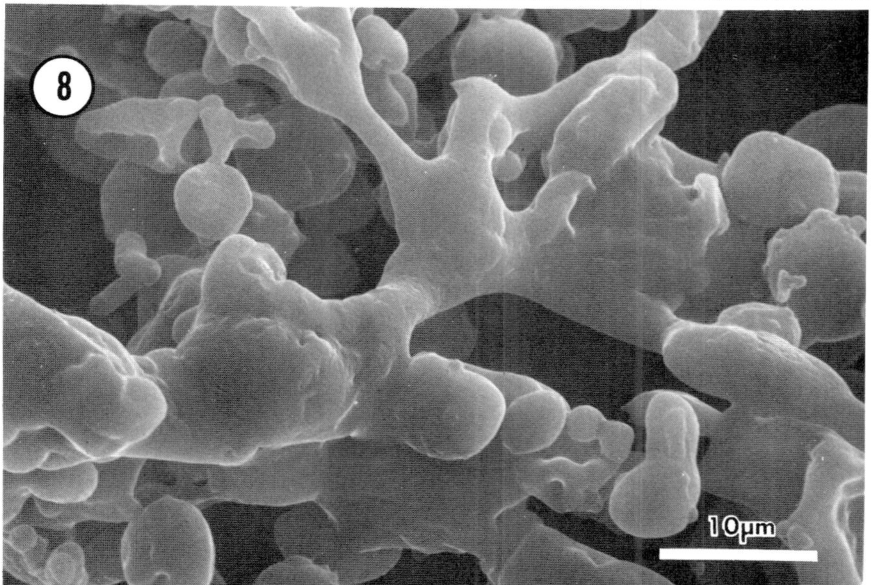

FIGURES 7 and 8. Scanning electron micrographs of amylomaize starch granules at 85°C. Fig. 7: 55% amylose, Fig. 8: 70% amylose.

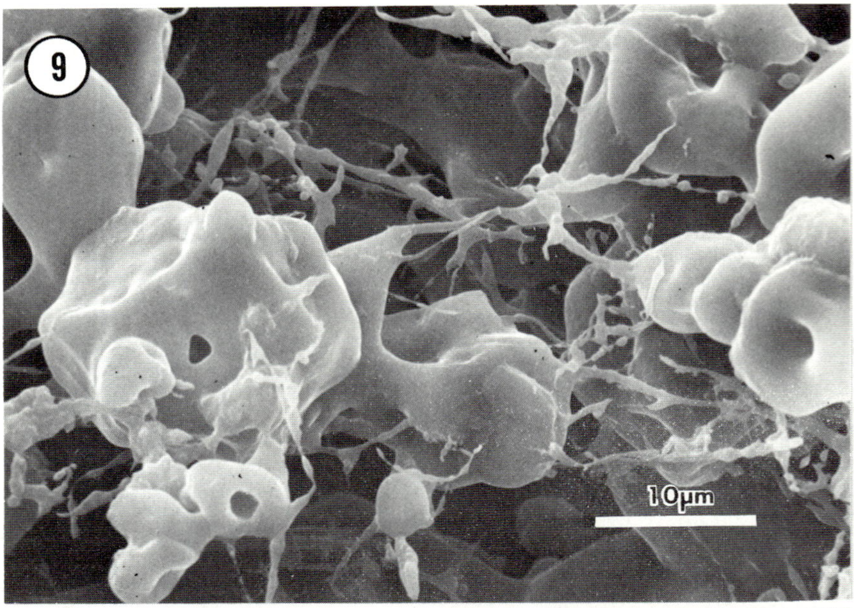

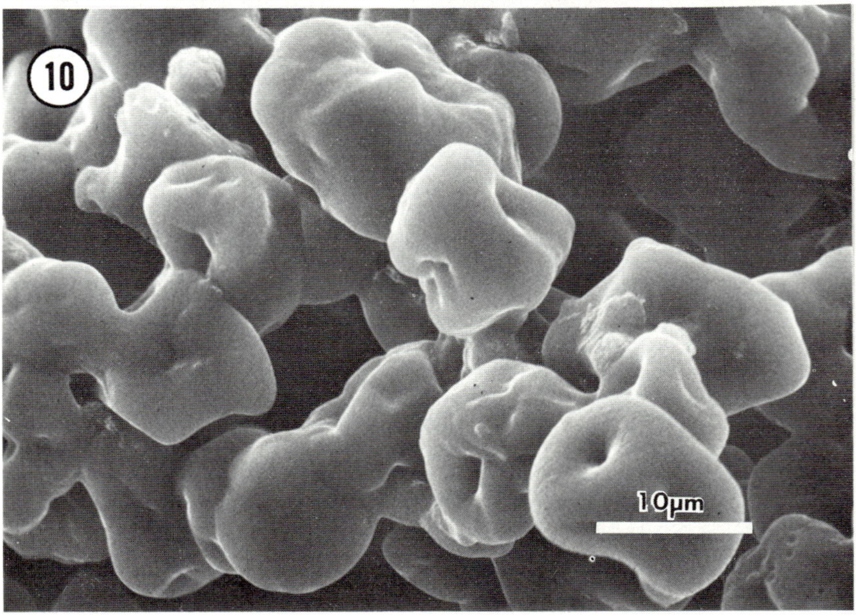

FIGURES 9 and 10. Scanning electron micrographs of unmodified (Fig. 9) and crosslinked (Fig. 10) corn starch granules at 65°C.

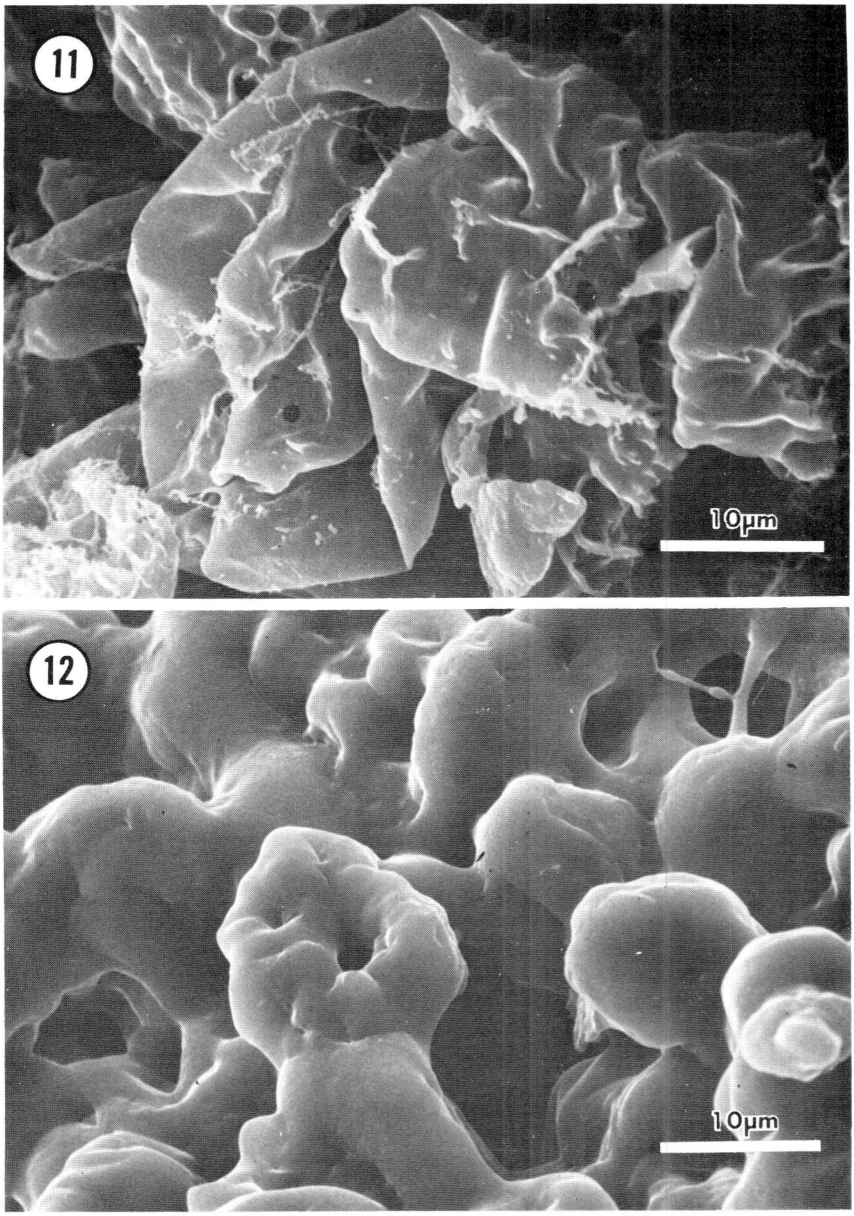

FIGURES 11 and 12. Scanning electron micrographs of unmodified (Fig. 11) and crosslinked corn starch granules at 80°C.

Hill and Dronzak (1973), Miller et al. (1973), and Allen et al. (1977) have observed similar extragranular material and have indicated that it appears before an increase in viscosity is discernable. This material is amylose (Chabot et al., 1976). As the starch suspension approaches peak viscosity, both the extragranular material and the swollen granule contribute to viscosity development.

Chemical modification of the starch can influence its structure and functional properties. Crosslinking retards granule swelling and reduces the amount of extragranular matrix (Fig. 10). Chabot et al. (1976) have shown that crosslinking inhibits the leaching of amylose from the granule during heating. At 80°C, crosslinked corn granules are swollen and deformed but still intact (Fig. 12). In contrast, when the unmodified corn starch is cooked to 80°C, granule rupture is common (Fig. 11). Obviously, the degree of swelling by crosslinked starches will be a function of the number of crosslinks, amount of shear and pH. The crosslinking reinforces the outer regions of the granule so that swelling can occur but granule rupture is inhibited (Hood et al., 1974, Allen et al., 1977).

Cooked unmodified waxy maize granules have a curley appearance with only a few fibrils evident between the granules. Cooking at moderate temperatures (65-70°C) results in almost total disruption of the granule. Crosslinked waxy maize granules are very convoluted and folded (Fig. 13). Practically no extragranular fibrular material is evident. As in the case of crosslinked corn starch, there is some alteration in granular structure during cooking but the modified granules are not disrupted.

Extrusion cooking can influence granule structure but does not appear to alter the molecular structure. Mercier and Feillet (1975) showed that all maize starches are solubilized at extrusion temperatures of 170 to 200°C but that no degradation into maltodextrin occurs. Charbonniere et al. (1973) indicated that a new structure, similar to the butanol-amylose complex, is formed in starch during extrusion cooking. The structure is not observed for waxy maize but is observed at cooking temperatures $\geq$ 135°C for corn and 90°C for amylomaize.

B. Molecular Structure

The molecular structure of starch has been of considerable research interest in recent years. This interest has been catalyzed by the discovery and availability of several enzymes; namely beta-amylase, pullulanase, isoamylase, and phosphorlase. The specificities and action patterns of these enzymes have

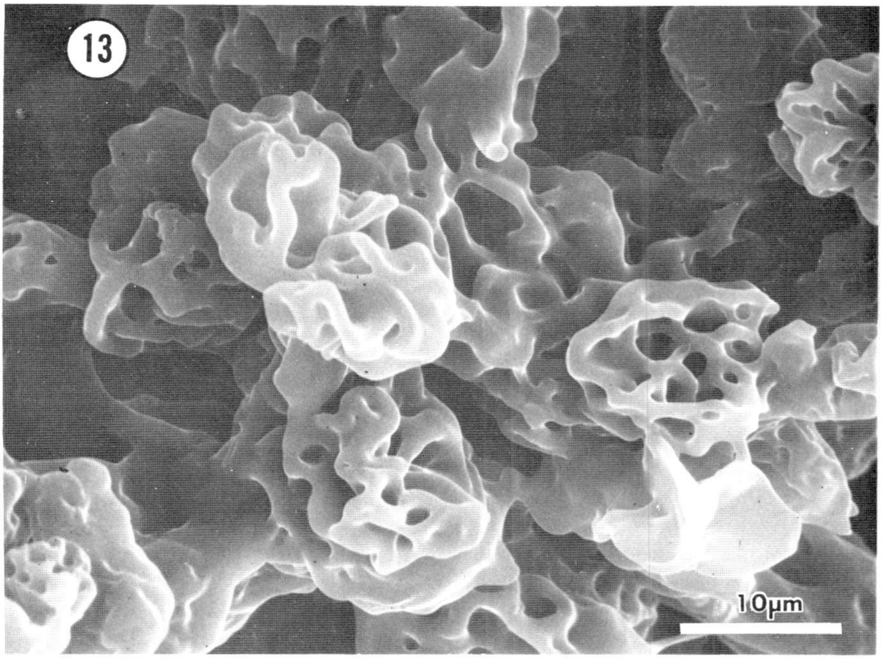

FIGURE 13. Scanning electron micrograph of crosslinked waxy maize starch granules at 85°C.

been reviewed (French, 1960; Lee and Whelan, 1966; Whelan, 1971; Marshall, 1975; Banks and Greenwood, 1975). Combined with gel permeation chromatography, these enzymes have enabled scientists to refine and extend the chemical studies of Haworth et al. (1937), Staudinger and Husemann (1937), and Meyer and Bernfeld (1940). Greenwood (1976) has reviewed the recent literature on enzymatic explorations of the molecular structure of starch. Except for amylomaize, amylopectin from most cereal and tuber starches appears to have a common structure (Lee et al., 1968; Mercier, 1973; Robin et al., 1974, 1975; Lii and Lineback, 1977; Hood and Mercier, 1978). Enzymatic studies on lintnerized (acid-treated) cereal starches has resulted in a new model for amylopectin (Robin et al., 1975). The model is based on the cluster models of Nikuni (1968) and French (1972) and contains aggregates of highly ordered $\overline{dp}$ 15 chains. The $\overline{dp}$ 15 polymers are attached to $\overline{dp}$ 45 chains through $\alpha(1\rightarrow6)$ linkages. Robin et al. (1975) have speculated that it is the ordered $\overline{dp}$ 15 chains that are primarily responsible for the crystallinity of starch.

The amylopectin fraction of amylomaize appears to differ from that of corn and waxy maize. It has higher iodine affinity, lower water solubility, lower number average molecular weight and longer average chain length (Wolff et al., 1955). Mercier (1973) showed that amylomaize amylopectin has longer inner chains ($\overline{dp} \geq 60$) than waxy maize and corn amylopectin ($\overline{dp} \sim 30$). Wolff et al. (1955) and others (Boyer et al. 1976; Mercier, 1973) proposed that amylopectin from amylomaize is intermediate in structure between amylopectin and amylose. However, Greenwood and Thompson (1962) suggested that amylomaize amylopectin is composed of two components: an amylopectin similar to normal amylopectin and a low molecular weight linear fraction. This theory was summarized and extended by Banks et al. (1974). The relative insolubility of the amylomaize has hampered progress on defining its molecular structure.

The stability of cooked maize starch dispersions is largely dependent upon the extent of retrogradation. It is the amylose component of the starch that is principally responsible for this phenomenon. Chemical modification techniques have been devised in order to provide starches that will resist retrogradation (Wurzburg and Szymanski, 1970). Bulky chemical groups are covalently linked to the starch molecule. These groups sterically interfere with retrogradation. While the modification processes are not new, some significant advances have been made in describing the molecular structures of the derivitized starches. Leegwater (1972) suggested that hydroxypropyl groups are distributed randomly on the starch molecules. Bollenback et al. (1968) demonstrated that the hydroxyethyl groups are predominately located on carbon two. The amount of multiple substitution by hydroxyethyl groups increase when the molar substitution (MS) is greater than 0.45 (Banks et al., 1973). Hood and Mercier (1978) have suggested that the modifying groups (hydroxypropyl) are concentrated in the amorphous regions of the amylopectin molecule, near the $\alpha(1\rightarrow6)$ branch points. While the work up until now has not been carried out on corn starch, it is logical to assume that the structure of modified corn starch will be similar to those starches studied to date.

Enzymatic modification can also influence retrogradation. Hofreiter et al. (1978) observed that the hydrolysis rate for amylose by immobilized alpha amylase is much greater than for amylopectin. In contrast, the soluble alpha amylase hydrolyzes the two starch fractions at approximately the same rate (Boundy et al., 1975). This preferential hydrolysis enables the preparation of a modified corn starch with a significantly reduced tendency to retrograde. The authors speculate that the amylose molecules that have a molecular weight in the range which normally retrograde are preferentially adsorbed onto the phenyl-formaldehyde resin and subsequently acted on by the immobilized enzyme.

C. Nutritional Aspects

The relative digestibilities of the various maize starches have been evaluated (Wolf et al., 1977). Alpha amylase susceptibility is related to its degree of gelatinization (Hood and Arneson, 1976; Kayisu and Hood, 1979). The relatively high temperature required to gelatinize amylomaize starch suggests that food products containing amylomaize would have lower caloric values than those containing other starches. Wolf et al. (1977) have shown that a portion of high amylose starch is indigestible and appears in the feces of both mice and humans fed diets containing the starch. No traces of corn starch were found in the feces during these studies. The nutritional significance of these observations needs to be explored.

Mineral binding by starch has been a subject of renewed study in recent years. In 1961, Leach et al. demonstrated that the absorptive affinity of starch for alkaline metals is not markedly affected by the species of the starch, the content of the linear fraction, the granule size, or the micellar organization within the granule. Various cations at relatively high concentrations induce gelatinization or lower the gelatinization temperatures of starch (Lancaster and Conway, 1968; Gough and Pybus, 1973). Hood and O'Shea (1977) demonstrated that corn, waxy maize and tapioca starches bind appreciable amounts of calcium. Since tapioca and corn starch have about the same amylose content, differences in degree of binding among starches is probably due to differences in granular structure and not to amylose:amylopectin ratios within the granules. In fact, gelatinization of the starch drastically reduces its ability to complex with calcium. Recently, Sukan et al. (1979) showed that the addition of calcium reduces the digestibility of uncooked corn starch. This effect is not the same for all starches since the digestibility of raw potato starch is increased by the addition of calcium. The interaction between metal ions and starch may turn out to be of greater consequence than currently envisioned because not only may it affect the bioavailability of minerals (Hood et al., 1976), but it may also influence metal-requiring chemicals and biochemical reactions in starch-containing foods.

IV. CELL WALL POLYSACCHARIDES

The growing interest in dietary fiber has focused new attention on the cell wall polysaccharides. Surprisingly, there are few reports in the literature on the composition and the relative physiological and nutritional effects of corn bran. Van Soest (1973) reported that corn bran is about 15-16% cellulose and 44% hemicellulose. Recently, Dintzis et al. (1979) applied a modification of the Van Soest method and obtained substantially higher values for apparent hemicellulose and cellulose in corn bran. Corn bran is high in hemicellulose relative to soy hulls and wheat bran (Dintzis, 1979). It has a comparatively high dietary fiber content of 92% (88% for soy hulls and 52% for wheat bran). When the corn bran is fed to humans, practically all of it is recovered in the feces (Dintzis, 1979). Furthermore, scanning electron micrographs reveal that the shape and dimensions of ingested corn bran is not altered by passage through the human alimentary tract. The physiological and nutritional significance of the high resistance of corn bran to digestion are unknown at this time.

V. ANALYTICAL METHODOLOGY

Significant progress has been made in recent years in the development of improved analytical methods for corn syrup and starch analysis. Much of this work has been carried out in industrial laboratories. In part, the motivation for the development of these methods has been the rapid growth in the high fructose corn syrup industry. Rapid, sensitive methods are required for on-line processing control and end product analysis. Scobell et al. (1977) have described an automated liquid chromatography method for the separation and quantification of sugar mixtures. The method is sensitive enough to separate the principle sugar components in high fructose corn syrup. Similar methods have been applied in the fractionation of starch hydrolyzates. Oligosaccharides of $\overline{\text{dp}}$ 1-15 can be separated within 75 minutes (Brobst, 1978).

Reliable enzymatic methods have been developed for quantification of starch in foods. One method is based on the hydrolysis of solubilized starch with glucoamylase and the subsequent quantification of the hydrolysis product, glucose, by glucose oxidase (Thivend et al. 1972) In addition the Corn

Refiners Association has over 30 methods for the analysis of unmodified and modified corn starch. Several procedures for identifying and quantifying modifying groups have been described (Stahl and McNaught, 1970; Mourtis et al., 1976). Newer techniques for determining residual reactants and minor reaction products from modified starch manufacture have recently been summarized by Brobst (1978).

ACKNOWLEDGMENTS

The assistance of Myra Liboff in preparation of the micrographs is greatly appreciated.

REFERENCES

Allen, J. E., Hood, L. F., and Chabot, J. F. (1977). *Cereal Chem.* 54, 783.

Banks, W., Greenwood, C. T., and Muir, D. D. (1973). *Br. J. Pharmacol.* 47, 172.

Banks, W., Greenwood, C. T., and Muir, D. D. (1974). *Starke* 26,289.

Banks, W., and Greenwood, C. T. (1975). "Starch and its Components". Halsted Press, Edinborough.

Bollenback, G. N., Golik, R. S., and Parrish, F. W. (1968). *Cereal Chem.* 46, 304.

Boundy, J. A., Smiley, K. L., Swanson, C. L., and Hofreiter, B. T. (1975). *Carbohydr. Res.* 48, 239.

Boyer, C. D., Garwood, D. L., and Shannon, J. C. (1976). *Starke* 28, 405.

Brobst, K. M. (1978). Seminar proceedings. Products of the Corn Refining Industry in Food. p. 62 Corn Refiners Association, Inc., Washington.

Cerning, J. (1970). These de doctorat d'Universite, Lille, France.

Chabot, J. F., Hood, L. F., and Allen, J. E. (1976). *Cereal Chem.* 53, 85.

Chabot, J. F., Allen, J. E., and Hood, L. F. (1978). *J. Food Sci.* 43, 727.

Charbonniere, R., Duprat, F., and Guilbot, A. (1973). *Cereal Sci. Today* 18(a), 57 (Abstr.).

Dintzis, F. R., Legg, L. M., Deatherage, W. L., Baker, F. L., Inglett, G. E., Jacob, R. A., Reck, S. J., Munoz, J. M., Klevay, L. M., Sandstead, H. H., and Shuey, W. C. (1949). *Cereal Chem.* 56, 123.

Earle, F. R., Curtis, J. J., and Hubbard, J. E. (1946). *Cereal Chem.* 23, 504.

French, D. (1960). *Bull. Soc. Chim. Biol.* 42, 1677.
French, D. (1972). *J. Jpr. Soc. Starch Sci.* 19, 8.
Gallant, D., Derrien, A., and Guilbot, A. (1973). *Starke* 25, 26.
Gentinetta, E., Zambello, M., and Salamini, F. (1979). *Cereal Chem.* 56, 81.
Gough, B. M., and Pybus, J. N. (1973). *Starke* 25, 124.
Greenwood C. T., and Thompson, J. (1962) *J. Chem. Soc. (London)* 222.
Greenwood, C. T. (1976). In "Advances in Cereal Science and Technology, Vol. II". (Y. Pomeranz, ed.) p. 119-157. American Association of Cereal Chemists. St. Paul.
Harness, J. (1978). Seminar proceedings. Products of the Corn Refining Industry in Food. p. 7. Corn Refiners Association, Inc. Washington.
Haworth, W. N., Hurst, E. L., and Isherwood, F. A. (1937). *J. Chem Soc.* 577.
Hill, R. D., and Dronzak, B. L. (1973). *Starke* 25, 367.
Hofreiter, B. T., Smiley, K. L. Boundy, J. A., Swanson, C. L., and Fecht, R. J. (1978). *Cereal Chem.* 55, 995.
Hood, L. F., Seifried, A. S., and Meyer, R. (1974). *J Food Sci.* 39, 117.
Hood, L. F., and Arneson, V. G. (1976). *Cereal Chem.* 53, 282.
Hood, L. F., Van Campen, D. R., House, W. A., and Szatkowski, E. (1976). *J. Nutr.* 106, 1768.
Hood, L. F., and O'Shea, G. K. (1977). *Cereal Chem.* 54, 266.
Hood, L. F. and Mercier, C. (1978). *Carbohyd. Res.* 61, 53.
Kayisu, K., and Hood, L. F. (1979). *J. Food Sci.* 44 (*In Press*).
Lancaster, E. B., and Conway, H. F. (1968). *Cereal Sci. Today* 13, 248.
Lasseran, J. C. (1973). *Starke* 25,257
Leach, H. W., Schoch, T. J., and Chessman, E. F. (1961). *Starke* 13, 200.
Lee, E. Y. C., and Whelan, W. J. (1966). *Arch. Bioch. Biophys.* 116, 162.
Lee, E. Y. C., Mercier, C., and Whelan, W. J. (1968). *Arch. Biochem. Biophys.* 125, 1028.
Leegwater, D. G. (1972). *Starke* 24, 11.
Lii, C. Y., and Lineback, D. R. (1977). *Cereal Chem.* 54, 138.
Marshall, J. J. (1975). *Starke* 27, 377.
Mercier, C. (1973). *Starke* 25, 78.
Mercier, C., and Feillet, P. (1975). *Cereal Chem.* 52, 283.
Meyer, K. H., and Bernfeld, P. (1940). *Helv. Chim. Acta* 23, 875.
Miller, B. S., Derby, R. I., and Trumbo, H. B. (1973). *Cereal Chem.* 50, 271.

Mourtis, J. W., Merkus, H. G., and de Galan, L. (1976). *Anal. Chem.* *48*, 1557.
Mussulman, W., and Wagoner, J. A. (1968). *Cereal Chem.* *45*, 162.
Nikuni, Z. (1969). Science of Cookery 2:6. Quoted in Nikuni, Z. (1978). Studies on Starch Granules. *Starke* *30*, 105.
Peplinski, A. J., Brekke, O. L., Griffin, E. L., Jr., Hall, G., and Hill, L. E. (1975). *Cereal Foods World* *20*, 145.
Robin, J. P., Mercier, C., Charbonniere, R., and Guilbot, A. (1974). *Cereal Chem.* *51*, 389.
Robin, J. P., Mercier, C., Duprat, F., Charbonniere, R., and Guilbot, A. (1975). *Starke* *27*, 36.
Scobell, H. D., Brobst, K. M., and Steele, E. M. (1977). *Cereal Chem.* *54*, 905.
Shannon, J. C., and Creech, R. G. (1973). *Ann. N.Y. Acad. Sci.* *210*, 279
Stahl, H. and McNaught, R. P. (1970). *Cereal Chem.* *47*, 345.
Staudinger, H., and Husemann, E. (1937). *Ann. Chem.* *527*, 195.
Sukan, G., Kearsley, M. W., and Birch, G. G. (1979). *Starke* *31*, 125.
Thivend, P., Mercier, C., and Guilbot, A. (1972). In Methods in Carbohyd. Chem. Vol VI. p. 100. Academic Press.
Van Soest, P. J. (1973). Proc. Cornell Nutr. Conf., Ithaca.
Vojnovich, C., Anderson, R. A., and Griffin, E. L., Jr. (1975). *Cereal Foods World* *20*, 333.
Watson, S. (1967) In "Starch:Chemistry and Technology", Vol. II. (R. Whistler and E. Paschall, eds). p. 1 Academic Press, New York.
Whelan, W. J. (1971). *Biochem. J.* *122*, 609.
Wolf, M. J., Seckinger, H. L., and Dimler, R. J. (1964). *Starke* *16*, 375.
Wolf, M. J., Khoo, U., and Inglett, G. E. (1977). *Starke* *29*, 401.
Wolff, I. A., Hofreiter, B. T., Watson, P. R., Deatherage, W. C., and Mac Masters, M. M. (1955). *J. Amer. Chem. Soc.* *77*, 1654.
Wurzburg, O., and Szymanski, C. D. (1970). *J. Agr. Food Chem.* *18*, 997.

ENZYME TECHNOLOGY IN THE MANUFACTURE OF SUGARS FROM CEREALS

Steen Helde Hemmingsen
Barrie E. Norman

Novo Research Institute
Denmark

I. INTRODUCTION

The term, sugar, is commonly related to the sweet tasting substance obtained from a gigantic tropical grass, sugar cane, or from sugar beets.

For many years cane was the only important source of sugar, then, as a result of serious shortages during the Napoleonic wars sugar production from beets was developed.

Attempts were also undertaken to exploit Kirchoff's discovery from 1811 that starch when boiled with acids can be converted into a sweet liquor which was later on demonstrated to contain glucose.

In this way the foundation was laid for a starch-based sweetener industry.

In nature starch serves as the prime energy reserve for most plants; it is found in seeds, bulbs and tubers, and it is from these sources that the commercial starches are obtained.

A variety of methods are used to separate the starch from other components such as protein, cellulose, hemicellulose and oil, which may also be present.

Fig. 1 lists some important commercial sources of starch products. By-products such as oil and protein from the corn wet milling process contribute significantly to the economic feasibility of the operation.

ISBN: 0-12-370960-1

Source	Processes	Products
Maize (corn) Wheat Potato Sago Cassava Rice	Milling Extraction Separation	Starch and by-products

G~0690

FIGURE 1. Manufacture of starches

Starch consists of two components, a linear glucose polymer, amylose, containing 1,4-alpha glucosidic links and a branched polymer, amylopectin, in which linear chains of 1,4-alpha glucose residues are interlinked by 1,6-alpha glucosidic bonds.

The glucosidic bonds in starch can by hydrolyzed to produce syrups containing glucose, maltose and other oligosaccharides.

By controlling the degree of hydrolysis products with desired properties may be obtained.

Traditionally, starch was hydrolyzed by means of acid catalysis. Acid hydrolysis is, however, associated with the formation of undesired by-products, especially at high degrees of conversion. The major limitation is the completely random nature of the acid catalyzed hydrolysis which makes the sugar composition solely dependent on the degree of hydrolysis.

This makes it difficult to adapt the properties to the highly diversified requirements of modern food processing.

G~0691~1

Conversion:	Acid		Acid/ enzyme		Enzyme			
% saccharides[1] DE	42	55	65	45	55[2]	65	98	98
DP_1, Glucose	20	31	39	5	-	39	96	54
DP_1, Fructose	-	-	-	-	-	-	-	42
DP_2	14	18	31	50	90	35	2	2
DP_3	12	12	7	20	6	11	-	-
Higher saccharides	54	39	23	25	4	15	2	2

[1]) on carbohydrate basis [2]) Laboratory experiment

Saccharide compositions of acid and enzyme converted starch

FIGURE 2. Saccharide compositions of acid and enzyme converted starch.

Besides sweetness, physical properties such as viscosity, osmotic pressure, crystallization retardation, gloss, humectancy, consistency and many others are considered important.

To obtain the desired properties, the manufacturer will look for specific saccharide compositions, and in this regard the specific action pattern of enzymes plays a key role.

II. ENZYMES FOR MANUFACTURE OF STARCH SYRUPS

The starch converting enzymes used industrially may be classified into four main groups (Norman, 1978):

Endo-Amylases
Exo-Amylases
Debranching Enzymes
Isomerases

A. Endo-Amylases

Endo-amylases are alpha-amylases which hydrolyze the 1,4-alpha bonds in the inner regions of amylose and amylopectin.

This results in a rapid decrease in the viscosity of gelatinized starch solutions and a decrease in the iodine staining ability of amylose.

The endo-amylases may be sub-classified into thermostable amylases which are used mainly for high temperature starch liquefaction and thermolabile amylases which are used mainly in saccharification.

1. Thermostable Alpha-Amylases. Enzymatic hydrolysis of raw (granular) starch is slow and therefore gelatinization prior to enzyme action is normally practised.

For many common starches which are converted at concentrations above 30% DS heating to temperatures higher than 100°C is necessary to achieve complete gelatinization and ideal filtration properties after the subsequent saccharification. This emphasizes the importance of the heat stability of alpha-amylases.

Amylases produced by _Bacillus amyloliquefaciens (subtilis)_ have been used widely for starch liquefaction, but a major improvement was achieved in 1973 when an even more heat stable enzyme from _Bacillus licheniformis_ was introduced under the trade name Termamyl$^{(R)}$ (Madsen et al, 1973). This enzyme tolerates temperatures which are 20°C higher than those used with the ordinary bacterial amylases. In addition its stability is less dependent on the presence of calcium ions.

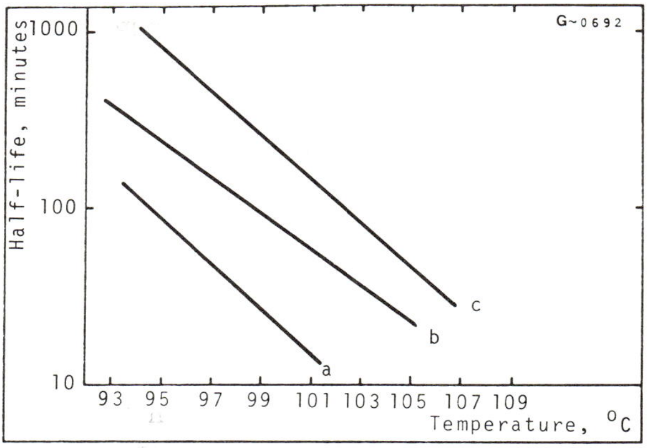

FIGURE 3. Temperature stability of Termamyl in corn starch slurry. Substrate concentration: 30% w/w; pH 6.5. {Ca^{++}}: a: 5 ppm, b: 20 ppm, c: 70 ppm.

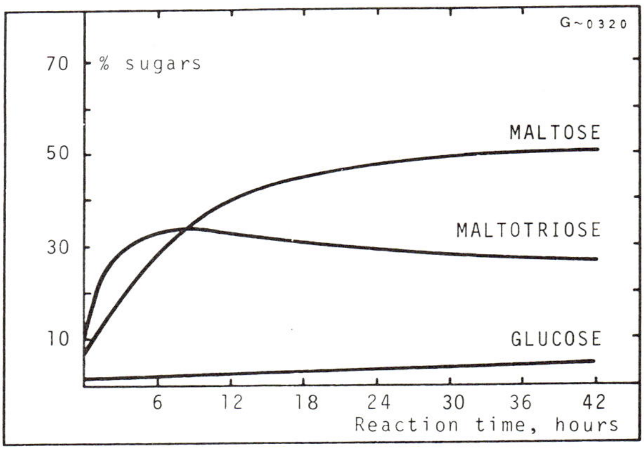

FIGURE 4. Saccharide composition obtained with Fungamyl. Substrate: 40% DS, 20 DE enzyme liquefied corn starch. pH 5.0, 50°C, Fungamyl dosage: 0.16 KNU/g DS.

2. Thermolabile Alpha-Amylases. Thermolabile alpha-amylases are not sufficiently heat stable for industrial starch liquefaction, but they find limited use in the production of syrups rich in maltose.

An example of maltogenic alpha-amylase is the fungal alpha-amylase produced by Aspergillus oryzae.

The carbohydrate spectrum obtained using an enzyme from this group is reproduced in Fig. 4.

If Fungamyl$^{(R)}$ is used alone, the reaction is virtually self-terminating. A maltose content up to about 60% can be achieved without the glucose concentration exceeding 7%. These so-called high maltose syrups inhibit crystallization in sugar blends and are relatively non-hygroscopic. They are therefore used in the manufacture of, for example, hard candy and frozen desserts.

Fungal alpha-amylase is also used together with glucoamylase for the production of so-called "High Conversion" (60-70 DE) syrups.

B. Exo-Amylases

Exo-amylases catalyze the hydrolysis of 1,4-alpha glucosidic bonds in starch by successively removing glucose or maltose units from the non-reducing chain end.

The glucogenic exo-amylases cleave the 1,4-alpha glucosidic bonds and also, though at a much slower rate, the 1,6-alpha bonds. The maltogenic exo-amylases such as cereal beta-amylase cleave only the 1,4-alpha bonds.

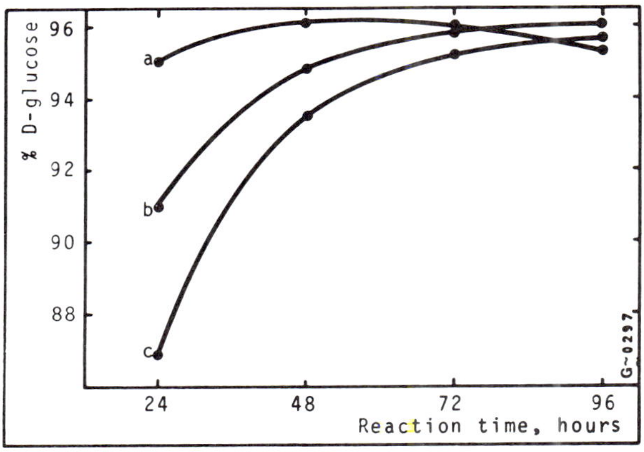

FIGURE 5. The influence of glucoamylase dosage and reaction time on D-glucose level. Substrate: 30% DS, 13 DE enzyme liquefied corn starch. pH 4.3, 60°C; a: 0.32, b: 0.16, c: 0.12 AG units/g DS.

1. Glucogenic Exo-Amylases. Glucogenic exo-amylases are commonly known under the designation glucoamylase or amyloglucosidase. Commercial products originate from either *Aspergillus niger* or *Rhizopus sp*.

Glucoamylase is able to effect a total conversion of starch into glucose. Yields are, however, affected by the ability of the enzyme to polymerize glucose mainly into maltose and isomaltose. This reversion reaction is favoured at the fairly high substrate concentrations used in the industrial process and complete conversion is therefore not achieved in practise.

2. Maltogenic Exo-Amylases (Beta-Amylases). Beta-amylases are found in cereals such as barley, wheat, rye and in soya beans and sweet potatoes. In recent years, beta-amylases from microbial sources have also been described (Marshall, 1974; Higashihara & Okada, 1976; Takasaki, 1974).

Beta-amylases may be used for the manufacture of maltose syrups. The saccharide spectrum will be different from the one obtained with the maltogenic alpha-amylases because of the difference in action pattern of the two types of enzyme.

While syrups obtained with maltogenic alpha-amylases contain a high amount of maltotriose, syrups obtained with beta-amylase contain large amounts of beta-limit dextrin.

C. Debranching Enzymes

The majority of industrial starches contain at least 70% amylopectin. The 1,6-alpha glucosidic linkages are resistant to attack by both alpha- and beta-amylases. Thus the prolonged action of these enzymes will result in the formation of limit-dextrins along with the sugars.

Enzymes which are able to hydrolyze the branch points in amylopectin and partially hydrolyzed amylopectin have been described and at least one type, pullulanase from *Klebsiella aerogenes* is used commercially to increase the yield in the manufacture of crystalline dextrose.

Another debranching enzyme is isoamylase which is not yet produced on an industrial scale operation.

In addition to the manufacture of dextrose, debranching enzymes may be used along with an exo-acting maltogenic enzyme such as barley beta-amylase for the production of syrups with maltose content above 80%.

The prevention of the formation of glucose polymers containing an odd number of glucose units is essential in this case because these would lead to formation of maltotriose during hydrolysis with beta-amylase.

G-0699

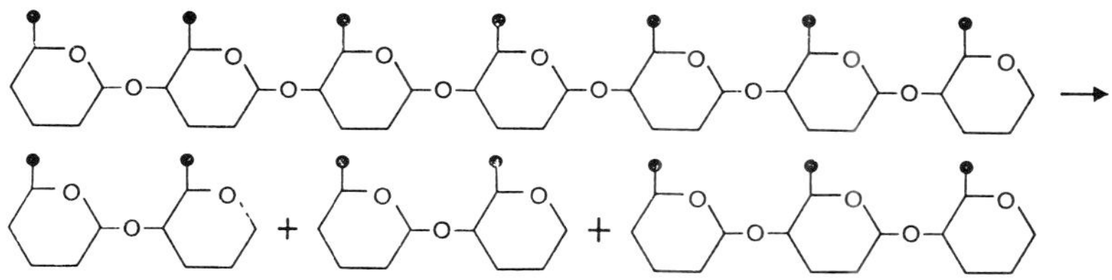

Reaction time hours	% saccharides, carbohydrate basis			
	Glucose	Maltose	Maltotriose	Higher
3	0	85.4	4.7	9.8
72	0	91.1	5.5	3.5

Substrate: 10% DS corn starch, gelatinized
pH 4.0 & 5.0, 50°C

FIGURE 6. Saccharide composition obtained by using isoamylase plus beta-amylase.

D. Glucose Isomerase

Glucose isomerase is quite distinct from the hydrolytic enzymes described in so far as it catalyses the isomerization of an aldose, glucose to the corresponding ketose, fructose, and vice versa.

Marshall and Kooi (1957) reported that an enzyme from Pseudomonas hydrophila, known as xylose isomerase was also able to isomerize glucose into fructose. Later investigations showed that similar isomerases are produced by other micro-organisms. The enzymes were found to be intracellular, produced in small quantities and were thought to require the presence of toxic salts for activity. The potential industrial process for conversion of glucose to fructose thus did not appear too promising for a number of years.

The industrial breakthrough came with the exploitation of the enzyme immobilization technique (Thomson et al, 1974).

Based on the work of Takasaki and his associates (Takasaki et al, 1969) the American company Standard Brands Inc. utilized glucose isomerase from Streptomyces immobilized by adsorption on DEAE cellulose. The isomerization was carried out in reactors having shallow beds of immobilized isomerase (Schnyder, 1974).

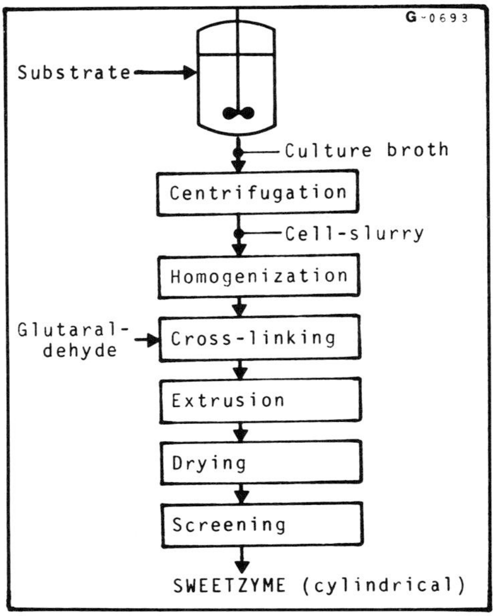

FIGURE 7. Survey of major steps in the manufacture of immobilized glucose-isomerase, Sweetzyme.

An immobilized enzyme suited for continuous column isomerization was introduced in 1975 under the trade name, Sweetzyme$^{(R)}$, (Zittan et al, 1975).

The enzyme is produced by a thermophilic strain of Bacillus coagulans in continuous culture fermentation. It is immobilized by crosslinking with glutaraldehyde (Hemmingsen, 1979).

The properties of this immobilized isomerase were thoroughly evaluated with the emphasis on applicability for a continuous column isomerization process.

Some of the important characteristics are listed in Fig.8. It is noted that half life for activity at 60°C is 1400 hours. In the industrial process running at about 60°C the enzyme is utilized over 2 or 3 half lives so that once a column is packed, it will run continuously for a period of 4 to 6 months before the enzyme is discarded. At this point 1 kg of enzyme will have produced about 3000 kgs of fructose syrup solids.

Initial activity (micromoles F/min./g, 65°C):
250 IGIC/g
Half-life in 40% DS dextrose solution at pH8.2-7.0
a) 2600 hours at 57°C
b) 1400 hours at 60°C
c) 500 hours at 65°C
pH-optimum: 8.2-7.0 (measured at 25°C)
Wet bulk density: 0.3 g/ml
G~0694

FIGURE 8. Sweetzyme characteristics.

III. MANUFACTURE OF FRUCTOSE SYRUP

The starch converting enzymes are utilized industrially for the manufacture of crystalline dextrose and a highly diversified range of syrups.

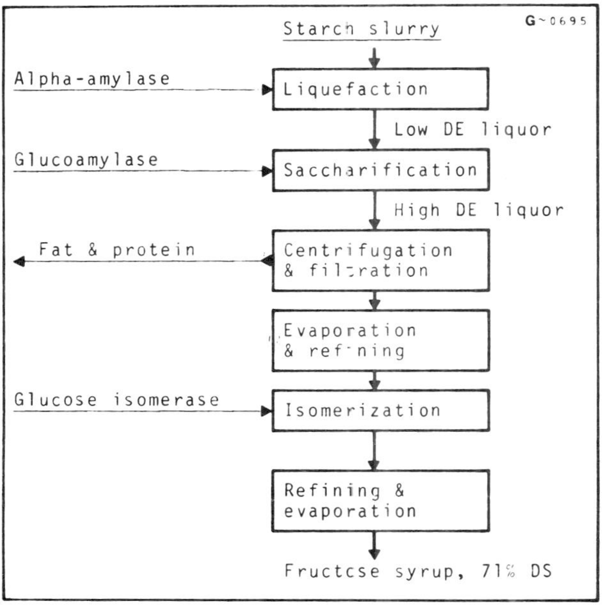

FIGURE 9. Survey of important steps in the manufacture of fructose syrup.

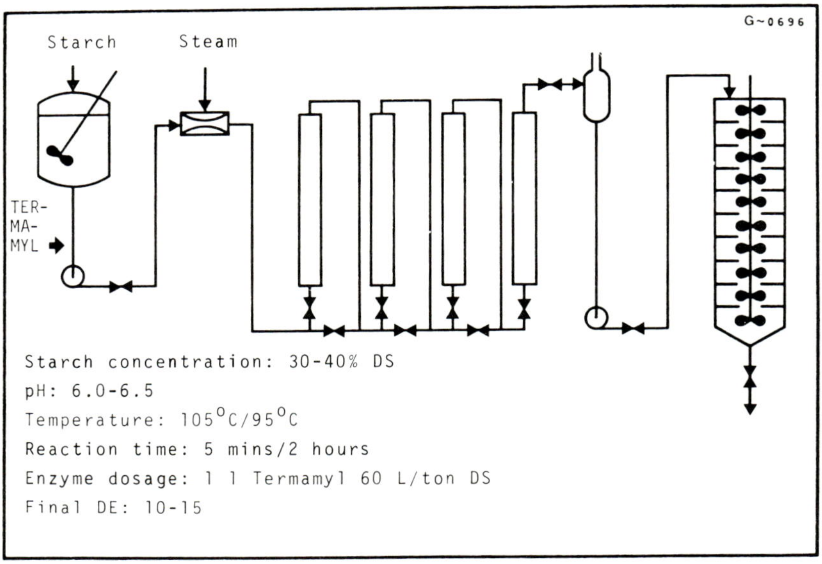

FIGURE 10. Starch liquefaction process.

An outstanding example of the utilization of enzyme technology is the production of fructose syrup.

The manufacturing process comprises three enzymatic steps, liquefaction, saccharification and isomerization and a number of purification and evaporation steps.

A. Liquefaction

The starch liquefaction process is outlined in Fig. 10. A starch slurry containing 30-40% DS is prepared in the feed tank and the pH is adjusted to 6-6.5 with sodium hydroxide.

An enzyme dosage of about 1 litre Termamyl(R) per ton of dry starch is fed into the starch stream and the temperature is raised to 105°C by direct steam injection in a jet cooker.

Some mechanical thinning is obtained by the high shearing effect of the jet cooking. Together with the viscosity reducing effect of the enzyme, this prevents the occurrence of peak viscosities.

After passing the jet cooker, the gelatinized starch is held in a series of pressurized holding cells at 105°C for a period of 5 minutes and then flashed to atmospheric pressure. The liquefaction is allowed to continue for about 2 hours at 95°C.

Saccharification process	
Dextrin concentration:	30-40% DS
pH:	4.5
Temperature:	60°C
Reaction time:	48 hours - 120 hours
Enzyme dosage:	[illegible].5-1.0 l/ton DS (AMG 150)
Final DE:	98
% dextrose:	~95

FIGURE 11. Saccharification process.

After this treatment the liquefied starch will have a dextrose equivalent, DE, of 10-15 (DE is defined as reducing sugars expressed as dextrose and calculated as percentage of dry substance).

B. Saccharification

After liquefaction, the pH is adjusted to 4.5 and the temperature quickly decreased to 60°C. Glucoamylase is added and the hydrolysis reaction allowed to continue for 48 hours to more than 100 hours depending on enzyme dosage.

The saccharification equipment usually consists of a number of large stirred tanks operated in series.

The aim of the saccharification process is to produce syrup with the maximum content of glucose.

When the required level is reached, the reaction should be terminated in order to avoid glucose polymerization. This is often accomplished as a result of subsequent refining with a strong cation-exchange resin which removes the enzyme activity.

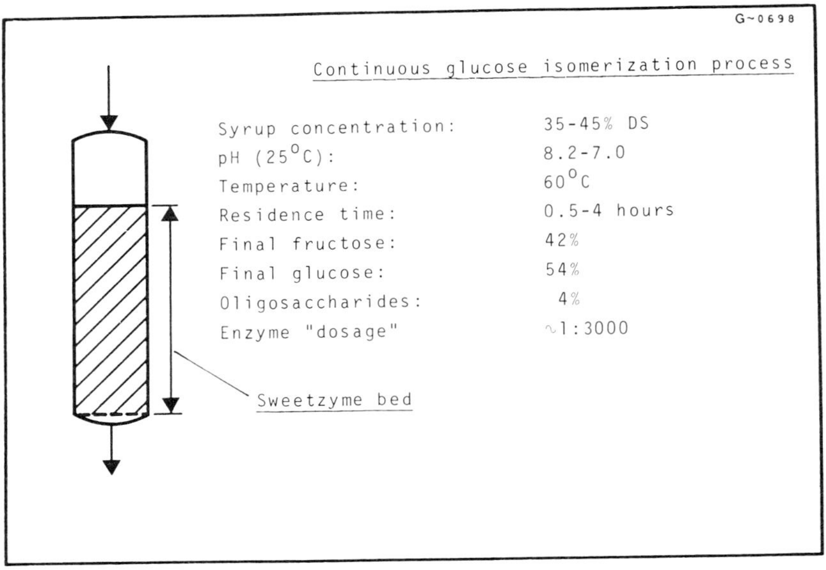

FIGURE 12. Continuous glucose isomerization process.

C. Isomerization

After saccharification the liquor is filtered to remove particulate matter which otherwise could cause clogging of the isomerization reactor. Further purification is done by means of activated carbon and ion-exchange treatment.

This pre-isomerization refining is important in order to achieve maximum productivity of the glucose isomerase. The syrup is then evaporated to the desired concentration and pumped to the isomerization plant.

Isomerization is accomplished by passing the purified syrup continuously through a fixed bed of isomerase at a proper flow rate.

In addition to the convenience of continuous operation, the process provides a short residence time whereby optimal conditions for productivity of the enzyme can be selected without risk of significant by-product formation.

For a given enzyme bed, residence time determines the degree of conversion and hence the fructose concentration.

The isomerization is a reversible reaction with equilibrium fructose concentration about 50% at 60°C. It has become industrial practise to run the isomerization to about 42% fructose where the sweetness of the reaction mixture is almost equivalent to that of sucrose in many applications.

An industrial isomerization plant normally comprises several reactors which are operated in a staggered mode to even out the decrease in flow caused by the gradual decrease of enzyme activity.

The design of the individual reactor is based on criteria which are the results of thorough evaluation of enzyme properties such as activity, wet bulk density and resistance to pressure drop (Norsker et al, 1979).

A plant with a capacity of 400 tons DS per day may thus have 6 isomerization reactors with enzyme bed heights of 5 m.

After isomerization, pH is adjusted to 4-5, the syrup is then refined and evaporated to 71% DS.

IV. CONCLUSION

A broad range of sweeteners produced by enzymatic converstion of starch is available to the food industry. Until recently the use of these sweeteners was related mainly to their functional properties rather than their sweetening power.

The introduction of the enzymatic process for isomerization of glucose changed this so that to-day a starch-based alternative to cane sugar exists as the consumption figures reflect.

	1976	1977	1978
U S A			
Industrial sugar	5700	6030	5600
High fructose syrup, DS	710	920	1200
Japan			
Industrial sugar	2100	2175	2250
High fructose syrup, DS	140	160	170
Western Europe			
Industrial sugar	6800	6900	7000
High fructose syrup, DS	60	80	100
Consumption in 1000 tons			G~0700

FIGURE 13. Consumption of sugar and fructose syrup for industrial use.

Syrups of virtually any combination of functional properties and sweetness can now be produced from starch by means of enzyme technology.

ACKNOWLEDGMENT

Figures and tables are used by permission from Novo Industri A/S, Copenhagen.

REFERENCES

Hemmingsen, S.H. (1979). In "Applied Biochemistry and Bioengineering" (L.B. Wingard, Jr., E. Katchalski-Katzir, L. Goldstein, eds.) Vol. 2, pp 157, Academic Press, New York, In Print.

Higashihara, M., Okada, S. (1976), Agricultural and Biological Chemistry 38, 1023.

Madsen, G.B., Norman, B.E., Slott, S. (1973), Stärke 25, 304.

Marshall, J.J. (1974), FEBS Letters 46, 1.

Marshall, R.O., Kooi, E.R. (1957), Science 125, 648.

Norman, B.E. (1978). Proceedings of the Society for General Microbiology 83rd Ordinary Meeting, Academic Press, In Print.

Norsker, O., Gibson, K., Zittan, L. (1979), Stärke 31, 13.

Schnyder, B.J. (1974), Stärke 26, 409.

Takasaki, Y., Kosugi, Y., Kanbayashi, K. (1969). In "Fermentation Advances" (D. Perlman ed.) pp 561, Academic Press, New York.

Thomson, K.W., Johnson, R.A., Lloyd, N.E. (1974), US Patent 3788945.

Zittan, L., Poulsen, P.B. and Hemmingsen, S.H. (1975), Stärke 27, 236.

ENGINEERING TECHNOLOGY FOR THE MANUFACTURE OF SUGARS FROM CEREALS

Cornelis Bos

DDS-KRØYER A/S
Copenhagen, Denmark

I. INTRODUCTION

For many years starch and starch containing materials have been converted by means of high temperatures in the presence of catalysts into soluble sweet products.

Historically Kirchoff in 1811 found, when potato starch slurry was treated with acid, a sweet tasting syrup was produced, from which dextrose crystallized.

Since his discovery the starch converting industry has made enormous technological progress, especially over the past 30 years.

DDS-KRØYER A/S a danish company is closely associated with the design and engineering of plants producing sweeteners from starch and cereals.

In this paper I propose to describe briefly the modern technology associated with the starch converting industry.

II. CHOICE OF RAW MATERIAL

This conference is primarily concerned with the production of sugars from cereals, I should mention, that other starch containing materials are also used in the industry, of which potato and manioca are the most important raw materials. Only

ISBN: 0-12-370960-1

maize and barley will be mentioned in the lecture.

Normally for the production of sweeteners the starch is extracted from the raw material, i.e. "maize wet milling" and refined before it is converted into sweeteners. However, a technology has been developed to put the raw material after only partial refining into the conversion system to produce sweeteners. We call it the "direct hydrolysis process".

I should also mention the conversion system for the production of barley syrup, based on the direct conversion of milled whole feed barley. A process, which we have developed, is now running for several years in Kenya.

Generally, the auxiliary materials used in the process, i.e. acids, alkalis, active carbons, filter aid etc. are of normal commercial technical qualities.

III. TECHNOLOGY OF STARCH CONVERSION

The starch conversion technology consists of the following sub processes:

preparation of raw material
conversion
filtration
refining
evaporation

and for crystalline products

crystallization/spray drying
centrifuging
drying

A. Preparation of Raw Material

Normally the starch is extracted from the cereal grain by means of "wet milling" and the starch product refined by means of separation techniques and counter current washing.

The relatively pure starch slurry for instance maize starch slurry is then fed to the conversion system.

A typical raw material analysis is as follows for maize starch:

RAW MATERIAL
MAIZE STARCH SLURRY

Slurry concentration - Baumé	- 21°
% Ash	- 0,12
% Protein	- 0,3
% Soluble Protein	- 0,02

ANALYTICAL INFORMATION ON
STARCH DRY SUBSTANCE

Table 1.

B. Conversion

There are 2 systems existing at present in the world to produce sugars from starch:

batch production of sugars
continuous production of sugars

The batch type system is gradually becoming obsolete, although in small plants the system is still operating.

1. Batch Conversion. Basically the batch system consists of an autoclave where a batch of starch slurry is heated in the presence of acid or enzyme to produce the desired degree of conversion. Acidity or enzyme quality, temperature and retention time of the slurry in the autoclave, effects the degree of conversion.

When the desired degree of conversion is obtained, checked by the operator, the autoclave is discharged and made ready for a new charge.

For plants with a sugar production greater than 10 tons per day a batch plant is not any more considered.

2. Continuous Conversion. This type of system converts a starch slurry continuous and various types are at present running in the starch converting industry.

Up to about 15 years ago only acid converted syrups were available.

Acid was used as a catalyst to convert starch slurry into glucose syrups the so called 42 DE confectioners glucose syrups. To replace the batch system an indirect heated conversion system was developed to convert continuously starch slurry into high quality glucose syrups.

Starch slurry was pumped by means of a high pressure positive displacement pump through a series of heat exchangers to convert the starch into glucose syrup. By varying acidity, temperature and holding time in the heat exchangers the degree of conversion could be controlled.

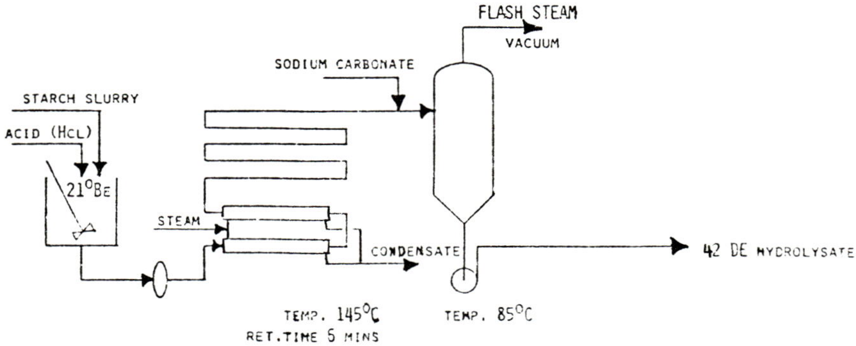

Diagram 1. Conversion system. Production of 42 DE glucose.

Typical conditions for the system are as follows:

CONDITIONS FOR PRODUCING STANDARD 42 DE GLUCOSE SYRUP

Slurry concentration - Baumé	- 21°
Slurry acidity (Hcl)	- 0,03 N
Temperature Heat Exchangers	- 145°C
Retention time in heat exchangers	- 6 mins.

Table 2. DEXTROSE EQUIVALENT RESULTANT SYRUP 42

The system is stable in operation and easy to control. Large quantities glucose syrup have been produced and are still produced using this system.

When enzyme became available for the industry, the indirect acid conversion system is still used but only for a limited number of products, i.e. for the production of total sugar, high dextrose/high maltose syrups and dextrose monohydrate.

After the initial acid conversion an additional alpha-amylase liquefaction step is added. If enzymes are used in the indirect system, instead of acid for the 1st liquefaction step very high viscosities are created and large back pressures are set up across the heat exchangers, creating mechanical pumping problems.

To overcome these difficulties the direct heating system was developed.

In this system a starch slurry is pumped through a jet cooker, at the same time steam is injected, retention time is short and due to high thermal and mechanical forces the starch in the slurry is rapidly gelatinised and the viscosity is reduced, the initial enzyme activity also contributes to the viscosity reduction.

It must be realised that the efficiency of breakdown and liquefaction has to be as high as possible, because if any unconverted gelatinised starch is present, it will create problems in the subsequent refining steps of the hydrolysate.

The jet liquefaction system runs as follows; see diagram 2. From diagram 2 can be seen the temperature profile and retention time for each step.

This temperature and retention profile is suitable for a bacterial heat stable alpha amylase. Positive displacement pumps are used for step 1 and step 2.

Typical feed conditions to the jet are as follows; see table 3.

When using normal bacterial alpha amylase the temperature and time profiles are as can be seen from diagram 3, also an extra jet is required.

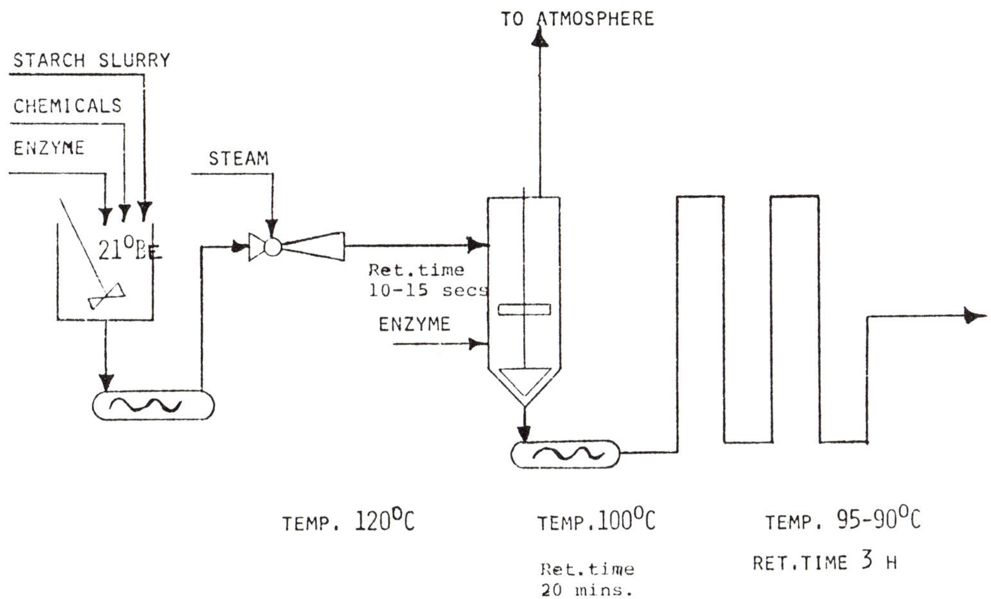

Diagram 2. Liquefaction system for high temperature stable bacterial alpha-amylase.

RAW MATERIAL
MAIZE STARCH SLURRY

Slurry concentration - Baumé	- 21°
pH slurry	- 6,5
Heat stable bacterial alpha amylase	- 1 ltr/ton D.S. starch
Ca^{++} concentration	- 100 ppm
DE after liquefaction	- 15-18
DS after liquefaction	- 34-36%

Note: Ca^{++} is added to improve the heat stability of the alpha amylase

Table 3.

Starch concentration and pH are as for heat stable bacterial alpha amylase.

Normal bacterial alpha amylase	- 2,5 kg/ton D.S. starch
C^{++}	- 250 ppm
DE after liquefaction	- 20-22
DS after liquefaction	- 34-36%

Table 4. Enzyme consumption.

A very important advantage using a total enzyme system is the final DX obtained in the hydrolysate after saccharification:

i.e. acid/enzyme DE 94 - DX 88
enzyme/enz. DE 98 - DX 95

This higher DX is important for instance with regard to yield of final products.

The increase in yield of dextrose monohydrate is demonstrated in table 5.

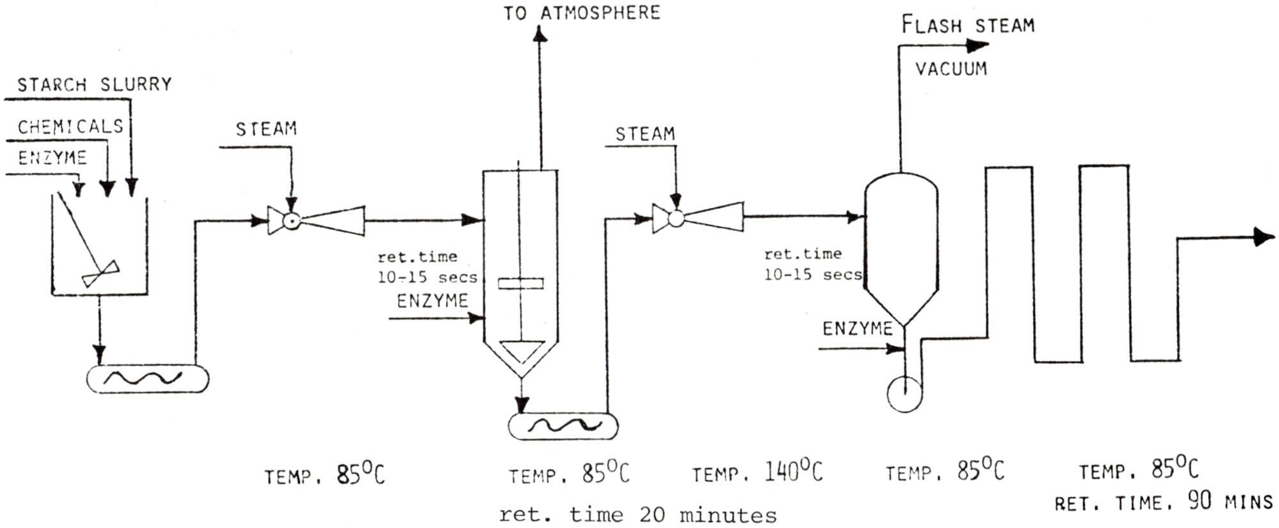

Diagram 3. Liquefaction system for normal bacterial alpha-amylase.

THE DEXTROSE YIELD AND THE DEXTROSE TO HYDROL PROPORTION
FOR DIFFERENT DE-VALUES OF HYDROLYSATE AND HYDROL.

Hydrolysate DE	98 (100 kgs)		97 (100 kgs)		96 (100 kgs)		95 (100 kgs)	
Hydrol DE	80	85	80	85	80	85	80	85
Dextrose monohydrate kg	98.9	95.3	93.4	87.9	87.9	80.5	82.4	73.3
Hydrol DS kg	10.0	13.3	15.0	20.0	20.0	26.7	25.0	33.3
Per cent hydrol DS of dextrose monohydrate	10.1	14.0	16.1	22.8	22.8	33.2	30.3	45.4

DE-values are based on ion-exchanged product
Process loss =0%

Table 5.

Temperatures in the various steps are controlled by temperature and pressure control systems and also, when a lower temperature is required after a higher temperature step. The hydrolysate is flash cooled either to atmosphere or under vacuum depending which temperature is required.

The 2 main advantages are:

a) collection of condensate for use in processing, i.e. condensate for anion exchangers

b) increase in the dry substance of the hydrolysate saving energy in the next processing steps.

The all enzyme liquifying starch process give a possibility of producing many types of sweeteners. Diagram 4 shows the number of products possible, see diagram 4.

Most of the products are made batch wise, i.e. high DE syrups, maltose syrups etc., the exception is isomerase syrup and the DDS-KRØYER Dextrose Monohydrate Process.

The conditions for producing high DE hydrolysates is shown in table 6.

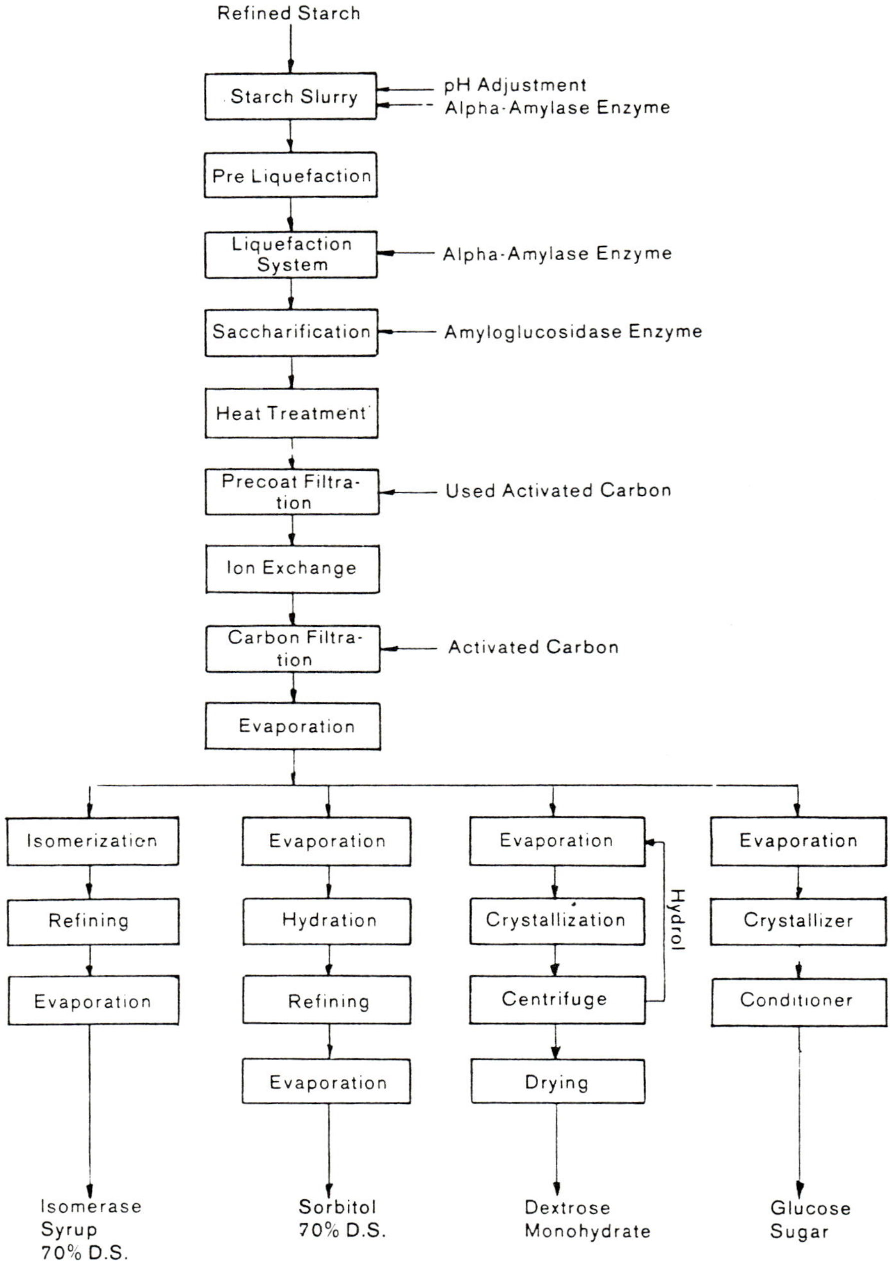

Diagram 4. Process: Continuous 3-step enzyme liquefaction.

Feed liquified hydrolysate	DE 15-22
pH	4.5
DS%	34-36
Temperature	60°C
Retention time	60 hrs.
Amyloglucosidase L 150	1.3 l/T DS sugar
Final DE	98_{+}
DX	95^{+}

Table 6.

During saccharification the tanks are slowly stirred and temperature is strictly controlled. By controlling time and enzyme dosage, the required degree of saccharification can be obtained.

C. Filtration

After conversion and/or saccharification the hydrolysates are filtered. The hydrolysates contain suspended matter and the product is filtered over precoat filters.

i.e. rotary precoat filters; filters operating under vacuum
precoat candle filters; filters operating under pressure
precoat rotary discs filters; filters operating under pressure

When large plants are designed, fat separation is carried out by means of centrifugal separators before filtration. In the smaller plants floatation tanks are used or a slight increase in filter aid is tolerated and the fat taken on the precoat filter.

The filtering surface is covered with filter aid prior to filtration.

The hydrolysates after filtration are clear and free from suspended matter. A typical consumption figure for filter aid is 5 kg/Ton D.S. starch.

D. Refining

If a normal 42 DE syrup or high maltose syrup is required the hydrolysate after 1st filtration is treated with active carbon and filtered over one of above filters and finally evaporated.

Syrups which are prepared enzymatically and need to be converted into crystalline products, sorbitol or iso syrup, additional refining is required.

These syrups are fully deionised before proceeding to the next process step.

The hydrolysate after precoat filtration is passed through a double set of cation and anion exchangers.

This system effectively removes inorganic and organic impurities from the sugar solutions and also improves the colours of the hydrolysates.

A 2 pass system is used to be able to have maximum use of the resin exchange capacity per operating cycle. The cation resin is of the strong acid type, the anion resin of the weak base type.

When the exchanger is exhausted, each set is regenerated, the cation exchanger with acid and the anion exchanger with alkali reagents. This operation provides advantages, for instance increase in yield and purity during DxH_2O manufacture and another example, a decrease in isomerase enzyme consumption during iso syrup manufacture. One serious disadvantage using ion exchange is the BOD produced during the regeneration time, which depending on the size of the plant can be considerable. Hydrolysates which do not need further processing into other products, for instance High DE Syrups, High Maltose Syrups are evaporated to about 80-82% dry solids in a continuous falling film evaporator. These syrups are now ready for consumption.

E. Evaporation

The dilute refined syrups are concentrated by means of falling film evaporators. The final concentration vary depending on the product produced.

F. Additional Operations Used in the Industry

The operations mentioned previously crystallization/drying, and crystallization, centrifying and drying refer to the production of special products.

These operations are described for the production of total sugar and dextrose monohydrate.

1. Total Sugar. Total sugar is produced by spray-drying high DE refined hydrolysate.

Two systems exist and are operating for producing total sugar:

spray dryer crystallizer, see figure 5
crystallizer dryer, see fig. 6

The diagrams are self explanatory.

The appearance of the product from each operation is different. The product from the spray-dryer crystallizer is a fine crystalline powder. The product from the crystallizer dryer consists of agglomerated spherical particles.

Chemically they are identical.

DE	98
H_2O%	1
Alpha/beta dx ratio	40/60

2. Dextrose Monohydrate. Ancther important sugar made in the starch conversion industry is dextrose monohydrate.

This sugar is extensively used in the food, pharmaceutical and beverage industries.

Dextrose monohydrate is crystalline alpha-d-glucose with one molecule of crystal water.

It is produced continuously from high DE hydrolysate in a cascade system of crystallizers. Crystallization time is about 90 hrs.

After crystallization the crystals are spun and washed in a centrifuge and finally dried in a rotary warm air dryer.

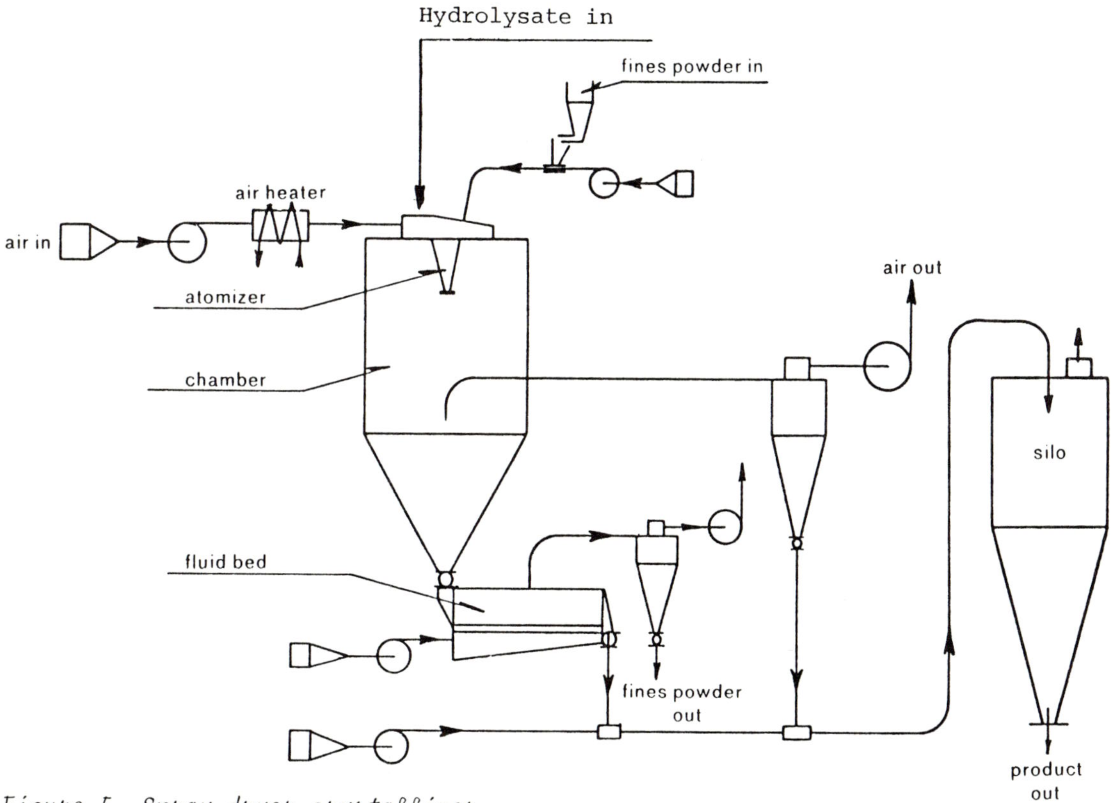

Figure 5. Spray dryer crystallizer.

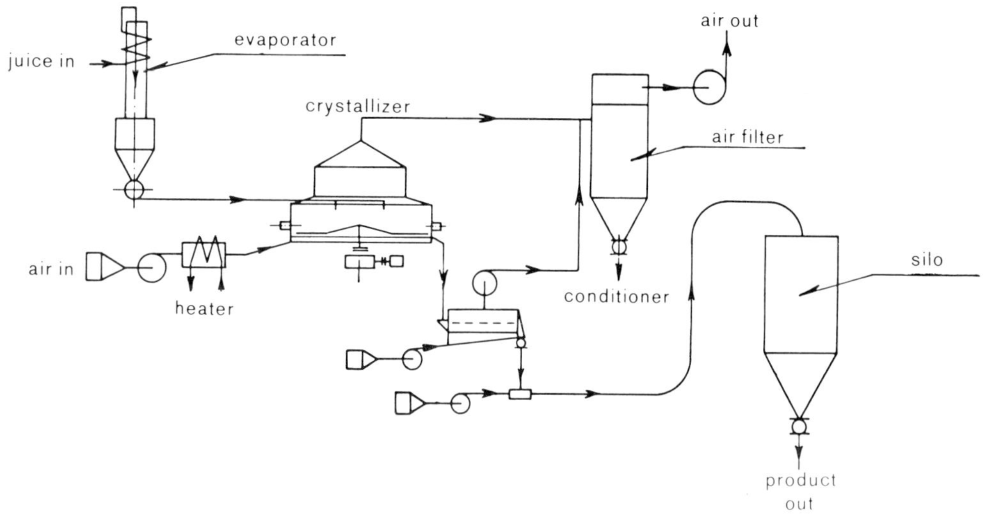

Figure 6. Crystallizer dryer.

The following is a typical analysis:

Appearance - crystalline
DE - 99.9
Dry substance - min. 90%
Quality conform to BP and USP specifications.

3. Fructose Syrups. Fructose syrups or the so called high fructose corn syrups are produced from high DE hydrolysates by means of isomerase enzyms. These syrups have in recent years obtained importance in the food industry.

The syrups have similar applications as liquid sugar in the food and beverage industries.

The syrups are made by passing high DE hydrolysates through columns containing immobilised isomerase enzymes.

The syrups before passing through the columns are highly refined and close process control is carried out during the isomerisation process, to obtain the correct fructose content in the final syrups.

Part of the alpha-d-glucose in the hydrolysates is isomerised to fructose.

After passing through the columns the syrups are fully deionised and evaporated to 71% dry solids. A typical fructose syrup has the following composition:

% fructose	- 42.0
% glucose	- 53.0
% higher sugars	- 5.0
% D.S.	- 71.0
pH	- 4.5

Figure 7. Composition of high fructose corn syrup.

4. High/High Fructose. These are the 2nd generation fructose syrups having more than 55% fructose on dry substance. Of interest are syrups containing 55%-60% and 90% fructose.

They are made at present by passing normal dructose syrup over a chromatographic column.

The column is loaded with a certain amount of fructose syrup.

When loaded the column is washed with water and the sugars eluted from the column.

The eluate consits of three fractions.

The 1st fraction contains the higher saccharides, i.e. maltose, triose etc.

The 2nd fraction contains most of the alpha-d-glucose and minor amounts of fructose.

The 3rd fraction contains mostly fructose. This fraction is blended with normal 42% fructose syrup to produce the required fructose concentration in the final high/high fructose syrup.

The high glucose fraction is send back to the isomerisation step.

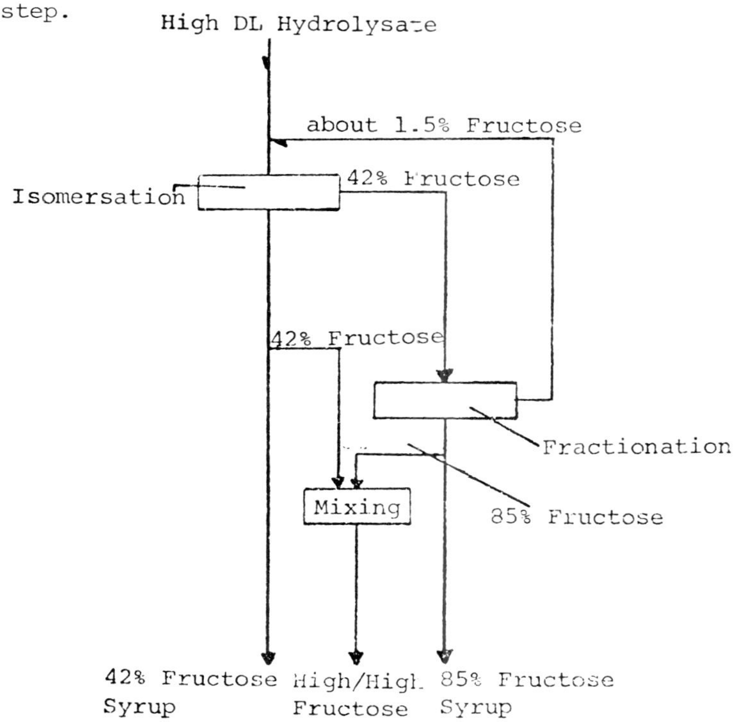

Figure 8. High/high fructose production.

IV. SPECIAL PROCESSES

Two special processes have been developed by DDS-KRØYER A/S.

The direct hydrolysis process using maize gritz as raw input material and the barley syrup direct hydrolysis process.

A. Direct Hydrolysis for "Maize Gritz"

A process running at present in Copenhagen for the past 10 years is the so called "direct hydrolysis of maize gritz".

The fundamental difference with this process with other enzymatic processes is the quality of the raw material.

The raw material is the "gritz" fraction from a dry milling operation of whole maize.

It contains a larger quantity of impurities, i.e. ash, fat and protein as compared with refined starch, see table 7.

The conversion of gritz is similar to converting starch, however the temperature profile across the jet liquefaction system is different.

Also the refining of the hydrolysate necessary is more stringent than when starting from starch, because of higher impurities in the raw material.

The quality of the final hydrolysate produced from gritz is identical to a pure starch hydrolysate.

The process produces a valuable feed cake by product, see table 8.

An advantage of using the gritz process is the much lower investment required compared with a wet milling operation, and an additional advantage is, if the operators of the plant are only interested in sugar and not in starch.

TYPICAL "MAIZE GRITS" ANALYSIS

		RANGE
% MOISTURE	: 11.5	11.0-12.0
% PROTEIN	: 9.3	9.0-9.5
% FAT	: 1.0	0.8-1.3
% PENTOSANS	: 1.3	1.2-1.4
% STARCH	: 81.5	81.0-83.0
% ASH	: 0.4	0.4-0.5
% OTHERS (BY DIFFERENCE)	: 6.0	5.0-6.0

RESULTS ON DRY BASIS.

MESH ANALYSIS

+ 1000 MICRONS	: 0.8%
+ 750 MICRONS	: 8.1%
+ 600 MICRONS	: 29.4%
+ 355 MICRONS	: 47.3%
+ 140 MICRONS	: 13.4%
- 140 MICRONS	: 1.0%

Table 7.

TYPICAL FILTER CAKE ANALYSIS

% MOISTURE	- 5.0
% CARBOHYDRATES	
(SOL. AND INSOL.)	- 37.3
% PROTEIN	- 40.0
% PENTOSANS	- 5.5
% FAT	- 6.7
% FIBRE	- 4.5
% ASH	- 0.4
% OTHERS	- 5.5

RESULTS ON DRY BASIS.

Table 8. Process direct hydrolysis of maize gritz.

B. Production of Barley Syrup.

A recent development in the starch converting industry is the production of Barley syrup from feed type Barley.

This syrup can and is replacing convertional malt/wort for the production of beer.

The process is continuous and is total enzymatic and runs as the DDS-KRØYER A/S "direct hydrolysis process".

Barley grain is partially dehulled, dry milled, slurried up with water and enzymatically liquified with normal bacterial alpha amylase. The liquified product is further saccharified with amyloglucosidase and a small percentage of milled malt in the presence of high temperature stable proteinase. The proteinase increases the alpha amino nitrogen in the syrup.

The resultant product is filtered and the syrup concentrated to 77% D.S.

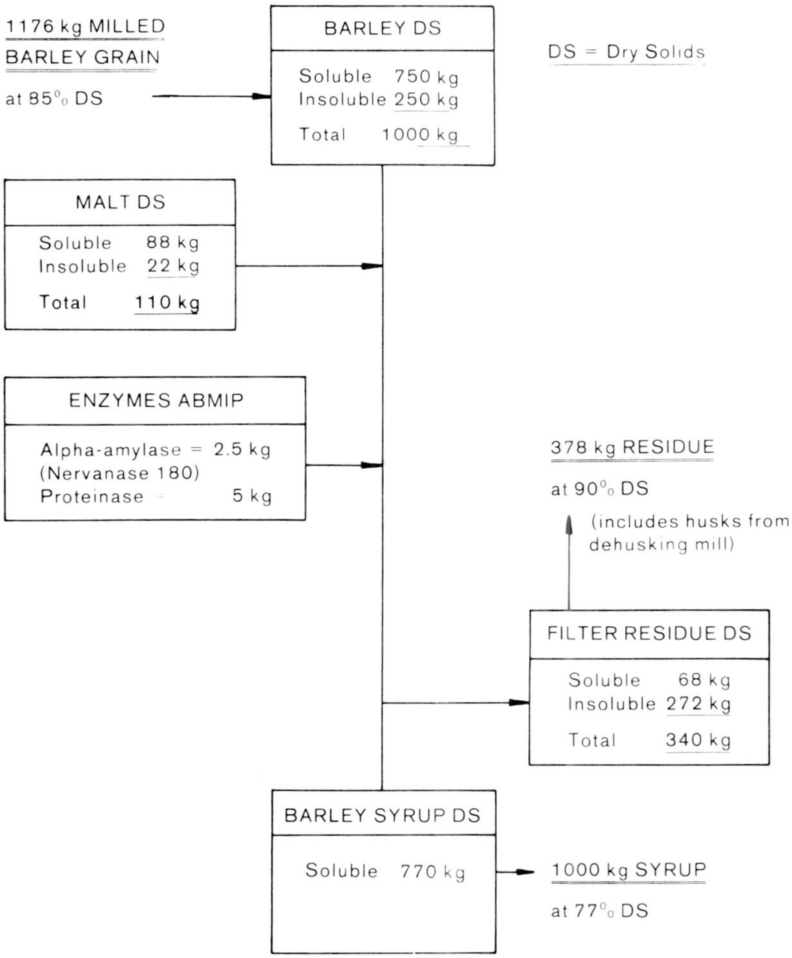

Figure 9. Barley syrup for brewing A.B.M.I.P. and DDS-KRØYER process approximate mass balance.

The hulls and filter cake are dried together to form a valuable animal feed product.

Finally I would like to mention that our basic liquefaction and saccharification systems for cereals and other starchy materials are used to provide the raw material for power alcohol production, which is becoming more and more important, because of the present world fuel situation.

WATER ADSORPTION IN CEREAL FOODS AND WATER ACTIVITY MEASUREMENT

Jean-Louis Multon[1]
Hervé Bizot

Laboratoire de Biophysique des Aliments
Institut National de la Recherche Agronomique
Nantes, France

Bernadette Savet[2]

Foss-Electric France

From the viewpoint of its functionnal properties, water is one of the more important components of biological and natural products, like cereals and derivatives. In fact, in such foods, water plays the role of a reactant, and of a diffusion and solvant medium.

The water behavior in foods can be analysed in terms of thermodynamic exchanges between atmospheric water vapour and condensed water, physically bound (adsorbed) to the substrate. This adsorbed water (from 0 to 40 g water/100 g wet matter) is of considerable interest, both theoretical (bonding mode, distribution) and applied (its role in alteration or transformation processes).

This is why we are going to discuss the adsorption state of water, and to describe the most suitable methods for measuring it in cereal foods.

[1] *Present address : Chemin de la Géraudière, 44072 Nantes Cedex, France.*
[2] *Present address : 80, rue Taitbout, 75009 Paris, France.*

ISBN: 0-12-370960-1

I. THE THERMODYNAMICAL PROPERTIES OF WATER IN FOODS

A. *The Concept of "Water Activity" (A_W)*

The term "Water activity" was introduced by Lewis and Randall (1923) and used in biological systems for the first time by Scott (1957). The theoretical discussion of the A_W concept is very difficult (Morin, 1979). Summarizing, one can say that in pure water, addition of a solute decreases the chemical potential of water, according to the expression (Moore, 1979) :

$$\Delta G = \mu_{w(T,P)} - \mu^{\circ}_{w(T,P)} = R\,T \ln X_w \quad (1)$$

where : $\mu_{w(T,P)}$ is chemical potential (or partial mole free energy) of water in the condensed phase at temperature T, and pressure P

$\mu^{\circ}_{w(T,P)}$ is chemical potential in pure water

X is mole fraction of water in system

R is the perfect gas constant.

If the solution is not ideal, the mole fraction must be corrected by an "activity coefficient" (Moore, 1979) :

$$\Delta G = \mu_{w(T,P)} - \mu^{\circ}_{w(T,P)} = R\,T \ln (X_w \gamma_w) \quad (2)$$

By definition, $X_w \gamma_w = A_W$ is the "water activity" in the system ; and then :

$$\Delta G = R\,T \ln A_W \quad (3)$$

Let us now consider the vapour phase in equilibrium with the condensed phase : in a non-ideal system, one must substitute the concept of "fugacity" to the concept of "pressure". The Raoult's law is then :

$$X_w \gamma_w = \left(\frac{f}{f'}\right)_T = A_W \quad (4)$$

where f is the fugacity of water vapour in equilibrium with the condensed phase
f' is the fugacity of water in equilibrium with pure water, considered a reference system.

However, under ordinary conditions of pressure and temperature, water vapour may be supposed to be a perfect gas, and then :

$$A_W = \left(\frac{f}{f'}\right)_T \simeq \left(\frac{p}{p'}\right)_T \qquad (5)$$

where p is the water vapour pressure in equilibrium with condensed system,
p' is the water vapour pressure of pure water

$$\Delta G = \mu_{w(T,P)} - \mu^{\circ}_{w(T,P)} \simeq RT \ln\left(\frac{p}{p'}\right) \qquad (6)$$

This approximation leads, according to Gal (1975), to an error of about 0.2 %.

By definition,

$$100 \left(\frac{p}{p'}\right)_T = \text{E.R.H.} \qquad (7)$$

relative humidity of atmosphere in equilibrium with solution.

Then :

$$A_W = \left(\frac{p}{p'}\right)_{T,P} = \frac{\text{E.R.H. \%}}{100} \qquad (8)$$

It must be emphasized, concerning the concept of water activity, that in solutions A_W is related to chemical potential of water bound to the system, and E.R.H. is related to partial water vapour pressure in the atmosphere ; then, even if there is a strict correlation between A_W and E.R.H., both concepts are quite different.

However E.R.H. is practically the only experimental way to measure A_W in food.

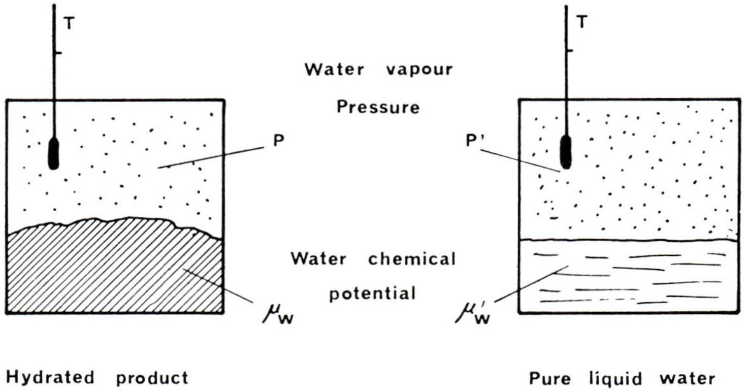

$$RH = \frac{P}{P'}$$

$$\mu_{w(P,T)} - \mu'_{w(P,T)} = RT \log A_w$$

FIGURE 1. Water activity definition.

B. The Relationship Between A_w and Moisture Content of the Food : Sorption Isotherms Curves

Several reviews have been made on this topic (Stitt, 1958 ; HEISS, 1968 ; Wolf *et al.*, 1973 ; Bizot *et al.* , 1978 ; Multon, 1978).The affinity between food products and water (sometime called hygroscopicity) is well depicted by sorption-desorption isotherms. Typical curves are shown in figure 1 in which the A_w of the product or the R.H. of air in equilibrium with the product is shown on the X-axis and the moisture content of the product, expressed as g water per 100 g dry or wet product is shown on the Y-axis.

It is not easy to define the moisture content of a product. The definition of Stitt (1958) is too theoretical, and we prefer to use that of Guilbot (1973) : "the moisture content is all water which could be removed from the product, when it is equilibrated exactly with a R.H. equal to zero".

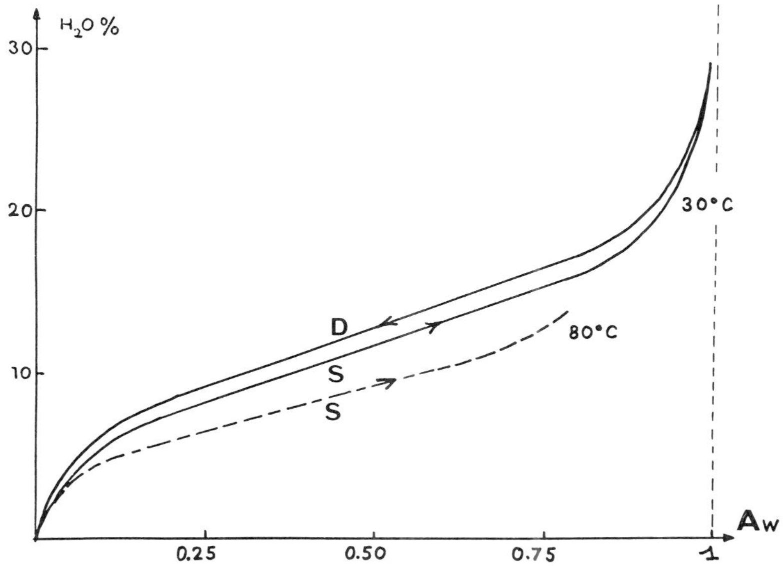

FIGURE 2. Sorption (S) and desorption (D) isotherms curves of maize, for different temperatures

One must emphizise that every point on these curves corresponds to an equilibrium state : a previously dried sample (sorption process) or humidified sample (desorption process) is placed in an atmosphere the R.H. of which is kept constant ; there is then a transfer of water from atmosphere to product (sorption) or from product to atmosphere (desorption), until equilibrium is reached.

There is an "hysteresis effect", corresponding to the fact that the desorption curve is higher than the adsorption curve. The magnitude of "hysteresis" is about 1 % on Y-axis for $A_W = 0.5$. Until now, no clear explanation has been given for this phenomenon.

An other characteristic of sorption isotherms is their variation with temperature ; in most cases increasing the temperature decreases the moisture content, when A_W is kept constant (fig. 2).

1. Sorption Curves. Water vapour sorption is only a particular case of so called physical gas adsorption by solids (Moore, 1979). The generally sigmoïdal shape (fig. 2) of water vapour sorption curves corresponds to the type II and III of the classification of Brunnauer *et al.* (1938). However, when there is a crystalline component, giving specific hydrates and saturated solutions, a different shape may be observed.

2. Physical Significance of Sorption Phenomena

a. Molecular level. The water molecule is highly polar because of the asymmetry of its electrical charges, and has a permanent dipolar moment (equal to 1.83 Debye). This is why this molecule may form up to four hydrogen bonds, oriented in space as a tetrahedral-like structure, which promotes the formation of molecular aggregates, more or less organized. These polar properties of water explain the interactions between water molecules and polar organic groups. Most active functionnal group which form hydrogen bonds with water during adsorption are the following :

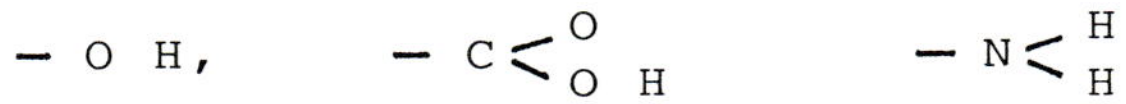

and to a lesser degree with :

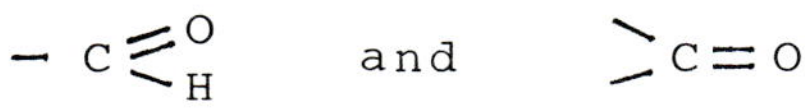

These polar groups organize reversibly the structure of water, orienting hydrogen bonds in a cooperative way, particularly if water molecules form several H-bonds.

This mechanism is the origin of the affinity of water for biological products, the amount of water bonded being a function of the number and the accessibility of polar sites in macromolecular structures i.e. carbohydrates such as starch (Guilbot *et al.*, 1979), and proteins (Eagland, 1975

Concerning proteins (Fig. 3), interactions with water may occur on two different kinds of polar sites : the side chains of polar amino acids (Mc Laren and Rowen, 1951), and the polar sites of polypeptide skeletons (Duckworth and Smith, 1968). The so-called "hydrophobic interactions", occuring in and around non-polar groups, involve the formation of intramolecular micelles (Nemethy and Scheraga, 1962) and minimize the organization of water around the repulsion area.

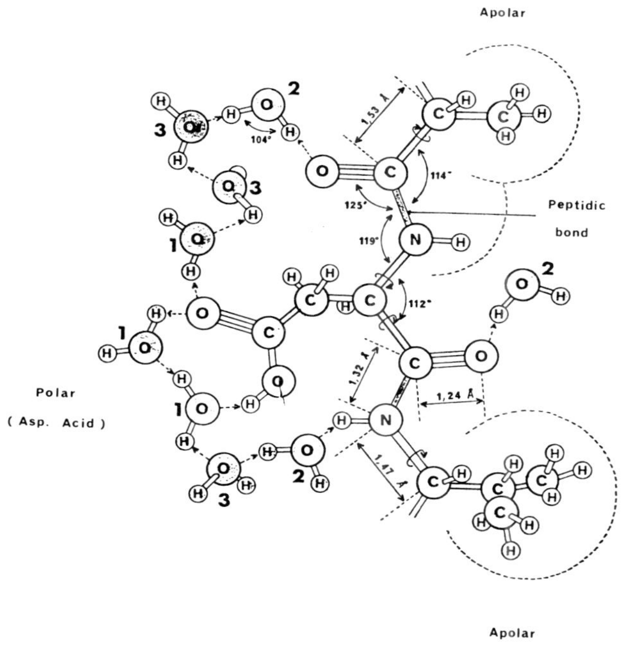

Figure 3. Possible polar sites of water adsorption in protein molecule.

In the hydration of carbohydrates, the stereochemistry plays an essential role. Water bonding sites are exclusively the - OH groups. When the macromolecule is partially crystallized, like starch or cellulose, water molecules are essentially adsorbed in amorphous zones ; in crystalline zones, interchain bonds are too strong, and steric accessibility too difficult for water molecules to be bonded (fig. 4).

On the other hand, water seems to act as a catalyst for crystallization, as shown by X-ray diffraction studies (Guilbot *et al.*, 1961, 1979).

Another type of carbohydrate macromolecules and water association is gel formation (Doublier *et al.*, 1975). Lipids have a low affinity for water, except at their polar extremities

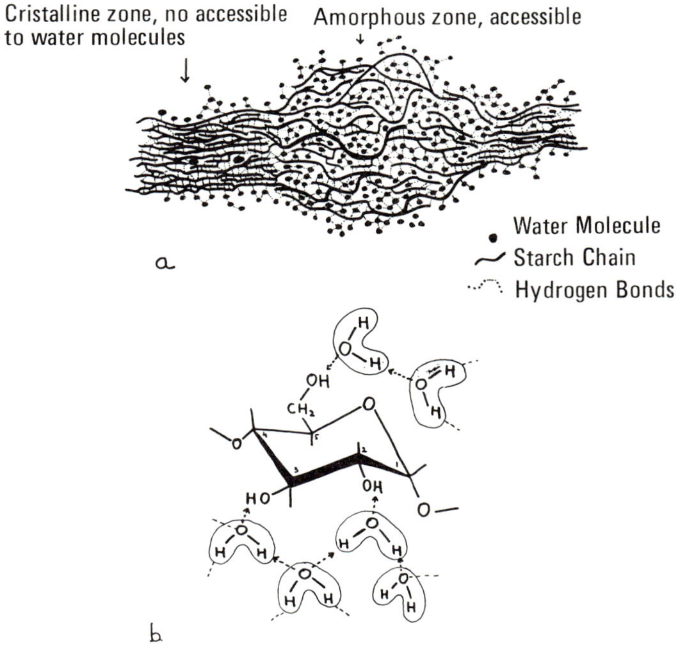

Figure 4. Model of starch hydration
a. Cristalline zone excluding water, and amorphous zone where water adsorption occurs
b. One glucose unity of amorphous zone, bonding up to six water molecules (corresponding to a moisture content of 40 %, dry basis).

b. Macroscopic level. Sorption curves, at least those of which the shape is sigmoïdal, are generally interpreted as the result of several phenomena, occuring successively and overlapping one another. One may distinguish three characteristic zones (Drapron and Guilbot, 1962).

- First, from 0 up to 10 % R.H., water is strongly and directly bonded on free and accessible polar sites. Heat of hydration (Guilbot, 1955), dielectric absorption (Guilbot, 1960) and size changes (Hermans 1949, 1952) show that water molecules have a specific orientation.
- Between 10 and 65 % R.H. (linear section of the curve) water molecules are bonded either on those previously bound or on polar sites previously hidden inside the structure and now accessible due to swelling. In this range water is not freezable.
- Over 65 % R.H., the curve deviates from linearity towards higher moisture content. Water molecules accumulate in small capillaries and intergranular spaces of the product ; a liquid phase may appear, with a vapour pressure slightly depressed. Bonding energies become weaker with increasing adsorption but are slightly stronger than in pure liquid water. Freezable water may appear. At this level, the water behavior may be interpreted in terms of the laws of capillarity and osmosis.

3. *Additivity*. When two products are mixed, the sorption curve of the mixture is the average by weight of the sorption curves of the two products, if there is no interactions between them (fig. 5).

On the contrary, if the products interact with each other, it is not possible to compute the sorption curve of the mixture. In this case it is also impossible to predict the A_W of the mixture from the A_W of the components, then the only way is to measure it.

4. *Functionnal Availability of Water and Its Role in Cereal Food Technology*. Sorption phenomena have been successfully interpreted from the point of view of mobility and functionnal availability of water. The concept of "solvent water", has been established by Guilbot and Lindenberg (1960) and more recently by Duprat (1975). These authors have demonstrated that water fraction comprised between the sorption curve and its extrapolated linear part, shows solvent properties ; its mobility is sufficient to play the role of a diffusion medium for biochemical reactions. There water molecules are weakly bound to the substrate, with bonding energies similar to those measured in pure

liquid water. The fraction of water above the linear part is strongly bound, and un-available, and is not solvent water.

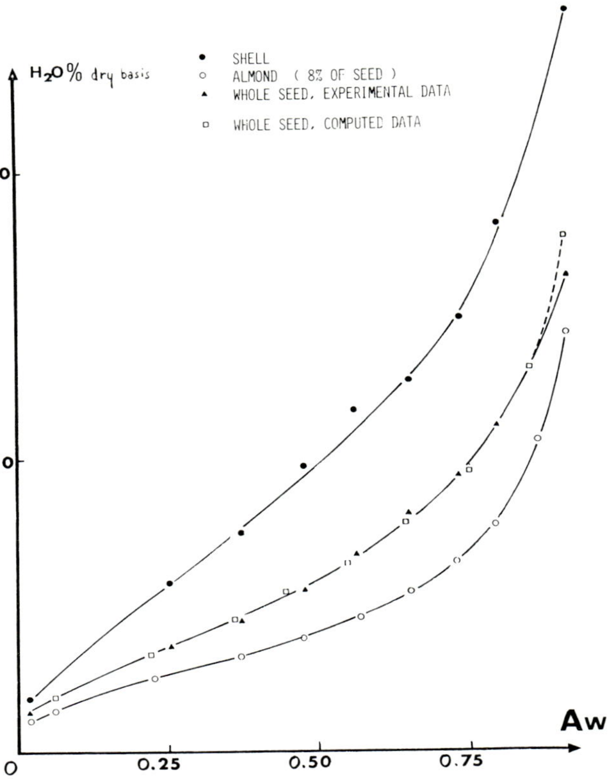

Figure 5. Additivity of sorption isotherms : shell, almond and complete seed of sunflower.

It has been shown that many reactions require a minimum quantity of solvent water to start and their speeds increase as a function of the amount of available solvent water, e.g., enzymatic reactions (Loncin *et al.*, 1968 ; Drapron, 1972) ; denaturation of proteins (Multon and Guilbot, 1975) carbohydrate gelification (Doublier *et al.*, 1975) ; growth of microorganisms (Scott, 1957).Figure 6 shows the A_w

threshold of the main types of alterations occuring in cereals.

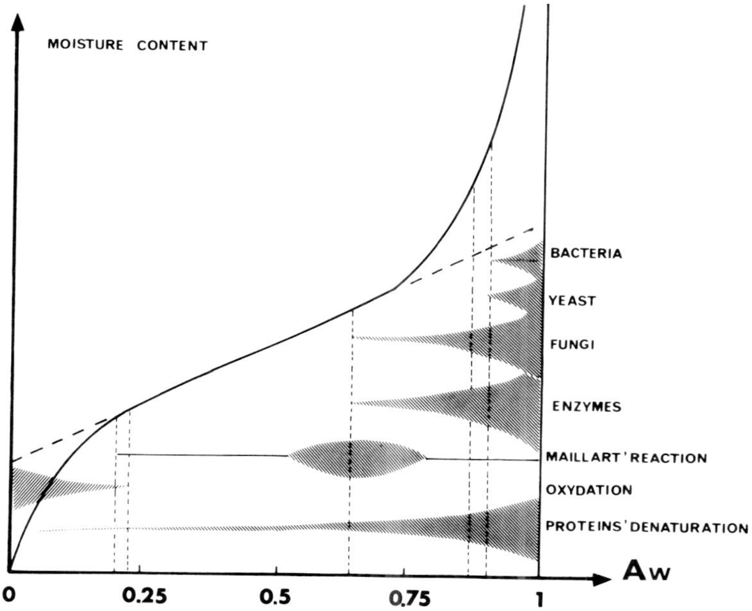

Figure 6. A_W threshold above or below which the biochemical reactions of alterations may occur in natural products.

On the other hand a lot of physical properties of grains and their components are also functions of the quantity of available water, and of A_W, e.g.crystallinity of starch (Guilbot *et al.*, 1979 ; Duprat, 1977) dielectric absorption (Guilbot, 1960), which is used for grain moisture-meter (Multon and Martin, 1979) ; rheological behavior of wheat grains (Multon *et al.*, 1978) ; thermal characteristics, heat capacity and heat transfer coefficient, heat of hydration (Hermans, 1949, 1952).

Consequently, almost all technological processing of cereals are mainly subordinate to water content and A_W. In particular,

- the preservation of cereals is mainly depending of moisture content and temperature of grains during storage ;
- the milling requires a precise tempering of grains ;

- cooking process (bread, biscuit) and extrusion processes need a precise adjustment of water content for controling Maillard reaction, protein and starch gel formation ;

- industrial pastry and all derived "intermediate moisture foods" need a good equilibration of A_W between different parts in order to avoid migration of water ;

- preservation of these products is highly dependant on A_W, particularly for microbiologic stability.

The great importance of moisture in cereal technology justifies the necessity to have good and precise methods for measuring the two parameters which determine the thermodynamic state of water : the moisture content, and the "water activity".

II. CEREALS MOISTURE CONTENT MEASUREMENT

As the measurement of moisture content is largely discussed elsewhere (Multon and Martin, 1978, 1979 ; Multon *et al.*, 1978, 1979) it is only necessary to recall here the different categories of methods commonly used.

A. Reference Methods

These methods are highly standardized at international level, thanks to I.C.C.[1], I.S.O.[2] and O.I.M.L.[3]. Two levels of accuracy are to be distinguished :

- fundamental reference methods (after grinding, drying over P_2O_5, under vacuum, at 50°C, until constant weight)

- practical reference methods (after grinding, oven drying at 130°C, during a time calibrated with respect to fundamental method, e.a. 2 hours for wheat, barley, oats, 4 hours for maize).

[1] *I.C.C. : International Association of Cereal Chemistry*

[2] *I.S.O. : International Standardization Organization*

[3] *O.I.M.L. : Organisation Internationale de Métrologie Légale.*

These methods are largely and commonly used, especially in Europe, where E.E.C.[1] has adopted the same standards.

However it must be pointed out that some other methods are also used : this is the case of A.A.C.C.[2] and A.O.A.C.[3] methods (oven drying at 100°C, which give slightly different results compared with ISO methods) and E.B.C. method for barley) which is empirical and incorrect.

Therefore, even if a large international consensus is now obtained concerning the determination of grain moisture content, some improvements are yet to be made in standardization.

B. *Moisture Meters*

For a quick measurement of moisture content in storage warehouses or factories, moisture-meters are commonly used. Several physical principles can be choosen for these apparatuses (conductivity, dielectric, NMR, neutron probe, infra-red spectrometer) but right now the dielectric measurement is the most common.

Concerning the construction, the calibration process and the control of these apparatuses, some countries have developped official rules, like in Germany, in France or in some states of U.S.A.

In addition, O.I.M.L. and of an E.E.C. standards concerning moisture meters are now in progress.

The best and most up-to-date apparatuses (fig.7, 8 and 9) now in use in the world are Dickey-John (Multigrain), CEDEM (HD 2000) and Foss-Electric (MK-II) (Multon and Martin, 1979).

[1] *E.E.C. : Economic European Community*

[2] *A.A.C.C. : American Association of Cereal Chemists*

[3] *A.O.A.C. : American Organization of Analytical Chemistry*

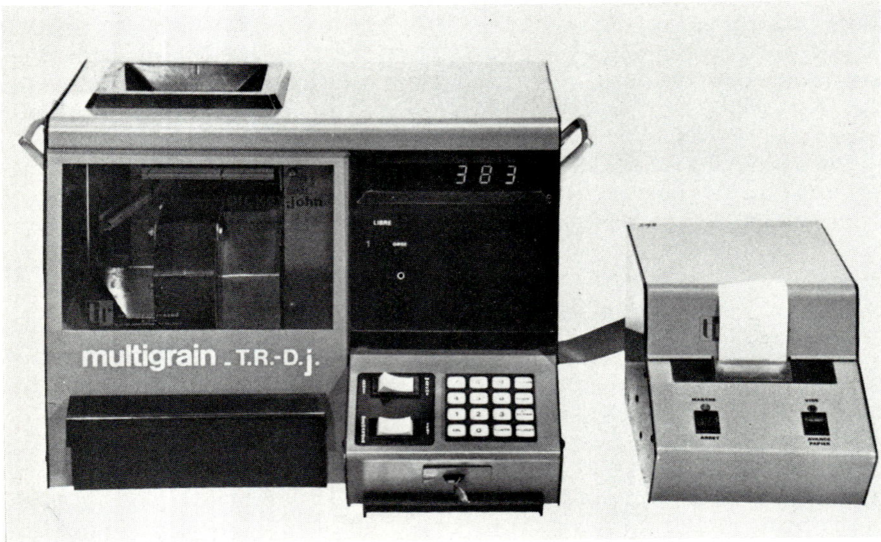

Figure 7. Moisture meter Dickey-John Multigrain

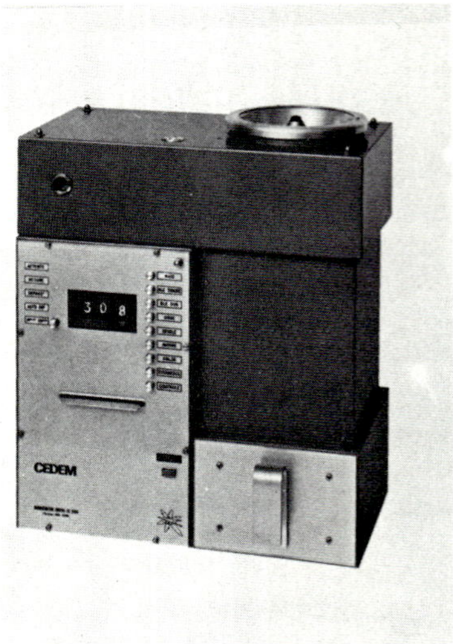

Fig.8. Moisture meter CEDEM HD-2000

Fig.9. Moisture meter Foss-Electric MK-II

III. METHODS USED FOR A_W MEASUREMENT

Concerning A_W measurement, first of all, we must emphasize that the sample must remain in thermodynamical equilibrium. If not, by definition, the measurement is meaning less.

Secondly, except in the case of a dilute solution where the chemical potential can be determined (Raoult's law), the only way to reach A_W value is to measure the Relative Humidity (R.H.) of air or gas in equilibrium with the sample. Therefore it is necessary to put the product in a small box, and to allow it to come to equilibrium with the controlled atmosphere. The attention must be focussed on : (i) the constancy of temperature, (ii) the realization as quickly as possible of the equilibrium, and (iii) the accurate measurement of the R.H. of the air.

In order to summarize the principle methods described, we must separate absolute methods (direct measurement of R.H.), methods using reference substance (research of equilibrium point) and calibrated methodes (Bizot *et al.*, 1978).

A. *Absolute Methods*

Under atmospheric pressure the gaseous mixture of air and water vapour is bivariant ; therefore, the determination of R.H. needs the measurement of two independant variables, with the sorption at equilibrium.

1. Manometric Method. The water vapour pressure P(t) is directly measured at equilibrium temperature with a manometric device, after evacuation of all other gases ; if P'(t) is the saturating water vapour pressure, at the same temperature (known in literature) then $A_W = P(t)/P'(t)$.

Therefore it is also necessary to measure very accurately the temperature of the sample.

Several manometric devices have been described :
- with a mercury manometer (Coppens and Wei, 1954 ; Taylor and Rowlinson, 1955 ; Acheson, 1965), particularly suitable for elevated temperature, when pressure is high ;

- with oil manometer, suitable for room temperature (Taylor and Rowlinson, 1955 ; Taylor, 1961 ; Sood and Heldman, 1974 ; Labuza, 1975 ; Multon and Bizot, 1978) ;
- with low pressure gauge, when water vapor pressure is small, e.g. over frozen products (Simatos, 1971).

Generally speaking, in order to evacuate non condensable gas, the sample is freezed at very low temperature (-80°C), then the temperature of sample is brought back to its initial temperature, water is defrozen and exert its pressure which is measured. Sometimes, the water evaporated from the product is trapped during evacuation but, by an appropriate procedure (Labuza *et al.*, 1976), the drying process may be minimized. We have developped (Multon and Bizot, 1978) such a method with an appropriate correction procedure, in order to take in account the residual pressure of non condensable and occluded gas (fig. 10).

The measure is long (1,5 hours for liquid and up to 5 hours for solids) but accurate, with a standard error of about 0,01 up to 0,02 A_W unity.

Concerning the apparatus, it must be built with the most extreme care, avoiding particularly any leak under vacuum.

This method cannot be used when there are other volatile components, the vapour pressure of which is similar to that of water. When ice crystal formation may change the bonding state of water (e.g. weak living cells structure), it is necessary to use an experimental procedure without freezing process.

In case of cereal grains, flour, dough and all cooked products (biscuit, bread, pastry), this method is very suitable.

2. *Dew Point Method*. The dew point temperature of air, in equilibrium with product, is the temperature at which water vapour begins to condensate on a brilliant surface (mirror) progressively cooled, generally by Peltier's effect.

Knowing the dew point temperature (t_c) and the temperature t of the sample, R.H. or A_W is :

$$A_W = P'(t_c) / P'(t)$$

$P'(t_c)$ and $P'(t)$ being respectively the water vapour pressure at temperature t_c and t.

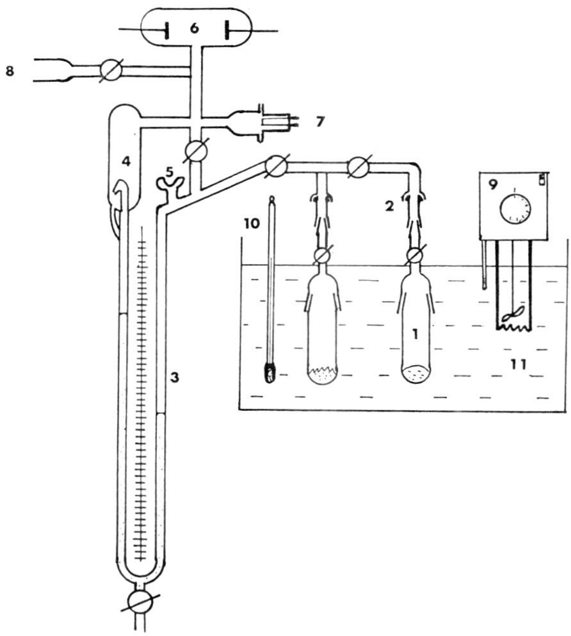

Figure 10. Measurement of A_W : manometric method 1- Sample ; 2- spherical junction ; 3- oil manometer 4- safety bottle ; 5- cup for freezing point ; 6- vacuum test ; 7- pirani vacuum gauge ; 8- vacuum 9- thermocouple ; 10- thermometer ; 11- constant temperature bath.

The dew-point detection can be made by eye (Wexler and Hasegawa, 1954 ; Ayerst, 1965 ; Cakebread, 1974), or automatically with a photoelectric cell (Rodel and Leistner, 1972).

The sample and the dew-point detection device can be in the same box, with a ventilation fan (Wexler and Hasegawa, 1954 ; Ayerst, 1965 ; Bizot, 1976) or not (Rodel and Leistner, 1972). But the temperature gradiant which remains necessarily between mirror and sample seems to be a source of errors. It seems better to put the sample and the dew-point device in two separate boxes, the circulation of air being ensured by a pump (Cakebread, 1974 ; Gough, 1975 ; Bizot, 1976 ; Multon, 1979). Accuracy, limitations, and equilibration time are similar to those for manometric techniques.

3. Psychrometry. The psychrometric method is extremely difficult to use in a small box, because of the necessity of a vigorous agitation of air around the wet thermometer (Bates, 1969).

B. Methods Involving a Reference or Standard Substance

1. Isopiestic Method. Knowing the sorption curve of a reference or standard substance, the method consists in placing the sample and the reference substance, side by side, in a very small box. When equilibrium is reached, the determination of water content of reference substance gives the A_W of the system. Fett (1973) used as reference substance proteins with small hysteresis, Vos and Labuza (1974) used micro-crystalline cellulose.

2. Microclimate Method. The interpolation method of Landrock and Proctor (1951) is simple and accurate. It consists in determining by interpolation, the R.H. of atmosphere for which there is no weight change of the sample. Practically, several samples are weighed and placed in a dessiccator (or in very small special devices, called Conway cells) with a saturated salt solution (or sulphuric acid or glycerol solutions) of well-known E.R.H.. After a certain time (several hours, up to 24 h), each sample is weighed again. The relative change of weight is plotted as a function of R.H. : the intersection of this curve with the abscissa (no weight change) give the A_W of the product.

In order to have good and reproducible results it is necessary to standardize all the steps of the procedure (Bem and Leistner, 1970 ; Thieme, 1972 ; Cheftel *et al.*, 1979).

C. Calibrated Methods

In these methods the sensitive part of a gauge equilibrates, by sorption or desorption process, with water vapour pressure of surrounding air. A physical property - electrical or mechanical - changes as a function of water content being choosen, the response of a gauge is calibrated in A_W, by comparison with a standard solution (saturated salt or glycerol). An appropriate device allows the gauge to be placed very close and just over the sample, in a closed box. The measurement is taken when equilibrium is reached between sample air and gauge (which needs from 2 up to 5 hours). Several such gauges are now commercially available.

1. *Change of an Electric Property*. The electrical resistance or capacitance of the gauge is generally used. The measurement of conductivity of a Lithium chloride solution has been used very often (Groninger, 1965, 1969 ; Greaves, 1976). Sina and Rotronic gauges are using this technique.

Some other material could be used, like a piece of plastic (Richard gauge), anodised aluminium (Cooke, 1976) or ceramic with titanium oxide (Norrish, 1966).

2. *Change of Mechanical Property*. Hair is the oldest material used for hygrometry determination. Instead of hairs, now it is often a polyamide fiber which is used.

The variation of length of the fiber is used to move a needle on a dial (Luft hygrometer), or is transformed into an electrical signal (Giroux Mesure gauge).

In order to have good results with calibrated gauges, both electrical and mechanical, it is absolutely necessary to respect the following precautions :

- to check very often (one or more time each day) the calibration, by using good and well known reference saturated salt solutions (Bizot, 1976 ; Labuza *et al.*, 1976 ; Greenspan, 1977 ; Cheftel, 1979) ; the difficulty to obtain a valid reference substance must not be underestimated ;
- to use the gauge always in sorption process, in order to avoid errors due to the hysteresis effect, because the sensitive part of the gauge is affected by sorption hysteresis ;
- to wait until equilibrium between sample, air and gauge is fully realized ;
- to avoid samples containing volatile components which could change the behavior of the gauge ; sometime it is possible to protect the gauge with a filter.

If all these precautions are carefully observed, then calibrated gauges can give good results.

They are particularly useful to follow with a potentiometric recorder, the change of A_W of a sample : for example the gauge can be introduced inside a stream of product on a production line, for the control of A_W.

CONCLUSION

The activity of water is now a very important parameter in cereal and food technology.

All alteration processes, both chemical and microbiological, and all industrial transformations are function of A_W, with regard to speed and intensity.

From a fundamental point of view, A_W reflects the thermodynamic state of water in products.

Therefore great attention must be devoted to the methods which could be used for measuring A_W. Each method, absolute or calibrated, involves large possibilities of errors :
- concerning the sample,which must be kept in the initial state, in order to avoid any change of A_W during the measurement ;
- concerning the method and apparatus ;
- concerning the experimental procedure, which needs a high level of standardization.

The consideration of the A_W concept and its determination is very helpfull for food scientists and technologists.

REFERENCES

Acheson, D. (1965). *In* "Humidity and Moisture. Measurement and Control in Science and Industry. Fundamentals and Standards., (Wexler A. ed.), Vol. 3, Reinhold Publishing Corporation, New-York, p. 521.

Ayerst, G. (1965).*J. of the Science of Food and Agriculture, 16,* 71.

Bates, D. (1969). *In* Relative Humidity in the Food Industry. B.F.M.I.R.A. Symposium Proceedings N°4, p.9.

Bem, Z., and Leistner, L. (1970). *Die Fleischwirtschaft, 50,* 1412.

Bizot, H. (1976). "Contribution à la mise au point de deux méthodes de référence pour la mesure de l'activité de l'eau dans les produits alimentaires". Rapport de D.E.A. de Sciences et Techniques des I.A.A., Université de Paris VII.

Bizot, H., and Multon, J.L. (1978). *Ann. Technol. Agric. 27,* 441.

Bizot, H., Tomé, D., Drapron, R., Guilbot, A., Multon J.L. (1978) "Les aliments à humidité intermédiaire. Série Synthèses Bibliographiques, CDIUPA et APRIA, 35, rue du Général Foy, 75008 Paris.

Brunauer, S., EMMETT, P., Teller, E. (1938). *J. of the American Chemical Society, 60,*309.

Cakebread, S. (1974). *Confectionery Production, 40, n° 2,*67 , *n° 3,* 104.

Cheftel, J.C. (1979). Personal communication.

Cooke, B. (1976). *Process Engineering, 8,* 51.

Coppens, R. and Wei, H. (1954). *Annales de Gembloux, 4,* 225.

Doublier, J., Lisch, J., Launay, B. (1975). "Agents de texture alimentaires : épaississants et gélifiants".Série Synthèses Bibliographiques, APRIA, Paris.

Drapron, R. (1972).*Ann. de Technol. Agric., 21,* 487.

Drapron, R., Guilbot, A., Barbe, F. (1962). *Ann. de Technol. Agric., 11,* 175.

Duckworth, R. and Smith, G. (1963). *Proceedings of the Nutrition Society, 22,* 182.

Duprat, F. and Guilbot, A. (1975). *In* "Water relations of foods" (Duckworth R.,ed.) Academic Press, London, p. 173.

Duprat, F. (1977). "Sur la détermination de quelques caractéristiques physiques du grain d'amidon et de sa phase cristalline". Thèse de Doctorat d'Etat, (in press).

Eagland, D. (1975). *In* "Water relations of foods", (R. Duckworth, ed.), p. 73, Academic Press, London.

Fett, H. (1973). *Journal of Food Science, 38,* 1097.

Gough, M. (1975). *Journal of Stored Products Research, 11,* 161.

Greaves, P. (1976). *Food Trade Review, 46,* 281.

Greenspan, L. (1977). *Journal of Research of the NBS, A.Physics and Chemistry, 81 A,* 89.

Groninger, K. (1965). *Mitteilungen aus dem Gebiete der Lebensmitteluntersuchung und Hygiene, 56,* 255.

Groninger, K. (1969). *In* Relative Humidity in the Food Industry. B.F.M.I.R.A. Symposium Proceedings n°4, p.16.

Guilbot, A., Charbonniere, R., Abadie, P., Girard,P. *Die Stärke, 12,* 327,(1960).

Guilbot, A. (1949). *C.R. Acad. Agric., 17,* 694.

Guilbot, A. (1965). Personal communication.

Guilbot, A., Duprat, F., Gallant D., Mercier, C., Robin J.P. (1979) "L'amidon" (in press).

Guilbot, A., Charbonniere, R., Drapron, R. (1961). *Die Stärke, 13,* 204.

Guilbot, A. and Lindenberg, A. (1960). *Biochemica et Biophysica Acta, 39,* 389.

Heiss, R. (1968). Springer-Verlag, Berlin, 163 p.

Hermans, P.H. (1952). *In* "Contribution à l'étude physique des fibres cellulosiques". Dunod, Paris

Labuza, T. (1975). *In* "Water relations of foods" (R. Duckworth, ed.) p. 155, Academic Press, London.

Landrock, A., Proctor, B. (1951). *Modern Packaging, 25,* 123.

Landrock, A. and Proctor, B. (1951). *Food Technology 5,* 332.

Ling, G. (1972). *In* "Water structure at the water-polymer interface" (H. Jellinek, ed.) p. 4, Plenum Press, New-York.

Loncin, M., Bimbenet, J., Lenges, J. (1968). *J. of Food Technology, 3,* 131.

McLaren, A. and Rowen, J. (1951). *J. of Polymer Sci., 7,* 289.

Morin, P. (1979). "Activité de l'eau dans un solide" C.N.E.R.N.A. Report.Personal Communication.

Moore, W.J. (1978). "Physical Chemistry", Longman, London.

Multon, J.L. and Bizot, H. (1976). *In* "Biodeterioration investigation techniques" (Walter, ed.) Applied Science Publishers, p. 149.

Multon, J.L. and Bizot, H. (1978). *Ann. Nutr. Alim., 32,* 631.

Multon, J.L. and Guilbot, A. (1975). *In* "Water relations of foods" (R. Duckworth, ed.) p. 379, Academic Press, London.

Multon, J.L., Gadet, M.H., Martin, G. (1978). *Revue de Métrologie Pratique et Légale, 5,* 342.

Multon, J.L., Martin, G. (1979). *Cereal Food World,* (in press).

Multon, J.L. (1979).Unpublished.

Nemethy, G. and Scheraga, H. (1962). *J. of Physical Chemistry, 66*, 1773.
Nemethy, G. and Scheraga, H. (1962). *J. of Chemical Physics, 36,* 3382.
Norrish, J.L. (1966). *J. of Food Technology, 1,* 25.
Rodel, W. and Leistner, L. (1972). *Die Fleischwirtschaft, 52,* 1461.
Scott, W. (1957). *Advances in Food Research, 7,*83.
Sood, V. and Heldman, D. (1974). *J. of Food Science, 39,* 1011.
Stitt, F. (1958). *In* "Fundamental aspects of the dehydration of foodstuffs" p. 67, Society of Chemical Industry, London.
Taylor, A. (1961). *Food Technology, 15*, 536.
Taylor, A. and Rowlinson, J. (1955). *Transaction of the Faraday Society, 51,* 1183.
Thieme, E. (1972). *Gordian, 72,* 132.
Vos, P. and Labuza, T. (1974). *Journal of Agricultural and Food Chemistry, 22,* 326.
Wexler, A. and Hasegawa, S. (1954). *J. of Research of the National Bureau of Standards, 53,* 19.
Wolf, W., Spiess, W., Jung, G. (1973). "Wasserdampfsorptionsisothermen von Lebensmitteln", 485 p., Maschinenbau Verlag, Frankfurt.

TECHNICAL EFFECTS IN CEREAL PRODUCTS OF LIPIDS - NATURALLY PRESENT AND ADDITIVES

Kåre Larsson

Department of Food Technology
University of Lund
Lund, Sweden

I. INTRODUCTION

The chemical composition of wheat lipids is well known, particularly from work by MacMurray and Morrison (1), and also lipids from other cereals have been well characterized (cf. Refs. 2,3). The physical chemistry of these lipids, however, has been neglected, although the significance of physical structure in relation to technical effects is obvious. The present review is limited to such aspects, and based on studies of association behaviour and molecular arrangements of cereal lipids in a water environment carried out at our Institute.

II. RELATIONS BETWEEN LIPID FUNCTIONALITY AND PHYSICAL STRUCTURE

General features of lipid-water interaction and lipid crystal and liquid-crystalline properties relevant to foods have earlier been reviewed (cf. Refs. 4,5,6). With regard to lipid functions in cereal products, particular attention should be given to the macroscopic properties of the phases. This is illustrated in Fig. 1, showing typical lipid crystals.

ISBN: 0-12-370960-1

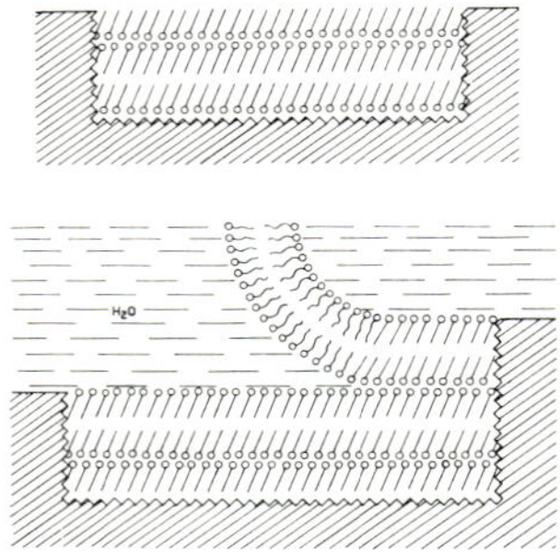

FIGURE 1. Lipid crystal with alternative hydrophobic or hydrophilic surfaces depending upon preparation procedure (7).

The crystals form thin plates parallel to the lipid bilayers, and the surface of the crystals can, depending upon preparation, consist of either chain methyl end groups or polar head groups. This means that the crystals as a whole phase can be surface active (7). The weak forces over the gap separating hydrocarbon chains means that the crystals have excellent lubrication properties, and also that it is mechanically easy to open the structure at the chain ends in order to expose a hydrophobic interface towards air or an oil phase.

The three types of lipid-water phases which we have observed in aqueous systems of cereal lipids are the lamellar and hexagonal liquid-crystalline phases shown in Fig. 2 and the liquid lipid-water phase of the so-called L2-type, see Fig. 3. The lamellar liquid-crystalline phase is formed for example by the phospholipid and digalactodiglyceride components of cereal lipids, and it has similar functionality as discussed above for lipid crystals. The hexagonal phase lacks these properties and behaves like a plastic fat. It is formed for example by monogalactodiglycerides in the presence of water. The structure of a pure L2-phase has been analysed in monoglyceride-water systems (8), and it is proposed to consist of disc-like water units separated by lipid bilayers. The L2-phase formed by cereal lipids should be expected to have the

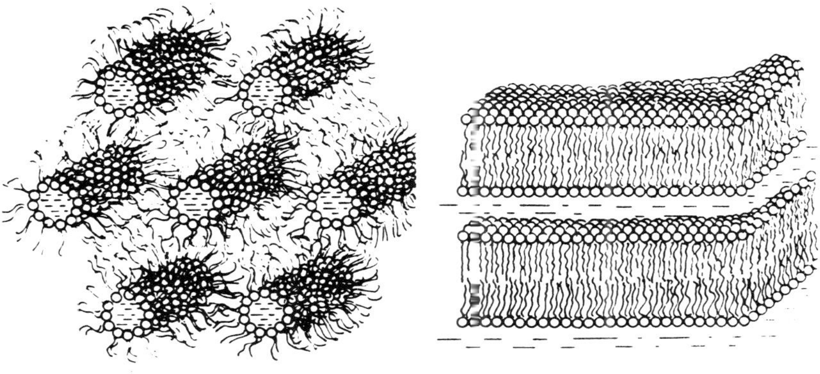

FIGURE 2. Lamellar and hexagonal liquid-crystalline phases, both observed in aqueous systems of cereal lipids (9).

FIGURE 3. Proposed structure of an L2-phase (8). This liquid is formed by cereal lipids in excess of water The analysis of the structure was carried out in monoglyceride--water systems (8)

same structure (9).

The functional properties of lipid-water phases, with special regard to cereal products, are summarized in Table 1. We can differentiate two levels. A colloidal level, where the lipid phase properties give certain technical effects, and a molecular level, where the amylose-lipid interaction is the only known example in cereals.

TABLE I. Schematic illustration of the most important lipid functions in cereal products. Most functions are related to a lipid phase, only starch interaction is known to occur on the molecular level.

Stabilise suspensions emulsions foams Influence adhesion (lubrication)	Lamellar structures Liquid-crystal, gel, crystal (L2)	The phase (not monolayer) is the functional unit (with hydrophilic or hydrophobic surface)
Starch interaction	Lipid monomer form amylose-lipid inclusion complex	The lipid monomer concentration is critical. (Ideal lipid distribution for lamellar structures)

III. INTERACTION BETWEEN CEREAL LIPIDS AND WATER

Separation of different phases when water is added to lipids (extracted in WSB) from triticale is shown in Fig. 4. There are two main types of lipids, non-polar ones with no water interaction which separates as an oil on the surface, and polar lipids which form aqueous phases. It is in this type of complex systems convenient to represent the phase equilibria in a ternary phase diagram, where the water/polar/non-polar lipids are the components. Such a mapping of lipids extracted from a wheat flour (9) is shown in Fig. 5. It can be seen that the three structures discussed above occur at different water composition. Although the relations are not known between this phase diagram and structural changes of lipids when a flour is worked in water, this phase diagram shows all possible structures. In a simplified way we can say from the association properties known in lipid systems that mechanical mixing in water means that membrane residues in the flour will associate in water into a lamellar liquid-crystalline structure. The degree of uptake of non-polar lipids determines whether the hexagonal phase is formed, and the degree of water uptake determines whether an L2-phase is formed. An illustration of the stepwise formation of ordered lipid-water phases from isolated membrane fragments and oil droplets from a cereal flour as indicated by our phase diagram results is given in Fig. 6.

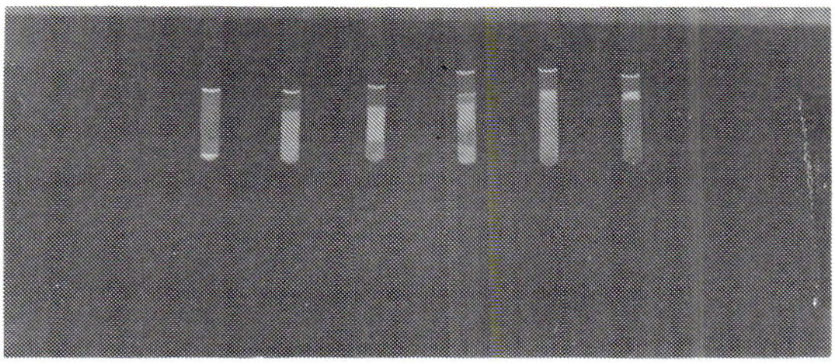

FIGURE 4. Illustration of phase separation after equilibria are reached between water and lipids extracted by water-saturated butanol (WSB) from triticale. The samples show separation by ultra-centrifugation of the pure liquid extract (to the left) and (towards the right) with increasing amounts of water (10, 20, 30, 40 and 50% (w/w)). The top phase at 50% (w/w) of water is non-polar oil, in the middle there is water, and at the bottom an L2-phase.

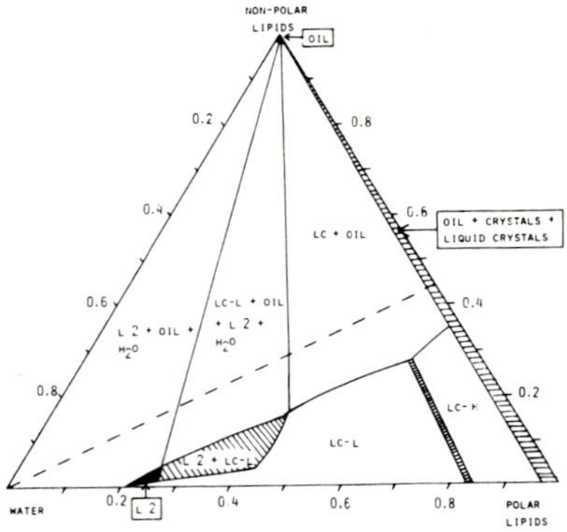

FIGURE 5. Phase diagram of lipids from wheat flour (Amy) illustrated as a ternary composition diagram where the corners are water, non-polar lipids and polar lipids (see Ref. 9).

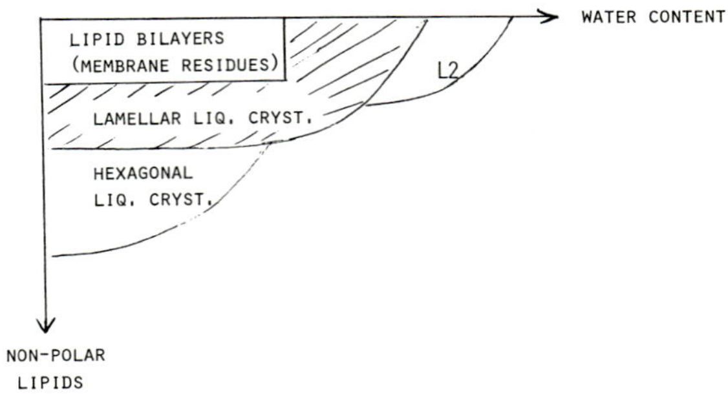

FIGURE 6. Schematic illustration of association structure formation when cereal flour is worked in water. Lipid bilayer units associate into liquid-crystalline domains. Depending upon uptake of water and of non-polar oil or a hexagonal structure can be formed.

There are a few minor differences between the phase diagram from wheat flour lipids (Fig. 5), and the diagram corresponding to gluten lipids (10), but the basic behaviour is the same with the same phases formed with increasing water content. We have also examined the corresponding phase equilibria with a salt solution corresponding to that of a dough (0.260 M NaCl, 0.070 M KCl, 0.016 M $MgCl_2$ and 0.004 M $CaCl_2$) instead of water, as it is known that the presence of ions drastically can change the water swelling properties of charged lipids. Also in this phase diagram (10) the behaviour is close to that of pure water shown in Fig. 5. The main differences are reduction in solubilization of non-polar lipids in the liquid crystalline phase, and reduction of the maximal water swelling. These effects should also be expected from shielding effects of the provided cations, as there are anionic lipid species among the polar lipids.

We have also studied different systems of extracted cereal lipids and found that the equilibria between these phases are different (11). One might therefore expect that they exhibit similar structural differences during dough-making. Systems of another wheat cultivar, rye lipids and triticale lipids are shown in Fig. 7. The phase properties are illustrated here in a simplified way by using binary systems, just showing the phases in the order they occur when water is successively added to the lipid. There are quite remarkable differences between these phase diagrams. Rye lipids (unable to form gluten) show no region of the lamellar liquid-crystalline phase at all, and the crucial role of this phase in gluten formation is discussed below. Triticale lipids show a phase behaviour in between rye and wheat lipids. Furthermore a wheat lipid with poor baking performance (M. Huntsman) shows coexistence of lamellar and hexagonal phases, i.e. there are less relative amounts of the functionally active lamellar phase compared to a good cultivar (Amy). Our present data indicate that the formation and existence range of the lamellar liquid crystalline phase in the actual cereal lipid-water system is related to the formation of this phase when the corresponding wheat flour is worked with water, and as discussed below, such a phase is one component of the gluten structure. The significance of cereal lipid-water interaction is also illustrated by possibilities for phase transitions due to small changes in lipid composition, such as hydrolysis giving free fatty acids known to take place during storage of cereals. As shown in Fig. 8 a very small amount of oleic acid added to cereal lipids can give such a phase transition.

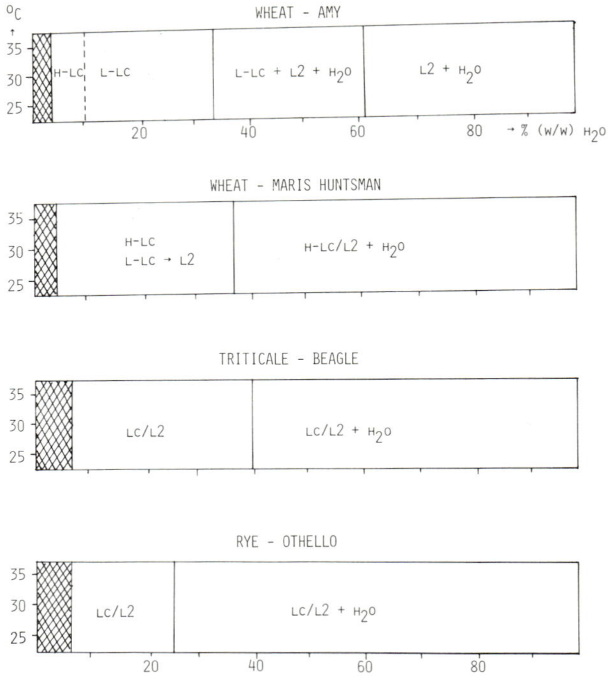

FIGURE 7. Phase properties in relation to water illustrated as binary phase diagrams of lipids extracted from flour from rye (Othello) and triticale (Beagle) compared to two wheat cultivars, M. Huntsman and Amy (11).

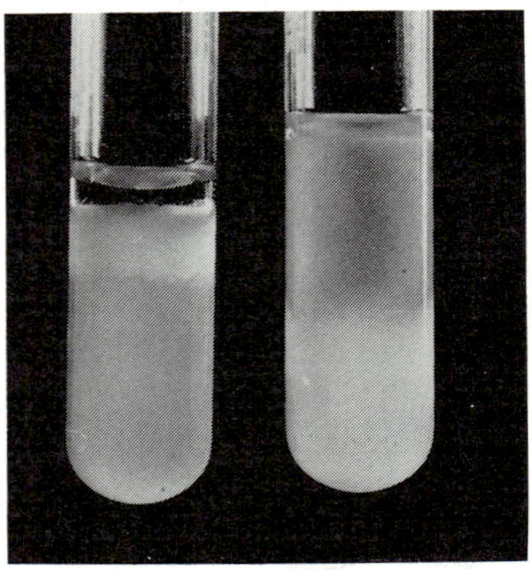

FIGURE 8. Samples of wheat flour lipids (Amy) to the left with 70% (w/w) of water showing from the top a liquid oil, a water layer (turbid) and an L2-phase. To the right effects when 3% (w/w) of oleic acid was added to the lipids.

IV. LIPIDS AND THE STRUCTURE OF GLUTEN

X-ray (or neutron) diffraction and low-angle scattering techniques are the only direct structural methods which can reveal the state of order and sizes of structural units of an aqueous gluten gel or a dough. Important X-ray studies have been reported by Hess (12), by Traub et al. (13) and by Grosskreutz (14). Of various published models on the gluten structure in a dough, the one by Grosskreutz (14) is based on the most thorough X-ray analysis, and the new results on lipid phase properties motivate a further discussion of this model. Based on X-ray low-angle scattering of diluted gluten protein solutions, it was concluded that gluten proteins form platelets with a diameter about 10 000 Å and a thickness of about 70 Å. A further feature of the proposed model was that lipid bilayers separate these protein platelets, so that the polar head groups of the lipid molecules interact with a polar surface of the protein platelets. The suggested lipid bilayer was based on the occurrence of an X-ray diffraction line near 50 Å. This interpretation, however, is not correct, and it is therefore surprising that this model is so often referred to when lipid function in gluten is discussed. A single lipid bilayer cannot give rise to a sharp diffraction line. Only a phase with strict periodicity in at least one dimension can give a sharp diffraction line. If the size of the ordered regions is reduced to a thickness below 1 000 Å perpendicular to the corresponding diffracting plane, the line is significantly broadened. The only reasonable explanation to the spacing near 50 Å is that it is produced by a lipid liquid-crystalline phase, and the following evidence support this:

1. The spacing is removed if most of the lipids are extracted from gluten (also used by Traub et al. (13) and Grosskreutz (14) in assigning the spacing to the lipid component).
2. Cereal lipids form liquid-crystalline phases with the same diffraction characteristics (similar line sharpness and thermal intensity fall-off).
3. Preparations consisting of isolated biomembranes spontaneously form multilamellar lipid-water phases, which is a direct consequence of the fact that the stable state of these lipids in an aqueous environment is just the multilamellar liquid-crystalline form. It should thus be expected that membrane residues from the endosperm will associate into a liquid crystalline lipid-water phase when wheat flour or gluten is mechanically worked with water. As an analogy it can be mentioned that the milk fat globule membranes, when butter is made, associate into

multilamellar aggregates (so-called liposomes, *cf.* (15).

4. If a lipid, which forms a lamellar liquid-crystal with water, is mechanically mixed with de-lipidized gluten in a proportion corresponding to native gluten, the X-ray diffraction and wet swelling resemble those of native wet gluten.

The first X-ray diffraction study of wet and dry gluten was reported by Hess (12). He reported a *d*-value of 43.2-47.7 Å for dry gluten and 52.1-55.7 Å for wet gluten. We have confirmed these values of about 55 Å for wet gluten with excess of water present, and about 44 Å for a gluten dried over P_2O_5. In our studies of extracted wheat lipids we observed a spacing of 49 Å at a ratio of 80:20 (w/w) polar/non-polar lipids and about 40 Å for polar lipids only. If the gluten X-ray diffraction data are assumed to be due to a lipid-water lamellar liquid-crystalline phase, such a phase contains about 90:10 polar/non-polar lipids, and its swelling from dry to completely swollen gluten corresponds to a water layer thickness of about 10 Å (or a water content in the liquid-crystalline phase of about 25%).

V. LIPIDS ADDED AS DOUGH CONDITIONERS

It was early observed that wheat lipids can improve breadmaking (16), and later it has been found from studies of different wheat lipid fractions that non-polar lipids have a negative effect whereas the improvement is due to the polar fractions (*cf.* 17). An interesting study of the effect of sucrose monopalmitate, ethoxylated monoglycerides and sodium stearoyl-2-lactylate (NaSSL) on bread made from defatted wheat flour was recently reported (18). Unfortunately, the physical state of the amphiphiles was not defined.

When technical effects of lipid additives are considered, the most relevant data are the bulk properties of lipid-water mixtures and the formation of air/water interfacial films. Self-association of lipid molecules is thus a characteristic feature of complex aqueous systems involving other biomolecules. This means that the same type of function units can be formed in complex systems as in the corresponding lipid-water system. Interaction between the natural wheat lipids and the lipid additives must also be considered. As most technically used lipid additives (monoglycerides or SSL-salts for example) exist in crystalline form when added to a dough one should not expect interaction with wheat lipids, which occur in liquid-crystalline states of order. We have also confirmed this experimentally by storing aqueous mixtures of wheat flour lipid WSB-extracts with monostearin and NaSSL at 25°C. Even when the monostearin was transformed into a gel state and it was stored with aqueous wheat lipids at 50°C, no interaction was observed.

An additive considered to be very effective in order to improve bread volume is NaSLL, and we have therefore studied its water interaction at room temperature in order to determine the functionally active structure. NaSLL forms a stable gel-state containing about 70% (w/w) of water. No other lipid additive is known which spontaneously gives a stable gel state at room temperature, and it is natural to assume that the function is related to this behaviour.

Any fat in the liquid state is known to function as a breaker of a protein foam, and this effect is due to their ability to displace the protein from the air/water interface (they give a higher surface pressure and can thus squeeze out the protein from the interface). Particularly from work by Krog and co-workers (19) it is known that α-tending emulsifiers are efficient in order to protect a foam against this destabilization tendency of liquid oils. The oil formed by non-polar lipids cannot be completely solubilized by polar lipids, as shown above, which means that there is always a non-polar oil phase able to act as foam destabilizer. For these reasons it is proposed that the effect of lipids as dough conditioner is to stabilize the structure of the air//water interface in a dough.

A secondary effect might be that the gel-phase might emulgate free oil formed from the wheat lipids during dough making, as it has been demonstrated that the gel-phase is more effective than any other lipid structure to stabilize oil-in-water emulsion (20). The functions to stabilize a foam and to encapsulate oil, which both are relevant for baking performance, are exhibited by lipid crystals, the lamellar liquid crystalline phase and the gel phase (see Table 1). The gel-phase, however, is most effective, and it is therefore not surprising that NaSSL, which forms a stable gel-phase, is superior compared to for example monoglycerides (crystals) or a lecithin dispersion (lamellar liquid-crystalline phase).

VI. LIPIDS AND STARCH

From work by Acker (cf. Ref. 21) and Arunga and Morrison (22) it is known that wheat starch contains about 1% (w/w) of lipids, most of which is lysolecithin. The lipids form an inclusion complex with amylose, which occurs in the helical V-conformation. This complex formed by amylose-monoglycerides has recently been studied (23), and the chains were shown to exist in the extended zig-zag conformation, see Fig. 9

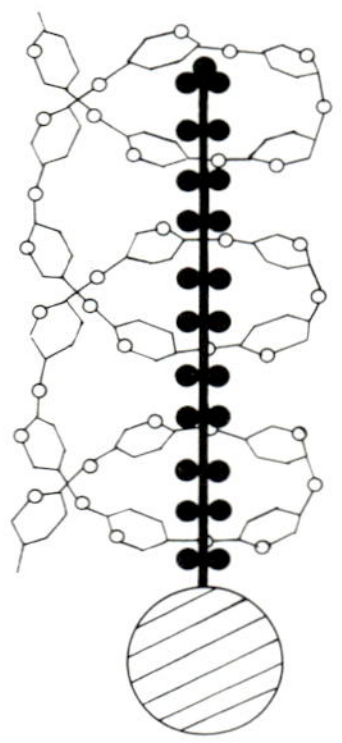

FIGURE 9. Proposed model of the amylose-monoglyceride inclusion complex with regard to conformation of the lipid (23).

One important application of lipids as functional additives in foods is to modify the texture of starch based products, for example reduce sticky consistency in pasta and potato-based products and reduce the aging velocity of bread (anti-staling) (cf. Ref. 24). These technical effects have also been discussed in relation to formation of the inclusion complex, but the actual mechanisms involved are really not known. An important study of cold gelatinization was recently reported (25), showing that gelatinization only can take place if amylose leaches out from the granule. Reasons are also given indicating that such gelatinization is closely related to thermal gelatinization (25), and thus that amylose leaching a general condition for starch gel formation. As lipid complexing of amylose can form insoluble precipitates (23), one might expect that lipids can block amylose leaching if the molecules can be supplied fast enough at the gelatinization temperature. This has also been shown to occur when the lipid is provided in excess and in the form of liposomes (26). If thus a starch suspension 2% (w/w) in water is slowly heated (below $10^{o}C/h$) in the presence of 1% (w/w) of 1-monolaurin in the liposomal dispersion state, no starch gelatinization occurs up to about $70^{o}C$, even if the temperature is kept there for 1 h. At higher temperature the lipid is successively transformed into the L2-phase and this phase, which separates from the starch suspension, cannot block amylose leaching. The liposomal form is the most effective lipid structure in order to supply the starch granules with bilayers; liposomes form easily lipid multilayers outside the granules. Kinetics of the lipid-amylose complex formation is directly related to the lipid monomer concentration, and monolaurin has a relatively high monomer concentration (about 10^{-5} M).

A characteristic property of cereal starch compared to other starches is the high lipid content, and Acker (21) has shown that the final stages of ripening involves incorporation of mainly lysolecithin. The role of starch in bread formation was recently studied (29) by examination of doughs prepared from wheat gluten and starch grains from wheat, potato, tapioka, rice and maize. Of these starches wheat starch gave the firmest bread with the granules most uniformly oriented in films in the gas cell walls. Also other studies indicate that cereal starch is superior in baking performance compared to starches from other sources. Can the reason be the high lipid content in cereal starches? For reasons discussed above, the lipid-amylose inclusion complex in cereal starch must be expected to result in reduced amylose leaching, and thus in reduced swelling compared to other starches. We have examined the physical properties of lipids physically adsorbed on starch granules after washing out starch from a dough (27). These lipids form spontaneously liposomes when exposed to excess of water. These lipids, roughly 0.5 mg/g starch, should also be expected to contribute the gelatinization properties by reduction of amylose transport from the granules, as discussed above.

Monoglycerides are certainly the additive most frequently used in order to modify the texture of starch products. In products where the gelatinized starch determines the texture, such as a pure starch gel, the amylose-lipid complex formation can fully explain the technical effects (cf. Ref. 24). It has been found that microcrystals of monoglycerides have a tendency to adhere to unswollen starch granules, and further that a 'scaly' structure is formed, which coats gelatinized starch (28). In products where other components have a dominating effect on the structure, the situation is more complicated. The mechanisms behind the aging of bread have been frequently discussed in the literature. It is known that the staling involves moisture transport from gluten regions to starch regions and starch retrogradation (27). If the gelatinized granules are covered by an amylose-lipid insoluble surface film, which must be assumed when a monoglyceride has been added, such a film should be expected to reduce the water transport involved in staling. This surface coating of starch granules might be the main antistaling effect of added lipids; i.e. they slow down the water transport towards the equilibrium distribution by forming a penetration barrier between the gluten matrix and starch granules.

ACKNOWLEDGMENTS

This review is based on work carried out in collaboration with T. Carlson and Y. Miezis. Financial support was obtained from the Swedish Board for Technical Development.

REFERENCES

1. MacMurray, T.A., and Morrison, W.R. (1970). J. Sci. Fd Agric. 21, 520.
2. Chung, O.K., and Tsen, C.C. (1974). In "The First Man Made Cereal" (C.C. Tsen, ed.). Amer. Ass. Cereal Chem., St. Paul, Minn.
3. Sahasrabudhe, M.R. (1979). J. Amer. Oil. Chemists´ Soc. 56, 80.
4. Luzzati, V., Gulik-Krzywicki, T., and Tardieu, A. (1968). Nature 218, 1031.
5. Larsson, K. (1976). In "Food Emulsions" (S. Friberg, ed.). Marcel Dekker Inc., New York.
6. Krog, N., and Birk Lauridsen, J. (1976). In "Food Emulsions" (S. Friberg, ed.). Marcel Dekker Inc., New York.
7. Larsson, K. (1968). In "Surface-Active Lipids in Foods". Soc. Chem. Ind., Monograph No 32. London.
8. Larsson, K. J. Coll. Interf. Sci., in press.
9. Carlson, T., Larsson, K., and Miezis, Y. (1978). Cereal Chem. 55, 168.
10. Carlson, T., Larsson, K., and Poovarodom, S. Cereal Chem., in press.
11. Carlson, T., Larsson, K., and Miezis, Y. J. Dispersion Techn., in press.

12. Hess, K. (1954). KolloidZ. 136, 84.
13. Traub, W., Hutchinson, J.B., and Daniels, D.G.H. (1957). Nature 179, 769.
14. Grosskreutz, J.C. (1961). Cereal Chem. 38, 336.
15. Mulder, H., and Walstra, P. (1974). "The Milk Fat Globule: Emulsion Science as Applied to Milk Products and Comparable Foods". Pudoc, Wageningen and Commonwealth Agric. Bureaux, Farnham Royal.
16. Pomeranz, Y., Rubenthaler, G.L., and Finney, K.F. (1965). Food Technol. 19, 1724.
17. McRitchie, F. (1977). J. Sci. Fd Agric. 28, 53.
18. Chung, O.K., Pomeranz, Y., Finney, K.F., and Sjogren, M.D. (1978). J. Amer. Oil Chemists´ Soc. 55, 635.
19. Krog, N. (1977). J. Amer. Oil Chemists´ Soc. 54, 124.
20. Larsson, K. (1978). Progr. Chem. Fats other Lipids 16, 163 (Pergamon Press).
21. Acker, L. (1977). Fette-Seifen-Anstrichmitteln 79, 1.
22. Arunga, R.O., and Morrison, W.R. (1971). Lipids 6, 768.
23. Carlson, T., Dinh, N., Krog, N., and Larsson, K. (1979). Stärke 31, 222.
24. Krog, N. (1971). Stärke 23, 206.
25. Lindqvist, I. (1979). Stärke 31, 195.
26. Larsson, K. Stärke, to be published.
27. Carlson, T., Eliasson, A.-C., Larsson, K., and Miezis, Y. Stärke, to be published.
28. van Lonkhuysen, H., and Blankestijn, J. (1976). Stärke 28, 227.
29. Willhaft, E.M.A. (1973). Bakers Digest 47, 14.

WHEAT BRAN AS A DIETARY FIBER

R. M. Saunders

Western Regional Research Center
Science and Education Administration
United States Department of Agriculture
Albany, California

I. INTRODUCTION

Dietary fiber can best be defined as that fraction of a food which is not broken down by enzymes in the human digestive tract to produce molecules which are then absorbed into the blood stream. Dietary fiber is generally acknowledged to include celluloses, hemicelluloses, pectins, gums, and lignins, but non-digestible fractions of protein, starch, lipids, minerals and sugars, and other non-metabolized compounds such as Maillard reaction products, in reality would qualify as part of the dietary fiber complex.

The intense current interest in dietary fiber has been generated by British workers Burkitt, Painter, and Trowell (1,2) who hypothesized that certain diseases in western civilization are due to failure by the population during early life to consume adequate amounts of fiber, particularly wheat bran. The diseases and disorders mentioned in their hypothesis are, in alphabetical order: appendicitis, atheroma, colon cancer, constipation, coronary thrombosis, dental caries, deep-vein thrombosis, diabetes, diverticulitis, gallstones, hemorrhoids, hiatus hernia, ischemic heart disease, peptic ulcer, polyps of the bowel (simple and malignant), and varicose veins.

Burkitt and his colleagues believe that a large part of the reason for decreased fiber intake can be blamed upon the wheat milling and baking industry. That is, highly refined white flour is almost devoid of fiber. These workers have stated "the greatest change in the British diet that has occurred in the past hundred years is a relative decrease in the intake of cereal fiber" (3). Their extrapolative suggestion that adding wheat bran to Western diets would prevent the disorders noted above has been refuted in a paper originating from the Flour Milling and Baking Research Association (4). Nevertheless, thanks to these workers, nutritionists, dietitians, physicians and food technologists now are acutely aware that the indigestible portion - or dietary fiber - of a food-

ISBN: 0-12-370960-1

stuff is a component deserving attention equal to that accorded the common nutrients.

In this paper on bran as a source of dietary fiber, the term dietary fiber includes the total indigestible fraction of the foodstuff, i.e., unavailable polysaccharides, lignin, protein, fat and micronutrients.

II. BRAN DIETARY FIBER: STRUCTURE AND COMPOSITION

The nature of the morphological entities within the wheat berry that comprise the bran fraction can be gleaned from Figures 1-3. In Figure 1 an artists rendition of the wheat kernel is illustrated; several cellular layers are present in the bran fraction. These layers in order from the exterior to the interior of the seed are: epidermis, hypodermis, intermediate cells, cross cells, tube cells, seed coat, nucellar epidermis (or hyaline layer) and aleurone layer. Distribution of these layers on a kernel weight basis is shown in Table I.

A cross-sectional light microscopy view of a wheat kernel section is shown in Figure 2, and a scanning electron micrograph of essentially the same section is illustrated in Figure 3.

When wheat is milled to provide white flour, the break occurs just within the endosperm, close to the junction of the aleurone layer and the endosperm (Figure 2). Thus bran contains all the layers mentioned above plus a small amount of starchy endosperm material. Commercial wheat bran contains germ, the matter of degree depending on the milling procedure. In normal milling practice, wheat is classified into several fractions (Table II). Commercial wheat bran refers to the larger particles which are removed from the kernel. These particles, which are platelets, range from 40 to 1600 u in size with over 90% being larger than 1500 u, and have an average thickness of 125 u.

The range in proximate composition of commercial bran from different types of wheat is shown in Table III (6). On a moisture-free basis, bran contains about 70% carbohydrates which consists of approximately 12% starch, 8% sugars, 45% hemicelluloses[1] and 35% cellulose (7, 8). The starch component may vary from 7% to as high as 16% depending on the variety of wheat and degree of milling (7, 9). It is noteworthy that the AACC certified food grade wheat bran contains 31% starch (or elsewhere (10) claimed to contain 23% starch), and thus is not typical of a bran. Rather it is bran plus about one third its weight of flour.

[1]In this text, hemicelluloses include pentosans.

WHEAT KERNEL

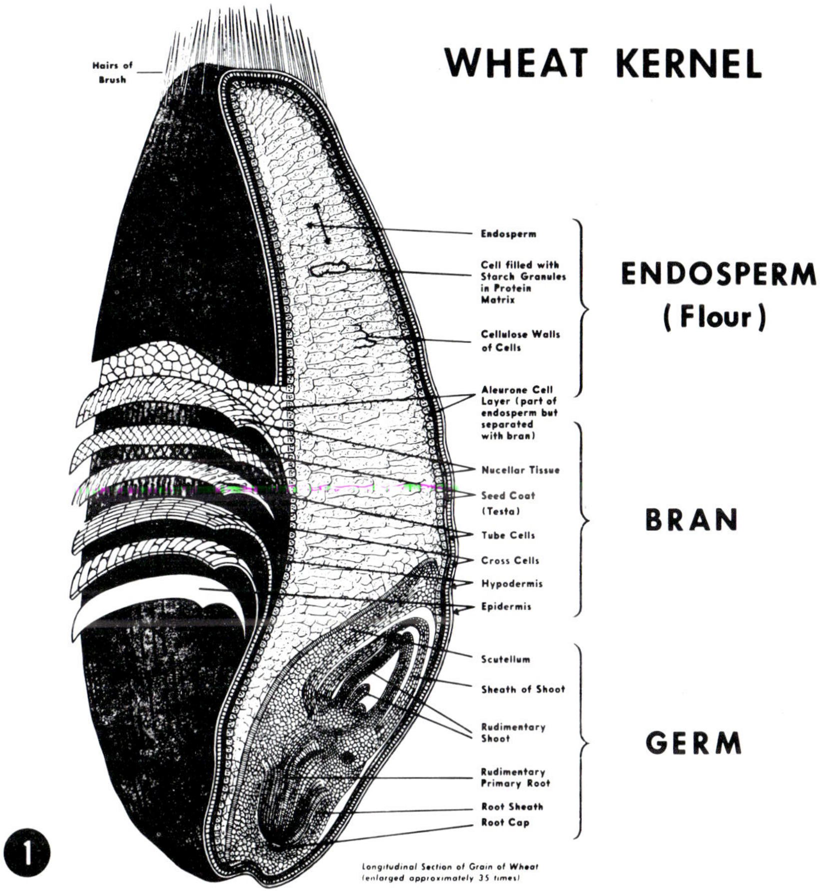

Longitudinal Section of Grain of Wheat (enlarged approximately 35 times)

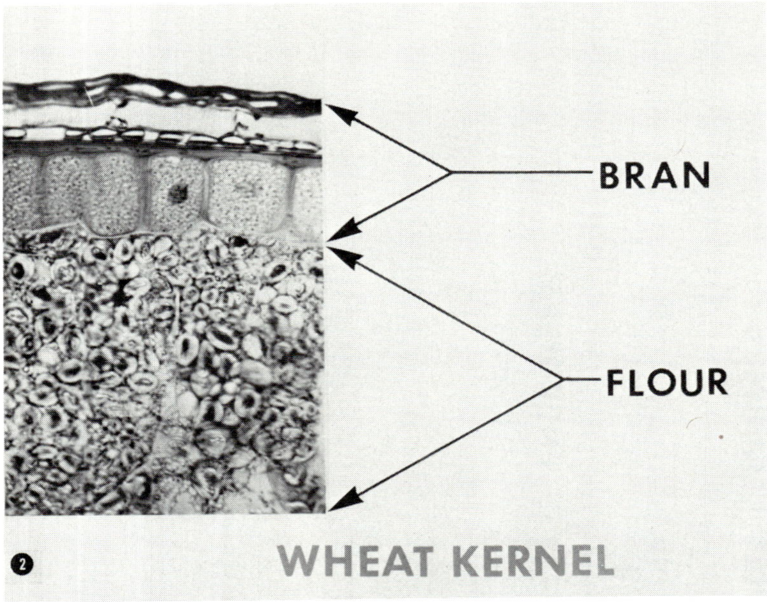
BRAN
FLOUR
2
WHEAT KERNEL

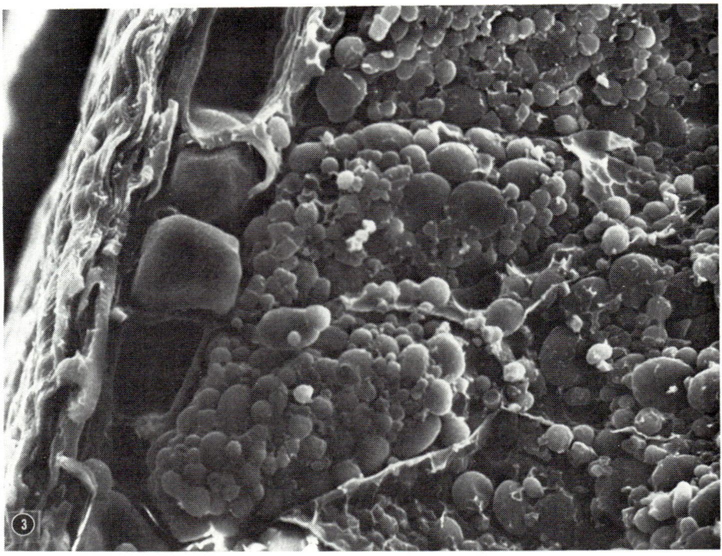
3

TABLE I. Composition of Component Parts of the Wheat Kernel[1] (Percentages)

Outer Pericarp	Cross cells	Seed coat	Nucellar layer & seed coat	Aleurone bran	Clean endosperm	Starchy Germ	Germ
2.0 4.4	0.5 1.5	0.3- 1.1	2.2- 3.1	4.6- 8.9	11.6- 21.4	74.9- 86.5	1.0- 3.8

[1]From data of Macmasters, Hinton, and Bradbury (5). Dry basis.

TABLE II. Commercial Milling Fractions Derived from Whole Wheat[1] (Percentages)

Wheat	Bran	Shorts	Red Dog	Germ	Flour
100	12.5-16.8	6.9-8.9	1.6-3.7	0.6-1.1	72.4-77.0

[1]Data from the Millfeed Manual (6).

TABLE III. Proximate Analysis of Wheat Bran[1] (Percentages)

Moisture	Protein (N x 6.25)	Fat	Ash	Crude fiber	Starch
3.7-17.7	11.9-22.9	3.0-6.8	3.8-9.6	6.8-17.5	7-16

[1]Data from the Millfeed Manual (6) and selected references.

Sugars are located for the most part within the aleurone layer (11) where sucrose, raffinose and neokestose account for approximately 75% of the total (7, 11, 12). Lesser amounts of other sugars include stachyose, fructosylraffinose, glycerol, xylose, arabinose, glucose, fructose, and low-molecular-weight fructosans.

No data exist on the composition or physical properties of wheat bran cellulose. Bran hemicellulose is more highly branched than the hemicellulose present in flour, straw or leaves, and has an arabinoxylan containing about 10% glucuronic acid as a major component (13).

With the exception of the known location of the sugars within the aleurone cells, little information is available on the precise layer(s) in which the polysaccharides are distributed. Bruckner (14) analyzed hand-dissected bran layers for polysaccharides and showed the epidermis and hypodermis were richer in hemicelluloses (50%) and crude fiber (including cellulose) (27%) than were the cross cells and tube cells (39 and 21% respectively). The seed coat is known to contain cellulose and pectin (15). The cell walls of the aleurone layer are believed to be cellulosic (16).

Since approximately 12% endosperm material is generally present in commercial bran, it should be noted that small quantities of endosperm hemicelluloses are present. The major sugar components of these materials are D-galactose and D-glucose (17). These polysaccharides in the form of glycoproteins are believed (18) to be responsible for the gelling power of certain wheat flour extracts. It is also likely that traces of water-soluble B-glucans are present (19).

Wheat bran contains less than 5% lignin. The quantity of lignin in different bran layers is not documented, nor has its chemical composition been established. Bran dietary fiber has been estimated to contain 7% lignin (20).

The cellulosic and hemicellulosic polysaccharides plus lignin are believed by most workers to form the major portion of bran dietary fiber, although the author believes that indigestible protein may be a significant dietary fiber component (21).

III. BRAN DIETARY FIBER: MEASUREMENT

Measurement of a value which is purported to be an index of dietary fiber is usually done chemically or enzymatically. Chemical methods include measurement of "unavailable carbohydrate" (22), or "neutral detergent fiber" (NDF), the residue left after treatment of a food with hot detergent with or without a pretreatment of the food with α-amylase (23). Enzymatic techniques are typically variations on digestion with protease and α-amylase (24, 25).

Values from this laboratory for crude fiber, NDF, and *in vitro* (enzymatic) techniques are compared with *in vivo* (rats) dietary fiber values for wheat milling fractions in Table IV (25, 26). The *in vitro* and *in vivo* methods have been described in detail elsewhere (25, 26). Analogous values for wheat-based breads and breakfast foods are listed in Table V (25, 26). Whilst it may not be immediately apparent that exacting linear relationships exist among these different fiber assessments, regression analyses indicate this to be the case. Correlations are listed in Table VI (25). Further evidence to support the concept of a linear relationship between crude fiber and *in vivo* (rats) dietary fiber in wheat bran is evident in Table VII. For nine different wheat brans, linear correlatiuon between dietary fiber (y) and crude fiber (x) calculated to be $y = 3.34x + 10.5$.

TABLE IV. Crude Fiber, Neutral Detergent Fiber, and Dietary Fiber in Wheat Milling Fractions Measured by *In Vivo* and *In Vitro* Techniques.

Material	NDF	Dietary fiber		Crude fiber
		in vitro	*in vivo*	
	%	%	%	%
Bran	49.66	52.41[a]	52.7[d]	12.2[d]
Shorts	41.97[a]	47.16[c]	41.2[e]	9.8[e]
Germ	26.47[b]	32.36[c]	23.9[e]	5.8[e]
Red Dog	19.33a	24.50[c]	18.8[e]	4.6[e]
Flour	1.84[a]	3.88[c]	2.1[d]	0.6[d]
AACC bran[f]	45.42[a]	43.85[a]	46.7[e]	10.0[e]

[a]Average value for 6 determinations.
[b]Average value for 5 determinations.
[c]Average value for 3 determinations.
[d]Average value for 2 determinations.
[e]Single determination.
[f]Certified food-grade wheat bran. American Association of Cereal Chemists, St. Paul, Minnesota.
All figures on moisture-free basis.

TABLE V. Crude Fiber and Dietary Fiber Measured by *In Vivo* and *In Vitro* Techniques in Wheat Breads and Breakfast Foods

Wheat food	Dietary fiber *In vitro*[a] %	Dietary fiber *In vivo*[b] %	Crude fiber[b]
Breads			
white 1	5.98	1.73	0.60
white 2	6.62	4.40	0.40
white/germ	5.84	2.29	0.74
mixed grains	6.57	3.00	1.06
whole wheat 1	10.46	7.78[d]	1.60[d]
whole wheat 2	11.83	7.78	1.47
whole wheat 3	12.14	9.27	2.45
whole wheat 4	12.85	7.77	2.34
Breakfast food			
wheat-processed	11.79	8.80	1.85
wheat-natural	9.37	4.35	1.17
bran	33.64	37.90	8.62

[a]Average value from 3 determinations.
[b]Single determination.
[c]Average value for 6 determinations.
[d]Average value for 2 determinations.
All figures on moisture-free basis.

TABLE VI. Linear Regression Equations and Correlation Coefficients for Relationships among *In Vivo* Dietary Fiber (Rats), *In Vitro* Dietary Fiber (pronase/amylase), NDF and Crude Fiber in Materials derived from Wheat.

Experimental values	Regression equation	Correlation coefficient
Wheat milling fractions		
In vivo dietary fiber (y) versus crude fiber (x)	y = 4.48x - 1.19	0.996
In vivo dietary fiber (y) versus *in vitro* dietary fiber (x)	y = 1.05x - 4.76	0.976
In vivo dietary fiber (y) versus NDF (x)	y = 1.05x - 1.29	0.996
In vivo dietary fiber (y) versus crude fiber (x)	y = 4.36x - 0.19	0.989
In vivo dietary fiber (y) versus *in vitro* dietary fiber (x)	y = 1.28x - 6.15	0.991

TABLE VII. *In vivo* Dietary Fiber and Crude Fiber of Wheat Brans

Bran No.	Dietary fiber %	Crude fiber %
1	52.2	12.4
2	50.3	12.4
3	53.4	12.2
4	52.1	12.2
5	50.7	13.2
6	61.0	14.3
7	53.8	13.4
8	52.2	12.3
9	49.3	11.6

All figures on moisture free basis.

The values for crude fiber and *in vivo* dietary fiber listed in Tables IV, V, and VII are shown graphically in Figure 4.

In wheat bran relationships clearly exist amongst these analytical values commonly associated with dietary fiber. In bran, dietary fiber measured *in the rat* is slightly more than four times the crude fiber value, or is fairly close to values determined by an *in vitro* digestive system, and by a detergent digest. What does this mean in the case of humans eating wheat bran? Will the human experience the same quantity of indigestible residue passing through the tract as does the rat, or will it be less because of intestinal microbial degradation of the bran? Which bran component is responsible for physiological changes? Is it conceivable that specific bran components can be associated with positive therapeutic effects and eventually become a prescribed food ingredient?

Such questions for the most part remain unanswered although research groups continue to look for the answers. One question which can be answered is that wheat bran in the

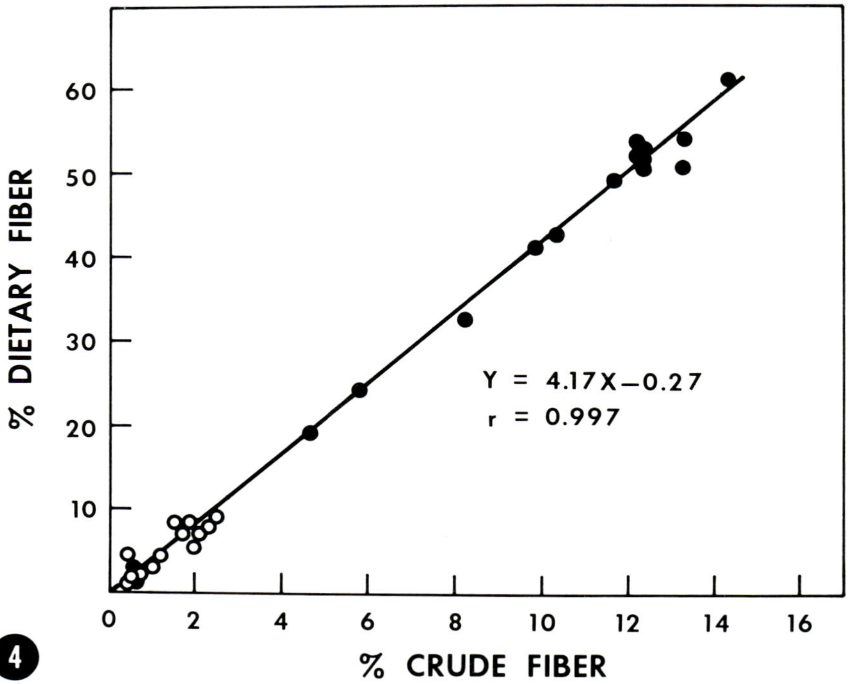
% DIETARY FIBER
60
50
40
30
20
10
0
2
4
6
8
10
12
14
16
% CRUDE FIBER
Y = 4.17X−0.27
r = 0.997

4

WHOLE WHEAT BREAD
WHITE BREAD

5

diet definitely increases the quantity of undigested food material entering the lower intestinal tract. This is illustrated in Figure 5 which shows the undigestible residue left after *in vitro* enzymatic (protease/α-amylase) digestion of equal quantities of whole wheat and white breads.

Even though the fraction of bran which is likely to remain undigested within the small intestine and, thus, enter the lower tract is thought to be known, the cellular components within this undigested fragment which are partly or wholly digested by gut microbial activity have not been clearly identified. A recent paper on this subject (10) lists for three human subjects a mean fecal recovery of bran cellulose of 83 $\pm$ 7%, hemicellulose 38 $\pm$ 6% and lignin, 92 $\pm$ 30%. Cell walls of the aleurone layer are likely to be less difficult to digest microbiologically. This is indicated in treatment of wheat bran with massive amounts of cellulolytic and proteolytic enzymes cause complete degradation of the aleurone cell wall matrix, but have no gross effect upon other bran layers when viewed by light microscopy (27). This hypothesis tends to be confirmed upon examination of fecal material recovered from different animals after ingestion of bran. The aleurone cell-wall matrix prevails virtually intact in chicks (28), is only partially intact after digestion by the calf (29), and disappears entirely when digested by the pig (29).

IV. BRAN DIETARY FIBER: HUMAN PHYSIOLOGICAL OBSERVATIONS

For purposes of this paper, observed physiological responses to ingestion of wheat bran have been tabulated briefly in Table VIII. The serious reader is referred to reference 30 and elsewhere for further information.

The difficulties of unequivocably establishing cause and effect relationships in broad physiological effects in humans is clearly illustrated in Table IX. In this experiment (31) fecal weight, plasma cholesterol (and low density lipoprotein cholesterol), were significantly affected by hard red spring wheat bran, but not by soft white winter wheat bran. No explanation has been postulated to explain these differences, but clearly brans from different varieties have different effects.

TABLE VIII. Observations and Reported Therapeutic Effects of Wheat Bran in Human Diets.

Stool volume increase
Positive effect upon constipation
Transit time decrease[1]
Bile acid excretion increase
Irritable bowel syndrome decrease
Diverticular disease decrease[1]
Serum cholesterol decrease[2]
Serum triglycerides decrease[1]
Glucose tolerance decrease[1]
Serum minerals decreased
Mineral absorption decreased
Energy absorption decrease
Fecal lipid, nitrogen, volatile fatty acids increase
Serum folate decrease

Other claims based only on epidemiological studies.

[1]Not observed in every study.
[2]Only observed in one study, not in other studies.

TABLE IX. Observed Human Responses to Ingestion of Wheat Brans

Observation	HRS bran	SWW bran
fecal weight	86% increase	no effect
plasma cholesterol	12% decrease	no effect
LDL cholesterol	21% decrease	no effect
plasma triglycerides	16% decrease	16% decrease

It is conceivable that bran components such as polyphenolics which can inhibit gut enzymes presumably could be involved in some of these and other physiological observations. It has been demonstrated on numerous occasions by animal nutritionists that certain polysaccharides or other cereal grain components exhibit growth depression effects in animals. Changes measured in human metabolic responses to bran may possibly be a manifestation of these same effects.

Acknowledgment:

The author thanks E. Hautala for measurement of NDF and *in vitro* dietary fiber recorded in Tables IV and V, D. Robbins and M. G. Gumbmann for assistance in rat feeding experiments, and D. W. Irving and M. A. Connor for microscopy work.

REFERENCES

1. Painter, N. S., and D. P. Burkitt. Brit. Med. J. ii: 450, 1971.

2. Travell, H. C. Revue Eur. Etude. Clin. Biol. 17: 345, 1972.

3. Burkitt, D. P., A. R. P. Walker, and N. S. Painter. Lancet, ii: 1408, 1972.

4. Eastwood, M. A., N. Fisher, C. T. Greenwood, and J. B. Hutchinson. Lancet, i: 1029, 1974.

5. MacMasters, M. M., J. J. C. Hinton, and D. Bradbury. In Wheat: Chemistry and Technology (ed. Y. Pomeranz). Amer. Assn. Cereal Chemists, St. Paul. 1971.

6. Millfeed Manual, The Millers National Federation, Chicago, Illinois, 1972.

7. D'Appolonia, B. L., K. A. Gilles, E. M. Osman and Y. Pomeranz. In Wheat: Chemistry and Technology (ed. Y. Pomeranz). Amer. Assn. Cereal Chemists, St. Paul 1971.

8. Fraser, J. R., and D. C. Holmes. J. Sci. Food Agr. 10: 506, 1959.

9. Saunders, R. M., A. L. Potter, M. A. Connor, R. M. McCready and H. G. Walker. Cereal Chem. 47: 147, 1970.

10. Dintzis, F. R., L. M. Legg, W. L. Deatherage, F. L. Baker, G. E. Inglett, R. S. Jacob, S. J. Peck, J. M. Munoz, L. M. Klevay, H. H. Sandstead and W. C. Shuey. Cereal Chem. 56: 123, 1979.

11. Stevens, D. J. J. Sci. Food Agr. 21: 31, 1970.

12. Saunders, R. M. and H. G. Walker. Cereal Chem. 46: 85, 1969.

13. Adams, G. A. Can. J. Chem. 33: 56, 1955.

14. Bruckner, G. In: Bread in Our Time, Detmold, Germany, 1966.

15. Bradbury, D., M. M. MacMasters and I. M. Cull. Cereal Chem. 33: 342, 1956.

REFERENCES (continued)

16. Bradbury, D., M. M. MacMasters and I. M. Cull. Cereal Chem. 33: 361, 1956.

17. Strobel, R. G., and J. Holme. Cereal Chem. 40: 361, 1963.

18. Udy, D. C. Cereal Chem. 33: 67, 1956.

19. Preece, I. A., and K. G. MacKenzie. J. Inst. Brewing, 58: 457, 1952.

20. Southgate, D. A. T. In Food and Fiber Symposium, Marabou, Sundbyberg,

21. Saunders, R. M., and E. Hautala. In Dietary Fibers: Chemistry and Nutrition (ed. G. E. Inglett and S. I. Falkehag). Academic Press 1979.

22. Southgate, D. A. T. J. Sci. Food Agric. 20: 331, 1969.

23. Southgate, D. A. T. J. Sci. Food Agric. 29: 979, 1978.

24. Hellendoorn, E. W., M. G. Noordhoff and J. Slagman. J. Sci. Food Agric. 26: 1461, 1975.

25. Saunders, R. M. and E. Hautala. Amer. J. Clin. Nutr. 32: 1188, 1979.

26. Saunders, R. M. Amer. J. Clin. Nutr. 31: 2136, 1978.

27. Saunders, R. M., M. A. Connor, R. H. Edwards and G. O. Kohler. Cereal Chem. 49: 436, 1972.

28. Saunders, R. M., H. G. Walker and G. O. Kohler. Poultry Sci. 48: 1497, 1969.

29. Saunders, R. M., M. A. Connor, G. O. Kohler and L. G. Blaylock. J. Animal Sci. 38: 1272, 1974

30. Kelsay, J. L. Amer. J. Clin. Nutr. 31: 142, 1978.

31. Munoz, J. M., H. H. Sandstead, . A. Jacob, G. M. Logan, S. J. Peck, L. M. Klevay, F. R. Dintzis, G. E. Inglett and W. C. Shuey. Amer. J. Clin. Nutr. 32: 580, 1979.

WHEAT GLUTEN

Marja-Leena Sarkki

Raision Tehtaat
Raisio, Finland

I. INTRODUCTION

Vital wheat gluten is the most common cereal protein which contains unique baking characteristics. Wheat contains 9-12 % of gluten which is uniformly distributed as a matrix throughout an endosperm. The specific property of gluten is to form an elastic mass when hydrated. The elasticity and thermosetting ability of gluten gives volume, texture, and apperance to our bread. In other food systems properties of gluten are utilized in manufacturing technology, texture, and flavor. Fractination of wheat into starch and gluten is made by old dough processes or by more sophisticated centrifugation processes. The manufacture of gluten does not demand any chemicals or complicated treatments. The separation occurs in a water medium utilizing mechanical and heat energy which yields all wheat components in their natural form. The most common utilization field of gluten is in baking products, but the use of fabricated foods has opened new markets for gluten. New economical, non-polluting manufacturing methods also permit the growth of gluten production in the near future. The food and techical uses of gluten have focused sharply on new perpectives, expanding techical ability and making use of gluten's unique properties. The specific functional properties of gluten protein adds to the knowledge of the behavior of gluten in different systems.

ISBN: 0-12-370960-1

II. THE PRODUCTION AND USES OF GLUTEN

Total production of wheat gluten is relatively small, ca. 120.000 tons a year. Only 0.2 % of the wheat on earth is processed to starch and gluten. The biggest consumer of gluten is the U.S.A., utilizing 40 % of the total production. Growth of convenience foods will provide an increased market for vital wheat gluten. Estimated annual market growth of vital gluten is 8 %. Vital gluten is the most desirable and most suitable form of gluten in most applications and because of this, production of chemical gluten modifications has remained small.

Table I shows the main utilization areas of gluten in the three largestest producing countries. The baking industry is naturally the largest utilizer in the United States and Australia. Japan uses traditional methods to fortify fish foods, processed meat, and sausages with gluten. (1, 2).

Table I

Production and utilization Distribution of Vital Wheat Gluten in the United States, Japan and Australia

	United States	Japan	Australia
Total Production, Million kgs	20	20	35
Utilization: %			
Bakery industry	69		75
Breakfast foods	12		10
Pet foods	8		5
Meat analogs	4	30	
Processed meat	1	50	10
Hydrolyzed gluten		20	
All other foods	6		

III. THE COMPOSITION AND NUTRITIVE VALUE OF GLUTEN

Wheat contains 10-14 % protein, and the insoluble protein part of the grain endosperm is called gluten which represents about 80 % of the grain protein. Gluten is actually a protein-lipid-carbohydrate complex and is characterized approximately as follows:

protein	75 %
fat	6 %
ash	0.8 %
carbohydrate	15 %

Gluten is composed from different protein components with regard to molecular weight and solubility, glutenin and gliadin. Both provide gluten with special characteristics; gliadin extensibility and glutenin elasticity. (Fig 1) (3).

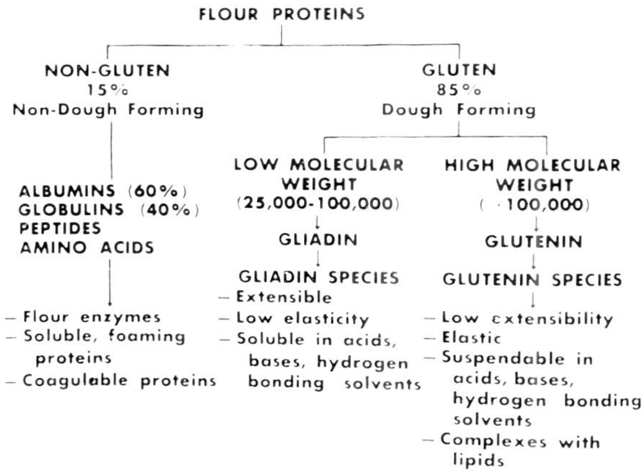

Fig. 1 Properties and composition of wheat flour proteins.

As a typical grain protein gluten suffers partial deficiencies of lysine, tryptophane, and methionine. This is not significant in a normal mixed diet and gluten can be regarded as a good quality protein for incorporation into foods. Table II shows the amino acid composition of gluten, soy protein, and FAO recommendations with regard to essential amino acids. The mixture of gluten and soy protein gives the ideal amino acid blance if required for nutritional purposes. The protein efficient ration of gluten (PER) is between 0.7-1.1 (4-5).

Table II

The contents of essential amino acids of wheat and soybean (g/16 g nitrogen).

	WHO/FAO PATTERN	WHEAT	SOYBEAN
Lysine	5,5	3,1	7,0
Threonine	4,0	3,1	4,2
Methionine, Cystine	3,5	4,3	2,8
Valine	5,0	4,7	5,3
Isoleucine	4,0	3,5	5,0
Leucine	7,0	7,2	8,5
Phenyalanine	6,0	8,1	8,9
Tryptophan	1,0	1,1	1,4

IV. PROPERTIES AND ANALYSIS OF GLUTEN

Vital wheat gluten retains its characteristics in processing. The most generally used methods of separation change neither chemically nor physically its characteristics. The method of dyring wet gluten causes only slight changes in the protein and if drying temperature is at a moisture level too high, the gluten looses its elasticity. In the developed world, where most gluten is consumed, the primary reason manufacturers pay for gluten and incorporate it into their products is that it improves appearance and overall palatability and gives economical advantages. In baking, gluten has unique properties compared to other vegetable proteins.

The question of how to indicate the quality of gluten is very complicated. As a product of an active raw material the quality of gluten can vary greatly. Gluten can be produced by different methods using different drying systems and the origin of the gluten may be strong or weak with good or poor gluten quality.

The consumer has to evaluate the acceptability of gluten with some simple analysis. Often gluten consumers have evaluated their own local methods testing the water absorbing capacity or hydration speed of gluten as a function of gluten vitality. The Farinographs and Extensographs are commonly used by adding dried gluten to a standard batch of flour and measuring the change at 1-3 % gluten addition. These alone are not very satisfactory ways of testing the baking properties of gluten. Booth and Timms have evaluated a new method which indicates the baking quality of gluten. Control baking in this test is made with dried gluten of good and poor quality with reconstructed flour. Good and poor quality gluten have remarkable linear differences in loaf volume as the protein (gluten) level increases. The slope of this linearity is called the protein response of gluten and it represents the standard of gluten quality. This technique is rather time consuming and the correlating method is based on the fact that although good and poor quality glutens develop elasticity fairly rapidly at low water additions, with larger additions the poor gluten takes a longer amount of time and requires more work to develop. A measure of this difference in development time at 200 % and 150 % water absorption is found to correlate quite well with the protein response of the gluten as shown in fig 2. (6).

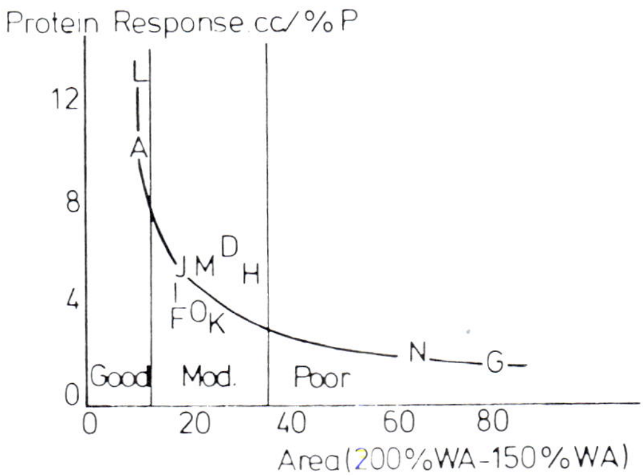

FIG 2. Gluten bread baking quality and work input to development.

In fortificating foods with gluten there is one disadvantage that must be discussed. A small part of the population, or about one in 2000-3000, are sensitive to the particular proteins present in cereals, wheat, rye, oat, and barley and this is called celiac disease. (= gluten sensitive enteropathy). The disease is characterized by malabsorption of nutrients as a consequence of damage to the absorptive epithelial cells that line the small intestine. Such persons have absolutly avoided gluten in their diets. The addition of gluten to products in which the public would not normally expect to find it, must be labeled with the amount of gluten contained in the product.

A. Functional Properties of Gluten

Functional properties of proteins for foods can be defined as the physicol chemical properties providing a certain amount of information on how a protein will act in a food system. The solubility, swelling, viscosity, foaming ability, and emulsifying ability are the common properties of proteins and are used to compare their effects in food systems. Those properties can be specified measuring them in varying pH, temperature, and ionic strength.

In solubility, swelling, and viscosity properties of gluten we have observed several differences, although not remarkable, in glutens made by different methods or from different origins. The analysis has to be examined very carefully because they give only basic ideas of how these gluten products behave in food systems, such as bread, sausages, cheese e.g.

The chemical modification of gluten causes drastic changes in functional properties, like solubility and swelling. In the following chapter the properties of succinylated and acetylated gluten are examined and compared to vital gluten.

1. Solubility. Vital wheat gluten in its natural form has very low solubility which is a limiting factor for use in some food systems. The low solubility in water can be attributed in part to the hydrogen bonds of the glutamine and asparagine side chains of gliadin and glutenin molecules. The low solubility of vegetable proteins usually indicates denaturation, but this is not valid for gluten. Solubilization of the gluten aggregate can be achieved by alterationing of the amino groups. Grant has described a 95 % solubilization of gluten when treating wheat flour with succinic anhydride (7). The succinylated gluten, with the amount of substitution at 55 %, increases the solubilization through the wide pH range. The acetylation of gluten to the same substitution grade has little effect. (Fig 3). The saltsensitivity of vital and modified glutens at pH 6 are not remarkable. (Fig 4).

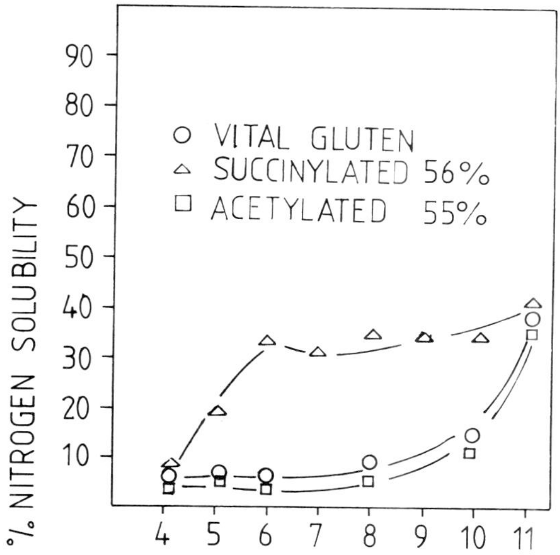

FIG. 3 Nitrogen solubility of vital gluten as a function of pH 0,2 M NaCl

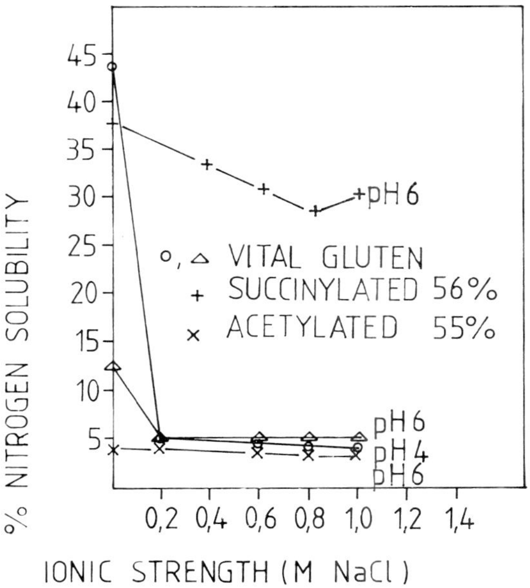

FIG. 4 Nitrogen solubility of vital gluten as a function of NaCl concentration

2. Swelling. The swelling determination of gluten differs from other proteins. The simplest and most common way is to make a gluten water mixture and let it stand for one hour at room temperature and then press the extra water out of gluten by hand and measure the wet weight of the gluten. This method is not very scientific and water absorption measurings with Baumann apparatus have been researched. This method also has certain disadvantages because water is sucked through the filter paper and the surface of the gluten against the filter paper forms a wet impenetrable layer. A good quality gluten develops the layer faster and the gluten remains partly dry. In the Baumann method a poor quality of gluten is more porous and absorbs water better than a good quality of gluten. The method is developed futher using one hour hydration time and centrifuging wet gluten in the tube equipped with iron net which is 1.5 cm over the bottom of tube standardizing water outpress. Vital wheat gluten, measured by this method, absorbs water 1.5-2 times its own weight and this absorption remains constant under any condition. Small concentrations of NaCl or CaCl2 add only a little water absorption. Succinylated gluten, on the contrary, increases water absorption to 4-5 times and acetylation effects are not so remarkable. (Fig 5).

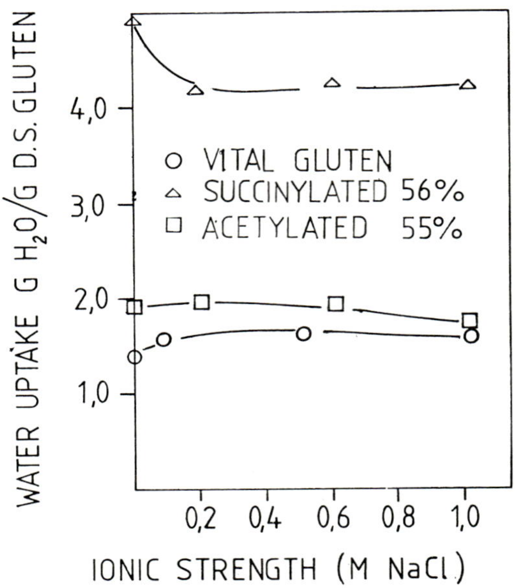

FIG. 5 Swelling of gluten as function of ionic strenght

The swelling ability of vital gluten increases at a temperature of over 60°C. This might be caused partly by starch gelatination and partly by protein denaturation. (Fig 6.)

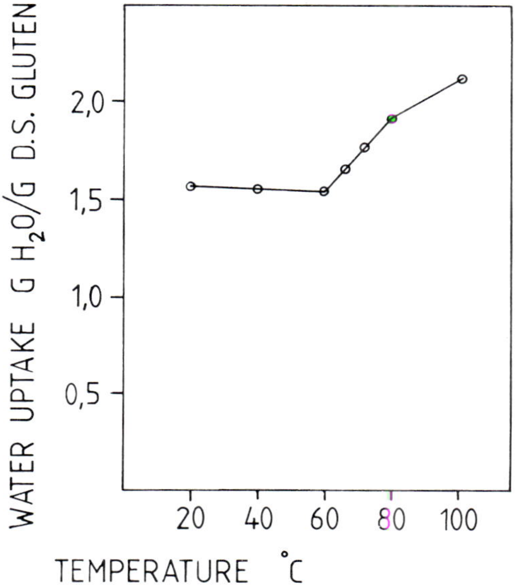

FIG. 6 Swelling of gluten as a function of temperature

3. Viscosity Measurements. The swelling of protein molecules in a dispersion results in a viscosity increase. Figure 7 shows viscosity as a function of protein concentration in pure water for vital gluten. In over 10 % concentration, the viscosity increases very rapidly due to the strong gel-forming property of gluten. Gluten doesn't disperse in water very easily, but forms large, insoluble gluten agglomerates (7).

At a temperature of about 40-50°C the gluten absorbs water and swells rapidly. In dough making and gluten prosessing this swelling ability is utilized, optimizing the fermentation and gluten forming time.

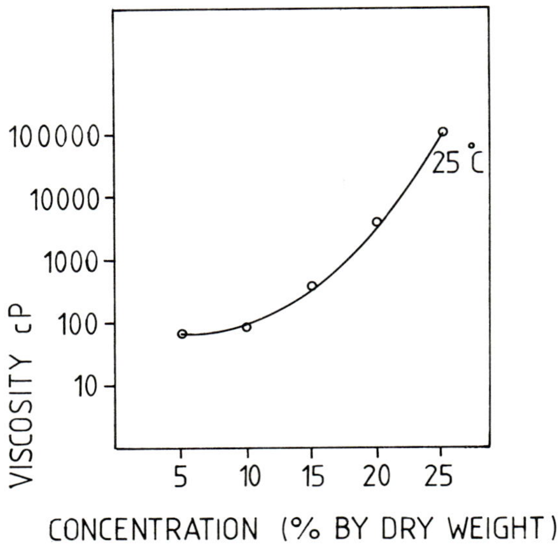

FIG. 7 Viscosity of gluten as a function of concentration

V. METHODS OF MANUFACTURE

The manufacturing methods of wheat gluten have developed greatly in the seventies because of the demand of enviroment protection. Disadvantages of the old processes, such as big waste water load and at the same time big material losses, have been avoided with modern technology. In principle the production of wheat gluten is harmless to the environment because the process needs no chemicals. Fractination of wheat into starch and gluten starts with a dry flour milling process which largely separates the skin, germ, and endosperm of the wheat. The object of the milling process is to produce as little damaged starch as possible, for which has been developed special short mills. Some gluten manufactures use certain flour fractions of wheat, for example second clear flour.

The Martin process is the oldest gluten production method and is still widely used. In the Martin process a fairly stiff dough is formed and thorouhly washed with water to remove the starch. After this process the gluten is recovered as an elastic mass. The A-starch and B-starch are then separated by sentrifuge. In this process the process water is lost as a waste.

The Raisio process is a new sophisticated method with a complete wheat milling and flour separation combination. The principle of the process is very simple (Fig 8). The wheat is milled in a short mill and as a by product a bran fraction is obtained. In the wet separation process the wheat flour is mixed with a specific amount of warm water in to a thick slurry which is then homogenized to free the starch granules from the gluten matrix. Subsequent centrifuging produces two fractions, a prime starch fraction with little secondary starch (= B-starch) and a fraction containing gluten and secondary starch, called proteinconcentrate, which contains a proteincontent of about 30 %. The prime starch fraction is refined by screening and separating to the desired purity. The gluten agglomorates during maturation under controlled temperature, time, and mechanical energy. This maturation stage permits strict regulation of the gluten agglomoration and seeks the optimum conditions for all wheats. The secondary starch is concentrated and yields a process water. This process water is then concentrated futher by evaporation with the soluble material remaining.

FIG 8. Process flow diagram of the Raisio wheat starch plant

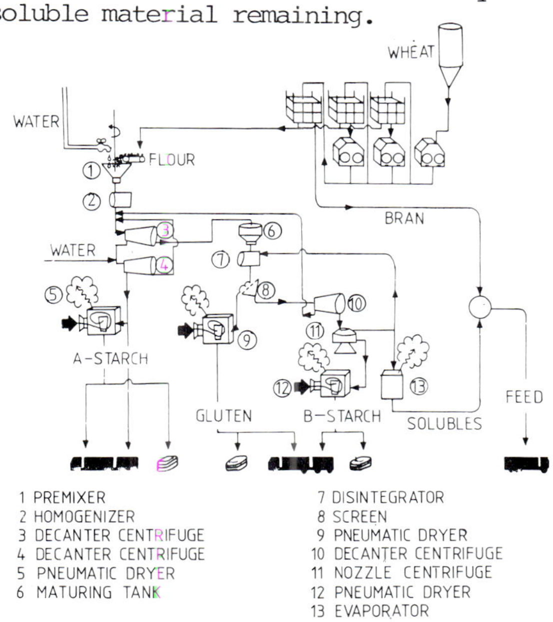

The Raisio process is versatile and flexible working with a wide spectrum of different wheats, including hard wheat and even mass wheats. All these wheat species yield a satisfactory amount of good quality gluten. Farinographs and baking results show the effect of gluten obtained from different wheat species in dough and bread. (Fig 9 and Fig 10.) Hybrid Triticale also yields a baking flour with the addition of gluten derrived from the Triticale wheat. The flexibility of the Raisio prosess is of great importance as the genetic research is constantly producing new varieties and hybrids with a continuously increasing yield per acre.

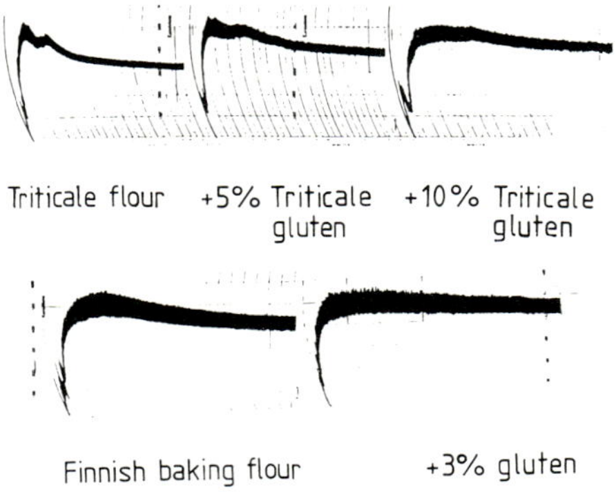

Fig. 9 The effect of gluten in the form of farinogram

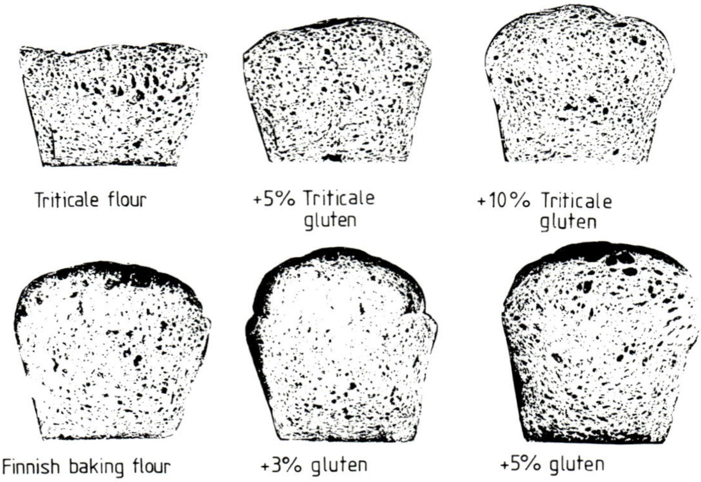

FIG.10 Baking test of Finnish flour and Triticale hybrid with added gluten

It is important to discuss the comparisons between the Martin process and the Raisio process in more depth due to economical, enviromental, and product quality reasons.

<u>Martin process</u>	<u>Raisio process</u>
Gluten is formed by strong mechanical mixing and gluten is recovered from the flour dough.	Gluten is recovered from the proteinconcentrat slurry and gluten is formed gently by heat, time, and slow mixing.
total amount of water used in process 10 m^3/1 tn flour	3 m^3/1 tn flour
waste water load 5-8 m^3/1 tn flour	Non-polluting. The process water with flour solubles is recovered, evaporated and used as a feed. Processwater 2 m^3/1 tn flour.

The quality of gluten differs in these two processes. In the Raisio process gluten retains its vitality better and water absorbtion and the speed of hydration are higher than in the Martin process. This might be due to the strong mechanical mixing of dought in the Martin process which disrupts the structure of gluten irreversible. In comparison, the standard quality gluten produced by the Martin process has the water absorption 150-160 %, when the Raisio gluten from hard and soft wheat has water absorption of 190-200 %, and even hybrid Triticale of 165 %.

New baking and extrusion techonology demand fast developing doughs before final treatment, and gluten with fast recombination has special advantages. Drying in a spray dryer yields a good quality of gluten with good hydration properties, but this method of drying consumes more energy.

VI. APPLICATIONS

A. Bakery Products

As a protein product processed from wheat, gluten is a natural raw material in baking. An addition of vital gluten improves the visco-elastic properties of dough and permits the formation of gas cells in bread which ensures proper configuration, volume, and texture in bread. The tolerance of inert ingredients of dough increases without unfavorable effects in baking quality. The water absorption ability of gluten keeps the products soft and extends their shelf life. The present demands for more wholesome nourishment has increased the use of gluten and the desire to optimize carbohydrate, protein, and fat levels in our diet.

The unique adhesive and film forming character of hydrated vital wheat gluten and its thermosetting properties are largely responsible for its use in meat and fish products. Actual applications range from extended ground meats, fabricated steaks, canned hams, and turkey rolls to fish and meat sausages, meat pies and others. In meat systems the water absorption ability of gluten is totally economized reducing water losses during cooking or roasting.

B. Non Food Applications of Gluten

The addhesive and filmforming characteristics of hydrated wheat gluten are largely responsible for its use in techincal fields. An adhesive modified gluten can be used in paper coatings and textile yarn-sizing. The foaming properties of gliadin may be utilized in foam drying and as a whipping agent. Solubilized gluten is also a good emulsifying agent in pharmaceutical pastes etc.

Film forming properties permit the production of edible films, chewing gum, sausage casing, and capsules from gluten.

The utilization of technical applications has been limited by the satisfactory markets in the baking field and the relatively small quantities of gluten produced. At the moment the situation is changing. The baking industries are saturated with increasing gluten production. As of yet gluten producers are seeking new applications in technical fields as well.

VII. SUMMARY

The demand of wheat gluten as well as other vegetable proteins with special functional properties is growing with the increased consumption of fabricated foods at home and at work. The natural origin and manufacturing of gluten confirm the safe use of gluten as a universal food ingredient. Totally new forms, flavors, and textures of foods are the future of gluten. Meat analog is a very good example of this type of food where new technology and nutrition knowledge are combined. As a renewed natural raw material, gluten opens new possibilities as a glue and binding material in the technical field where the adhesive and film forming properties of gluten are utilized. Chemical modification widens the application field of gluten, changing its solubility and water uptake properties.

In the food application the use of wheat gluten and other vegetable proteins should optimize in their specific fields and the synergetic characteristics should be used in fields where each type of protein has something of its own to contribute.

REFERENCES

1. Hall, J.B., Cereal Foods World. 22, 144 (1977)
2. Burrows, V.D., et al. in "Food Protein from Grains and oilseeds A Development Study Projected to 1980". p. 120. Office of the Minister Responsible for the Canadian Wheat Board (1972).
3. Pomeranz, Y., Finney, L.F. Food, Eng. 45, 134 (1973)
4. Amino Acid Content of Foods and Biological Data on Proteins, p. 24. FAO Publication (1970)
5. Hulse, J.H., and Laing, E.M., Monograph IDRC-021e, Canada, 1974
6. Booth, M.R., Timms, M.F., Sixth International Cereal & Bread Congress, 1978
7. Grant, D.R., Cereal Chem., 50, 417 (1973)
8. Henriksnäs, H., Weckström, P.E., Unpublished study i Åbo Academi 1978

THE FUNCTIONALITY OF WHEAT PROTEINS

Aris Graveland
Pieter Bosveld
Willem J. Lichtendonk

Institute for Cereals, Flour and Bread TNO
Wageningen
The Netherlands

I. INTRODUCTION

Animal or vegetable material used in the production of foodstuffs must be suitable both for the processing technology and for human consumption.

Technological suitability is determined by the volume, structure, smell, and taste of the final product. The acceptability of a product is also largely determined by its smell and taste. All these properties are in turn determined by the denaturation (irreversible chemical and physical changes) of the raw material during processing, when gelatinization, emulsification, hydration/dehydration, etc. can occur.

Suitability for human consumption is determined by the digestibility, nutritional value, toxicity, and acceptability of the product.

Both technological and consumptive suitability are ultimately determined by the components that make up the raw material: starch, protein, lipids, minerals, vitamins, etc.

I should like to confine myself in this paper to the (technological) function of the protein in the production of bread from wheat flour.

The protein of a given raw material generally consists of a large variety of components, which may differ greatly in their physical and chemical properties. Even more marked are the differences found when the proteins of different raw materials are compared.

The protein of each raw material thus has its own specific properties, and with them largely determines the type

ISBN: 0-12-370960-1

of product that can be prepared from the raw material.

Which properties of a given protein and which chemical and physical factors can be important in the production of a given foodstuff can be seen from Fig. 1.

Aspects of food preparation from proteins			
Primary functional properties	Secondary functional properties	Tertiary functional properties	Food
Amino acid composition Amino acid sequence Sub-units Mol. weight Conformation	Solubility Hydration Swelling Viscosity Elasticity Flexibility Film formation Heat stability Enzymatic activity Bonds: H-bonds Hydrophobic bonds Ionic bonds	Gelatinization Emulsification Fixation Interactions with: Proteins Carbohydrates Lipids Emulsifiers Reactions: SH / SS interchange Oxidation / Reduction	Texture Colour Flavour Digestibility Nutritive value Safety

FIGURE 1. Functional properties of proteins.

Wheat protein has a very complex composition and posses-ses moreover the unique properties that are necessary for bread making. During dough mixing the proteins must on the one hand form into a continuous network and on the other hand interact with the starch granules, so that the CO_2 which is formed during the fermentation period can be retained. The network must be strong enough and at the same time elastic enough to continue to retain the CO_2 throughout the baking process, thus yielding a product with a light, expanded structure. This requires the protein network to have both viscous and elastic properties. The three-dimensional protein-starch network must be capable of forming a thin film around the CO_2 bubbles while the CO_2 is forming and, as a result of the rise in temperature, expanding. Furthermore, the proteins must have the property of forming a fairly stable structure (fixation) as a result of heat denaturation (concluding the baking process).

At the molecular level this means that there must be sufficient large - but not too large - protein molecules with elastic properties as well as smaller proteins with viscous properties.

II. FLOUR PROTEIN FRACTIONS

In order to understand why bread with the typical light, expanded structure can be made from wheat and not from other cereals such as rye, barley, oats and maize it is desirable to separate the protein as completely as possible into discrete fractions, determine the characteristics of these, and

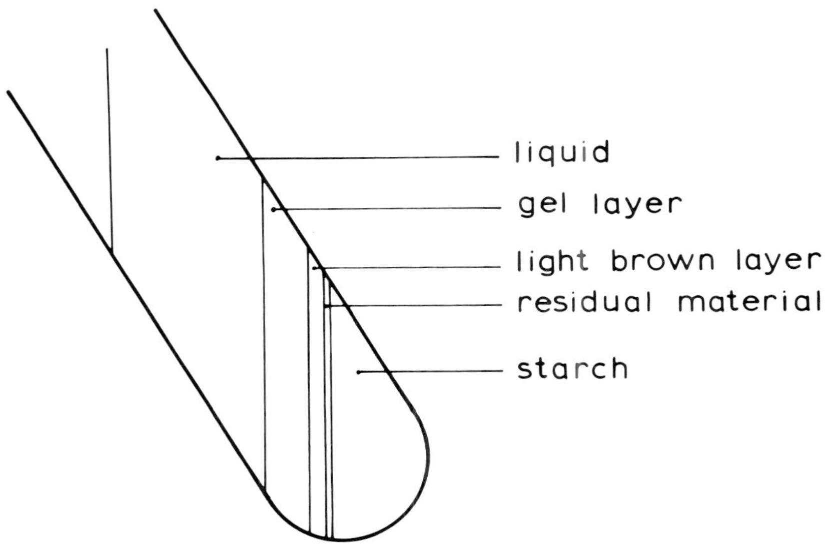

CENTRIFUGATION OF SDS-FLOUR SUSPENSION.

FIGURE 2.

to establish what contribution each fraction makes to the structural formation of the end product. For this purpose we have developed a new protein extraction and fractionating method at our institute (1). Use is made of a sodium dodecyl sulphate solution (SDS). Flour is suspended in SDS and then centrifuged at high speed. Approximately 40 per cent of the protein is precipitated and at the same time separated into two layers: a light brown layer made up of pentosan-like sub-

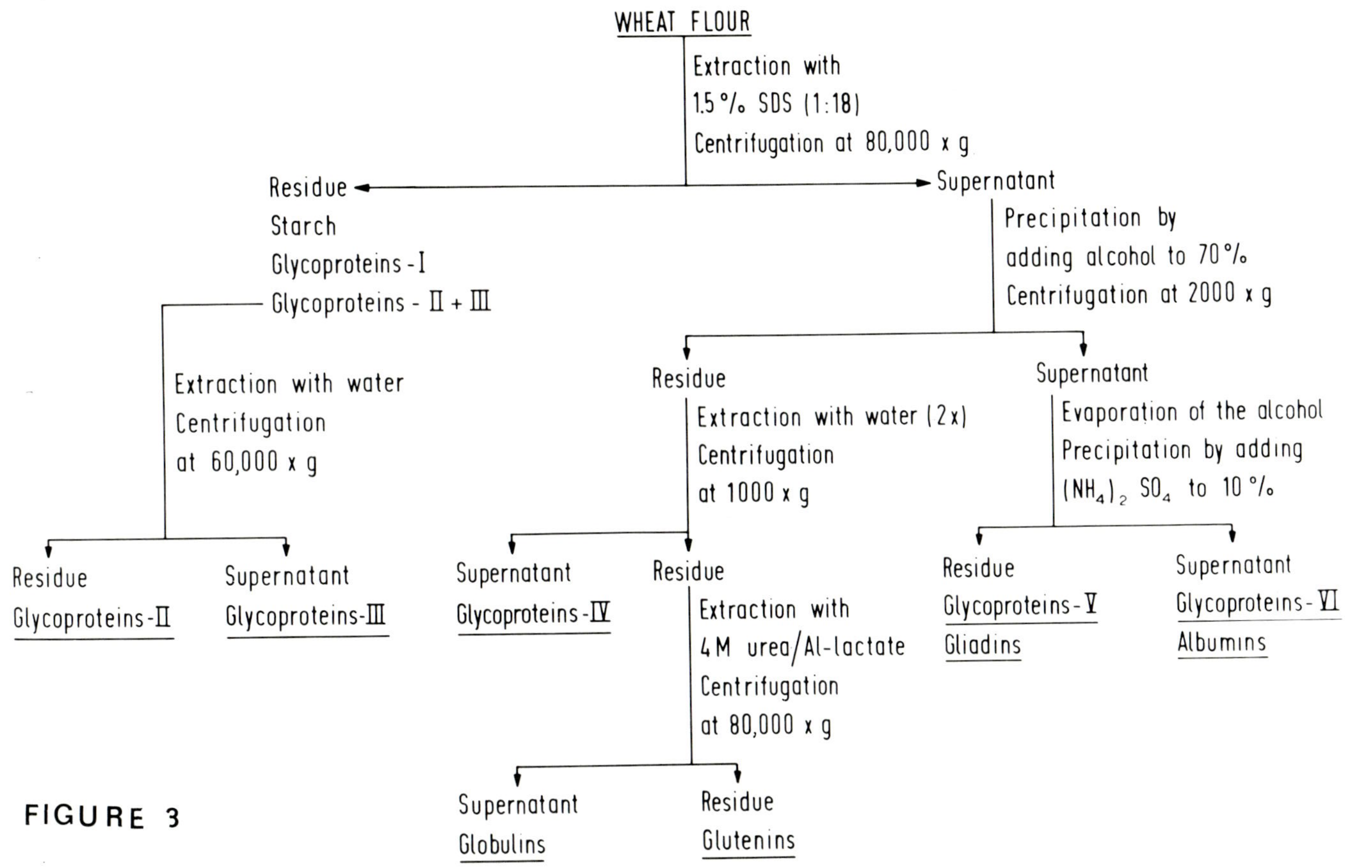
WHEAT FLOUR
Extraction with
1.5 % SDS (1:18)
Centrifugation at 80,000 x g
Residue
Starch
Glycoproteins - I
Glycoproteins - II + III
Extraction with water
Centrifugation
at 60,000 x g
Residue
Glycoproteins - II
Supernatant
Glycoproteins - III
Supernatant
Precipitation by
adding alcohol to 70 %
Centrifugation at 2000 x g
Residue
Extraction with water (2x)
Centrifugation
at 1000 x g
Supernatant
Glycoproteins - IV
Residue
Extraction with
4 M urea/Al-lactate
Centrifugation
at 80,000 x g
Supernatant
Globulins
Residue
Glutenins
Supernatant
Evaporation of the alcohol
Precipitation by adding
$(NH_4)_2 SO_4$ to 10 %
Residue
Glycoproteins - V
Gliadins
Supernatant
Glycoproteins - VI
Albumins

FIGURE 3

stances (glycoprotein I) and a gel layer that consists of a complex of high-molecular-weight glycoproteins II and glycoproteins III (Fig. 2).

That part of the protein that dissolves in SDS, 60 per cent of the total, is fractionated by the addition of alcohol (Fig. 3). The residue this produces consists of glutenins, globulins, and glycoproteins. The supernatant liquid that remains after the addition of alcohol can be further fractionated by adding $(NH_4)_2SO_4$. The fractions that precipitate are glycoproteins and gliadins. The fraction that remains in solution contains glycoproteins as well, together with albumins. The quantities of these fractions are presented in Table I.

TABLE I.

Distribution of Nitrogen in the protein fractions (percentages of total Nitrogen)

SDS - insoluble	Manella	DNS
Glycoproteins - I	6 %	7 %
Glycoproteins - II	15 %	20 %
Glycoproteins - III	11 %	7 %
Residual Material	1 %	1 %
Total	33 %	35 %
SDS - soluble		
Glycoproteins - IV	7 %	6 %
Glutenins	14 %	15 %
Globulins	1 %	1 %
Glycoproteins-V	30 %	32 %
Gliadins	3 %	3 %
Glycoproteins - VI	10 %	6 %
Albumins	2 %	2 %
Total	67 %	65 %

This fractionation shows that the gliadins, albumins, and globulins together make up only 6 to 8 per cent of total protein. These fractions can be identified with the aid of starch gel electrophoresis (1). The glutenins, approximately 15 per cent of total protein, can not be identified by electrophoresis. It is noteworthy that a large proportion of the protein contains chemically bound sugars.

Three fractions are of great importance for the formation of the continuous protein network: the glycoproteins II and III, which in the presence of SDS make up the gel layer in the fractionating sequence described, and the glycoproteins V, which are soluble in SDS and 70 % alcohol.

The glycoproteins II and III together form a complex with strongly elastic properties and the formation of the protein network and its consistency depend upon the quantities in which they are present. Although the glycoproteins III are soluble in SDS they are precipitated together with GP_{II} during centrifugation, as enclosed compounds. GP_{II} and GP_{III} are made up of the same subunits and have the same amino-acid composition. They differ, however, in that GP_{II} has a higher molecular weight than GP_{III}, which makes it insoluble in SDS. The presence of SDS enables GP_{II} to expand massively; all intramolecular hydrophobic and electrostatic interactions are broken and the GP_{II} polymer attains its greatest volume. On the grounds of the great similarity between GP_{II} and GP_{III}, which also exists in respect of their physical properties and their appearance on examination with a scanning electron microscope, it is reasonable to assume that GP_{II} is constructed from GP_{III}.

From an investigation of a series of wheat varieties with different baking properties we have been able to establish that there is a correlation between the amount of GP_{II} per gramme of flour and the volume of bread obtained. This is in agreement with results obtained with the SDS sedimentation value (2). After all, the sediment obtained with this method consists of glycoproteins I and II.

III. PROTEIN FRACTIONS IN DOUGH

During dough mixing the GP_{II} fraction is converted into smaller fragments which are soluble in SDS. The mixing results, moreover, in the GP_{III} fraction being released from the GP_{II} + GP_{III} complex. In consequence of these processes the gel layer diminishes during mixing (3). It has been established that the conversion of GP_{II} is due to SS bonds being split. The number of SS bonds in this fraction decreases, while the number of SH groups increases. GP_{II} contains approximately 50 μmol SS per gram of material. The conversion of GP_{II} into SDS-soluble material takes place under the influence of a component of the GP_V and GP_{VI} fractions.

A study conducted to identify this component has established that it is an SS reductant. The SS bonds are split by a radical mechanism and only a limited number (20 to 30 per cent) of the SS bonds of GP_{II} and GP_{III} are reduced according to the equation:

$$P_r^1 S - S P_r^2 \xrightarrow{\text{Reductant}} P_r^1 S\cdot + P_r^2 S^-$$

The conversion of GP_{II} to SDS-soluble material can also be brought about by adding cysteine or glutathione, which again split the SS bonds of GP_{II} and GP_{III} probably by SH/SS exchange reactions.

Under the influence of cysteine and glutathione the GP_{II} and GP_{III} molecules are broken down into smaller fragments; that is, cysteine and glutathione react with SS bonds with which the abovementioned SS reductant does not react.

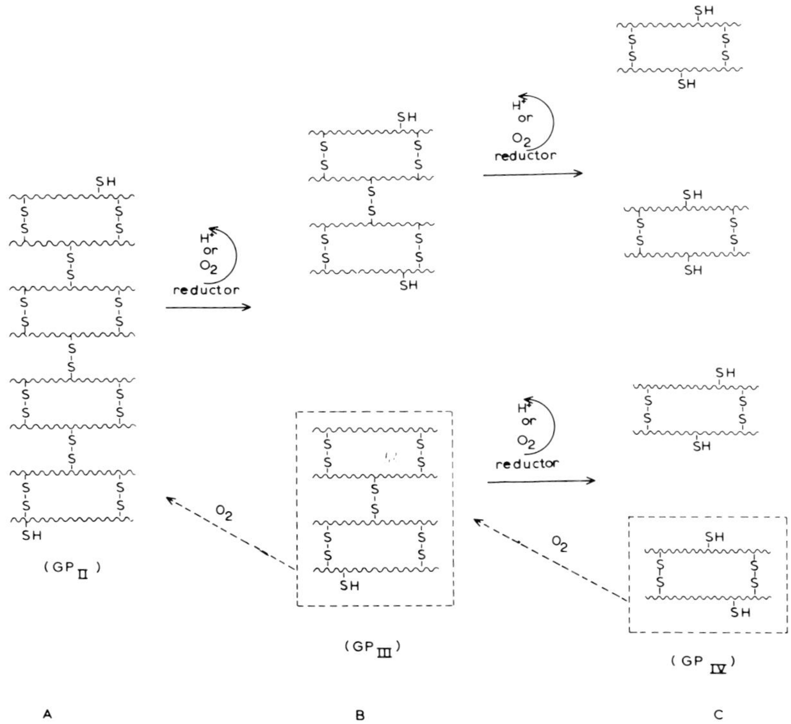

FIGURE 4. Reduction of GP_{II} and oxidation of the reaction products.

Adding SDS to a dough also promotes the conversion of GP_{II}. This can be explained by its unfolding action, which makes the SS bonds of GP_{II} more accessible to the reductant so that the conversion is more complete. The addition of $(NH_4)_2$-SO_4 inhibits the conversion of the gel layer; this is in consequence of the salt stabilizing action of that substance, which diminishes the accessibility of SS bonds. $CaCl_2$ and KCNS

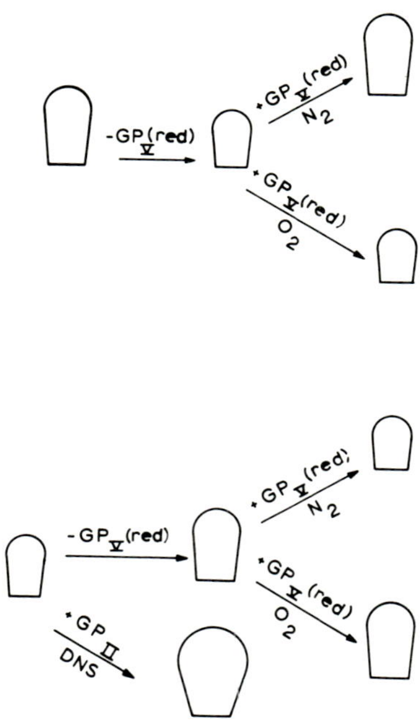

FIGURE 5. Influence of reductor on loaf volume.

are further agents that accelerate the conversion. This can be explained by the destabilizing effect of these salts on GP_{II}, which enables the reductant to react more easily with more SS bonds.

Increasing mixing energy and raising temperature also enhance conversion.

Another interesting phenomenon is the fact that during the fermentation period after mixing the gel layer re-forms from the reaction products formed during mixing from GP_{II} and GP_{III} (3). The increase in the gel layer during the resting period is due to the oxidation of SH groups into SS bonds. This reaction can be retarded by adding NEMI or fumaric acid. These substances react with the Pr S^- or Pr S· and thus retard the oxidation into SS bonds. The oxidation of the SH groups can be accelerated, on the other hand, by adding oxidizing agents such as KIO_3, $KBrO_3$, or hydroperoxides of polyunsaturated fatty acids. The conversion of GP_{II} and GP_{III} and the re-formation are illustrated in Fig. 4.

The difference between a flour that bakes well and one that bakes poorly can be explained by reference to Fig. 5.

Extraction of the SS reductant from a flour that bakes well, using an isopropanol-water mixture, causes the volume to diminish, while removal of the reductant from a flour that bakes poorly produces an improvement. With flour that bakes well complete reconstitution is possible following the addition of non-oxidized reductant. With flour that bakes poorly the improving effect is retained after the addition of the SS reductant, if this is oxidized. The volume of bread obtained

TABLE II. Distribution of Nitrogen in the protein fractions of several cereals (percentages of total Nitrogen).

	Wheat			Barley		Rye
	DNS	Manella	Clement	Pirouette	Banting	
Glycoproteins I	7	6	6	19	15	16
II	20	15	10	11	7	5
III	7	11	6	3	4	4
Glycoproteins IV	6	7	18	9	11	18
Glutenins	15	14	15	32	38	16
Globulins	1	1	1	1	1	1
Glycoproteins V	31	30	28	5	4	26
Gliadins	3	3	3	1	1	2
Glycoproteins VI	8	10	10	18	16	8
Albumins	2	2	2	2	2	2

from a flour that bakes poortly can be greatly increased by adding GP_{II} from a good flour. The flour that bakes poorly contains enough natural reductant to ensure the required conversion. These tests show that the volume of bread obtained is chiefly determined by the quantity of GP_{II} present in the flour and the reaction between GP_{II} and the SS reductant. It should be noted further that a major part is played in the formation of the protein network not only by the quantity of

GP_{II}, but also by its composition (the number of SS bonds and the molecular weight). The GP_{II} polymers of a flour that bakes poorly are smaller than those of one that bakes well. Conversion of such polymers by reduction of SS bonds results in units that are too small for the formation of a good protein network. Presumably the GP_{II} polymers of a wheat flour that bakes well are too large for optimum participation in the protein network, so that partial conversion by reduction of SS bonds in necessary.

TABLE III. Amino acid composition of glycoproteins II, III, V and glutenins (Amino Acid Residues per 100,000 g Protein).

Amino acid	Glycoproteins				Glutenins
	II	III		V	
Lysine	9	10		4	34
Histidine	12	13		12	18
Arginine	19	19		13	39
Tryptophan	5	5		4	12
Aspartic acid + asparagine	18	22		17	60
Threonine	24	24		15	33
Serine	58	56		45	54
Glutamic acid + glutamine	260	261		300	131
Proline	104	98		135	58
Glycine	60	51		19	65
Alanine	26	27		19	59
Valine	32	34		27	52
Methionine	10	8		7	14
Isoleucine	24	26		27	30
Leucine	52	54		49	61
Tyrosine	22	20		12	22
Phenylalanine	26	29		39	31
Cysteine	2	2		2	9
Cystine	6	6		13	3

The baking tests have also shown that the glutenins have a negative effect on the volume of bread obtained. Addition of extra glutenins to a given flour caused the volume to diminish substantially. As is evident from their amino-acid composition

(Table III), the glutenins are too hydrophobic to participate well in the formation of the protein network.

The answer to the question of why bread with a light, expanded structure can be made from wheat and not from barley or rye can now be obtained from table II.

The percentage of glutenins in barley is high, and we have seen that this fraction has poor baking properties. Moreover, barley has very little of the two fractions that are essential for the formation of the protein network needed: GP_{II} and GP_{V}. Rye, on the other hand, contains approximately the same amount of glutenins as does wheat, bus has little GP_{II} and a lot of GP_{I} and GP_{IV}. Both GP_{I} and GP_{IV} are fractions whose contribution to the formation of a good protein network is only moderate. Rice and maize are also very rich in GP_{I}, but they both have no GP_{II} at all and so can not be used for making raised bread.

REFERENCES

1. Graveland, A., Bongers, P., and Bosveld, P., J. Sci. Food Agric. 30, 71 (1979).
2. Axford, D.W.E., McDermott, E.E., and Redman, D.G., Milling Feed and Fertiliser 161, No. 5, 18 (1978).
3. Graveland, A., Bongers, P., and Bosveld, P., J. Sci. Food Agric. (1979) in Press.

WHEAT PROTEINS
EVALUATION AND MEASUREMENTS OF WHEAT QUALITY

Pierre Feillet
Laboratoire de Technologie des Céréales
Institut National de la Recherche Agronomique
Montpellier, France

I. INTRODUCTION

A considerable amount of work has been directed to correlate wheat protein composition and wheat quality. As summarized in Figure 1, wheat proteins are a highly heterogeneous material including albumins (soluble in water), globulins (soluble in neutral salt solutions), gliadins (soluble in 70 % ethanol, acids) and glutenins (soluble in acids, bases, hydrogen and hydrophobic bonding solvents). Albumins and globulins are cytoplasmic proteins with enzymatic activities, foaming and emulsifying properties ; they account for 15-20 % of total proteins. Gliadins, known for their medium molecular weight (25 000 - 100 000) and their high extensibility, and glutenins, known for their high molecular weight (100 000 - > 3 000 000) and their high elasticity, are associated together with lipids and other minor components (minerals, carbohydrates) to form a unique viscoelastic protein complex referred to as gluten. The relative importance of disulfide cross-links (Bloksma, 1975 ; Greenwood and Ewart, 1975) and of polar and hydrophobic bonds in the structure of this complex is rather contested (Kasarda et al., 1976). Recent findings by Kobrehel and Bushuk (1977, 1978) who have found that glutenins are dissolved in distilled water in presence of Na-salts of fatty acids are in favor of the prominent role of secondary bonding forces.

Both protein quantity and quality are considered to be the main factors of the suitability of wheat flour or wheat semolina to be processed into bread, pastry or macaroni having a suitable quality. Early works of Finney (1943) have clearly shown that increased level of gluten and improved gluten quality are mostly responsible for increases in loaf volume.

It is well established that the quantity of proteins is primarily influenced by growing conditions and that their quality is essentially under the control of genetic factors.

ISBN: 0-12-370960-1

GROUPS	SOLUBILITY	STRUCTURE	PROPERTIES	CONTENT %	ORIGIN
ALBUMINS	WATER		ENZYMIC	15-20	CYTOPLASMIC PROTEINS
GLOBULINS	NEUTRAL SALT		FOAMING, EMULSIFYING		
GLIADINS	70 % ETHANOL	 LOW MOLECULAR WEIGHT	EXTENSIBILITY	40	STORAGE PROTEINS
GLUTENINS	ACIDS, BASES, HYDROGEN AND HYDROPHOBIC DISRUPTING AGENTS	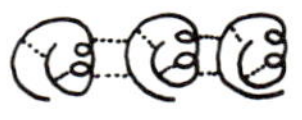 HIGH MOLECULAR WEIGHT (> 100,000)	ELASTICITY TENACITY	40	

FIGURE 1 . WHEAT PROTEIN PROPERTIES

With regard to the major effect of protein composition on wheat quality, breeders anxiously attempted to evaluate the quality of new lines and varieties from the results of protein composition analysis. To develop new biochemical screening test is challenging for wheat protein chemists and cereal genetists.

In this presentation, I would like to outline the extremely promising results which have been recently reached by several groups in the world. A special attention will be give to those obtained in our laboratory by Autran, Jeanjean and Kobrehel. Before, it might be useful to specify what we mean by wheat quality.

II. CONCEPT OF VARIETAL INTRINSIC QUALITY

By itself, the word quality does not mean many things and cannot be expressed in terms of specific characteristics. Its significance is different for millers, bakers, pasta processors and consumers, or for nutritionits, technologists and genetists. Generally speaking, and from the view point of wheat breeding, it is fruitful to distinguish between the concepts of varietal intrinsic quality (VIQ) and wheat commercial quality (WCQ).

VIQ is a genetic characteristic which takes account for the potential quality of utilization of a variety ; by definition, it is no influenced by the growing conditions. WCQ is a phenotypic property resulting from the interactions between VIQ and environmental factors ; processors are primarly concerned about its value.

The most frequently, as illustrated Figure 2, breeders who are interested in knowing VIQ of a new line use to grow this line in different locations and to measure WCQ of samples with conventional methods as alveograph, farinograph, mixograph, viscoelastograph, Pelshenke tests, Zeleny tests or baking tests ; by averaging all data, they get a mean value which is an estimate of VIQ. An accurate measurement of VIQ needs to replicate the same experiment at least two years. Within one variety, VIQ (which is a constant) cannot be correlated to WCQ values of several samples while VIQ values of different varieties must be correlated to WCQ mean values of each variety.

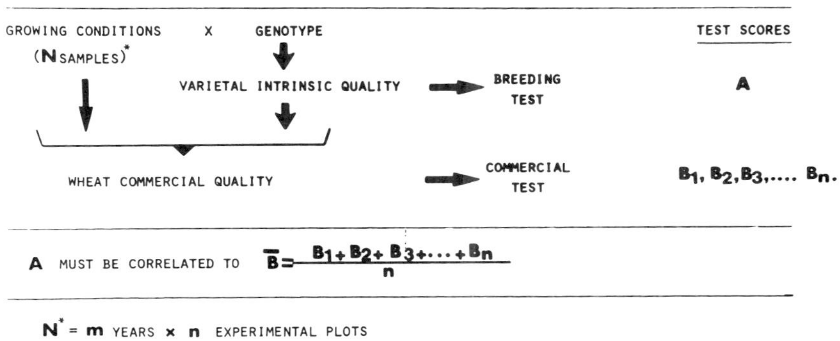

FIGURE 2 . CONCEPT OF VARIETAL INTRINSIC QUALITY

It is clear that such a determination is tedious and time consuming. Therefore, an easier evaluation of VIQ is requested by breeders. However, the history of the development of breeding tests indicates that most of them derive from methods which were primarly developed to evaluate WCQ in terminals, mills and bakeries ; until very recently, only few studies were devoted to the development of methods which could allow a direct assessment of VIQ value of varieties.

Our increasing knowledge on the biochemistry and genetic of wheat proteins - excellent and recent reviews have been published by Kasarda et al. (1976), Konzak (1977), Mecham et al. (1978) - opens new fields of investigations to clear the way toward a solution of this problem. In this direction, different ways have been explored : content in specific protein classes, gliadin electrophoregrams, protein subunits composition.

III. CONTENT IN SPECIFIC PROTEIN CLASSES

While the influence of gliadin content and gliadin properties on rheological behavior of a dough is far from being negligible (Finney, 1971, Preston et al., 1975), glutenins are frequently recognised as the most important of wheat flour proteins in giving strength and elasticity to dough (Bietz et al., 1973).

The fact that the amounts of glutenins extracted with various solvents are related to the breadmaking quality of wheats has been recognized for almost one hundred years (Guthrie, 1896). For instance linear relationship between Zeleny test and glutenin content was found (Feillet, 1965) (Figure 3).

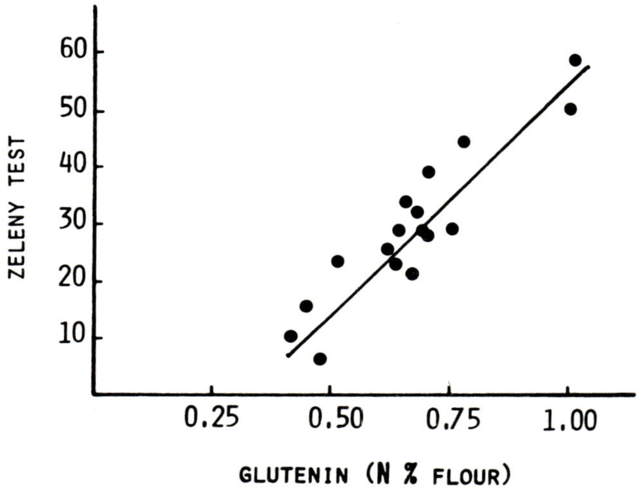

FIGURE 3 . RELATION BETWEEN GLUTENIN CONTENT OF FLOUR AND ZELENY TEST VALUE (FEILLET, 1965)

In the last ten years, many authors (Godon, 1969 ; Bietz and Wall, 1975 ; Tanaka and Bushuk, 1973 ; F. Mac Ritchie, 1978) underlined the importance of the acetic acid insoluble gluten and glutenin fraction. The importance of the so-called gel protein fractions, which are those proteins that occur in the highly hydrated insoluble viscous residue remaining when wheat flour is extracted with dilute acetic acid was suggested by Meredith (1961) and stressed by Mecham *et al.* (1972) and Cole *et al.* (1976).

Based upon these biochemical studies, new tests have been developed by Orth (1976) and by Feillet *et al.* (1976) to appreciate the varietal intrinsic quality of wheat : (a) Orth and Bushuk (1972) have shown that bread making quality is negatively correlated with the proportion of the acetic acid-soluble glutenin and positively with the proportion of the residue protein (acetic acid-insoluble glutenin). A pratical test was then developed by Orth (1976) : one gram of flour was extracted with 250 ml of 0.05 M acetic acid. After shaking and centrifuging, residue proteins were determined and expressed as a percentage of total flour protein. The greater the proportion of residue protein, the stronger is the baking strength ; and (b) in the gel protein test (Feillet et al., 1976), 5 grams of flour are extracted twice by 30 ml water for 15 min and centrifuged (15 min ; 38 000 g) ; the supernatant is discarded and the residue is extracted once again by 30 ml. Chloro-2-ethanol 70 % (pH = 1.8) for 15 min. By centrifugation (15 min. ; 38 000 g), three layers are formed in the centrifuge tube : the gel protein layer is located in the middle part ; its thickness is measured (Figure 4).

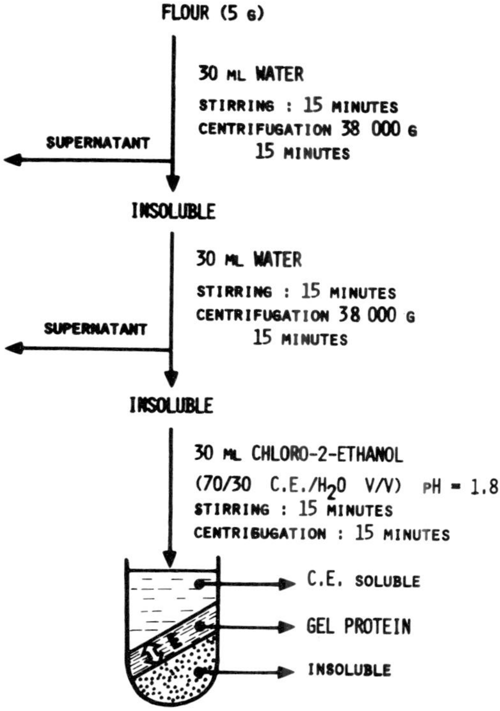

FIGURE 4 . DETERMINATION OF GEL PROTEIN CONTENT OF WHEAT

From the analysis of a very large number of french varieties covering a wide range in quality and grown several years in many different locations, Jeanjean and Feillet (1979) reached the following conclusions : (a) gel protein contents (i.e. the gel layer thickness) of wheats range from 4 to 18 (Table 1), (b) within one variety, standard deviation of gel thickness is usually lower than 1 mm, (c) there is a high significant correlation between wheat breadmaking quality (as appreciated by baking score, Zeleny test and alveograph test) and gel protein content when all samples are examined (Table II), (d) there is no correlation between breadmaking quality and gel protein content of samples of the same variety (Table III), (e) varietal intrinsic quality of varieties (which is, for each variety, the baking quality mean value) and gel protein content are very highly correlated as illustrated Figure 5, (f) consequently, the gel protein test appears to be a good tool in the hand of breeders to appreciate the varietal intrinsic quality of wheat. But it cannot be of any use to measure the wheat commercial quality.

TABLE I. Gel protein composition of wheat

	Gel Protein Thickness (MM.)		
	Number of samples	Mean value	Standard deviation
Maris Huntsman	19	5.2	0.89
Clément	7	6.2	1.07
Maris Hobbit	4	6.6	0.54
Joss	15	6.6	1.19
Talent	24	7.5	0.87
Champlein	34	7.6	0.74
Top	45	9.5	0.81
Capitole	41	9.5	1.05
Kolibri	4	9.6	0.48
Hardi	34	11.2	1.57
Darius	3	12.2	0.29
Florence Aurore	1	14.5	-
Magdalena	1	15.0	-
Poncheau	1	15.0	-
Rex	1	15.0	-

TABLE II. Correlation coefficients between gel protein content and flour quality parameters

Year	1975	1976		1977	
Number of samples	69	37	125	117	348
Baking score	0.83***	0.76***	0.50***	-	-
Bread volume	-	-	-	0.49***	-
Zeleny test	0.80***	0.68***	-	0.54***	0.53***
Alveograph W	0.75***	-	0.61***	0.65***	0.59***
Flour protein content	0.08	-0.01	-0.08	0.00 (1)	

*** Significant at 0.1 % level of confidence

(1) Protein content of wheat

TABLE III. Intravarietal correlation coefficients between gel protein content and baking score

Variety	1975		1976	
	N	R	N	R
Champlein	13	0.07	16	0.58 *
Top	18	0.22	17	-0.28
Capitole	15	0.07	22	-0.14
Hardi	17	0.02	17	0.10

N = Number of samples
R = Correlation coefficient

* Significant at 5 % level of confidence

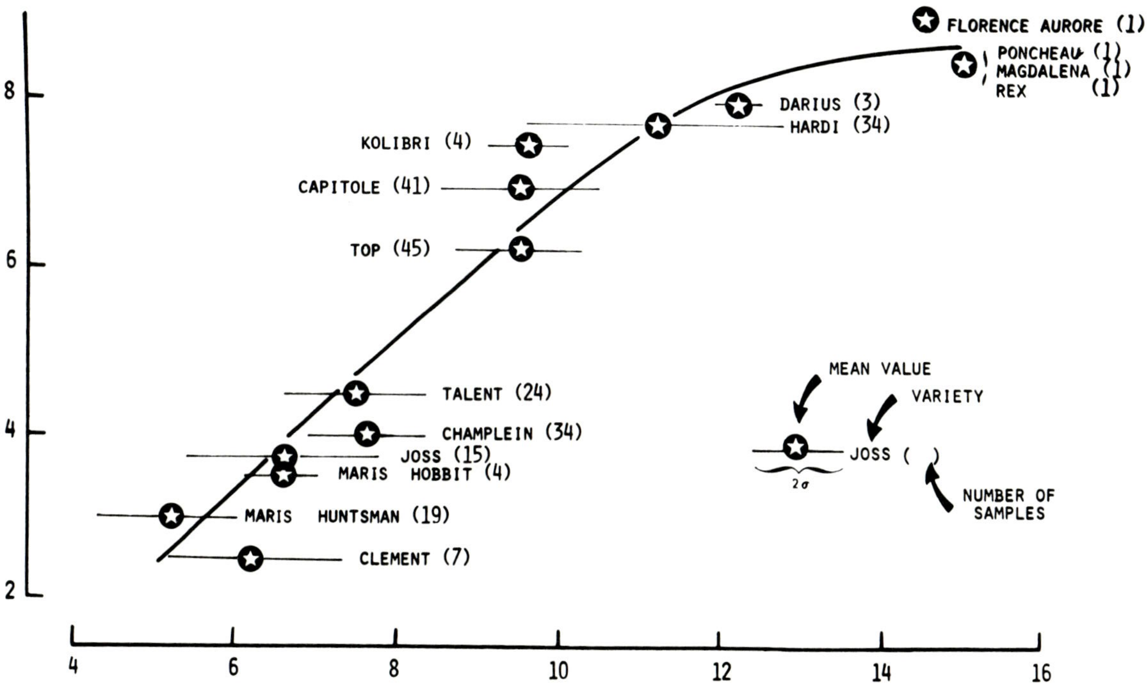

FIGURE 5 . RELATION BETWEEN GEL PROTEIN CONTENT AND BAKING QUALITY SCORE

IV. GLIADIN ELECTROPHOREGRAMS

Wheat gliadins are a very heterogeneous material. Upon gel electrophoresis, 20 to 30 individual components can be detected ; patterns are distinctive for each wheat cultivar (Coulson and Sim, 1961 ; Elton and Ewart, 1962 ; Bourdet *et al.*, 1963 ; Lee and Wrigley, 1963).

Wrigley and Shepherd (1973) identified 46 different fractions in a single variety by combining gel electrophoresis and isoelectrofocusing. Slight changes in experimental conditions allowed to obtain very clear and reproducible gliadin fractionation either by starch gel electrophoresis (Autran and Bourdet, 1975) or by polyacrylamide gel electrophoresis (Bushuk and Zillman, 1978).

It has been clearly demonstrated that gliadin components of wheat are varietal characteristics not modified by environmental factors (Feillet, 1965 ; Feillet and Bourdet, 1967 ; Doekes, 1968 ; Wrigley and Shepherd, 1973) ; they are coded by chromosomes of homoeologous groups 1 and 6 (Shepherd, 1968 ; Wrigley and Shepherd, 1973).

Varietal identification of wheats was the first application of those findings (Autran and Bourdet, 1975 ; Autran, 1975 ; Wrigley, 1976 ; Bushuk and Zillman, 1978). In the scope of this paper, I would like to comment another promising use of the genetic controlled gliadin electrophoregram in breeding programs.

In a pioneered work, Doekes (1970) claimed that α, β, γ and ω gliadins occur in only a few different configurations ; accordingly, wheat varieties can be characterized with the aid of a simple code, e.g. 1423 for a combination of $\alpha 1$, $\beta 4$, $\gamma 2$ and $\omega 3$; he hypothesized that this configuration is connected with baking quality of wheat varieties.

The occurence of gliadin components which are inherited together and constitute gliadin blocks were corraborated by Sozinov *et al.* (1974). From their very extensive study, they found that baking quality of wheats can be predicted from the knowledge of the gliadin block formula.

More recently, Rousset and his group (Rousset and Branlard, 1977 ; Branlard, 1979) confirmed that the varietal intrinsic quality of wheats is linked to the occurence of some specific gliadin bands in the electrophoretic pattern. They examined 50 different wheat cultivars and compared the qualitative and quantitative composition of their gliadin patterns (according to Autran and Bourdet, 1975) to the results of technological tests (Zeleny test, alveograph, farinograph, baking test). This experiment was performed on samples grown in different locations, some for two years. From correlation studies, they concluded that presence of 15 bands (mobilities:

28, 44, 49, 52, 55, 68, 72, 74, 81, 83, 85, 88, 93, 98 and 100) and absence of 13 bands (mobilities : 22, 34, 37, 41, 50, 53, 59, 60, 62, 71, 77, 91, 96) were related to a high baking quality. Furthermore, they observed that some bands were associated together to constitute gliadin groups similar to the gliadin blocks of Doekes (1970) and Sozinov _et al_. (1974).

Relationships between durum wheat gliadin patterns and gluten viscoelasticity, which is known to be related to the pasta making quality (Damidaux and Feillet, 1978), were studied in our laboratory (Damidaux _et al_., 1978 ; Feillet, 1979 ; Damidaux, 1979). 117 durum wheat varieties of different genetical origins were examined. After extraction by 70 % ethanol, gliadins were fractionated by polyacrylamide gel electrophoresis according to Bushuk and Zillman (1978) ; component mobilities were established by reference to the standard 51 band in agreement with the common wheat gliadin nomenclature of Zillman and Bushuk (1979). By giving a special attention to the gamma gliadin bands, durum wheats were classified into two main groups (Figure 6) : (a) group 45 is characterized by presence of a strong band 45 and absence of band in the 38-42 region and (b) group 42 is characterized by presence of a strong band 42 and absence of band 45.

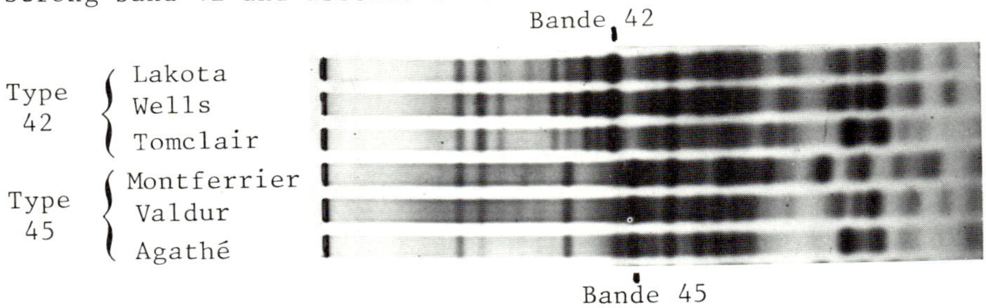

FIGURE 6. CLASSIFICATION OF DURUM WHEATS INTO TWO GROUPS ACCORDING THE PRESENCE OF GLIADIN BANDS 42 AND 45 (polyacrylamid gel electrophoresis in aluminium lactate buffer)

Sixty six varieties belong to group 45 ; 47 varieties to group 42 ; 4 varieties could not be classified into one or other of these two groups.

Gluten viscoelasticity of these varieties was determined as follows : after hand washing extraction, 1 g gluten was hold into a molding tube by a tightening device ; the tube was put in boiling water for 90 seconds and cooled during 2 min in 1 liter of water at 15°C. Viscoelastic parameters were then measured with a viscoelastograph by placing the gluten disk between two plates and by applying a 500 g constant force for 40 seconds, at right angles to the plate ; the change in thickness of the sample was related to the time. Difference

(e_2-e_1) between final thickness e_2 of the sample 20 seconds after loading off and residual thickness e_1 of the sample after the load has been applied and just before it is removed was called absolute elastic recovery (AER) ; it expresses the elasticity of gluten (Damidaux and Feillet, 1978).

According to the AER values, all varieties were distributed into two classes around the mean values 0.6 and 1.8 mm.

The most striking conclusion of this work is the occurence of a very narrow correspondance between electrophoretic pattern types and gluten AER values, as illustrated Figure 7 : within 45 gliadin group, 59 out of 66 varieties own a gluten AER higher than 1.2 mm : within 42 gliadin group, 46 out of 47 varieties own a gluten AER lower than 1.2 mm.

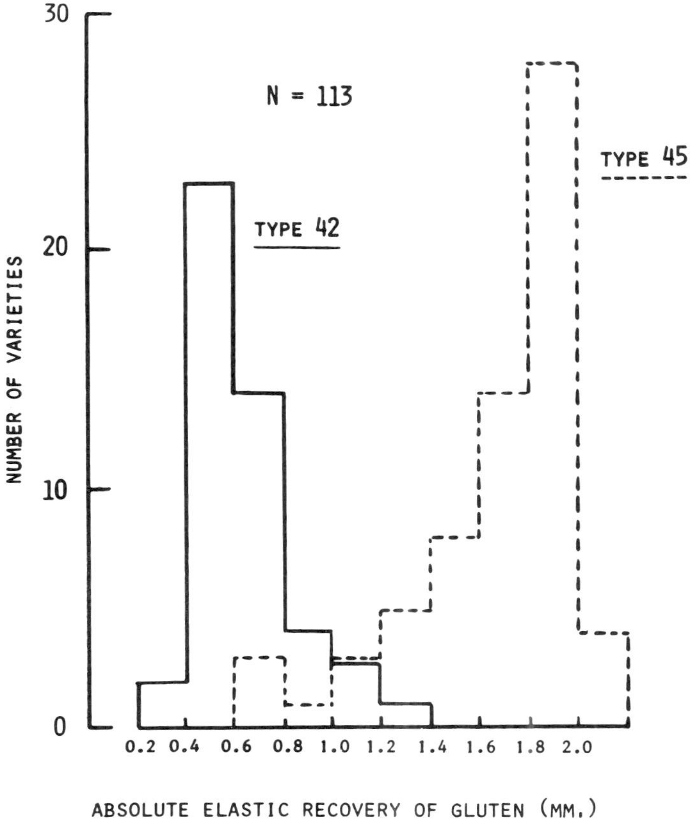

FIGURE 7 . DISTRIBUTION OF DURUM WHEAT VARIETIES ACCORDING TO THEIR GLUTEN ABSOLUTE ELASTIC RECOVERY AND THEIR GLIADIN TYPE (DAMIDAUX, 1979)

One can wonder about the nature of the relations which were demonstrated between gliadin electrophoregrams and baking strength or gluten viscoelastic properties. It might be either a genetic linkage : genes coding for mixing and baking properties of wheat flour dough or for pasta making ability of durum wheat semolina are close by genes coding for specific gliadin bands ; or, more possibly, a functionnal-type relation : individual or genetic linked gliadin components would have specific characteristics which would act on the dough properties even at low concentration (1 to 3 % of total proteins). This hypothesis is supported by recent findings of Jeanjean et al. (1979) that some gliadin fractions can participate to the formation of an insoluble protein network which give high viscoelasticity to the gluten upon heating and that others cannot. It is not contradictory with the recognized major effect of the glutenins since gliadin-type fractions participate to the glutenin protein complexes (Bietz and Wall, 1972 ; Jeanjean and Feillet, 1978).

Whatever are the basic explanation of these relationships, it turns out that these results supply breeders already with new methods to screen wheat lines for technological characteristics. For instance, when breeding durum wheat for cooking quality, we suggest to screen for the presence of electrophoretic gliadin component 45 as soon as the F3 (or even F2 on half a kernel) generation ; then to screen from F5 generation (all lines will belong to 45 type) for high gluten elastic recovery. Later on, cooking tests will have to be performed to check the cooking quality and to evaluate the stability of this quality when changing growing conditions. Simultaneously, it would be advisable to screen for high protein content in the kernel.

V. PROTEIN SUBUNITS COMPOSITION

Sodium dodecyl sulfate polyacrylamide gel electrophoresis (SDS-PAGE) in basic buffer is commonly used to separate subunits of reduced proteins and to estimate their molecular weight (MW). Using this technique, Bietz and Wall (1972) found that glutenins are composed of subunits of at least 15 distinct MW's, ranging from 11,600 to 133,000 while most gliadin proteins are single chained and have MW's near 36.500. In spite of small discrepancies, which could be attributed to slight differences in experimental conditions (Cole et al., 1976), the main conclusions of this study were confirmed later on by other authors (Orth and Bushuk, 1973a ; Jeanjean and Feillet, 1978 ; Khan and Bushuk, 1979).

Variations in the patterns of glutenin subunits between wheat varieties were noted by Bietz and Wall (1972) but differences were small. Using 26 hexaploid bread wheats of diverse baking quality, Orth and Bushuk (1973a, 1973b) showed that there was no obvious relationship between their quality differences and the SDS-PAGE patterns of their reduced glutenins. By comparing hexaploid and tetraploid wheats, they found that the absence of high MW subunits can be correlated to the lack of breadmaking quality in durum wheats and speculated that these subunits could play a dominant role in baking quality.

Wasik and Bushuk (1975) observed that glutenin of durum wheat varieties with excellent spaghetti-making quality had more of subunit 6 (MW 53,000) than subunit 5 (MW 60 000) ; varieties with low or mediocre spaghetti-making quality appear to have less subunit 6 relative to subunit 5.

Possibly, Kobrehel (1979) could be credited with a new breakthrough in the knowledge of relationships between glutenin subunits composition and wheat baking quality. Pursuing his study on the solubility of wheat proteins by soaps, he came to the conclusion that varietal intrinsic quality (VIQ) of wheats might be related to the strength of polar and hydrophobic bonds between native flour proteins. In this work, proteins were extracted by stirring overnight various amount of sodium hexadecanoate (0 to 100 mg), 20 ml distilled water and 2 g flour ; after centrifugation, soluble proteins (20 to 95 % total proteins) were reduced with β-mercaptoethanol and fractionnated by SDS-PAGE in SDS-Tris-borate buffer (pH = 8.9) without previous alkylation. Proteins were more rapidly solubilized from low quality than from high quality wheat flours when increasing the amount of soaps in the extracts.

As shown Figure 8, protein subunits with MW higher than 90.000 were not detectable when using small quantities of soaps ; furthermore, high-MW components were more easily extracted when VIQ was low. From his study, Kobrehel concluded that confirmation of these results would allow to develop a breeding test for bread wheats based upon the determination of the minimum amount of soaps which is necessary to solubilize high-MW subunits of wheat proteins : higher would be the amount of soaps, better would be the VIQ of bread wheats.

VI. CONCLUSION

For many years, most of the wheat protein studies aimed to isolate protein components, mainly gliadins and glutenins, and to relate their physicochemical characteristics (amino acid content, subunit composition, molecular weights, carbohydrate-protein and lipid-protein interactions) to their functionnal properties (breadmaking and pasta making ability).

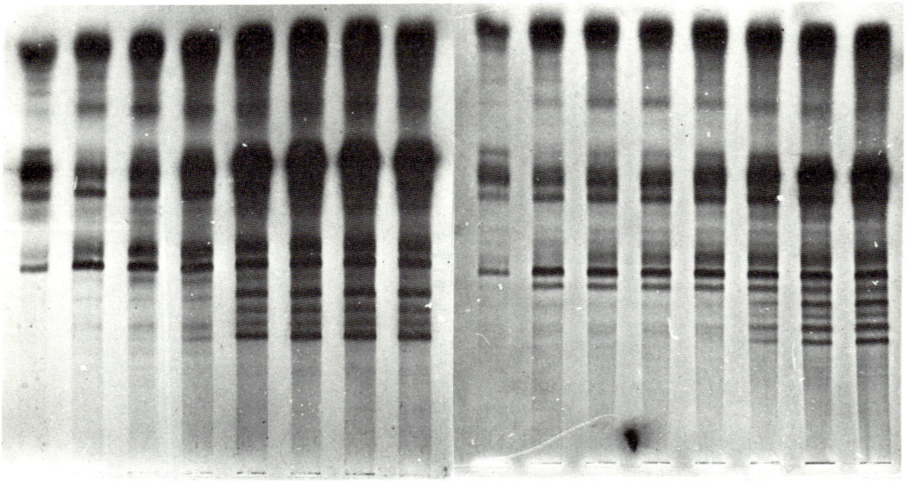

0 40 50 60 70 80 90 100 0 50 60 70 80 90 100 110

(mg. Sodium Palmitate in 2 g. Flour)

Varietal Intrinsic Quality	Low	High
Variety	Maris Huntsman	Magdalena

FIGURE 8. SDS-PAGE ELECTROPHORESIS OF REDUCED PROTEIN EXTRACTED WITH SODIUM PALMITATE FROM LOW QUALITY AND HIGH QUALITY WHEAT (KOBREHEL, 1979)

Recent advances in methods of protein separation have allowed to reduce the amounts of material which are necessary to characterize proteins of wheat grains. Polyacrylamide gel electrophoresis methods are now available to separate gliadin components and glutenin subunits from as little as half a kernel, keeping the germ for further studies.

These methods are simple, fast, highly specific and reproducible. They are now applicable by breeders to screen for quality in early generations. Nevertheless, because of their independance toward the environmental factors, it is clear that these techniques have no value to evaluate the commercial quality of wheat sample.

We think that the development of biochemical methods, as protein electrophoresis, to evaluate the varietal intrinsic quality of wheats could be of considerable value to wheat breeders and genetists. The study of the genetic control of wheat proteins has just started ; quite possibly, it is the only way to know some day the genetic regulation of wheat quality.

REFERENCES

Autran, J.C. 1975. Nouvelles possibilités d'identification des variétés françaises de blé par électrophorèse des gliadines du grain. *Ind. Agric. Alim.* 92: 1075-1094.

Autran, J.C. and A. Bourdet 1975. L'identification des variétés de blé : établissement d'un tableau général de détermination fondé sur le diagramme électrophorétique des gliadines du grain. *Ann. Amelior. Plantes* 25: 227-301.

Bietz, J.A. and J.S. Wall 1972. Wheat gluten subunits : molecular weights determined by sodium dodecyl sulfate polyacrylamide gel electrophoresis. *Cereal Chem.* 49: 416-430.

Bietz, J.A., P.R. Huebner, and J.S. Wall 1973. Glutenin - the strength protein of wheat flour. *Baker's Dig.* 47: 26.

Bietz, J.A., and J.S. Wall 1975. The effect of various extractants on subunit composition and associations of wheat glutenin. *Cereal Chem.* 52: 145.

Bloksma, A.H. 1975. Thiol and disulfide groups in dough reology. *Cereal Chem.* 52: 170r-183r.

Bourdet, A., P. Feillet, and F. Mettavant 1963. Sur le comportement électrophorétique des prolamines du blé en gel d'amidon. *C.R. Acad. Sci.* 256: 4517-4520.

Branlard, G. 1979. Relation entre le polymorphisme des gliadines et la force boulangère des génotypes de blé tendre. *C.R. Journée Biochimie-Génétique, INRA, Paris.*

Bushuk, W., and R.R. Zillman 1978. Wheat cultivar identification by gliadin electrophoregrams. I. Apparatus method and nomenclature. *Can. J. Plant Sci.* 58: 505-515.

Cole, E.W., H. Ng, and D.K. Mecham 1976. Some chemical and physical properties of the mercuric chloride-solubilized gel protein from different wheat varieties. *Cereal Chem.* 53: 250.

Coulson, C.B., and A.K. Sim 1961. Starch gel electrophoresis of isolatedwheat gluten. *Biochem. J.* 80: 46-47.

Damidaux, R. 1979. Nouveaux critères de sélection pour l'amélioration de la qualité culinaire du blé dur. *Thesis, Université des Sciences et Techniques du Languedoc, Montpellier.*

Damidaux, R., J.C. Autran, P. Grignac, and P. Feillet 1978.

Relation applicable en sélection entre l'électrophorégramme des gliadines et les propriétés viscoélastiques du gluten. *C.R. Acad. Sci.* 287: série D, 701-704.

Damidaux, R., and P. Feillet 1978. Relation entre les propriétés viscoélastiques du gluten cuit, la teneur en protéines et la qualité culinaire des blés durs (T. durum). *Ann. Techn. Agric.* 27: 799-808.

Doekes, G.J. 1968. Comparison of wheat varieties by starch gel electrophoresis of their grain proteins. *J. Sci. Food Agric.* 19: 169-176.

Doekes, G.J. 1970. Electrophoresis of grain proteins as an aid in wheat breeding. *Proc. 5th Word Cereal and Bread Congress, Dresden.*

Elton, G.A.H., and J.A.D. Ewart 1962. Starch gel electrophoresis of cereal proteins. *J. Sci. Food Agric.* 13: 62-72.

Feillet, P. 1965. Contribution à l'étude des protéines du blé. Influence des facteurs génétiques, agronomiques et technologiques. *Ann. Techn. Agric.* 14: HS 1, 1-94.

Feillet, P. 1979. Evaluation de la qualité culinaire intrinsèque des variétés de blés durs (T. Durum) par examen des électrophorégrammes des gliadines et par mesure des propriétés viscoélastiques du gluten. *Symposium ICC, matières premières et pâtes alimentaires. Rome, 30-31 Mai.*

Feillet, P., and A. Bourdet 1967. Composition protéique et caractéristiques génétiques des blés. *Bull. Soc. Chim. Biol.* 49: 1273-1283.

Feillet, P., E. Fevre, and M.F. Jeanjean 1976. Recherche d'un nouveau test d'appréciation de la force boulangère des blés tendres. Résultats préliminaires. *Bull. Anc. Elèves ENSMIC* 231: 22.

Finney, K.F. 1943. Fractionating and reconstituting techniques as tools in wheat research. *Cereal Chem.* 20: 281.

Finney, K.F. 1971. Fractionating and reconstituting techniques to relate functionnal (breadmaking) to biochemical properties of flour components. *Cereal Sci. Today* 16: 342-356.

Godon, B. 1969. Etude physicochimique des associations du gluten de blé. *Thesis, Faculté des Sciences de Paris.*

Greenwood, C.T., and J.A.D. Ewart 1975. Hypothesis for the structure of glutenin in relation to rheological properties of gluten and dough. *Cereal Chem.* 52: 146r-153r.

Guthrie, F.B. 1896. The absorption of water by gluten of various wheats. *Agric. Gaz. N.S. Wales* 7: 583.

Jeanjean, M.F., and P. Feillet 1978. Physichemical properties of wheat gel proteins : effects of isolation conditions. *Cereal Chem.* 55: 864-876.

Jeanjean, M.F., R. Damidaux, and P. Feillet 1979. Effects of thermic treatments on protein solubility and viscoelastic properties of wheat gluten. *Cereal Chem.*, submitted for publication.

Jeanjean, M.F., and P. Feillet 1979. Eine neue biochemishe Methode zur Ermittlung der Backfähigkeit von Weichweizen. *Getreide U. Mehl* 33: 127-130.

Khan, K., and W. Bushuk 1979. Studies of glutenins. XII. Comparison by sodium dodecyl sulfate polyacrylamide gel electrophoresis of unreduced and reduced glutenin from various isolation and purification procedures. *Cereal Chem.* 56: 63-68.

Kobrehel, K. 1979. Solubilization of proteins with soaps in relationship with the breadmaking properties of wheat flour. *Cereal Chem.*, submitted for publication.

Kobrehel, K., and W. Bushuk 1977. X. Effects of fatty acids and their sodium salts on solubility in water. *Cereal Chem.* 54: 833.

Kobrehel, K., and W. Bushuk 1978. Studies of glutenin. XI. Note on glutenin solubilization with surfactants in water. *Cereal Chem.* 55: 1060.

Konzak, C.F. 1977. Genetic control of the content, amino acid composition and processing properties of proteins in wheat. *Adv. Genet.* 19: 407-582.

Lee, J.W., and C.W. Wrigley 1963. The protein composition of gluten extracted from different wheats. *Aust. J. Exp. Animal Husb.* 3: 361-384.

Macritchie, F. 1978. Differences in baking quality between flours. *J. Fd. Technol.* 13: 187-194.

Mecham, D.K., E.W. Cole, and H. Ng 1972. Solubilizing effect of mercuric chloride on the "Gel Protein" of wheat flour. *Cereal Chem.* 49: 62-71.

Mecham, D.K., D.D. Kasarda, and C.O. Qualset 1978. Genetic aspects of wheat gliadin proteins. *Bioch. Gen.* 16: 831-852.

Meredith, P. 1961. A gel fraction of wheat gluten ; mixing, oxidation and lipid relationships. *New Zealand. J. Sci.* 4: 66-77.

Orth, R.A., and W. Bushuk 1972. A comparative studies of the proteins of wheats of diverses baking quality. *Cereal Chem.* 49: 268.

Orth, R.A., and W. Bushuk 1973a. Studies of glutenin. II. Relation of variety, location of growth, and baking quality to molecular weight distribution of subunits. *Cereal Chem.* 50: 191.

Orth, R.A., and W. Bushuk 1973b. Studies of glutenin. III. Identification of subunits coded by the D-Genome and their relation to breadmaking quality. *Cereal Chem.* 50: 680-687.

Orth, R.A., and L.O'brien 1976. A new biochemical test of

dough strength of wheat flour. *J. Aust. Inst. Agric. Sci.* 42: 122-124.

Preston, K.R., N. Woodbury, and V. Bendelow 1975. The effects of gliadin fractions of varying molecular weight on the mixing properties of a synthetic dough system. *Cereal Chem.* 52: 427-430.

Rousset, M., and G. Branlard 1977. L'électrophorèse des gliadines appliquée à la sélection des blés tendres de haute valeur en panification : travaux préliminaires. *Proc. Symposium Franco-Soviétique, Odessa, URSS.*

Shepherd, K.W. 1968. Chromosomal control of endosperm proteins in wheat and rye. In : *Proc. 3rd. Int. Wheat Genet. Symp.* p. 86. *Aust. Acad. Sci., Canberra.*

Sozinov, A.A., F.A. Poperela, and R.J. Stacanova 1974. Application du fractionnement électrophorétique des gliadines à la sélection qualitative des blés. *Proc. Symposium Franco-Soviétique, Amélioration génétique de la qualité des blés durs, Montpellier, France,* 47-56.

Tanaka, and W. Bushuk 1973. Changes in flour protein during dough mixing. I. Solubility results. *Cereal Chem.* 50: 590-596.

Wasik, R.J., and W. Bushuk 1975. Sodium dodecylsulfate-polyacrylamide gel electrophoresis of reduced glutenin of durum wheats of different spaghetti-making quality. *Cereal Chem.* 52: 328-334.

Wrigley, C.W. 1976. Single-seed identification of wheat varieties : use of grain hardness testing, electrophoretic analysis and a rapid test paper for phenol reaction. *J. Sci. Fd. Agric.* 27: 429-432.

Wrigley, C.W., and K.W. Shepherd 1973. Electrofocusing of grain proteins from wheat genotypes. *Ann. N.Y. Acad. Sci.* 209: 154-162.

WHEAT FLOUR COMPONENTS IN BREADMAKING[1]

Y. Pomeranz

U.S. Grain Marketing Research Laboratory
Science and Education Administration
Agricultural Research
U.S. Department of Agriculture
Manhattan, Kansas

I. SUMMARY

This is a review of investigations designed to determine the role of wheat flour components in breadmaking. The investigations involved fractionation of the components without impairing their functional properties, appropriate characterization of the separated fractions (and, whenever possible, synthesis of purified components), and performance tests. Those studies (combined with examination of dough and bread by light microscopy and scanning and transmission electron microscopy) have established the roles of gluten proteins, water-solubles, and lipids in breadmaking. They have provided insight into the interaction among various flour components, the significance of the complexes formed in breadmaking, and the mechanism of the shortening response. In addition, they have stimulated research on making foreign proteins functional and compatible with wheat flour proteins, thereby enhancing consumer acceptance of nutritionally improved, protein-enriched bread.

[1]Mention of a trademark or proprietary product does not constitute a guarantee or warranty of the product by the U.S. Department of Agriculture and does not imply its approval to the exclusion of other products that may also be suitable.

ISBN: 0-12-370960-1

II. INTRODUCTION

Much of the biochemical research on foods is designed to answer the following four questions: (1) What are the identities of the food components? (2) How much of each component does the raw material contain? (3) Where are the specific components concentrated in the plant or animal? and (4) What are the functions of the components? The fourth question varies with the research: the theoretical researcher generally deals with the physiological functions of components; the applied investigator deals with the roles of components in producing desirable foods. This review summarizes investigations, mostly in the author's laboratory, that were conducted to answer the above four questions with regard to the wheat flour components of bread.

It has been known for many years that flours milled from certain wheat varieties grown in some locations consistently produce bold loaves with good internal crumb grain and color and with excellent shelf life. On the other hand, some wheat flours produce compact bread that has a crumbly interior and stales rapidly. Researchers surmised that those differences result from qualitative and quantitative compositional variations. The problems of relating chemical composition and structure of wheat flour components to functional properties in breadmaking is complicated by several factors: the large number of components; the high molecular weight, limited solubility, and difficulty of separating or isolating pure components without altering them; and the interaction of the components during dough mixing, fermentation, and baking. Some classical fractionation methods make it possible to separate flour components; some modern biochemical techniques enable the researcher to characterize those components and to follow interactions among macromolecules; and the combination of separation and characterization techniques can be used to elucidate the roles of wheat-flour components in breadmaking.

III. BAKING TESTS

For establishment of a causative rather than causal relation between chemical composition and structure of wheat-flour components and their functional properties in breadmaking, it is necessary to (1) extract, fractionate, and characterize flour components, (2) reconstitute the isolated moieties, and (3) ascertain that neither the isolation nor the reconstitution procedures impair the functional properties of the components.

Historically, and logically, the last requirement, was met first. An optimized baking test, rather than test bakings under fixed and arbitrarily selected conditions, was developed to determine the characteristics inherent in wheat and flour components. In the optimized baking test, four factors -- mixing time, water absorption, yeast activity, and fermentation time -- are optimized and balanced (Finney, 1978). Oxidation requirements are met by use of the optimum amount of either potassium bromate ($KBrO_3$) or a mixture of $KBrO_3$ and ascorbic acid (Shogren and Finney, 1974). Because flours from commercial wheats show a relatively narrow range of oxidation requirements, 50 ppm ascorbic acid alone effectively replaces $KBrO_3$ as a dough oxidant and developer. The added ingredients -- shortening, sugar, and malt -- are supplied in excess because an insufficient quantity of any one would limit the performance of the wheat-flour ingredients. Figure 1 illustrates the effects and importance of dough ingredients on loaf volume and crumb; and Figure 2, some differences between good and poor quality flours (from Finney, 1978). When 100 g flour were mixed with water, as needed, the loaf volume of 230 cc equaled the dough volume after mixing. Loaf volume was increased to 583 cc by the addition of 6% sugar and yeast; and to 732 cc by the further addition of 1.5% salt. Still, none of the crumb grains was satisfactory. When 4% soy flour, 0.25% malt, 50 ppm ascorbic acid, and 3% shortening were added to the five basic ingredients (wheat flour, water, sugar, yeast, and salt), loaf volume increased to 1006 cc and crumb grain was excellent. The optimized baking test can be performed with the precision of most biological assays, and it establishes several parameters (the most important being loaf volume) as standards of comparison.

Loaf volume is highly correlated, over a relatively wide range, with dough-handling properties, consumer acceptance -- as defined by crumb texture and freshness retention -- and with technological versatility. For a given type of commercial bread, taste and overall freshness retention depend on the types and amounts of wheat flour components as well as on the method used to produce the bread, optional ingredients, skill of the baker, and storage conditions. The last-mentioned factors are beyond the control of the plant breeder and biochemist and were not included in investigations described in this review.

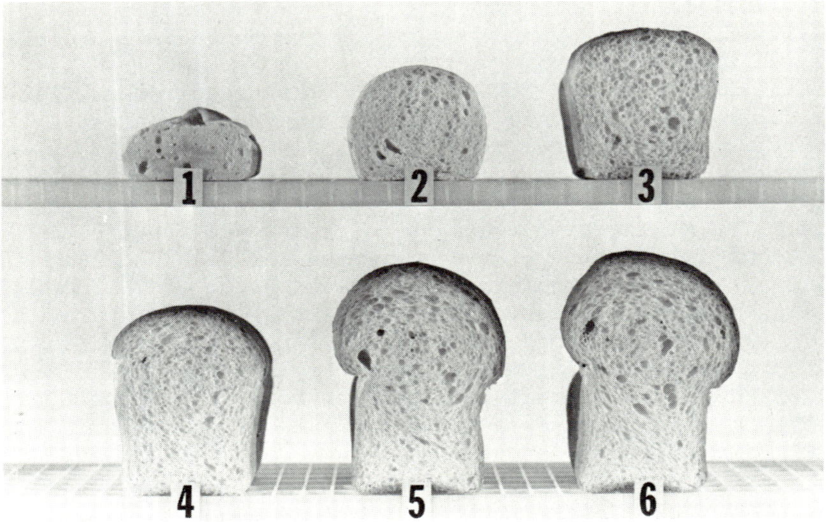

FIGURE 1. Relative loaf volumes and crumb grains of breads made from flour and water (1) plus: yeast (2), yeast and sugar (3), yeast, sugar, and salt (4), yeast, sugar, salt, malt, ascorbic acid, and shortening (5), and yeast, sugar, salt, malt, ascorbic acid, shortening, and soy flour (6) (from Finney, 1978).

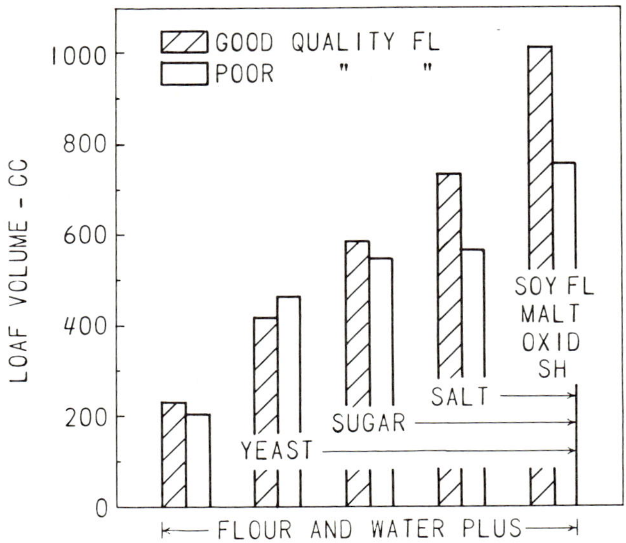

FIGURE 2. Loaf volumes of breads made from good- and poor-quality wheat flours (100 g) and very lean to optimized formulations (from Finney, 1978).

III. WHEAT FLOUR COMPONENTS

A. Protein Content

In initial investigations, bread was baked from flours milled from wheat varieties grown under a widely different climatic and soil condition. Protein content was the major factor that accounted for variation in loaf volume within a single variety. Indeed, the relation between loaf volume (provided the bread was baked by the optimized test) and protein content is linear (Fig. 3 from Finney and Barmore, 1948). Because the protein content-loaf volume relation is linear within a single variety, the breadmaking quality of new wheat varieties can be easily determined.

In practice, a plant breeder can determine the slope of the regression line from two or three samples of a new variety at different protein levels. As shown in Figure 3, the regression of loaf volume with protein quantity (content) differs widely among different varieties. This variation

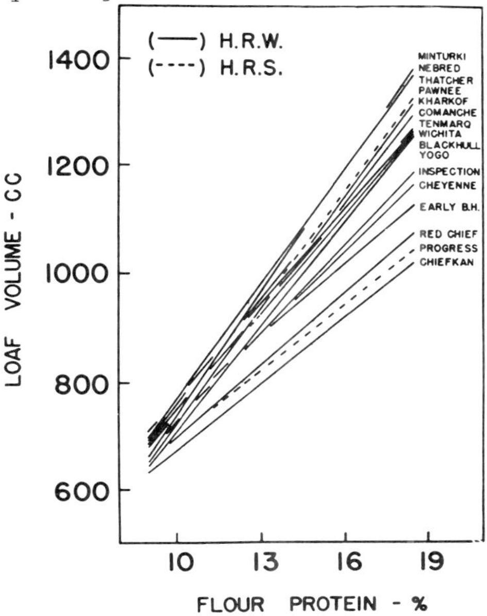

FIGURE 3. Loaf volume-protein content regression lines for hard red winter and hard red spring wheat varieties. Each varietal regression line represents many samples harvested throughout the Great Plains region of the United States during several crop years (from Finney and Barmore, 1948).

results in differences in the slopes of regression lines for individual varieties: the greater the slope, the greater the increase in loaf volume per 1% increase of protein. Those differences in slope reflect differences in breadmaking quality of the protein in the different varieties of wheat.

The significance of protein content as a factor that governs breadmaking quality of wheat and wheat flour was confirmed recently in a study of the functional properties of dark hard and yellow hard red winter wheats (Pomeranz *et al.*, 1976). In that study, functional properties of 14 commercial and 6 varietal hard red winter wheat samples, which varied widely in protein content and percentages of dark, hard, vitreous kernels (DHV), and of yellow hard kernels, were evaluated by laboratory milling and baking tests. Simple correlation coefficients were calculated for the relationships between protein content (of wheat or flour), DHV, baking absorption, mixing time, and loaf volume. The results are summarized in Table I. The correlation coefficients were constantly higher for the commercial samples alone than for all samples. That difference probably resulted from the relatively small effects of varietal differences in the commercial samples, which were mixtures of several varieties from several locations. *Protein content* explained about 90% (0.95^2) of the variability in loaf volume of all samples and about 96% (0.98^2) of the variability in loaf volume of the commercial samples; consequently, only 10 and 4% of loaf volume variability, respectively, was attributable to differences in *protein quality* and to experimental errors. The correlation was high between baking absorption and protein content, but was not significant between mixing time and protein content.

Because the correlation between protein or DHV and breadmaking was so high, we calculated several partial correlation coefficients, keeping either protein content or DHV constant. The results are given in Table II. Wheat protein and loaf volume were highly correlated ($r = 0.901$), irrespective of the DHV contents; the correlation between DHV and loaf volume ($r = 0.219$) was not significant when the protein content was constant.

In Table I, the correlation coefficient between baking absorption and mixing time was not significant. However, when the protein content was kept constant (Table II), the correlation between baking absorption and mixing time was highly significant (0.842). For flours with less than about 12% protein, mixing time increases with decreasing protein; above 12%, increases in protein content have little effect on mixing time. Consequently, when the effect of protein was eliminated, a relationship between baking absorption and mixing time was demonstrated.

TABLE I

SIMPLE CORRELATION COEFFICIENTS FOR PROTEIN AND BAKING CHARACTERISTICS OF COMMERCIAL AND COMMERCIAL PLUS VARIETAL SAMPLES

	Simple Correlation Coefficients for All Samples[b]					
	DHV[a]	WHEAT PROTEIN	FLOUR PROTEIN	BAKING ABSORPTION	MIXING TIME	LOAF VOLUME
DHV		0.832**	0.829**	0.640**	-0.117	0.841**
WHEAT PROTEIN	0.898**		0.995**	0.713**	-0.298	0.951**
FLOUR PROTEIN	0.916**	0.996**		0.743**	-0.310	0.949**
BAKING ABSORPTION	0.859**	0.841**	0.861**		0.092	0.625**
MIXING TIME	-0.270	0.420	-0.391	0.060		-0.353
LOAF VOLUME	0.900**	0.981**	0.985**	0.842**	-0.425	
	Simple Correlation Coefficients for Commercial Samples[c]					

[a]Dark, hard, and vitreous.

[b]n = 20, [c]n = 14.

**Denotes significance at the 1% level.

(from Pomeranz *et al.*, 1976)

TABLE II. PARTIAL CORRELATION COEFFICIENTS FOR WHEAT PROTEIN AND BAKING CHARACTERISTICS OF COMMERCIAL WHEAT SAMPLES

	PARTIAL CORRELATION COEFFICIENTS-DHV CONSTANT			
	DHV[a]	BAKING ABSORPTION	MIXING TIME	LOAF VOLUME
WHEAT PROTEIN		0.311	-0.419	0.901**
BAKING ABSORPTION	0.434		0.591**	0.312
MIXING TIME	0.268	0.842**		-0.432
LOAF VOLUME	0.219	0.161	-0.073	
	PARTIAL CORRELATION COEFFICIENTS-PROTEIN CONSTANT			

[a]Dark, hard, and vitreous. **Denotes significance at the 1% level (from Pomeranz *et al.*, 1976).

In addition, the six varietal wheat samples were separated into DHV and yellow hard kernels, and the functional properties of each were determined. The separated DHV contained more protein and produced larger loaves than the yellow hard kernels. However, expressed on an equiprotein basis, flours from the DHV and yellow hard kernels were comparable in breadmaking quality. Figure 4 (from Finney, 1979) shows the regressions of loaf volume with protein content for different wheat varieties. The regression curves form a family of lines for predicting loaf volumes associated with any desired protein content.

I do not imply that the quality of wheat flour proteins is constant; there are certainly significant differences among wheat types, classes, and varieties grown in different locations and under different environments. However, if a sound plant breeding program is established, in an area that is suited for growing bread wheats (such as the Great Plains of the United States), with approved varieties that have been selected also on the basis of their acceptable milling and baking properties, the effects of differences in

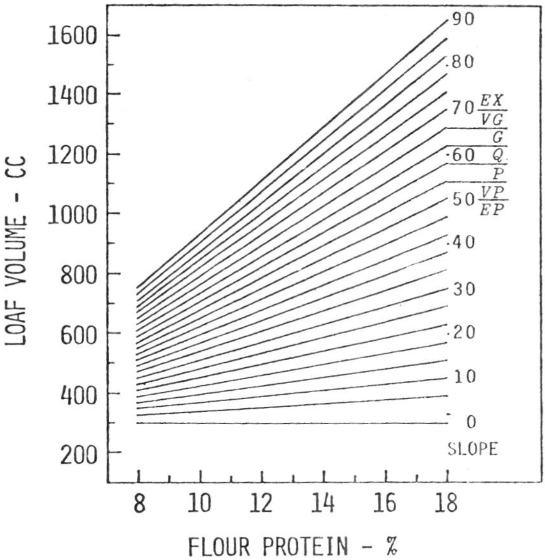

FIGURE 4. Loaf volume (100 g flour) versus protein content regression lines for correcting loaf volumes of wheat varieties to a constant protein basis. Slope is rate of change in loaf volume per 1 percent protein. EX = excellent; VG = very good; G = good; Q = questionable; P = poor; VP = very poor; EP = extremely poor (from Finney, 1979).

protein quality should be relatively small. The effects of differences in protein quality should then be minimal in commercial samples, for their protein quantity is already a consistent and satisfactory index of breadmaking quality.

B. Fractionation

After the linear relation between protein content and loaf volume was established in the optimized baking test, wheat flours representing a wide range in quality (as indicated by loaf-volume response per one percent of flour protein) were fractionated into starch, gluten, and water-soluble components. Fractions from one variety were recombined in their original and in different proportions, and various fractions of different varieties were interchanged (Finney, 1943). Invariably, flours that were reconstituted in their original proportions gave bread that was identical to bread made with the original, unfractionated flour; this established that neither the fractionation nor reconstitution techniques impaired breadmaking qualities.

When wheat gluten, starch, and water-soluble fractions from different varieties were interchanged and the flours so formed were baked, gluten proteins accounted for differences in bread quality of the varieties studied, as defined by loaf-volume increase per one percent protein. Although these results indicated that gluten proteins govern differences in breadmaking potential of wheat varieties, bread of acceptable volume, texture, and freshness retention cannot be produced unless other flour components, namely starch, water-solubles, and lipids, are also present in approximately the same quantities found in normal, unfractionated wheat flours.

C. Gluten Proteins

Once the key role of gluten proteins in governing varietal differences in breadmaking potential was established, the next logical step was to identify the gluten component(s) responsible for those differences.

Fractionation of gluten components on the basis of solubility damages the breadmaking properties of those components. Reconstitution studies showed that separation by ultra-centrifugation of the two main fractions of gluten into a centrifugate containing glutenin and a supernatant rich in gliadin did not impair the breadmaking properties of the gluten components. The gliadin fraction controls loaf volume and varies in flours that differ in breadmaking potential (Hoseney *et al.*, 1969a; Finney *et al.*, 1978). The factor responsible for mixing time and dough development is in the glutenin fraction. During dough mixing, the protein mass is converted from granular protein bodies into a homogeneous network in which starch granules are imbedded. Proper dough development is essential to optimum performance in breadmaking.

D. Amino Acids

Automated techniques to determine the amino acid composition of proteins, have facilitated the assay of amino acids in gluten proteins. Gluten proteins are characterized by high concentrations of glutamic acid (about 32%) and proline (about 10%). The acidic amino acids, glutamic and aspartic acids, occur mainly as amides. Thus far, studies on the amino acid composition of cereals have been disappointing because they have failed to explain the differences in breadmaking characteristics of wheat varieties. However, many studies have shown that both ionic and nonionic groups govern the unique viscoelastic properties of wheat gluten. The four types of groups/bonds that are known to contribute to viscoelastic properties of dough are (1) amide groups, (2) sulfhydryl groups and disulfide bonds, (3) hydrogen bonds, and (4) hydrophobic interactions.

When the acidic moieties of amides in gluten, glutenin, or gliadin are methylated, solubility, intrinsic viscosity, and cohesion are changed significantly (Beckwith *et al.*, 1963). Dough-handling properties can be modified, either beneficially or adversely, by the addition of minute amounts of reducing agents or sulfhydryl-blocking reagents. Also, the performance of a flour in breadmaking can be improved significantly by addition of an appropriate amount of oxidizing agent, such as 20-50 ppm potassium bromate.

The oxidation requirement (the amount of oxidant needed to produce the best loaf of bread in terms of volume, crumb texture, and freshness retention) is related to total protein content and to protein sulfhydryl groups and disulfide linkages. Continuous interchange reactions between sulfhydryl and disulfide groups and the reactivity of protein sulfhydryl groups affect oxidation requirements. Some studies indicate that protein quality may be affected significantly by the disulfide-to-sulfhydryl ratio (Belderok, 1966). That hydrogen bonds are involved in stabilizing dough structure was demonstrated by use of deuterium oxide instead of water in dough mixing (Vakar *et al.*, 1965). The role of hydrophobic interactions is discussed later in this review.

E. Water Solubles

The three fractions obtained when wheat flour dough is washed under a gentle stream of water are gluten, starch, and water-soluble components (Figure 5). The water-soluble components of flour (Figure 6) can be separated, in turn, by dialysis and centrifugation into four fractions: (1) membrane-permeable dialyzate (a mixture of soluble carbohydrates, amino acids, peptides, minerals, and growth factors essential for yeast fermentation); (2) globulins, which precipitate in the dialysis bag in the absence of minerals and can be removed after centrifugation; (3) heat-coagulable albumins in the supernatant; and (4) residual supernatant (Fig. 6, from Hoseney *et al.*, 1969b).

Baking experiments showed that the role of water-soluble components in breadmaking is two-fold. The membrane-permeable dialyzate is essential to panary fermentation because it contributes to gas formation of yeast; however, that contribution can be replaced by synthetic yeast food. Neither the globulins nor the albumins are essential to produce a

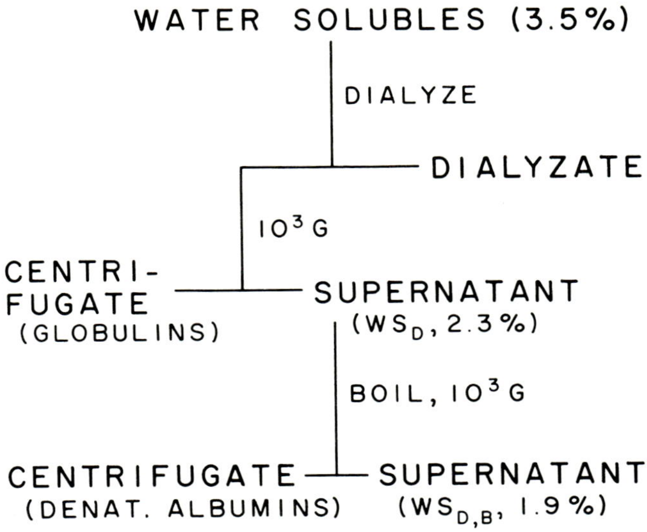

FIGURE 5. Scheme for fractionating wheat flour into gluten, water solubles, and starch.

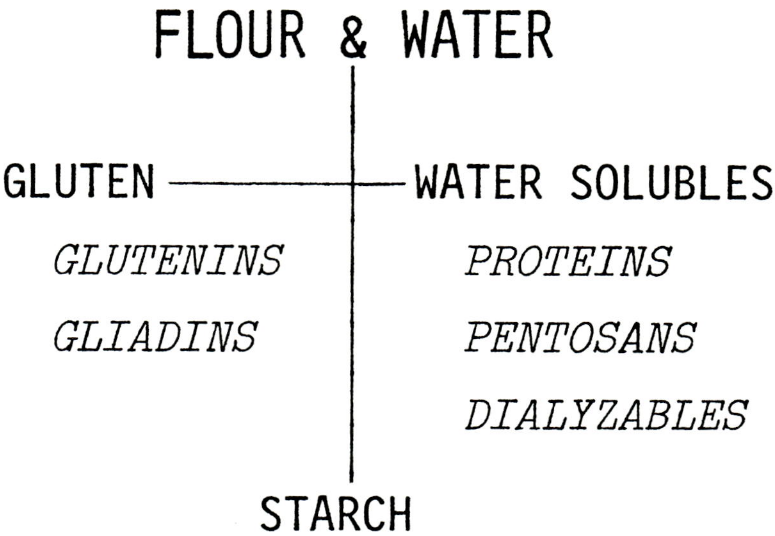

FIGURE 6. Fractionation scheme for certain water-soluble fractions of flour. Percentages are based on total flour weight. Abbreviations: WS, water-solubles; D, dialyzed; and B, boiled (from Hoseney *et al.*, 1969b).

normal loaf of bread. The residual supernatant is rich in water-soluble pentosans and glycoproteins, and contributes to gluten extensibility and to retention of gas in a fermented dough.

F. Lipids

Wheat flour lipid composition is shown schematically in Figure 7 (adapted from data reported by MacMurray and Morrison, 1970). Total wheat flour lipids consist of about equal amounts of nonpolar and polar components. Triglycerides (TG) are a major component of nonpolar lipids, digalactosyldiglycerides (DGDG) of glycolipids; and lysophosphatidylcholines (LPC) and phosphatidylcholines (PC) are major components of phospholipids. Differences in solubility provide a convenient and useful means of separating wheat flour lipids into major categories: free and bound (Figure 8). Free lipids can be extracted with nonpolar solvents such as ether or petroleum ether (PE). For extraction of bound (mainly to protein) lipids, polar solvents, such as water-saturated butanol (WSB) or a mixture of chloroform-methanol-water, are required. Lipids extracted by PE are arbitrarily defined as free and those by WSB, following PE extraction, as bound lipids.

The free lipids can be fractionated according to their elution from a silicic acid column. About 70% of the total free lipids can be eluted with chloroform; and they form what is arbitrarily called the nonpolar fraction, containing TG as a major component. The residual 30% free lipids can be eluted from the column with a more polar solvent, such as methanol, and comprise a mixture of free polar lipids. Among the free polar lipids, about two thirds are glycolipids, containing DGDG as a major component, and one third are phospholipids, with PC as a major component.

About 0.6 to 1.0% bound lipids can be extracted from flour with WSB after PE extraction. Bound lipids contain about 30% nonpolar and 70% polar lipids. Bound polar lipids are rich in phospholipids, with LPC as a major component. Because glycolipids are important in breadmaking, it is important to distinguish clearly as to whether they are free or bound. Although percentagewise the free polar lipids are richer in glycolipids than the bound polar lipids are, the actual amounts of both glycolipids and phospholipids are higher in the bound polar than in free polar lipids (Fig. 8).

Recently the lipids of wheat starch have received new attention. Morrison *et al.* (1975) divided flour lipids into those inside starch granules, which are true starch lipids, and all other lipids outside the starch granules, which are called non-starch lipids. Even polar solvents consisting of

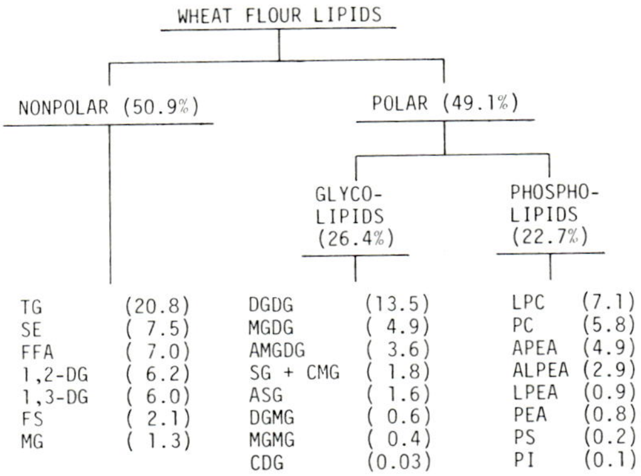

FIGURE 7. Composition of total wheat flour lipids (extracted with water-saturated butanol). The abbreviations are: TG = triglycerides; SE = steryl esters; FFA = free fatty acids; 1,2-DG = 1,2-diglycerides; 1,3-DG = 1,3-diglycerides; FS = free sterols; MG = monoglycerides; DGDG = digalactosyldiglycerides; MGDG = monogalactosyldiglycerides; AMGDG = O-acylmonogalactosyldiglycerides; SG = steryl glucoside; CMG = ceramide monoglycerides; ASG = 6-o-acylsteryl glucosides; DGMG = digalactosylmonoglycerides; MGMG = monogalactosylmonoglycerides; CDG = ceramide diglycosides; LPC = lysophosphatidylcholines; PC = phoshatidylcholines; APEA = N-acylphosphatidylethanolamines; ALPEA = N-acyllysophosphatidylethanolamines; LPEA = lysophosphatidylethanolamines; PEA = phosphatidylethanolamines; PS = phosphatidylserines; PI = phosphatidylinositols. (Adapted from MacMurray and Morrison, 1970).

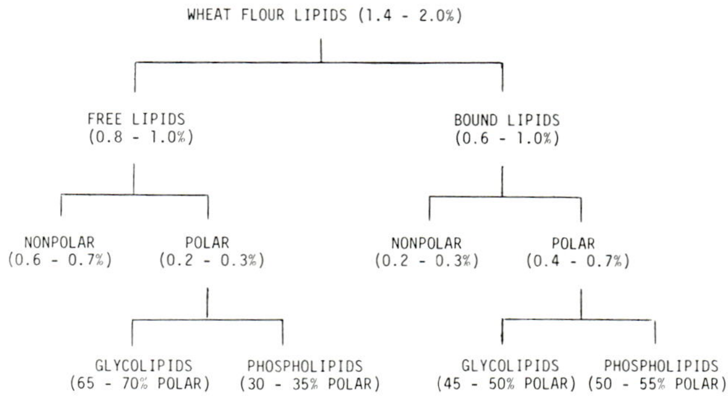

FIGURE 8. Classification and distribution of the main lipids in wheat flour. Percentages, unless stated otherwise, are based on total flour weight. Free lipids are extractable with petroleum ether; bound lipids require more polar solvents (i.e., alcohols). The terms "polar" and "nonpolar" relate to the solubility and reactivity of the lipids. (From Chung and Pomeranz, 1978).

alcohol-water mixtures, such as WSB, extract mainly non-starch lipids at room temperature. Starch lipids can only be extracted efficiently with hot N-butanol-water (65:35) or hot WSB. Starch lipids are almost exclusively monoacyl lipids, and 86 to 94% of the total starch lipids are lysophospholipids, with LPC as the chief constituent.

G. Lipids in Breadmaking

In tests to demonstrate the role of lipids in breadmaking, gluten was isolated from both untreated and defatted flours and then remixed with starch and water-soluble components in different proportions. Loaves were then baked from these doughs. At each gluten level, loaf volume was higher in bread baked from untreated (i.e. lipid-containing) than from defatted flour (Figure 9) (Chiu *et al.*, 1968). Studies with defatted flours have shown that nonpolar lipids are deleterious, and that polar lipids are beneficial in breadmaking (Figure 10)(from Daftary *et al.*, 1968). The deleterious effects of nonpolar lipids can be counteracted by polar lipids, and the effect on loaf volume has been shown to depend on the levels of nonpolar and polar lipids as well as on their ratio. The effects of lipids demonstrated that they are minor components that have major importance in breadmaking.

The significance of protein-lipid interactions was examined recently by Chung and Pomeranz (1978). Using 0.05N acetic acid, they determined the extractability of proteins from untreated flours of good and poor baking quality and from the same flours after removal of fat by acetone and isopropanol. Next, they dialyzed, lyophilized, and fractionated the protein extracts by gel filtration on a Sephadex G-100 column. The whole extracts and their fractions were characterized by starch and sodium dodecyl sulfate polyacrylamide gel electrophoresis. Both the non-defatted and the defatted poor (Chiefkan/Tenmarq) flours showed higher protein extractability than did the good flours (Shawnee). More proteins were extracted from the untreated good and poor control flours than from the corresponding defatted flours. In elution profiles from the gel column, all protein extracts showed similar retention times for the glutenin, gliadin, and water-soluble fractions. Quantitatively, the glutenins were the most variable among the gel-separated fractions in the acetic acid extracts from the flours. Whereas concentrations of gliadins were similar in all extracts, concentrations of glutenins depended on the

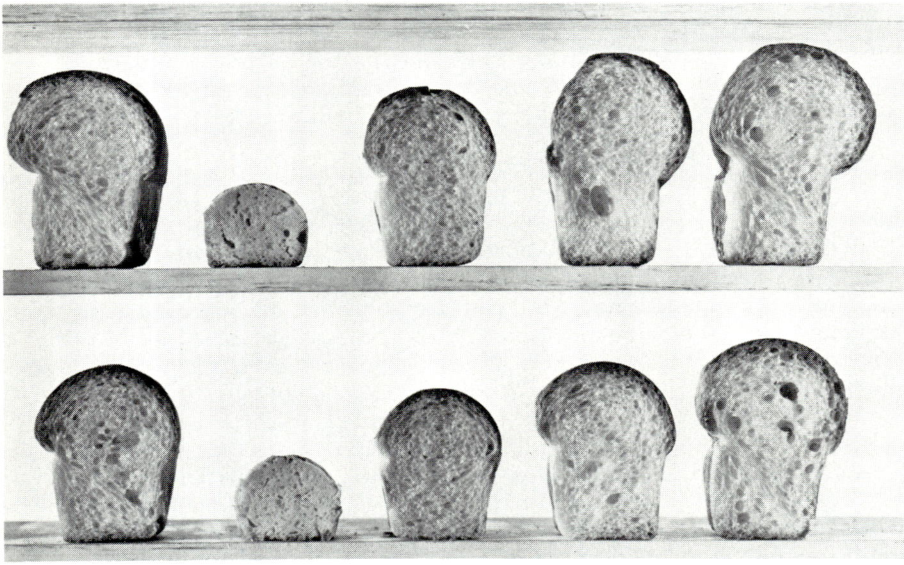

FIGURE 9. Microloaves (10 g flour) baked from untreated (top) and defatted (bottom) flours. From left to right: (1) Bread from original flour (13% protein); (2) bread made from flour minus gluten (a mixture of starch and water solubles); and (3-5) bread from a mixture of starch and water solubles plus 10%, 13%, and 16% gluten protein, respectively. (From Chiu *et al.*, 1968).

flour samples and the defatting solvent. Defatting decreased glutenin/gliadin ratios. The glutenin and gliadin fractions of strong and weak flours and of nondefatted and defatted flours differed in number and intensity of electrophoretic bands.

In a subsequent study, Chung and Pomeranz (1979) examined the binding of acid soluble proteins from untreated and defatted good (Shawnee, C.I. 14157) and poor (Chiefkan/Tenmarq, KS501097) baking quality flours to Phenyl-Sepharose CL-4B by batch- and column-elution techniques. Batchwise elution of the proteins from the hydrophobic gel with five solvents showed that total absorbance at 280 nm of the eluates was higher for the poor than for the good baking quality flour. Acid soluble proteins of defatted good and poor baking flours differed little from those of nondefatted flours in hydrophobic binding capacity. Glutenins from isopropanol-defatted

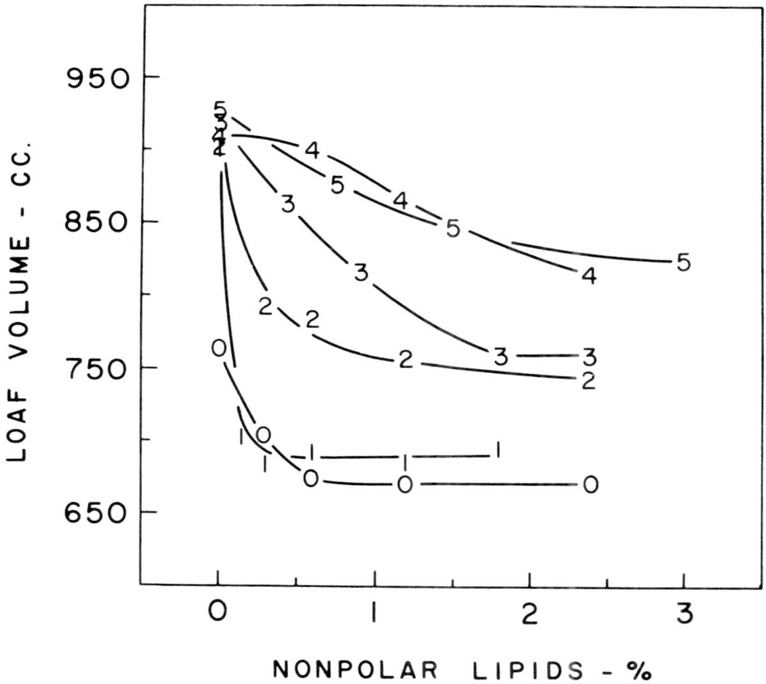

FIGURE 10. Effects of nonpolar and polar lipids on loaf volume of bread from petroleum ether-extracted flour baked without shortening and with various combinations of nonpolar and polar lipids (Numbers 1 to 5 denote 0.1 to 0.5 g. polar lipids per 100 g. flour) (from Daftary *et al.* 1968).

poor baking quality flour (KS501097) were less hydrophobic, and gliadins were more hydrophobic, than glutenins and gliadins from the good baking quality flour. The difference in apparent hydrophobic interaction was more pronounced for glutenin than for gliadin. The relationship between the amount of protein eluted and the amount of protein absorbed by the hydrophobic gel varied with source and concentration of protein. The results indicate that acid-soluble proteins from good and poor breadmaking flours differ in apparent hydrophobic properties.

H. Glycolipids

Olcott and Mecham postulated in 1947 that a gluten-lipid complex is formed during dough mixing or gluten preparation. However, the precise structure of that complex and the manner in which it contributes to breadmaking have been elucidated only recently as a result of intensive investigations of wheat flour glycolipids, which, as the name implies, are complexes of carbohydrates and lipids. Structural formulas of glycolipids are given in Figure 11. Because glycolipids combine the polar features of polyols (carbohydrates) with the lipophilic behavior of long aliphatic chains (fatty acids), some of them show considerable solubility in lipid solvents, but they also form aqueous solutions. The combination of polar and nonpolar properties suggests that glycolipids might be associated with lipid-aqueous interfaces (Law, 1960). Noncharged surface-active glycolipids apparently interact with structural proteins by hydrophobic bonding, and, together with other complex lipids, may be responsible for interactions between lipid micelles and proteins in chloroplasts (Benson, 1964).

Glycolipids are of great interest to the cereal chemist. Historically, certain types of glycolipids were first isolated from wheat flour by Carter and co-workers (1961). Practically, studies in the author's laboratory have shown that glycolipids, whether natural or synthetic, are essential to the production of acceptable quality bread.

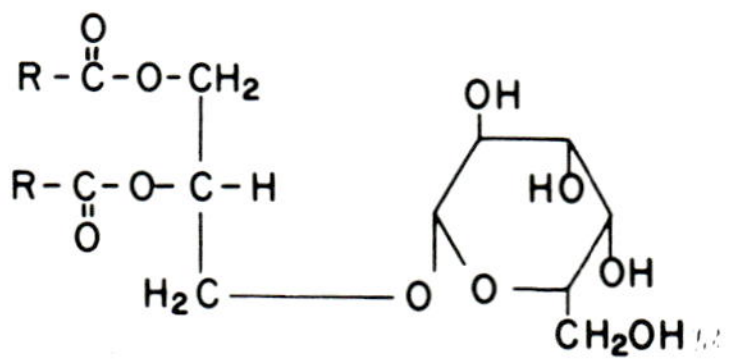

(I)

1,2-Di-O-acyl-3-β-D-galactopyranosyl--sn-glycerol
(= Monogalactosylglyceride)

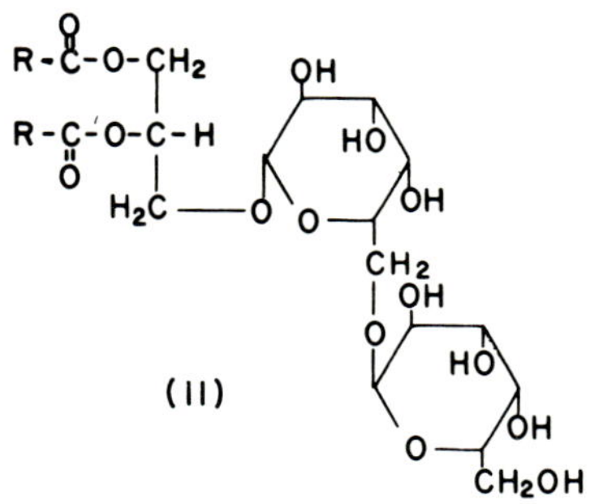

(II)

1,2-Di-O-acyl-3[O-α-D-galactopyranosyl-(1→6) β-D-galactopyranosyl]--sn-glycerol
(= Digalactosylglyceride)

FIGURE 11. Structural formulas of glycolipids.

I. Glycolipid Interaction

The interactions of glycolipids with wheat flour macromolecules and the complexes that glycolipids form with starch, gliadin, and glutenin were studied by infrared and nuclear magnetic resonance spectroscopy (Wehrli and Pomeranz, 1970a). Infrared spectroscopy indicated the presence of hydrogen bonds between glycolipids and gelatinized starch and between glycolipids and gluten components. NMR spectra showed hydrophobic bonding between glutenin and glycolipids. Those spectroscopic studies and subsequent investigations (Hoseney *et al.*, 1970) on differences in solubility and migration of gluten proteins in an ultracentrifuge have shown that, in gluten, glycolipids are bound to glutenin proteins by hydrophobic bonds and to gliadin proteins by hydrogen bonds. In unfractionated gluten, the glycolipid is apparently bound to both proteins at the same time and forms a complex that can be pictured as units of gliadin and glutenin proteins bound together by glycolipids.

Those spectroscopic, solubility, and centrifugation studies concerned the interaction at room temperature of glycolipids with isolated starch or gluten. In the dough, conditions are more complicated because of the competition for interaction by other components, chiefly carbohydrates and proteins, with glycolipids. Another complication arises during baking as a result of increased temperatures. Additional studies were therefore conducted to determine which interactions take place in dough and bread containing both starch and proteins. Tritium-labeled glycolipids were synthesized, and sections prepared from dough and bread containing the labeled glycolipid were studied by autoradiography (Wehrli and Pomeranz, 1970b). In the dough, most of the glycolipid was in the gluten; in the bread, most of it was in the oven-heated gelatinized starch granules. The findings on the macromolecular interaction between glycolipid and wheat flour components are summarized in Table III.

Basically, the important increase in loaf volume during breadmaking is controlled by the strength of interaction and of complex formation between glycolipids and gluten proteins, principally the gliadin fraction. To form the complex, adequate amounts of glycolipids are required. However, the strength of interaction apparently depends on the composition of gliadin proteins, which is genetically controlled. Recent studies indicate that gliadins from various flours differ consistently in their electrophoretic mobilities. Thus, the key to breadmaking strength seems to be in the gliadin fraction and the mode of its interaction with other flour components.

TABLE III

BONDS IN GLYCOLIPIDS AND WHEAT FLOUR MACROMOLECULE COMPLEXES

Method of study	Type of bond between glycolipid and: Starch	Gliadin	Glutenin
Solvent extraction of gluten proteins	-	Hydrogen	Hydrophobic
Lipid binding in starch dough	Hydrogen	-	-
Infrared	Hydrogen	Van der Waals, hydrogen	Van der Waals, hydrogen
NMR	Hydrogen, some induced dipole interaction	-	Hydrophobic and hydrogen
Autoradiography	Strong interaction in bread	-	Interaction in dough
Baking test	Hydrophobic and hydrogen bonds are essential for improvement in breadmaking.		

(from Wehrli and Pomeranz, 1969)

The occurrence of a gliadin-glycolipid-glutenin complex explains, at least in part, the importance of hydrophobic and hydrogen bonding in gluten structure. The complex is of special significance in view of the high sensitivity of hydrogen bonds to heat and the increased stability of hydrophobic bonds at elevated temperatures (gliadin is bound to glycolipids by hydrogen bonds). The complex also explains the relatively large effect of small amounts of glycolipids on the breadmaking performance of wheat flours (Wehrli and Pomeranz, 1969a).

IV. WHEAT FLOUR LIPIDS AND THE SHORTENING RESPONSE

Interaction between lipids and wheat flour macromolecules was followed in recent years in studies on the mechanism of the "shortening effect", i.e., the increase in loaf volume and improvement of crumb grain from the addition of 1 to 3%

shortening or hardened vegetable fat. The shortening effect was reviewed by Bell *et al.* (1977), who visualized two mechanisms of shortening effects: chemical and physical. The chemical effect would involve lipid oxidation; the mechanism was considered to be inoperative, or at least insignificant, in breadmaking. The following physical effects were reviewed: lubrication, sealing, foam formation, involvement of hydrogen and hydrophobic bonds, and delayed carbon dioxide release. Bell *et al.* (1977) showed that the rate of carbon dioxide release was faster in doughs baked without than in doughs baked with shortening. Fat in dough increased gas retention in the initial stage of rapid expansion. The final loaf volume depends on the permeability of the dough to carbon dioxide in the earliest stages of baking. They suggested that the difference in carbon dioxide release might explain the shortening response. They postulated that physical mechanisms account for the increased loaf volume due to shortening in the dough. Loaf volume increases when sufficient solid shortening components remain free in the dough. The free components are especially important in the initial stage of rapid dough expansion -- the so-called oven spring. The free shortening components presumably facilitate the production of oriented structures in dough. The structures persist even when the temperature exceeds the melting point of the fat, and seem to favor gas retention in the early stages of baking. The concept of delayed carbon dioxide release mechanism agrees with several well-established facts, such as the significance of free lipids and the critical oven spring. It seems, however, that attributing the differences in shortening effect among wheat flours to delayed carbon dioxide release alone is an oversimplification.

Wheat flour lipids and their role in breadmaking have been the subject of several comprehensive reviews (for a detailed list, see Chung and Pomeranz, 1977, and Chung *et. al.*, 1978). It is well established that in PE-defatted flours, nonpolar lipids are detrimental, and that polar lipids, especially glycolipids, are effective improvers. The roles of polar and nonpolar lipids are modified, however, by the addition of shortening and/or surfactants. MacRitchie and Gras (1973) emphasized that if baking formulations include shortening or other lipid additives, the net effect of natural flour lipids may be obscured. On the other hand, it is difficult to determine the effects of adding lipids to an untreated flour because of the presence of natural lipids. Recent studies in the author's laboratories indicate that interactions among wheat flour components are important in the shortening response (Pomeranz and Chung, 1978).

TABLE IV

CHARACTERIZATION OF FLOURS[a]

Flour	Flour Yield (%)	Protein (%)	Ash (%)	Free lipids (%) Total	Nonpolar	Polar
RBS-75	72.8	12.4	0.42	1.00	0.70	0.30
Shawnee	75.9	13.0	0.40	1.03	0.66	0.37
KS501097	73.4	13.4	0.43	0.93	0.69	0.24

[a]Flour yield, protein and ash contents are expressed on 14% moisture basis. Free lipid (Skelly B extract) contents were averages of two replicates and expressed on dry flour weight basis (from Pomeranz and Chung, 1978).

The results reported in this section were obtained for three experimentally milled flours (untreated or extracted with Skellysolve B). The flours are described in Table IV. First, a flour composite (RBS-75) made from many varieties grown at many locations in the Great Plains, U.S., was prepared. The flour had a medium-long mixing time, contained 12.4% protein, and was of satisfactory breadmaking quality. When the flour (10 g) was baked into bread with 3% shortening, water absorption was 68.3%, mixing time 4 7/8 min, and loaf volume 81.0 cc (Table V). The significance of the free flour lipids in the shortening effect is shown in Figure 12. In bread baked with 3% shortening (top row), extracting free flour lipids reduced loaf volume from 81.0 to 67.5 cc; in bread baked without shortening (bottom row), however, extraction of free lipids actually increased loaf volume from 64.5 to 71.0 cc.

The picture becomes even more complicated when we consider the shortening response and interaction effect in flours varying in breadmaking quality. Two additional flours used are described in Table IV. The flour milled from the 'Shawnee' wheat and the flour milled from the experimental KS501097 contained comparable amounts of proteins; Shawnee contained more free polar lipids than KS501097 flour. Defatting the flours increased their water absorptions; the increase was reduced somewhat by the addition of 3% shortening (Table V). The increase in water absorption was accompanied, generally, by an increase in mixing time.

TABLE V

BAKE ABSORPTION, MIXING CHARACTERISTICS, LOAF VOLUME, AND CRUMB GRAIN OF BREAD BAKED WITH OR WITHOUT 3% SHORTENING FROM THREE UNTREATED OR SKELLY B DEFATTED FLOURS (10 g)[a]

Parameter	RBS-75		Shawnee		KS501097	
	Untreat.	Defat.	Untreat.	Defat.	Untreat.	Defat.
Bake absorption (%)						
Without shortening	69.8	73.5	67.3	73.0	65.8	69.1
With 3% shortening	68.3	70.6	65.3	71.5	64.4	67.6
Mixing time (min)						
Without shortening	$4\frac{3}{4}$	$5\frac{3}{8}$	$4\frac{7}{8}$	$7\frac{5}{8}$	$\frac{7}{8}$	1.0
With 3% shortening	$4\frac{7}{8}$	$5\frac{1}{8}$	$5\frac{3}{8}$	$7\frac{1}{4}$	$\frac{7}{8}$	$\frac{7}{8}$
Loaf volume (cc)						
Without shortening	64.5	71.0	73.3	71.8	51.6	46.3
With 3% shortening	81.0	67.5	91.7	71.7	47.3	53.2
Crumb grain[b]						
Without shortening	U	O-U	O-U	Q	U	U
With 3% shortening	S	U	S	Q	U	U

[a] Averages of three replicates; overall standard deviation: 0.4% for bake absorption, 1/8 min for mixing time, and 1.3 cc for loaf volume.

[b] S = Satisfactory; Q = Questionable; U = Unsatisfactory (from Pomeranz and Chung, 1978).

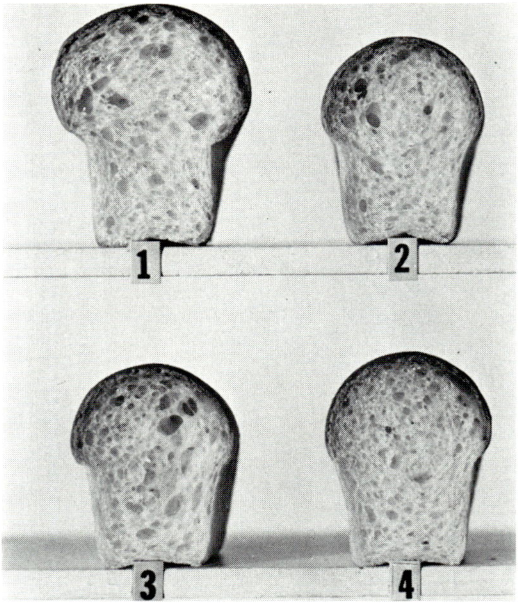

FIGURE 12. Loaves baked with 3% shortening (top row) or without shortening (bottom row) from untreated (1 and 3) and Skellysolve B extracted (2 and 4) composite flours (RBS-75) (from Pomeranz and Chung, 1978).

Shawnee flour had a long mixing time and produced excellent bread (in loaf volume and crumb grain); KS501097 produced inferior bread (Table V and Figure 13). The flours varied widely in shortening response, loaf volume, and crumb grain. Thus, the untreated two good breadmaking flours (RBS-75 and Shawnee) were improved by shortening; untreated KS501097 produced somewhat better bread when it was baked without, than with, shortening. For the defatted flours, however, the shortening response was reversed. Crumb grain was highly correlated with loaf volume, i.e., the higher the loaf volume the better the crumb grain.

Those studies show clearly that the shortening response is affected to a large extent by wheat flour quality, which is governed by the inherent wheat flour components (presumably gluten proteins), and is modified by the removal of wheat flour lipids. According to Wehrli (1969), since shortening, in the absence of glycolipids is detrimental to loaf volume, shortening probably interfers with the formation of a stable membrane between starch and proteins. Such a membrane,

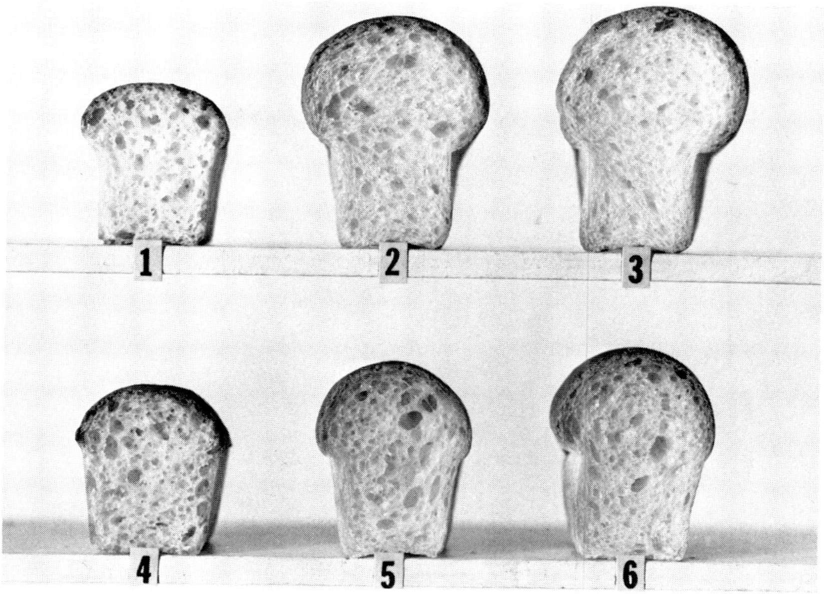

FIGURE 13. Loaves baked with 3% shortening (top row) or without shortening (bottom row) from KS501097 (1 and 4), RBS-75 (2 and 5), and Shawnee (3 and 6) flours (from Pomeranz and Chung, 1978).

presumably, can form only if the starch surface is covered with glycolipids. Probably, shortening is beneficial when the quantity and quality of gluten proteins and free lipids, especially polar glycolipids, in wheat flour are favorable.

Studies underway in our laboratories indicate that the shortening response is further affected by the nature of wheat flour lipids that are removed (Chung *et al.*, 1979). A composite hard red winter wheat flour was defatted at 4 to 75 °C by four solvents. The solvents extracted the following amounts of lipids: Skellysolve B, 0.89 to 1.05%; benzene, 0.96 to 1.12%; acetone, 1.02 to 1.13%; and 2-propanol, 0.96 to 1.32%. The increase in total extracted lipids was almost entirely from increase in polar lipids. The defatted flours contained little residual nonpolar lipids but various amounts of polar lipids. An increase in mixing time of the defatted flours depended mainly on the quantity and the types of lipids in the flour and on the specific effect of 2-propanol on flour components other than lipids. Loaf volume was affected by the quantity of flour lipids and shortening. Without shortening, removing up to about 1% total lipids

(0.3% polar) had little effect on loaf volume, but removing more than 1% total lipids increased loaf volume significantly: loaf volume increased 13% with removal of 1.32% total (0.64% polar) lipids. With 3% shortening, removing lipids had a deleterious effect on loaf volume: the most significant deleterious effect was shown at 1.13% total (experimentally, 0.46% polar) lipid removal. Shortening had a detrimental effect on loaf volume and crumb grain in the absence of native flour lipids, and severity of the detrimental effects was linearly related to the amount of polar lipids removed from the defatted flour. It was concluded that in flour containing no nonpolar lipids, loaf volume was governed by the quantity of polar lipids, shortening, and interaction of shortening and polar lipids.

V. PROTEIN-ENRICHED BREAD

Experiments with glycolipids synthesized in the author's laboratory (Wehrli and Pomeranz, 1969b) indicated that a certain hydrophilic-hydrophobic balance is required for best effects in breadmaking (Pomeranz and Wehrli, 1969). The significance of those findings with synthetic compounds was three-fold: they confirmed the role of natural flour glycolipids in breadmaking; they opened the possibility of synthesizing glycolipids superior to those found in flour; and they provided the basis for the production of protein-enriched bread through the addition of small amounts of glycolipids (Pomeranz *et al.*, 1969).

Wheat flour contains fairly large concentrations of proteins, but the proteins are not well balanced in amino acids and are particularly low in lysine. The quality of protein poses practically no problem if bread is part of a diet rich in protein foods of high biological value, such as milk, eggs, meat, fish, and legumes. In countries where high-protein foods are not generally available, however, wheat flour enriched with plant and animal proteins, including defatted oilseed flours, edible yeasts, and fish flour, might be used to improve nutrition in low-priced protein-enriched breads and other baked products.

Without the addition of extra glycolipids even relatively low levels, of about 5%, of lysine-rich protein supplements (i.e., soy flour) decrease loaf volume, impair crumb texture, and reduce shelf life. If small amounts of glycolipids are added, however, at least 10% soy flour, containing about 50% protein, can be included in the formula without impairing consumer acceptance of the bread. Soy flour costs about 12¢ per pound, and wheat flour, about 8¢. Soy protein contains about 6% lysine (compared to only 2% in wheat flour), so the addition of 10% soy flour easily doubles the amount of nutritionally limiting lysine that would be supplied by flour. The use of glycolipids in the production of acceptable, nutritionally improved bread is particularly promising because little or no changes would be required in breadmaking processes, formulations, schedules, or equipment. The processes for production of protein-enriched bread and for the synthesis of glycolipids have been granted public patents (Pomeranz and Finney, 1972; Pomeranz and Wehrli, 1973); and the patents are assigned to the U.S. Department of Agriculture, which is authorized to grant royalty-free licenses for their use.

Recently, the functional (breadmaking) properties of commercially-available sucrose esters were improved (Chung *et al.*, 1976). As much as 4% sucrose tallowate previously was required to restore breadmaking quality of high-protein breads (Pomeranz *et al.*, 1969); but now, acceptable high-protein breads can be made with as little as 0.5% sucrose esters. The effect on loaf volume of as little as 0.25 to 0.50% of the improved sucrose esters, added to wheat flour alone, is equivalent to the effects of 3% shortening. On the other hand, addition of unimproved sucrose tallowate, even to wheat flour alone, is deleterious to loaf volume and crumb grain. Effectiveness of sucrose esters is proportional to their monoester contents (Chung *et al.*, 1976). Their effectiveness can be further enhanced by removal of undesirable nonpolar and other components together with, presumably, di- and tri-ester fractions.

An "ideal" nutritionally improved high-protein bread should be low priced; require no special formulations, schedules, and equipment to produce; and be acceptable to the consumer with regard to overall eating qualities, including freshness retention. Although it is possible to produce today a high-protein bread that is superior to the one that was produced about 10 yr ago, it has not been possible to meet satisfactorily all the above requirements for our ideal bread. A recent study from our laboratories (Pomeranz *et al.*, 1977) showed that the use of flour from germinated soybeans (in the presence of small amounts of surfactants such as sucrose esters or sodium stearoyl lactylate) overcomes some of the difficulties. Scanning electron micrographs indicated differences in the size and shape of particles constituting soy milk and flours from chemically treated, high temperature treated, and germinated-heat treated soybeans. The last-mentioned treatment modified the particles the most. Of the three flours, nitrogen solubility index was highest for the flour milled from germinated soybeans. When bread was baked with each of the four soy products in a no-sugar formula, soy milk and high temperature treated soy flour produced unacceptable bread with regard to loaf volume and crumb grain. Bread baked from 90 g wheat flour and 10 g of flour from the chemically treated or germinated product, in the presence of 0.50 g sucrose palmitate or 0.50 g sodium stearoyl lactylate plus 3 g vegetable shortening, was consumer-acceptable with regard to loaf volume, crumb grain, crumb color, freshness retention, taste, and flavor. Wheat flour enriched with flour from germinated soybeans could well be the solution to the problems associated with producing low-priced, nutritionally improved, protein-enriched bread.

VI. SEEING IS BELIEVING

Recent studies from our laboratories indicate that microscopy provides a powerful tool and dimension in studies on dough structure. Specifically, it can help us to understand differences between flours that vary in breadmaking potential and to learn how to improve nutritional value (such as by the addition of protein-rich supplements) without adversely affecting consumer acceptance of bread.

Bechtel *et al.* (1978) used light and transmission electron microscopy to 1) elucidate microscopic changes associated with undermixed, optimally mixed, and overmixed flour-water doughs, 2) determine differences among doughs formed from poor- and good-quality flours, and 3) examine the ultrastructural changes that take place in complete bread dough during mixing, fermentation, proofing, and baking. They examined the starch structure of a water-flour dough, a fermented bread dough and baked bread. They found that protein strands provided a matrix network in a mixed dough, and that adequate mixing was required for matrix formation Excessive mixing, however, destroyed the matrix. The effects of overmixing and undermixing depended on mixing requirements and mixing tolerance of the flour. Fermentation of a complete dough produced gas vacuoles. After oven spring, protein strands were thin and had small vacuoles. Starch granules in the bottom center of a loaf varied widely in degree of gelatinization after oven spring; starch started to gelatinize from the interior of the granule and appeared fibrous. In the baked bread, most of the starch was gelatinized into fibrous strands interwoven with thin protein strands (Figure 14).

It was concluded that an optimally mixed dough formed from a good breadmaking composite flour results in even and continuous distribution of inclusions within the protein. The undermixed composite-flour dough does not have all these

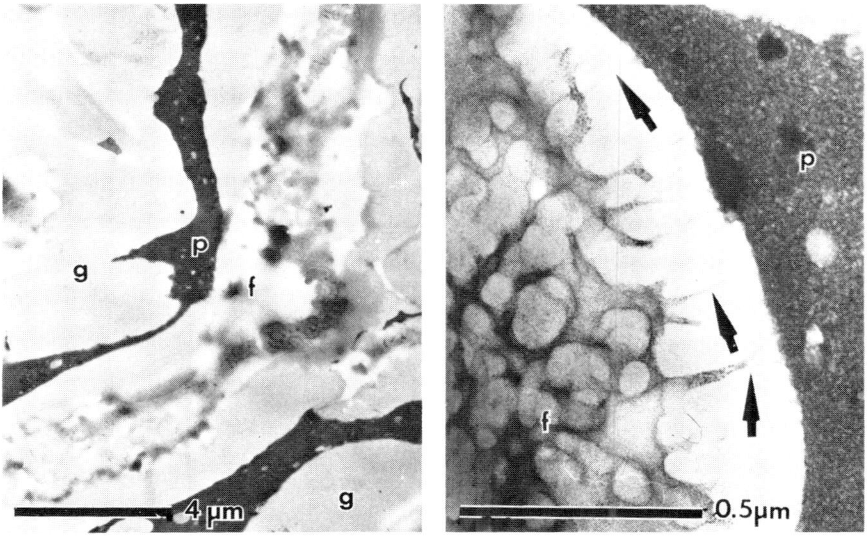

FIGURE 14. Left. Micrograph showing top center of loaf immediately after baking. Note gelatinized starch (f) between thin protein strands. Note lack of vacuoles in protein (p). Gas vacuoles (g). Right. Highly magnified protein (p)-gelatinized starch (f) interface. Note fine connections between starch and protein (arrows).

characteristics. In the overmixed dough, distribution of inclusions within the protein is even, but the protein is not continuous; moreover, protein strands are disrupted easily. Those results agree with the findings of Baker and Mize (1946), who measured (by changes in dough density) vacuole formation in doughs and effects of vacuole formation on bread quality. Doughs that are inadvertently overmixed can be relaxed and then remixed to form an optimally mixed dough (Bloksma, 1971). We found (Bechtel *et al.*, 1978) that when an overmixed dough was allowed to relax, the protein vacuoles decreased in size and number. Possibly relaxation and gentle remixing "mends" the broken protein strands so that an optimally mixed dough is restored.

Although both flour lipids and shortening contribute to the size and overall quality of the bread, the mechanisms of their contributions differ (Pomeranz, 1971). We could differentiate between wheat-flour lipids and shortening, but only the flour lipids were consistently associated with either protein or starch. It is possible that our fixation and dehydration techniques caused the extraction of some lipids and disrupted the association among bread dough components, or perhaps the interaction is of a type not visible by electron microscopy.

The poor-quality flour dough differed strikingly from the good-quality flour dough. The protein strands of the poor flour dough broke easily, even before the protein inclusions were uniformly distributed. In good-quality flour, the tensile strength of the hydrated proteins apparently governs the structure of the dough and the bread. That was quite apparent when a good-quality flour was grossly overmixed; the protein strands remained continuous, with relatively few being broken. This "stability" to overmixing (and even undermixing) is one of the most important and desirable characteristics of good-quality flours.

Our results point to the significance of starch and starch-protein interaction in the baked bread. They are in agreement with the findings of Yasunaga *et al.* (1968) on factors that govern gelatinization of starch during baking, of Derby *et al.* (1975) on the significance of limiting amounts of water in gelatinization of starch in baked or cooked products, and of Wehrli and Pomeranz (1970a,b) on the significance of starch-protein interactions in dough and bread.

REFERENCES

Baker, J. C., and Mize, M. D. (1946). *Cereal Chem. 23,* 57.

Bechtel, D. B., Pomeranz, Y., and deFrancisco, A. (1978). *Cereal Chem. 55,* 392.

Beckwith, A. C., Wall, J. S., and Dimler, R. J. (1963). *Arch. Biochem. Biophys. 103,* 319.

Belderok, B. (1966). *Inst. Cereal Flour Bread (TNO) Netherlands News 21,* 279.

Bell, B. M., Daniels, D. G. H., and Fisher, N. (1977). *Food Chem. 2,* 57.

Benson, A. A. (1964). *Ann. Rev. Plant Physiol. 15,* 1.

Bloksma, A. H. (1971). In "Wheat Chemistry and Technology" (Y. Pomeranz, ed.), Am. Assoc. Cereal Chem., St. Paul, Minn.

Carter, H. E., Hendry, R. A., and Stanacev, N. Z. (1961). *J. Lipid Res. 2,* 223.

Chiu, Chien-Mei, Pomeranz, Y., Shogren, M. D., and Finney, K. F. (1968). *Food Tech. 22,* 1157.

Chung, K. H., and Pomeranz, Y. (1978). *Cereal Chem. 55,* 230.

Chung, K. H., and Pomeranz, Y. (1979). *Cereal Chem. in press.*

Chung, O. K., and Pomeranz, Y. (1977). *Baker's Dig. 5(15)*, 32.

Chung, O. K., Pomeranz, Y., and Finney, K. F. (1978). *Cereal Chem. 55*, 598.

Chung, O. K., Pomeranz, Y., Goforth, D. R., Shogren, M. D., and Finney, K. F. (1976). *Cereal Chem. 53*, 615.

Chung, O. K., Pomeranz, Y., Finney, K. F., and Shogren, M. D. (1977). *Cereal Foods World 22*, 481.

Daftary, R. D., Pomeranz, Y., Shogren, M. D., and Finney, K. F. (1968). *Food Technol. 22(3)*, 79.

Derby, R. I., Miller, B. S., Miller, B. F., and Trimbo, H. B. (1975). *Cereal Chem. 52*, 702.

Finney, K. F. (1943). *Cereal Chem. 20*, 381.

Finney, K. F. (1978). *In* "Cereals '78: Better Nutrition for the World's Millions" (Y. Pomeranz, ed.), Am. Assoc. Cereal Chem., St. Paul, Minn.

Finney, K. F. (1979). "Wheat Proteins: What They Do", Proc. Conf. *In press.*

Finney, K. F., and Barmore, M. A. (1948). *Cereal Chem. 25*, 291.

Finney, K. F., Jones, B. L., and Shogren, M. D. (1978). *Cereal Foods World 23*, 479.

Hoseney, R. C., Finney, K. F., and Pomeranz, Y. (1970). *Cereal Chem. 47*, 135.

Hoseney, R. C., Finney, K. F., Shogren, M. D., and Pomeranz, Y. (1969a). *Cereal Chem. 46*, 126.

Hoseney, R. C., Finney, K. F., Shogren, M. D., and Pomeranz, Y. (1969b). *Cereal Chem. 46*, 117.

Law, J. H. (1960). *Ann. Rev. Biochem. 29*, 131.

MacMurray, T. A., and Morrison, W. R. (1970). *J. Sci. Food Agr. 21*, 520.

MacRitchie, F., and Gras, P. W. (1973). *Cereal Chem. 50*, 292.

Morrison, W. R., Mann, D. L., Wong, S., and Coventry, A. M. (1975). *J. Sci. Food Agr. 26*, 507.

Olcott, H. S., and Mecham, D. K. (1947). *Cereal Chem. 24*, 207.

Pomeranz, Y. (1971). *In* "Wheat Chemistry and Technology" (Y. Pomeranz, ed.), Am. Assoc. Cereal Chem., St. Paul, Minn.

Pomeranz, Y., and Chung, O. K. (1978). *J. Am. Oil Chem. Soc. 55*, 285.

Pomeranz, Y., and Finney, K. F. (1972). U.S. Patent 3,679,433.

Pomeranz, Y., and Wehrli, H. P. (1969). *Food Tech. 23(9)*, 109.

Pomeranz, Y., and Wehrli, H. P. (1973). U.S. Patent 3,729,461.
Pomeranz, Y., Shogren, M. D., and Finney, K. F. (1969). *Cereal Chem. 46,* 503, 512.
Pomeranz, Y., Shogren, M. D., and Finney, K. F. (1977). *J. Food Sci. 42,* 824.
Pomeranz, Y., Shogren, M. D., Bolte, L. C., and Finney, K. F. (1976). *Baker's Dig. 50(1),* 35.
Shogren, M. D., and Finney, K. F. (1974). Abstr. 59th AACC Mtg., Montreal, Canada.
Vakar, A. B., Pumpyanski, A. Ya., and Semenova, L. V. (1965). *Appl. Biochem. Microbiol. 1,* 1.
Wehrli, H. P. (1969). "The Synthesis of Glycolipids and Their Role in Breadmaking", Ph.D. Dissertation, Kansas State University, Manhattan.
Wehrli, H. P., and Pomeranz, Y. (1969a). *Baker's Dig. 43(6),* 22.
Wehrli, H. P., and Pomeranz, Y. (1969b). *Chem. Phys. Lipids 3,* 357.
Wehrli, H. P., and Pomeranz, Y. (1970a). *Cereal Chem. 47,* 160.
Wehrli, H. P., and Pomeranz, Y. (1970b). *Cereal Chem. 47,* 221.
Yasunaga, T., Bushuk, W., and Irvine, G. N. (1968). *Cereal Chem. 45,* 269.

BREAD AND SOFT WHEATS: RECENT MILLING PROGRESS

John Wingfield

Department of Grain Science and Industry
Kansas State University
Manhattan, Kansas

INTRODUCTION

Progress is more often noted by a series of subtle advances than by one giant leap. In the past 100 years we have seen such major milling advances as the advent of rollermills, purifiers, stacked sieve sifters, pneumatic conveying and air classification. Recent progress has utilized innovative redesign and application of proven machinery, taking advantage of all the new technologies available to us today. Progressive endeavors have been in the areas of flow sheet application, machinery efficiencies, building techniques and mill control.

The modern flour mill is at once complex and highly efficient. Millions of individual wheat berries are processed each hour, wherein the outer hull is separated from the inner endosperm material, and this endosperm is then gradually reduced and purified to white flour. The efficiency of this separation is quite high and the system requires very little manpower, making it one of the highest ton per man-hour manufacturing processes known.

The demands made by the modern high speed bakery for increased flour uniformity have triggered many innovations that improve mill control, which in combination with uniform wheat supply, is the key to product uniformity. The bonus feature of improved mill control is improved milling results through

ISBN: 0-12-370960-1

higher extractions and reduced downtime. This can result in favorable reductions in conversion costs to combat the ever increasing costs of such things as capitalization, interest, energy, and manpower.

With these thoughts in mind, we will now take a look at what is new in the milling industry. The subject matter will be divided into three main areas:

I. Milling Equipment
II. Milling Techniques
III. Mill Construction

One should keep in mind that the overall business philosophy of each mill or milling group will determine to a great degree the type and arrangement of its milling equipment, its mill design and layout, and the manner in which the entire milling complex will be managed.

There is no such thing as a standard mill.

I. MILLING EQUIPMENT

A. Grain Preparation

Before wheat can be milled into flour it must be cleaned of all foreign materials, brought to its best milling condition by adding moisture and sometimes heat, and then allowed to rest. Additional cleaning after the rest period is common and a small amount of water is often added just prior to milling.

1. Cleaning. The most common alignment of cleaning equipment and their use is as follows:

a. Magnet. This removes all tramp iron.

b. Receiving separator. This multiple sieve machine removes material larger and smaller than the wheat being cleaned. Air aspiration is often used at this point to remove the light chaff and dust.

c. Destoners. An inclined deck, through which air is drawn, takes advantage of differences in density to make a wheat/stone separation. (Also see washers).

d. Oats and barley separators. Discs or cylinders with especially designed indentations or pockets will pull out the wheat leaving the foreign grains.

e. Seed separators. Discs or cylinders with especially designed indentations or pockets will pull out the seeds, leaving the wheat.

f. Reseed separators. Discs or cylinders with especially designed indentations or pockets will separate broken wheat from seeds. This allows for a rich cut to be made on the seed separators to insure full seed removal. The small or broken kernels removed with the seeds are then reclaimed.

g. Gravity tables. Seed and foreign grain removal can also be accomplished on machines similar in operation to stoners, i.e. an inclined perforated deck through which air is permitted to flow.

h. Scourers. These machines are used to remove crease dirt, brush hairs, and loose outer bran layers. The wheat is thrown into a cylindrical screen by beaters. In dry scourers, the scourerings are removed by aspiration. In wet scourers, small amounts of water assist in removal of crease dirt.

i. Washers. Large quantities of water are used to float the wheat, allowing the stones to settle out. A whizzer section acts as a wet scourer while at the same time throwing off excess moisture.

It should be noted that wet scourers and washers have lost favor due to problems with sanitation, efficient solids disposal and control of water absorption.

h. Infestation destroyers. Aspiration will remove external insects present in wheat. When the berry has been bored into by certain insects, the structurally weakened kernel is broken open by impact. Aspiration can then remove the insect. These machines are often used in series as well as being used on either tempered or dry wheat or both.

RECENT IMPROVEMENTS. Improved thruput quantities and better control of results are featured in almost all new cleaning machines. Notable in the innovations introduced in recent years are the following:

Receiving separators. Countersunk sieve openings, higher speeds, greater deck inclinations and improved sieve cleaning have all contributed to higher efficiencies for this key machine.

Aspirators. Aspirators may be part of a more complex cleaning machine, or may be used by themselves. Improved control of separation is being accomplished by using adjustable volume aspiration channels. This allows for highly selective separation by terminal velocity. The use of clear plastic facilitates the adjustment by allowing visual inspection of separation.

Closed circuit aspirators reuse the precleaned aspirator air to conserve energy.

Concentrators. This is a recent introduction to the wheat cleaning lineup. Working on the same principles as a middlings purifier, a combination of sieving, stratification and aspiration results in three separations. The normal wheat mix will separate as follows:

Approximately 75%	Heavy wheat plus stones.
Approximately 20%	Light wheat and broken wheat plus foreign grains and seeds.
Approximately 5%	Light chaff and foreign material.

The heavy wheat goes to destoners then direct to temper. The stone-free light wheat continues through a full cleaning cycle. The lightest fraction goes direct to screenings. The end result of this technique is a net reduction in the amount of cleaning equipment needed.

Gravity selectors. Combining the results obtained on destoners with those of a gravity table, these versatile machines make the three separations described above under concentrators while at the same time removing the stones as a fourth separation.

Destoners. Substantial improvements in the design of dry destoners has resulted in much better separation accuracies, higher capacities and less operator attendance.

Scourers. Gentle action is the key to improved scourer action. This reduces the production of broken kernels and minimizes damage to the bran coat.

2. Tempering/Conditioning. To gain maximum flour yields with a minimum of bran intrusion, the bran coat must be toughened for milling by the addition of water. Heat is optional, and is often applied to facilitate the entrance of the water into the berry while at the same time bringing the wheat to the optimum milling temperature. Gluten modification by heating is sometimes desired. A rest time after moisture

addition will vary in length of time with types of wheat, degree of heat application and method of milling. This rest time permits uniform, full penetration of the moisture to occur.

RECENT IMPROVEMENTS

Intensive Mixers. These machines combine high speed agitation with scientifically placed water atomizers to improve the water addition operation. One inclined model features backflow to further insure intimate water-to-wheat mixing. These mixers have steam jackets and direct steam jets as options.

The improved milling results claimed for this method of tempering include better flour ash, color and yields. Tests involving the measurement of individual wheat kernel moistures before and after tempering indicate that a lessening of the spread of kernel moistures from the average contributes to these results. Up to 7% water addition can be obtained with no evidence of free or unabsorbed moisture.

Moisture detection and control. Many new and sophisticated control systems have been developed for the automatic and continuous addition of temper moisture. For cold or warm conditioning, the system will consist of wheat flow control and measurement, continuous detection of either the dry or wetted wheat moisture content and a control loop signal to a controlled water meter. For hot conditioning, additional instrumentation is needed to measure heat rise and evaporative cooling effects and integrate these quantities into the overall moisture addition equation. Additional controls can be used to control either water flow, steam flow or cooling air flow.

Among the many innovative devices being applied to moisture control are Impact Scales to detect wheat flow. These devices are compact and simple in construction. Radioactivity is being used at safe low levels to detect both wheat flow rates and freshly wetted wheat moisture content. Electronic detectors of several types are giving continuous, accurate dry wheat moistures. Near Infrared Reflectance instruments are also being utilized for quick moisture detection.

Bin flow control. Much attention has been given to obtaining laminar flow in temper bins. It is recognized that uniform rest or lying time is essential to uniform milling results and that multiple hoppered bins are one of the most successful methods of obtaining first-in, first-out flow. Some mills are taking advantage of this flow control procedure to better utilize temper bin space by tempering a new mix directly on top of another mix with only a very short period of intermixing occuring.

B. Grain Processing

The basic flour mill flow sheet contains an allotment of rollermills for separating the bran from the endosperm. Another allotment of rollermills is used to gradually reduce the endosperm to flour. Sifters are used to separate, grade and purify the various milled products. Purifiers may be used to further grade the larger-than-flour size endosperm particles and make further bran/endosperm separations through the use of air.

Additional equipment may consist of pin mills to assist the rollermills, brush and beater type dusters for specialized bran/endosperm separation, auxiliary equipment for air purification and stabilization, additive addition, insect egg destroying and the necessary transport equipment.

RECENT IMPROVEMENTS

Rollermills. Many improvements have been made to rollermills that result in higher specific loadings, better control of feeding and grinding and reduced maintenance.

Load sensing devices will automatically engage and disengage the movable roll to protect the roll surface from damage under no-load conditions and to facilitate the starting and stopping process of large mills. On start-up, it is only necessary to bring wheat to the first separating (breaking) rolls and the rollermills down stream will engage automatically as soon as sufficient material is flowing. Shut-down works in reverse of this process.

A recently introduced rollermill features a totally enclosed gear box, air cylinder support of the rolls (no springs), direct acting micro-adjustment of the roll gap or grinding zone and a reduced angle positioning of the rolls towards the horizontal.

A recently introduced U.S. mill features the horizontal roll position favored by some millers.

Internal pneumatic pickup of roll stocks within the rollermill adds much versatility to the design and operation of flour mills.

Improvements in feeder design, higher speeds of the rolls, improved bearing and drive systems and micro-control of roll gaps have all contributed to higher specific loadings combined with more uniform grinding results.

Flake disrupters. These are specialized devices to eliminate flakes made on the rollermills and thus facilitate the subsequent sifting operation. New design disrupters, or flake busters, can be installed directly into the pneumatic conveying line with on-off operation optional without affecting the conveying rate of the pneumatic lift.

Pin mills. A new design pin mill allows control of grinding action during operation by easy removal of pins from the outside. The pins are reversed and reinserted to close the hole opening.

Sifters. High Intensity sifting is being utilized in several areas of the mill flow. By forcing the mixed size materials onto rotating and/or vibrating screens higher thruputs in a greatly reduced space can be achieved.

Air sifting. The use of controlled air with or without screens is gaining favor as an auxiliary method of particle size separation.

Stacked sieve sifters. Although the basic idea is quite old, new improvements in design, application and materials of construction have contributed to increased use of stacked sieve designs as opposed to the fixed tray plansifter. Greater flow versatility, combined with higher thruputs, are among the benefits obtained.

Dusters. Improved design dusters have led to increased use of these machines which do a dual job of separation and classification, thus supplementing both the rollermill and sifter.

Air Stabilization. Although this item is not strictly a milling machine, the application of controlled temperature and humidity air to the mill has been greatly improved and is considered essential to uniform milling. The modern pneumatic mill moves air weighing ten times more than the weight of the wheat being ground. It is therefore, essential that this air be controlled. Air stabilization can conserve energy by utilizing processing heat also. Air from the dust filters is blended with outside air, then optionally heated or cooled and optionally dried or humidified to arrive at a constant temperature and relative humidity before being forced back into the various milling levels. Air purity is also improved due to the air washing effect.

C. Ancillary Operations

A modern mill consists of many segments indirectly related to the milling or grain conversion operation. Grain must be received and stored. Power must be received or generated and distributed. Finished products must be binned for further processing, packaging or bulk delivery. Each of these and other areas are a study in themselves. Some highlights are given below.

RECENT IMPROVEMENTS

Grain receiving. High speed unloading and transfer of grain has spawned a number of innovative designs in pneumatic and mechanical transport devices. The receiving of grain by unit trainload - 100 or more hoppered cars per train - has lowered freight costs for many terminals.

Product storage and delivery. The use of aerated or mechanically vibrated bin bottoms has proven to be a reliable, high speed, space saving method of holding and discharging flour and millfeed.

Pressurized truck vans and rail hopper cars are an improvement that allows fast, clean movement of bulk flour.

Bulk flour barges are also being utilized on inland rivers.

Automation. All new mills are using the technique of electrical interlocking to allow full or semiautomatic start-up and shut-down of mill segments. This control is centralized into panels or boards which allow the operator to have manual override and visual awareness of the total process. Sensing devices for flow and storage control will automatically control surges in flow from upset conditions, warn the operator and shut down preceding sections if necessary.

Automatic process manipulation is still quite limited due to the complexity of the milling system. However, some mills are now using continuous or semi-continuous monitoring of flour color, protein and cellulose content to automatically control flour stream selection for various grades through the use of micro-computers and servo mechanisms in conjunction with Near Infrared Reflectance analyzers.

Continuous yield reporting is now accomplished by integrating product weights obtained from scales or load cell suspended storage bins. Feedback signals from these systems could control various processing functions by using existing computer hard ware.

II. MILLING TECHNIQUES

The manner in which the wheat is converted to flour and feed is documented into what is called a flow sheet. This flow sheet can indicate all of the special notations of machine specifics as well as in-process product descriptions and distributions. It is not uncommon to have over 100 separate and distinct products or stock passages in process simultaneously. This is necessary to produce the whitest flour possible while optimizing yields and controlling quality.

RECENT IMPROVEMENTS

Flowsheet simplification. The striving for uniformity of product through ease of process control has resulted in many mills simplifying their flow sheet by reducing the number of passages required.

Use of purifiers. In past years, many flour mills used as many as 15 or more purifiers in their flow. With improved performance purifiers available now, combined with newer milling techniques, the need for purifiers has been greatly reduced and some millers are satisfied with the results obtained with no purifiers at all. The average new mill will use five to seven purifiers. Mills making a farina product may require more.

Mill control. Electrical, mechanical and applied techniques are being innovatively used to improve mill control and milling uniformity. This will result in improved uniformity of flours produced and improved milling results through reduction of downtime.

Flour granulation. Some shifting to coarser granulation flours has occurred in recént years and will undoubtedly continue in that direction. Energy conservation and improved baking quality are the primary motivations. Coarse flour is also being produced for use in starch/gluten extraction processes.

Air classification of flours. Sometimes called protein shifting, this use of controlled particle size separation of finished flours allows many mills to tailor make flours from wheats not normally usable due to their natural protein level. Control of the protein level is accomplished by taking advantage of the differences in the aerodynamic properties of starch and protein.

Sizings protein diagram. By capitalizing on the natural protein gradient within the wheat berry, this flow sheet adaptation produces in-process flours having greater spreads in protein than those obtained in the more orthodox methods. These streams can then be combined to produce various grades of finished flours. Results similar to those obtained by air classification are claimed.

Dry milling. Some soft wheat mills are milling wheats considerably below the moisture levels considered normal. This results in better sifting and promotes higher mill thruputs. Machines are available to add moisture to the resultant flours for improved baking results.

Wet milling. Limited use of water suspended milling has been attempted. This approach has appeal in the production of starch and gluten due to the high starch yields and low starch damage obtained when compared to conventional milling.

III. MILL CONSTRUCITON

The methods, applications and materials used in mill construction are as many as there are mills. Some highlights of recent trends are reported below.

RECENT IMPROVEMENTS

Low profile mills. A trend towards building mills of only one, two, or two and one-half floors is evident in recent years. This approach has promoted the use of internal pick-up rollermills, which eliminate the need for a pick-up floor below the rolls. The objectives of this technique are to reduce building costs and improve operator observation of the mill equipment.

Prefabricated construction. The use of prestressed concrete panels and prefabricated metal panels has reduced building costs.

Windowless mills. For purposes of reducing building costs and improving process air control, the use of windowless mills has been on the upswing in recent years. The lack of windows simplify building techniques and reduce heat transmission from mill to atmosphere.

Upside down mills. A mill design called "Superior" places the rollermills above the sifters. This allows the use of a single supporting structure for both of these machines which can be independent of the building walls. This configuration can result in lowered building costs.

Sanitation. Much emphasis is now being given to the design of mill equipment, mill buildings and supporting structures to eliminate cracks, crevises and ledges where insect infestation can progress. Ease of cleaning equipment is also given consideration in mill design. New materials of construction utilize plastics, epoxies, stainless steel and other smooth, non-toxic substances. Internal vacuum cleaning systems help pick up casual dust. Good design and use of pneumatics and general exhaust systems combined with stable flow characteristics allow new mills to operate virtually dust free.

Occupational safety and health. The concern for improved working conditions has resulted in many design and application innovations. Guarding of all exposed moving machinery, protection against slips and falls, emergency lighting and first aid and lowered noise levels are among the many items the modern mill designer must include in the overall scheme.

Environmental control. The discharge of particulate matter into the atmosphere and the discharge of solids in waste water have received a great deal of attention in recent years. The use of high efficiency cloth-type air filters is highly desirable to meet air quality standards.

Mills using the wet milling technique must look to solids recovery through evaporation, filtration, centrifugation or bacteriological methods to meet water quality standards.

Explosions. The prevention of explosions and minimizing the damage from explosions has received emphasis in recent years. Magnets, fabrication of moving parts from non-sparking materials, motion detectors and alarms, heat detectors and alarms, dust concentration detectors and alarms, dust tight and/or explosion proof electrical controls and lighting, motor overload circuit breakers, and explosion venting of dust collectors, bins and buildings are among the many steps being taken for explosion prevention.

Generous suction on all equipment prevents the escape of explosion producing dust. Room pressurization prevents dust migration and is being used in elevator tunnels and motor control rooms.

Explosion detection and suppression systems are available.

SUMMARY

The advances in flour milling in recent years have been substantial. Greater reliability of results, higher machine loadings, reduced space requirements, improved control and reduced operator attendance are among the prime factors built into the latest equipment.

Considerations for economy, sanitation, accessibility and power usage will continue to dominate the overall mill design.

Mills of the future will undoubtedly continue to move towards a fully automated process control.

A better understanding of the many changes the wheat berry undergoes in the milling process and how these are affected by the multitude of variables present will lead to more efficient mills of the future.

CEREAL GERMS USED IN BAKERY PRODUCTS: CHEMISTRY AND NUTRITION[1]

Cho C. Tsen

Department of Grain Science and Industry
Kansas State University
Manhattan, Kansas

I. INTRODUCTION

Cereal germs, long known for their excellent nutritive value, are high in protein, minerals, oil, and vitamins, as shown in Table I. The protein efficiency ratios (PER) for wheat germ range from 2.16 to 2.86 (1-4) and for corn germ, from 2.04-2.56 (2-6). Recently Miladi et al. (7) determined the relative nutritive value of mill fractions of wheat germ to be 80%, which compares favorably with 74% for casein. Canolty et al. (8), using the relative protein value (RPV) method, studied defatty corn-germ flour consisting of 20% protein and 2% fat on a dry basis. The RPV determined for casein and corn-germ flour were respectively, 0.71 and 0.62. All those data indicate that cereal germs contain protein that equals or approaches casein, a high-quality protein generally used as a reference.

Cereal germs can be used as valuable nutrient-fortifiers in both food and feed. World population increases the demand for protein of high nutritive value to be used in foods. As shown in Table II, the world's total wheat, corn, and rye germ productions in 1976 were expected to reach 12.4, 40.0, and 0.9 million tons, had all been dry milled (9). Obviously the huge grain crops, together with a substantial portion of

---

[1]Contribution No. 80-66-A, Department of Grain Science and Industry, Kansas Agricutural Expt. Station, Kansas State University, Manhattan, Kansas 66506

ISBN: 0-12-370960-1

TABLE I. Chemical Compositions of Defatted Corn and Wheat Germs, and Full-Fat Rye Germ

Germ sample	Moisture %	Protein (Nx6.25) %	Ash %	Fat %	Fiber %
Corn	5.9	23.6	10.0	1.0	3.3
Rye	15.0	34.6	4.3	9.7	2.7
Wheat	10.5	32.9	4.9	3.8	-

the grain kernel, represented by the germ (10), should supply large amounts of germs for food and feed every year.

Of all foods, bread, in various forms, is most popular. Its high consumption, wide acceptability, and low price make it an ideal staple to be fortified for various feeding programs and for general public use. Cookies and cakes are particularly popular snack items for children. All can be fortified to increase their nutritive values. Although wheat germ has been used in a variety of bakery foods (11, 12), corn, rye, and other cereal germs have not been extensively explored despite their great potential. Corn germ has been examined only recently as a nutrient fortifier for breads, cookies, muffins, and patties (Tsen et al., 13-15; Gardner et al., 16, 17; and Blessin et al., 18-20). Little information on the utilization of rye germ in bakery foods, however, is available.

TABLE II. World Wheat, Corn, and Rye Productions, and Estimated Germ Yields, 1976

	Production million tons	Germ	
		Yield %	Amount million tons
Wheat	413.4	3	12.4
Corn	336.3	12	40.0
Rye	29.4	3	0.9

II. EFFECT OF CEREAL GERMS ON RHEOLOGICAL PROPERTIES OF WHEAT FLOUR-WATER DOUGH

The rheological properties of dough change drastically when wheat flour is supplemented with cereal germs. Table III presents farinographic data on various wheat flour-water doughs containing rye and wheat germ flours[2]. Germ supplements increased absorption but reduced developing time and stability of wheat flour-water dough. Heated rye germ flour had less effect on dough stability than an unheated sample. Although different wheat flours were used to test the effects of rye and wheat germs in two studies, the data in the table indicate that wheat germ flour exerts less weakening than rye germ flour on wheat flour-water dough.

To alleviate the weakening effect of cereal germs, surfactants have been used to strengthen dough structure and baking performance of wheat flour fortified with cereal germ flours (14). As shown in Table IV, sodium stearoyl-2 lactylate (SSL), a surfactant, added at 0.5% substantially

TABLE III. Effects of Original or Heated and Rye-Germ Flour (RGF) and Heated Wheat-Germ Flour (WGF) on Indicated Farinographic properties

Dough	Absorp., %	D.T., min	Stability, min
Wheat flour	57.2	5.8	12.7
+ 6% Orig. RGF	59.0	4.8	5.8
+12% Orig. RGF	60.0	4.6	3.6
+18% Orig. RGF	60.4	4.3	3.6
+ 6% Heated RGF	57.8	5.5	7.8
+12% Heated RGF	58.6	4.9	6.8
+18% Heated RGF	59.2	5.0	6.8
Wheat flour	62.0	6	12.0
+12% Heated WGF	73.0	6	9.0
+24% Heated WGF	80.6	6	6.0
+36% Heated WGF	87.0	7	6.5

[2]Unpublished data

TABLE IV. Effects of Corn Germ Flour (CGF) and SSL on the Indicated Farinographic Properties of Wheat Flour (WF)-Water Dough

	No SSL			0.5% SSL		
	Absorp. %	D.T. min	Stab. min	Absorp. %	D.T. min	Stab. min
WF	67.2	8.5	12.5	67.2	18.3	30.0
88%WF+12%CGF	72.0	8.0	8.5	71.4	8.5	16.0
82%WF+18%CGF	74.2	5.8	7.0	73.4	8.8	9.5
76%WF+24%CGF	76.0	6.3	5.0	74.8	8.3	7.0
70%WF+30%CGF	77.4	6.5	3.8	76.2	9.0	5.5

increased developing time and stability of doughs prepared from wheat flour supplemented with 12, 18, 24, or 30% corn germ flour.

III. BAKERY FOODS FROM WHEAT FLOUR SUPPLEMENTED WITH CEREAL GERMS

The strengthening effect of a surfactant, such as SSL, on dough is also reflected by improvement in baking performance. As shown in Table V, the specific volume and grain score of

TABLE V. Baking Quality of Wheat Flour Fortified with Indicated Levels of Corn-Germ Flour

Corn-germ flour %	No SSL		0.5% SSL	
	Sp. vol. cc/g	Grain score	Sp. Vol. cc/g	Grain score
0	6.65	9	7.17	10
12	4.16	4	6.69	8
20	3.19	3	5.11	6
28	2.76	2	3.65	4

corn-germ breads lowered steadily as corn-germ flour increased. When 0.5% SSL was added, however, acceptable bread could be prepared from wheat flour fortified with up to 12% corn germ flour (Tsen et al; 13). Not only SSL, but also other surfactants such as ethoxylated mono-glycerides, sucrose monopalmitate and tallowate, each improved the baking quality of corn-germ bread (13).

In the U.S. specific loaf volume of marketable bread (1-lb. loaf) should be at least 6.00 with acceptable appearance, crumb texture, and grain. We regard a grain score below 5 on a 1-to-10 scale for finished bread as unsatisfactory. With such standards, acceptable breads were prepared from wheat flour fortified with 12% of either wheat-germ flour or rye-germ flour[3].

Cereal germs affect the quality of cakes, mainly cake crust color, when used instead of nonfat dry milk. As shown in Table VI, the crust color of white layer cakes prepared according to the AACC Method 10-90 (21) darkened substantially when cereal-germ supplements were used. The darkening was more pronounced with wheat germ than corn germ.

The effect of germs on cookies is similar to that on bread. Spread and top grain of sugar cookies gradually diminish as corn-germ flour fortification increases. Germs such as corn germ are particularly suitable for deposit on

TABLE VI. Average Weight, Volume, and Crust Color of (NFDM) Layer Cakes with Nonfat Dry Milk or Cereal Germ

Protein Fortifier	Wt. g	Vol. cc	Specific vol. cc/g	Agtron[A] unit
NFDM (Nonfat dry milk)	390.5	887.5	2.27	70.8
Wheat germ	394.5	887.5	2.25	19.5
Corn germ	393.5	910.0	2.31	36.5

[A]Agtron multicomatic abridged reflectance spectrophotometer model M-300A with monochromatic spectral line (blue: 436nm) was used. It was standardized with standardized with standard discs M68 and 24 to read 100 and 0, respectively.

[3]Unpublished data

wire-cut cookies and may be incorporated as high as 48% (on wheat flour basis) without detrimental quality effects, Tsen and Weber (15) reported. Acceptable germ-fortified foods including breakfast cereals, muffins, snack foods, and meat extenders were developed by Gardner et al. (16, 17) and Blessin et al. (18-20).

IV. NUTRITIVE VALUE OF CEREAL GERMS AND GERM-FORTIFED BREADS

A. Amino Acid Contents and Chemical Scores of Germ Proteins

Amino acid contents and chemical scores of corn, rye, and wheat germs are listed in Table VII. The quality of germ protein can be evaluated by its essential amino acid contents. Our data on corn germ agree with those of Garcia et al. (22), Neilsen et al. (23), Wall et al. (6), and Tsen et al. (13) and the wheat germ data, with those of Pomeranz (12). Chemical scores in Table VII indicate that the essential amino acid pattern of cereal germs are lower in isoleucine, methionine, and cysteine than recommended by the World Health Organization (24). However, the germs contain adequate amounts of lysine so they can be advantageously used as fortifiers to improve the nutritive value of wheat flour in bakery foods.

B. Protein Efficiency Ratio (PER) of Cereal Germs and Germ Breads

The excellent quality of germ proteins is demonstrated by their high PER values[4]:

Flour	PER
Corn germ	2.64
	2.55
Wheat germ	2.62
Wheat	0.93
Soy	2.32
Casein	2.50

The PERs of corn and wheat germs range from 2.55 to 2.64 slightly higher than that of casein, and considerably higher

[4]data

TABLE VII. Amino Acid Contents and Chemical Scores of Corn, Rye, and Wheat Germ

Amino acid	WHO Req. Pattern content mg/g	Corn germ content (Chem. Score) mg/g	Rye germ content (Chem. Score) mg/g	Wheat germ content (Chem. Score) mg/g
Lysine	55	56(102)	74(135)	63(115)
Threonine	40	42(105)	47(118)	44(110)
Methionine and Cysteine	35	18 7 (71)	20 3 (66)	19 4 (66)
Leucine	70	73(104)	69(99)	66(94)
Isoleucine	40	21(53)	26(65)	26(65)
Valine	50	38(76)	43(86)	39(78)
Phenylalamine and Tyrosine	60	39 28 (112)	38 28 (110)	40 26 (110)
Tryptophan	10	—	—	—

than 0.93 and 2.32, respectively for wheat and soy flours. Obviously the nutritive quality of germ proteins is excellent.

Our results from feeding rats to ascertain the PER of breads fortified with cereal germs (presented in Table VIII) show that 12% wheat-germ bread and 12% corn-germ bread have PERs equal to each other statistically and significantly more nutritional than white bread. A linear relationship exists between PER and germ supplementation; PER increases significantly with increasing germ supplements in bread. The digestibility of germ breads is slightly higher than that of white bread.

V. SUMMARY

Cereal germs are readily available in large quantities from the dry milling industry. They are not only rich in oil and vitamins, but also are excellent sources of proteins and minerals. They are suitably used as nutrient fortifiers in bakery foods. However, the fortification can adversely affect dough properties and baking quality. But modifying the bread making process and adding a surfactant such as SSL or a combination of surfactants such as SSL and EMG effectively improve the quality of germ-fortified bread and cookies.

Cereal germs are low in isoleucine, methionine, and cysteine but contain adequate lysine, which makes the germs attractive for fortifying cereal products.

Wheat and corn germs have PER values equivalent to that of casein, but higher than soy and wheat flours. The germ-supplementation significantly improves the PER value of bread.

TABLE VIII. Protein Efficiency Ratios and Digestibilities of Germ-Fortified Bread--Diets Fed Rats 28 Days

Bread	PER (Adj.)	Digestibility, %
Wheat (white)	0.90	82
Corn germ, 12%	1.16	88
24%	1.43	86
36%	1.73	85
Wheat germ, 12%	1.21	86
24%	1.63	87
36%	1.84	84

REFERENCES

1. Hove, E. L., and Harrel, C. G., Cereal Chem. 42:141 (1943).
2. Hove, E. L., Carpenter, L. E., and Harrel, C. G., Cereal Chem. 22:287 (1945).
3. Jones, D. B. and Widness, K. D., J. Nutr. 31:675 (1946).
4. Beeson, W. M., Lehrer, W. P., and Woods, E., J. Nutr. 34:587 (1947).
5. Reussner, G., and Thiessen, R., J. Nutr. 62:575 (1957).
6. Wall, J. S., James, C., and Cavins, J. F., Cereal Chem. 48:546 (1971).
7. Miladi, S., Hegsted, D. M., Saunders, R. M., and Kohler, G. O., Cereal Chem. 49:119 (1972).
8. Canolty, N. L., Schoenborne, B. M. Gregg, V. A., and Haring, R. V., J. Food Sci. 42:269 (1977).
9. U.S.D.A., Agricultural Statistics 1978.
10. Matz, S. A. The Chemistry and Technology of Cereals as Food and Feed 194 (1959).
11. Pomeranz, Y., Carvajal, M. J., Shogren, M. D., Hoseney, R. C., and Finney, K. F., Cereal Chem. 47:429 (1970).
12. Pomeranz, Y., Baker's Dig. 44:30 (1970).
13. Tsen, C. C., Mojibian, C. N., and Inglett, G. E., Cereal Chem. 51:262 (1974).
14. Tsen, C. C., Baker's Dig. 49:42 (1975).
15. Tsen, C. C., and Weber, J., Food Product Development 11:27 (1977).
16. Gardner, H. W., and Inglett, G. E., J. Food Sci. 36:645 (1971).
17. Gardner, H. W., Inglett, G. E., Deatherage, W. L., Kwolek, W. F., and Anderson, R. A., J. Food Sci. 36:640 (1971).
18. Blessin, C. W., Garcia, W. J., Deatherage, W. L., Cavins, J. F., and Inglett, G. E., J. Food Sci. 38:602 (1973).
19. Blessin, C. W., Garcia, W. J., Deatherage, W. L. and Inglett, G. E. Cereal Sci. Today 19:224 (1974).
20. Blessin, C. W., Inglett, G. E., Garcia, W. J. and Deatherage, W. L. Food Product Development 6:34 (1972).
21. American Association of Cereal Chemists. Cereal Laboratory Methods. The Association St. Paul, MN (1962).
22. Garcia, W. J., Blessin, C. W., and Inglett, G. E., Cereal Chem. 49:158 (1972).
23. Nielsen, H. C., Inglett, G. E., Wall, J. S., and Donaldson, G. L. Cereal Chem. 50:435 (1973).
24. World Health Organization. Tech Rep. Series 522 (1973).

RECENT RESEARCH ON MAJOR MAIZE PROTEINS : ZEINS AND GLUTELINS

Jacques Mossé
Jacques Landry

Laboratoire d'Etude des Protéines
Département de Physiologie et Biochimie végétales
Centre INRA Versailles, France

I. INTRODUCTION

Several authors have reviewed the different kinds of problems concerning seed proteins with more or less attention devoted to maize ones. We just mention some of them.

From a biochemical point of view, a few papers treat of whole cereal proteins : Waldschmidt-Leitz and Metzner (1962), Wall (1964), Mossé (1968) and Inglett (1977). Other ones deal only with cereal prolamins (Mossé, 1966) or glutelins (Dimler, 1966) or even zein (Mossé, 1961). The most recent and completed one of Wall and Paulis (1978) concerns maize and sorghum proteins.

From structure and localization point of view, Inglett (1970) reviews our knowledge on maize grain. Ory (1972), Mascherpa (1975), Miège (1975b), Briarty (1978) and Pernollet (1978) discuss protein bodies.

From a physiological point of view, Miège (1975c) considers seed development and Miège (1975d), Ashton (1976) and Tomos and Laidman (1979) analyze germination and storage mobilization processes.

From a genetic point of view, Nelson (1969), Mertz (1972), Nelson and Burr (1973), Kaul (1973), Mossé (1978), Röbbelen (1979) and Thomson and Doll (1979) treat of plant breeding, particularly for better protein quantity and quality, while Mossé (1973) analyzes the nature and origin of plant protein heterogeneity and polymorphism. Such a polymorphism indeed enables the use of storage proteins as genetic markers in

ISBN: 0-12-370960-1

taxonomy and phylogeny topics as underlined by Miège (1975a) and Stegemann (1975).

The present paper deals with prolamins (zeins) and glutelins, the two major protein groups[1] of maize grain. They mainly belong to storage proteins and roughly correspond to 80 % or more of total seed nitrogen.

It will be divided into three parts. First, a short analysis of our present knowledge about these two protein groups : their large complexity explains many difficulties and discrepancies in their study. Then a review about localization and biosynthesis of their protein components. Endly a few results concerning heterogeneity and polymorphism of these storage proteins.

Except for the first part, we will take into account exclusively the results reported during the last three or four years and not yet mentioned by Wall and Paulis (1978) in their noteworthy survey. Indeed they give a list of references which is practically exhaustive up to 1975 or 1976.

II. GLUTELIN SUBGROUPS AND ZEIN : ANALOGIES AND DISSIMILARITIES

It can be expected that main zein components will be separated, identified and even sequenced (Bietz *et al.*, 1979) in a rather near future. Such is not the case for glutelins which constitute a very complex group, a part of which is often coextracted with zeins. This is one of the reasons which turn hard the comparisons of literature data and for which we wish to be helpful in giving a brief digest on the question.

Preliminarily, it seems important to have in mind two comments.

Firstly, extractibility and solubility do not correspond accurately to the same characteristics. Because of multiplicity and diversity of seed chemical compounds, it is possible to extract a pool of seed proteins with a solvent mixture in which they are insoluble and reciprocally.

Secondly, the seed proteins of one group are often able to be partially cosolubilized by those of another group, which results in partial contaminations.

As noticed by Wall and Paulis (1978), it is possible to distinguish in maize grain three glutelin subgroups called G_1, G_2 and G_3 by Landry and Moureaux (1970) who proposed a fractionating extraction procedure for their separation.

[1]The word "group" will be systematically used in order to call the Osborne proteic fractions of maize grains like albumins, globulins, prolamins and glutelins.

After extraction of salt-soluble proteins and zeins by the classical Osborne method, 2-mercaptoethanol is systematically added to solvents used for glutelin subgroup extraction (Table I). G_1 is extracted by an alcohol water mixture. The two other subgroups are then extracted by an alkaline buffer solution (pH 10) containing either some salt (0.5 M NaCl) for G_2, or some detergent (0.5 % in SDS) for G_3.

As evidenced by these authors and in more recent investi-

TABLE I. Main Characteristics of Zeins, G_1 and G_2 Glutelins

Proteins		Zeins	G_1	G_2
Extracting medium components	1)[a,b]	---- Water ----	---- Water ----	---- Water ----
	2)[a,b]		---- Reductant ----	---- Reductant ----
	3)[a]	---- Alcohol ----	---- Alcohol ----	Salt
	or[b]	---- Alcohol + Salt ----	---- Alcohol + Salt ----	---- Alcohol + Salt ----
Fractionation	3[b]	-	Water insoluble	Water soluble
Composition : typical richness		Gln[a,c]	Met[a,b]	His[a,b]
Main size (Number and *Kilodaltons*)		2 or 3: *23*[d] or *25*[e]; *21*[d] or *22*[e]; *22*[b]	2 to 4: *23*[d] or *26*[e]; *21*[d] or *23*[e]; *14*[b,d,e]; *10*[b,d]	1 *26*[e]; 3 *22*[b]
Electrophoretic mobility pH 3.5[b]		slowest bands	slowest + faster bands	faster bands

[a]Landry and Moureaux (1970)
[b]Paulis and Wall (1977b)
[c]Righetti *et al.* (1977)
[d]Gianazza *et al.* (1976)
[e]Misra and Mertz (1976)

gations, a few typical and specific characteristics differentiate the three subgroups between themselves and with zein (see Table I concerning zeins, G_1 and G_2):

1) Constancy of their level when the order of extraction is changed (G_2 can be extracted before dissolving zeins).

2) Constancy of their level also among total endosperm nitrogen (8 to 10 % for G_1 and G_2 and 15 % for G_3[1]) as well for ripened grain as during seed development (Landry and Moureaux, 1976).

3) Specificity of their aminoacid composition (Table II). G_1 subgroup also termed "zein 2" by Fonzo *et al.* (1977) and by Miflin and Shewry (1979) has a composition rather close to that of whole zeins but differs from them by a high methionine content (54 per 1 000 total residues). G_2 is typified by a high histidine amount (62 per 1 000). On another hand G_3 has a composition rather similar to that of salt-soluble proteins. This suggests that it could contain membrane proteins, all the more as detergent is necessary to its dissolution. Moreover the specificity of composition of the three subgroups, when isolated from endosperm, was again found to be the same among several mutant maizes by Misra *et al.* (1975), with a level which of course can be changed by mutant genes.

4) Occurence of a few component sizes evaluated by SDS PAGE in presence of reducing agent. Misra and Mertz (1976) and Gianazza *et al.* (1976) reported for zein the presence of two major sizes, of respectively 21,000 to 23,000 daltons and 23,000 to 25,000 daltons. For G_1 the first workers observed the presence of two sizes slightly different from those of zeins and one of 14,000 daltons, whereas the second found 13,500 and 9,600 daltons. The occurence of this light compound appears to depend on genetic factors. At last for G_2 glutelins prominent component(s) of 25,000 daltons, found by Misra and Mertz (1976) could correspond to the heaviest sizes detected in G_1.

5) Exhaustivity of the procedure. Only about 5 % of total nitrogen remains unextracted (R in Table II), so that the three subgroups G_1, G_2 and G_3 can be considered as representative of total glutelins, as confirmed by Miflin and Shewry (1977) in a comparative study between different laboratories.

By the use of a partially different procedure, Paulis and Wall (1977a and b) have fractionated zein and glutelin groups. After discarding salt-soluble proteins, zeins are extracted by means of an alcoholic solution with 0.5 % sodium acetate. With a reducing agent added to the same mixture, they obtain

[1]In fact, because it partially arises from salt-soluble proteins insolubilization, G_3 does not remain quite constant during ripening (Landry and Moureaux, 1976).

Table II. Amino Acid Composition of Protein Groups and Subgroups of Maize Grain[a]. (Number of Residues per 1 000 Recovered Residues[b])

Protein[c]	SSP	Z	G_1	G_2	G_3	R
Asx	84	50	40	21	80	66
Thr	47	30	32	46	50	61
Ser	66	67	64	52	64	56
Glx	128	211	191	198	122	111
Pro	48	97	115	195	68	128
Gly	104	19	41	78	92	114
Ala	98	141	128	69	106	86
Cys	(22)	(3)	-	(2)	-	(7)
Val	98	41	37	65	70	70
Met	(16)	(7)	54	8	22	(13)
Ile	35	36	28	22	40	31
Leu	67	190	159	92	95	82
Tyr	24	35	43	26	25	29
Phe	34	51	41	21	38	30
Lys	63	1	1	11	51	42
His	27	9	10	62	25	36
Arg	68	10	14	30	50	38

[a]From Landry and Moureaux (1970).

[b]Numbers in parentheses indicate approximative values because of absence of specific assays.

[c]SSP : salt-soluble proteins ; Z : zeins ; G_1, G_2, G_3 : glutelin subgroups ; R : non extracted proteins.

glutelins. These are fractionated by dialysis in water insoluble and water-soluble glutelins which are respectively rich in methionine and histidine and can be considered as rather near of G_1 and G_2.

In spite of recent progress on these glutelin subgroups, many questions remain unsolved. Can we consider G_1 as "zein-like glutelins", or are there only a few common compounds between zeins and G_1 ? Are G_1 and G_2 localized inside protein bodies or outside or both ? The answer can be expected within a few years.

III. PROTEIN BODIES (STRUCTURE, ORIGIN) AND STORAGE PROTEIN BIOSYNTHESIS

Everybody now agrees that zeins are mainly — perhaps exclusively — accumulated inside endosperm protein bodies, the diameter of which is around one or a few microns. Already evidenced in seeds for over a century, these organelles (sometimes called aleurone grains, proteoplasts, protein granules.. and so on) were recognized in maize seed by Mottier (1921), then by Duvick (1955 and 1961) who suggested that they contain zeins. Pernollet (1978) points out that these are not only encountered inside plant seeds but also among animal oocytes.

In maize, their ultrastructure was being studied during the last ten years (Wall and Paulis, 1978). It can be different according to cereal genus (Adams *et al.*, 1976), tissue nature : scutellum, aleurone layer, endosperm (Pernollet, 1978) and perhaps also according to physiological stage. Aleurone layer protein bodies show phytin inclusion called globoid but seem devoid of zeins, contrary to endosperm ones (Kyle and Styles, 1977). These last authors suggest that aleurone layer protein bodies have a vacuolar origin, while endosperm ones develop from protrusions of the rough endoplasmic reticulum : this last point seems now unquestionable (Larkins and Hurkman, 1978 ; Pernollet, 1978).

Larkins and Dalby (1975) and Viotti *et al.* (1975 and 1976) have probably been the first to recognize that, during seed development, the protein body membrane is surrounded by bound polyribosomes inside which zein polypeptide chain biosynthesis is carried out. By isolating and translating *in vitro* the membrane polyribosomes mRNAs, Larkins *et al.* (1976) obtain two proteins differing by their size and quite similar to major zein group proteins. They study the influence of opaque-2 gene upon a major zein component (Jones *et al.*, 1977a and b). Simultaneously Burr and Burr (1976) quite confirm these results by a slightly different way, *i.e.* by first isolating protein bodies and then dissociating polyribosomes from their surface by detergent treatment. Viotti *et al.* (1977) followed the same procedure.

In a noteworthy publication, Burr *et al.* (1978) describe the isolation of at least two mRNAs. They show that zeins behave just like secretory proteins : the *in vitro* translation of mRNAs coding for zeins produces polypeptide chains roughly 10 % heavier — that is to say longer — than the seed extracted ones. In other words, nascent zein polypeptides going out from membrane-free polyribosomes are zein precursors, one can call "prezeins". They have N-terminal sequence (10 to 20 aminoacid residues long) which enables them to cross the membrane and penetrate the vesicle which is turning into a protein body. A specific protease could then split this particular N-terminal sequence and give rise to native zein chains, which constitutes a first kind of post-translational modification. Another one is found to be a glycosylation of polypeptide chain. In conclusion, Burr (1979) applies to zein the "signal hypothesis" of Blobel and Dobberstein (1975) : the additional N-terminal sequence is considered to be a "signal peptide" the structure of which allows the association with and penetration through the protein body membrane. And he suggests that mutations which increase seed lysine level by a decrease of zein amount without altering the protein structure, as proposed by one of us (Mossé, 1966), could be explained by modifications of this kind of secretory protein synthetic machinery. Besides the teams in Purdue (Jones, Larkins and coworkers) and in Upton (Burr and coworkers), three other teams in Milano, Freiburg and Stillwater have studied zein *in vitro* synthesis.

Wienand and Feix (1978) corroborate the occurence of two mRNAs differing by their size and coding for zeins of molecular weight respectively 22,000 and 19,000. Moreover, as far as we know, they are the first to get immune-serums containing zein antibodies. This enables them a complementary checking of the *in vitro* synthetized zein polypeptide nature.

Melcher (1979) finds 5 different mRNAs and points out a third size of zein polypeptide (MW = 14,500). In the mean time, Soave *et al.* (1979) claim the existence of 20 to 25 different zeins, arising from a mRNA family, the sizes of which are rather close : their demonstration is performed by the use of cDNA (DNA copies). Melcher and Fraij (1979) also suggest the existence of multiple zein polypeptide chains, like Viotti *et al.* (1978).

All these very interesting works show a few discrepancies which should imply a thorough discussion. We will make very few comments.

Maize strains are often differing, being either hybrids or inbred lines which already constitute a more homogeneous material from a genetic point of view.

Extraction processes of authentic zeins, protein body preparation, polyribosome and mRNA isolation involve the use of either reductants or detergents or both, which facilitate some G_1 or G_2 glutelin coextraction besides zeins : this can result in some confusing interpretations.

The frequent use of the SDS PAGE[1] alone does not allow any distinction between different polypeptides of the same size. It's a matter of fact that Milano and Stillwater teams who

[1]SDS PAGE : Sodium dodecylsulfate polyacrylamide gel electrophoresis.

work with IEF[1] find many more constituents.

But two main conclusions can be drawn from these numerous works :

1) the occurence of prezein, that is to say zein precursors with a signal peptide. First assumed, then demonstrated by Burr *et al.* (1978), it was recently confirmed by the use of batracia oocytes (*Xenopus laevis*) for *in vitro* synthesis (Burr and Burr, 1979 and Hurkman *et al.*, 1979). The question of course is : what becomes the signal peptide ? Are there special amino peptidases to hydrolyze it or does it remain unaltered inside protein body ?

2) in agreement with the results of next chapter, the existence of multiple kinds of zein polypeptide chains. This already resulted of zein starch gel electrophoresis of Landry and Sallantin (1966) and Mossé (1966) but we did not insist on this interpretation thirteen years ago. It appears now to be confirmed by the use of IEF and cDNA techniques : contrary to the only post-translational modifications suggested by Burr *et al.* (1978) as only (or main) origin of heterogeneity, zeins are made of a population of between 10 to 30 distinct polypeptide chains the sizes of which belong to a very small number of classes. Soave *et al.* (1976 and 1979), Viotti *et al.* (1978) and Valentini *et al.* (1979) are presenting arguments in favour of structure gene polymorphism.

So that the question is : are there any chemical facts able to give a direct demonstration of such an heterogeneity ?

IV. ZEIN HETEROGENEITY AND POLYMORPHISM

A number of investigations have recently contributed to characterize zein samples obtained by more or less significantly differing procedures. Several techniques were used :

- electrophoresis : PAGE, SDS PAGE, SGUE[2], by Gianazza *et al.* (1976), Fonzo *et al.* (1977), Paulis and Wall (1977a and b), Righetti *et al.* (1977), Viotti *et al.* (1978), Bietz *et al.* (1979), Handa *et al.* (1979). Moreover, Miflin and Shewry (1979) used two-dimensional electrophoresis.
- aminoacid analysis by Gianazza *et al.* (1977), Paulis and Wall (1977a and b) and Righetti *et al.* (1977).
- primary structure determination of N-terminal segments by Bietz *et al.* (1979).
- ion exchange chromatography by Esen (1979), and Landry (1979a).

[1]IEF : isoelectrofocusing
[2]SGUE : starch gel urea electrophoresis

All these results are in agreement with the heterogeneity of zein preparations. But comparisons and discrepancies between data result in the idea it is first necessary to describe accurately appropriate conditions of extraction. One of us (Landry, 1979a and b) reported a procedure which allows to get purified zein practically devoid of glutelins.

The main steps can be summarized. Zein is extracted from a defatted grain meal with a 55 % (w/w) isopropanol water mixture free of salt and reducing agents. After washing the extract with petroleum ether, zein is salted out, chromatographied on Sephadex LH 20 and lyophilized from a solution of aqueous tertiary butyl alcohol.

On SDS PAGE (Fig. 1), the purified zein, isolated from an hybrid maize (INRA 260), resolved into several bands which are localized in two distinct regions : a region consisting of 3 or 4 slow moving bands with molecular weight around 45,000 and a region with 2 fast moving bands of 24,000 and 22,000 daltons. When the sample is pretreated with 2-mercaptoethanol, slow moving bands in 45,000 dalton region disappear and the relative intensity of the 24,000 dalton band increases.

On SGUE (Fig. 2) at pH 3.5 and with a low ionic strength (0.01), the purified zein gives rise to five typical bands, that are in order of decreasing mobility :

- 1 faintly visible band, the fast moving, designated as O2
- 2 major bands of close mobility (doublet), designated as 11 and 12
- 1 prominent band 21
- and 1 minor band 31 located at the interval ℓ from band 21 which was equal to the interval between the bands 21 and 11.

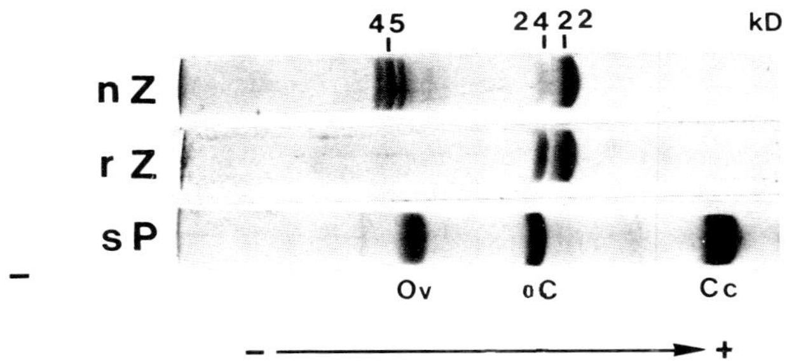

FIGURE 1. SDS PAGE of native (nZ) and reduced (rZ) zeins. sP, standard proteins : Ov, ovalbumin ; αC, α chymotrypsinogen ; Cc, cytochrome c.

In addition to these bands, several minor bands are also detectable. Moreover the reduction of the sample before electrophoresis does not result in increase of number or change in proportion of major bands (not shown), suggesting that the multiplicity of the band patterns is mainly bound to charge heterogeneity.

With respect to aminoacid composition, only the presence of minute amounts of lysine (one residue per thousand residues) and of 2 N-terminal residues (threonine and phenylalanine with predominance of the first one, Landry, 1979b) needs to be pointed out.

These results corroborate the data recently found by Bietz *et al.* (1979) and emphasize several types of heterogeneity in purified zein. Additional information was obtained by the determination of composition and physicochemical properties of components isolated after gel filtration and ion exchange chromatographies.

Gel filtration chromatography of crude as well as purified zein performed on Sephadex G 200 in 8 M urea at pH 3.5 yielded a 280 nm absorbance elution profile consisting of 2 major peaks D and M (Fig. 3). The characteristic properties of D and M fractions are the following :

1) their existence is independent of medium in which they were isolated : SDS or aqueous ethanol.

2) native D and M had aminoacid composition and N-terminal residues typical of those of unfractionated zein.

3) native D and M presented different molecular weights with approximate values of 45,000 and 22,000 respectively (Fig. 4).

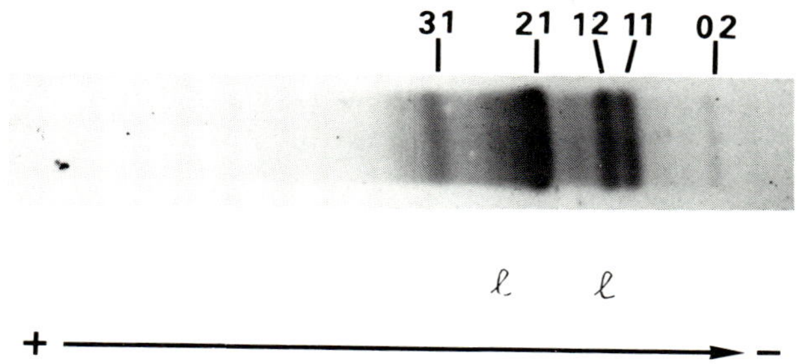

FIGURE 2. SGUE of native zeins (6M urea, pH 3.5). For ℓ see text.

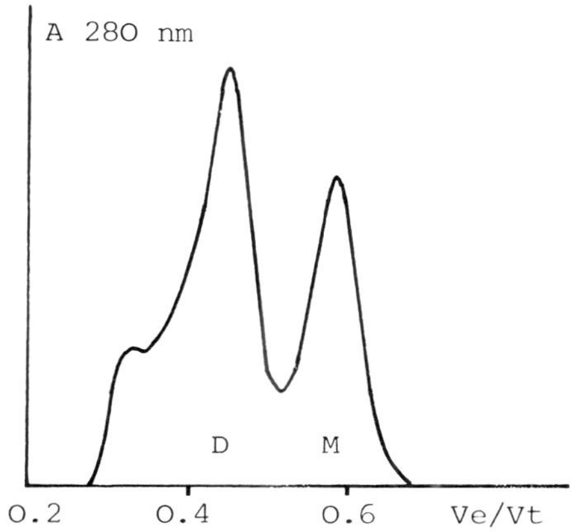

FIGURE 3. Gel filtration of native zeins on Sephadex G 200 equilibrated in 8 M urea and 0.01 N HCl. Ve : elution volume ; V_t : total volume of gel bed.

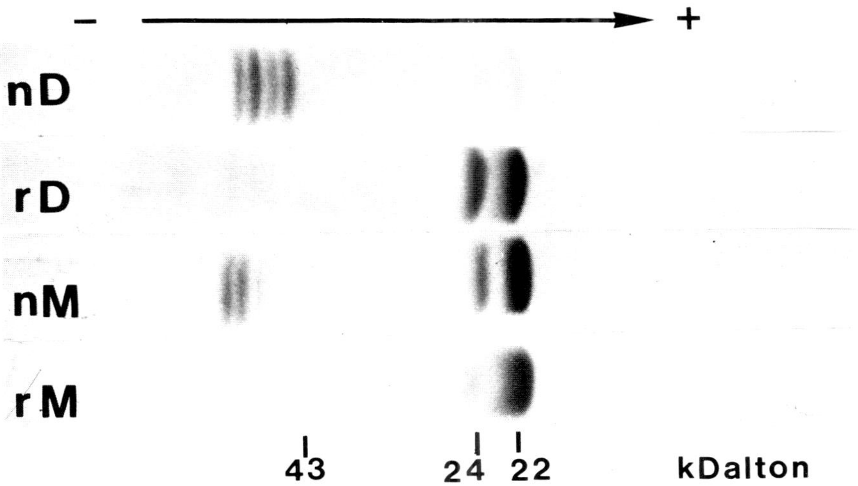

FIGURE 4. SDS polyacrylamide gel electrophoresis of D and M fractions. n, native ; r, reduced.

4) native D and M were also different in their electrophoretic behaviour on SGUE at pH 3.5. The pattern of D was a series of regularly spaced minor bands whereas that of M consisted of typical bands as detected with unfractionated zein (not shown).

5) reduced D was similar to native and reduced M with respect to molecular weight mean value which was about 22,000 and to SDS electrophoretic patterns (Fig. 4). Nevertheless on SDS PAGE reduced D differed from reduced M by a nearly equal intensity of 22,000 and 24,000 dalton components and by the presence of an additional fast moving band (Fig. 4). Likewise in SGUE, the existence of several minor bands with slower mobilities than that of major bands differentiates reduced D from reduced M.

These findings suggest the existence of the quaternary structure which may occur *in vivo* in crude or in purified zein : monomeric form (M fraction) or dimeric species, and homo- or hetero-dimers (D fraction). Moreover some subunits absent in M fraction are present in D in the associated state. Consequently since M appeared less complex than reduced D, we chose to focus attention on subunits of M instead of taking into account the subunits originating from reduced zein as most authors do.

Fraction M, chromatographied on sulfoethyl cellulose (SEC) as described by Huebner *et al.* (1966), was separated into 4 subfractions I, II, III and IV (Fig. 5). They all contained minute amounts of lysine (Table III).

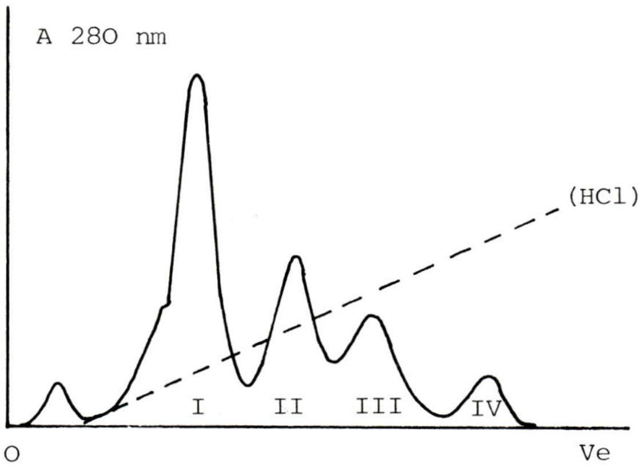

FIGURE 5. Chromatography of M fraction on sulfoethylcellulose (8 M urea ; 0.01 N HCl ; 0.03 N acetic acid, pH 3.5 with linear concentration gradient in HCl).

TABLE III. Amino Acid Composition of M (Zein Monomers) Subfractions Isolated by Sulfoethyl Cellulose Chromatography (Number of Residues per 1 000 Recovered Residues[a])

Fractions	I	II	III	IV
Asx	52	50	50	63
Thr	24	28	24	24
Ser	57	63	51	62
Glx	203	203	201	243
Pro	107	99	111	123
Gly	27	22	26	25
Ala	130	147	137	128
Val	33	34	29	40
Ile	42	42	47	35
Leu	193	192	198	141
Tyr	37	37	39	31
Phe	61	56	61	50
Lys	2	4,3	0,9	1,1
His	10	8	12	14
Arg	11	12,5	11	12

[a]Because of absence of specific assays, proportions of sulfur aminoacids are not mentioned.

The separation effectiveness was checked by SGUE at pH 3.5 (Fig. 6). Thus fractions I, II and III all migrated as a

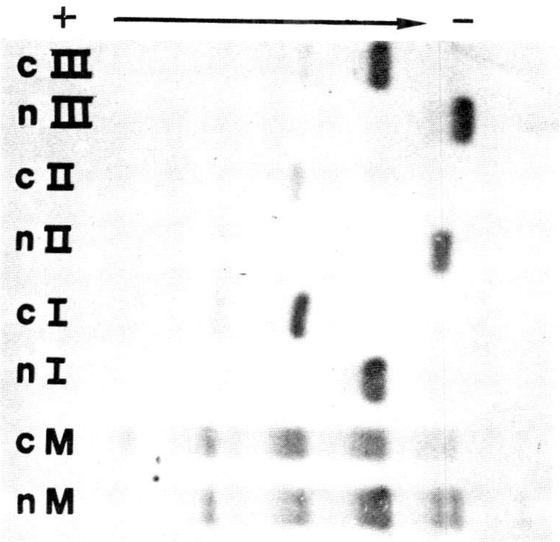

FIGURE 6. SGUE (6M urea, pH 3.5) of M and its SEC subfractions ; n : native ; c : carbamylated.

single band. But after a two days'incubation with potassium cyanate, fractions I and II both formed electrophoretic patterns with two new bands. With respect to native bands the new fast moving band had undergone an anodic shift equal to ℓ ; it corresponds to a subunit having lost one charge unit because of carbamylation of its N-terminal residue. The new slow moving band which had undergone a 2 ℓ shift toward the anod, is a subunit having lost two charge units because of carbamylation of N-terminal and lysyl residues. Fractions I and III are therefore heterogeneous : each one consists of at least two subunits with the same number of basic residues, one of them possessing one lysyl residue. Fraction II, when it was carbamylated, appeared to consist chiefly of a single new band, localized at 2 ℓ length from the native bands. The subunits of this fraction, if any, all contained some lysine.

On the other hand subunits of fractions I, II, III and IV contain respectively 4, 5, 5 and 6 basic residues per polypeptide chain.

The different behaviours of fractions I, II and III at pH 8.9 in SGUE and in PAGE support their heterogeneity, which comes from the variability in number and quality of basic residues. The results presented here provide critical evidence of discrete variations within the primary structure of zein subunits.

V. CONCLUSION

In the present paper we underlined the fact that zeins are deposited inside protein bodies in order to play a nutritional function with respect to the young seedling when it is still heterotrophic. This first feature of storage protein allows a comparison of zeins and possibly glutelins in maize seed with animal proteins like lactoproteins in mammals or ovalbumin in ovipara, proteins which all provide aminoacids to the young being.

Second feature : zeins appear as secretory proteins with signal peptides at the N-terminal end. Such is the case of lactoproteins, whichever they are (Mercier and Gaye, 1979) : caseins (Gaye *et al.*, 1977) ; α-lactalbumin (Mercier *et al.*, 1978b) ; β-lactoglobulin (Mercier *et al.*, 1978a).

The signal peptide is a kind of "passport" for the secreted protein, in order to cross the frontier represented by the endoplasmic reticulum membrane.

Any untimely cutting of the signal peptide could prevent the protein from crossing the membrane. This may occur anywhere, for instance in a seed, and hinder or diminish the accumulation of the protein(s) involved.

Another possibility is that no splitting occurs after passing through the membrane. Such is the case for ovalbumin (Gagnon *et al.*, 1978) which keeps its N-terminal signal peptide.

An intermediary event may correspond to a non systematic cutting, which could explain a two-sized family of polypeptides with practically the same aminoacid composition : it is a matter of fact that zein belongs to the proteins which are the richest in hydrophobic aminoacid residues. Now a recent comparison of some 20 different amino terminal extension sequences of secreted proteins (Austen, 1979) reveals the high level of these kinds of residues.

The third feature of likeness between prolamins and lactoproteins is their multiplicity, their polymorphism and their genetic variability. Lactoproteins are known to show a few main components by electrophoresis, with a most important number of minor components. The same thing is found among zeins and possibly will be found among glutelins. Zeins show a post-translational heterogeneity due both to a possible *in vivo* dimerization and to polypeptide modifications by losses of segments like signal peptides and by any *in vivo* alkylation (glycosylation, phosphorylation... and so on) : a few lactoproteins are known to be phosphorylated. But we think that the most simple polymorphism arises from a few point mutations which only result in discrete differences in primary structure and aminoacid composition : zeins partially consist of "isoproteins" (Mossé, 1973) corresponding to a structural gene polymorphism. In any case they appear as one of the best genetic markers.

ACKNOWLEDGMENTS

We wish to express our appreciation to Dr J.-C. Pernollet for helpful suggestions and for reviewing the manuscript. We also wish to thank Isabelle Pernollet and Danielle Vivier for their kind and efficient assistance during preparation of this text.

REFERENCES

Adams, C.A., Novellie, L., and Liebenberg, N.v.d.W. (1976). *Cereal Chem. 53,* 1.

Ashton, F. (1976). *Annu. Rev. Plant Physiol. 27,* 95.

Austen, B.M. (1979). *Febs Lett. 103,* 308.

Bietz, J.A., Paulis, J.W., and Wall, J.S. (1979). *Cereal Chem. (in press).*

Blobel, G., and Dobberstein, B. (1975). *J. Cell Biol. 67,* 835.

Briarty, L.G. (1978). *In* "Plant Proteins" (G. Norton, ed.), p. 81. Butter Worthes, London.

Burr, F.A. (1979). *In* "Seed Protein Improvement in Cereals and Grain Legumes" (Proc. internat. Symp. IAEA/FAO, Neuherberg, 1978), p. 159. IAEA, Vienna.

Burr, B., and Burr, F.A. (1976). *Proc. natl Acad. Sci. USA 73,* 515.

Burr, F.A., and Burr, B. (1979). *In* "Genome Organization and Expression in Plants" Abstracts (NATO/FAO, Edinburgh, 1979), p. 22.

Burr, B., Burr, F.A., Rubenstein, I., and Simon, M.N. (1978). *Proc. natl Acad. Sci. USA 75,* 696.

Dimler, R.J. (1966). *Fed. Proc. 25,* 1670.

Duvick, D.N. (1955). *Am. J. Bot. 42,* 717.

Duvick, D.N. (1961). *Cereal Chem. 38,* 374.

Esen, A. (1979). *Plant Physiol. suppl. n° 5,* 93.

Fonzo, N. di, Fornasari, E., Salamini, F., and Soave, C. (1977). *Maydica 22,* 77.

Gagnon, J., Palmiter, R.D., and Walsh, K.A. (1978). *J. biol. Chem. 253,* 7464.

Gaye, P., Gautron, J.P., Mercier, J.C. and Haze, G. (1977). *Biochem. Biophys. Res. Commun. 79,* 903.

Gianazza, E., Righetti, P.G., Pioli, F., Galante, E., and Soave, C. (1976). *Maydica 21,* 1.

Gianazza, E., Viglienghi, V., Righetti, P.G., Salamini, F., and Soave, C. (1977). *Phytochemistry 16,* 315.

Handa, A.K., Harmodson, M., Tsai, C.Y., and Larkins, B.A. (1979). *Plant Physiol. suppl. n° 5,* 94.

Huebner, F.R., and Wall, J.S. (1966). *Cereal Chem. 43,* 325.

Hurkman, W.J., Pedersen, K., Smith, L.D., and Larkins, B.A. (1979). *Plant Physiol. suppl. n° 5,* 94.

Inglett, G.E. (1970). *In* "Corn : culture, processing, products" (G.E. Inglett, ed.), p. 123. Avi publishing company, Westport.

Inglett, G.E. (1977). *In* "Food Proteins" (J.R. Whitaker, and S.R. Tannenbaum, ed.), p. 267. Avi Publishing Company, Westport.

Jones, R.A., Larkins, B.A., and Tsai, C.Y. (1977a). *Plant Physiol. 59,* 525.
Jones, R.A., Larkins, B.A., and Tsai, C.Y. (1977b). *Plant Physiol. 59,* 733.
Kaul, A.K. (1973). *In* "Nuclear Techniques for Seed Protein Improvement" (Proc. Meet. FAO/IAEA, Neuherberg, 1972), p. 1. IAEA, Vienna.
Kyle, D.J., and Styles, E.D. (1977). *Planta 137,* 185.
Landry, J. (1979a). *Thèse de Doctorat d'Etat, Paris.*
Landry, J. (1979b). *Biochimie 61,* 549.
Landry, J., and Moureaux, T. (1970). *Bull. Soc. Chim. biol. 52,* 1021.
Landry, J., and Moureaux, T. (1976). *Qual. Plant. 25,* 343.
Landry, J., and Sallantin, M. (1966). *C.R. Acad. Sci. 262,* 156.
Larkins, B.A., and Dalby, A. (1975). *Biochem. Biophys. Res. Commun. 66,* 1048.
Larkins, B.A., and Hurkman, W.J. (1978). *Plant Physiol. 62,* 256.
Larkins, B.A., Jones, R.A., and Tsai, C.Y. (1976). *Biochemistry 25,* 5506.
Mascherpa, J.M. (1975). *In* "Les Protéines des Graines" (J. Miège, ed.), p. 125. Georg, Genève.
Melcher, U. (1979). *Plant Physiol. 63,* 354.
Melcher, U., and Fraij, B. (1979). *Plant Physiol. suppl. n° 5,* 94.
Mercier, J.C., and Gaye, P. (1979). *Proc. natl Acad. Sci. USA (in press).*
Mercier, J.C., Haze, G., Gaye, P., and Hue, D. (1978a). *Biochem. Biophys. Res. Commun. 82,* 1236.
Mercier, J.C., Haze, G., Gaye, P., Petrissant, G., Hue, D., and Boisnard, M. (1978b). *Biochem. Biophys. Res. Commun. 85,* 662.
Mertz, E.T. (1972). *In* "Symposium : Seed Proteins" (G.E. Inglett, ed.), p. 136. AVI Publishing Company, Westport.
Miège, J. (1975a). *In* "Les Protéines des Graines" (J. Miège, ed.), p. 305. Georg, Genève.
Miège, M.N. (1975b). *In* "Les Protéines des Graines" (J. Miège, ed.), p. 31. Georg, Genève.
Miège, M.N. (1975c). *In* "Les Protéines des Graines" (J. Miège, ed.), p. 55. Georg, Genève.
Miège, M.N. (1975d). *In* "Les Protéines des Graines" (J. Miège, ed.), p. 101. Georg, Genève.
Miflin, B.J., and Shewry, P.R. (1977). *In* "Techniques for the Separation of Barley and Maize Proteins" (Workshop CEE, Harpenden, 1976) (B.J. Miflin and P.R. Shewry, ed.), p. 105. Commission of the European Communities, Luxembourg.

Miflin, B.J., and Shewry, P.R. (1979). *In* "Seed Protein Improvement in Cereals and Grain Legumes" (Proc. internat. Symp. IAEA/FAO, Neuherberg, 1978), p. 137. IAEA, Vienna.
Misra, P.S., and Mertz, E.T. (1976). *Cereal Chem. 53,* 705.
Misra, P.S., Mertz, E.T., and Glover, D.V. (1975). *In* "High-Quality Protein Maize" (Proc. CIMMYT - Purdue Symp., Dowden, 1975), p. 291. Hutchinson and Ross, Stroudsburg.
Mossé, J. (1961). *Ann. Physiol. vég. 3,* 105.
Mossé, J. (1966). *Fed. Proc. 25,* 1663.
Mossé, J. (1968). *In* "Progrès en Chimie Agricole et Alimentaire", p. 47. Hermann, Paris.
Mossé, J. (1973). *Physiol. vég. 11,* 361.
Mossé, J. (1978). *Cah. Nutr. Diet. 13,* 305.
Mottier, D. (1921). *Ann. Bot. 35,* 349.
Nelson, O.E. (1969). *Adv. Agron. 21,* 171.
Nelson, O.E. Jr, and Burr, B. (1973). *Annu. Rev. Plant Physiol. 24,* 493.
Ory, R.L. (1972). *In* "Symposium : Seed Proteins" (G.E. Inglett, ed.), p. 86. AVI Publishing Company, Westport.
Paulis, J.W., and Wall, J.S. (1977a). *J. Agric. Food Chem. 25,* 265.
Paulis, J.W., and Wall, J.S. (1977b). *Cereal Chem. 54,* 1223.
Pernollet, J.C. (1978). *Phytochemistry 17,* 1473.
Righetti, P.G., Gianazza, E., Viotti, A., and Soave, C. (1977). *Planta 136,* 115.
Röbbelen, G. (1979). *In* "Seed Protein Improvement in Cereals and Grain Legumes" (Proc. internat. Symp. IAEA/FAO, Neuherberg, 1978), p. 27. IAEA, Vienna.
Soave, C., Righetti, P.G., Lorenzoni, C., Gentinetta, E., and Salamini, F. (1976). *Maydica 21,* 61.
Soave, C., Viotti, A., Fonzo, N. di, and Salamini, F. (1979). *In* "Seed Protein Improvement in Cereals and Grain Legumes" (Proc. internat. Symp. IAEA/FAO, Neuherberg, 1978), p. 165. IAEA, Vienna.
Stegemann, H. (1975). *In* "The Chemistry and Biochemistry of Plant Proteins" (J.B. Harborne, ed.), p. 71. Academic Press. London, New York, San Francisco.
Thomson, J.A., and Doll, H. (1979). *In* "Seed Protein Improvement in Cereals and Grain Legumes" (Proc. internat. Symp. IAEA/FAO, Neuherberg, 1978), p. 109. IAEA, Vienna.
Tomos, A.D., and Laidman, D.L. (1979). *In* "Recent Advances in the Biochemistry of Cereals" (Annu. Proc. Phytochem. Soc. Europe n° 16) (D.L. Laidman and R.G. Wyn Jones, ed.), p. 119. Academic Press, London, New York, San Francisco.
Valentini, G., Soave, C., and Ottaviano, E. (1979). *Heredity 42,* 33.
Viotti, A., Sala, E., Alberi, P., and Soave, C. (1975). *Maydica 20,* 111.

Viotti, A., Soave, C., Gianazza, E., and Righetti, P.G. (1976). *In* "Colloques internationaux CNRS : Acides Nucléiques et Synthèse des Protéines chez les Végétaux" (Strasbourg, 1976), p. 511. CNRS, Paris.
Viotti, A., Balducci, C., Sala, E., and Soave, C. (1977). *In* "Techniques for the Separation of Barley and Maize Seed Protein" (Workshop CEE, Rothamsted, 1976), p. 81. Commission of the European Communities, Luxembourg.
Viotti, A., Sala, E., Alberi, P., and Soave, C. (1978). *Plant Sci. Lett. 13,* 365.
Wall, J.S. (1964). *In* "Symposium on Foods : Proteins and their Reactions" (Oregon State University) (H.W. Schultz and A.F. Anglemier, ed.), p. 315. AVI Publishing Company, Westport.
Wall, J.S., and Paulis, J.W. (1978). *In* "Advances in Cereal Science and Technology" (Y. Pomeranz, ed.), p. 135. American Association of Cereal Chemists, St Paul.
Waldschmidt-Leitz, E., and Metzner, P. (1962). *Z. Physiol. Chem. 329,* 52.
Wienand, U., and Feix, G. (1978). *Eur. J. Biochem. 92,* 605.

MAIZE PROTEIN CONCENTRATE

M. Sternberg [1]
R. D. Phillips
L. H. Daley

Biochemistry Department
IPG Research
Miles Laboratories, Inc.
Elkhart, Indiana

Maize, Indian corn or in plain American, corn, is a valuable source of oil, starch and protein. Its origin in the new world is not undisputed. Some sources show that it was introduced to Europe in the 13th century by Arabs who brought it from Asia (1). Archeological findings in the new world trace the origin of maize to Mexico about 7,000 years ago; subsequently it played a major role in the rise of Meso-American and Andean civilizations (2). Today, maize bred into high yield varieties is the largest single crop in the United States and an important one worldwide (Table 1).

Most maize is directly used in foods and feeds yet in the U.S. 6% is processed by industrial plants into starch, oil and syrups. This significant amount may grow further on account of steady replacement of sucrose by high fructose syrups prepared from glucose and ultimately starch. The industrial processing of maize aims to separate in one form or another the major constituents of the kernel, that is oil, starch, bran, and protein. While the proteins end up in animal feeds, some fractions are rich in sulfur amino acids suggesting the possibility of use for human nutrition. In the typical wet milling process, the protein distributes into several products of which the largest and most concentrated is known as Gluten Meal (Figure 1). This protein from the maize endosperm, can also be obtained by a dual enzymatic saccharification of dry milled maize flour. A typical composition of gluten meal is given in Table II.

---

[1] Present address: The University of Georgia College of Agriculture, Department of Food Science.

ISBN: 0-12-370960-1

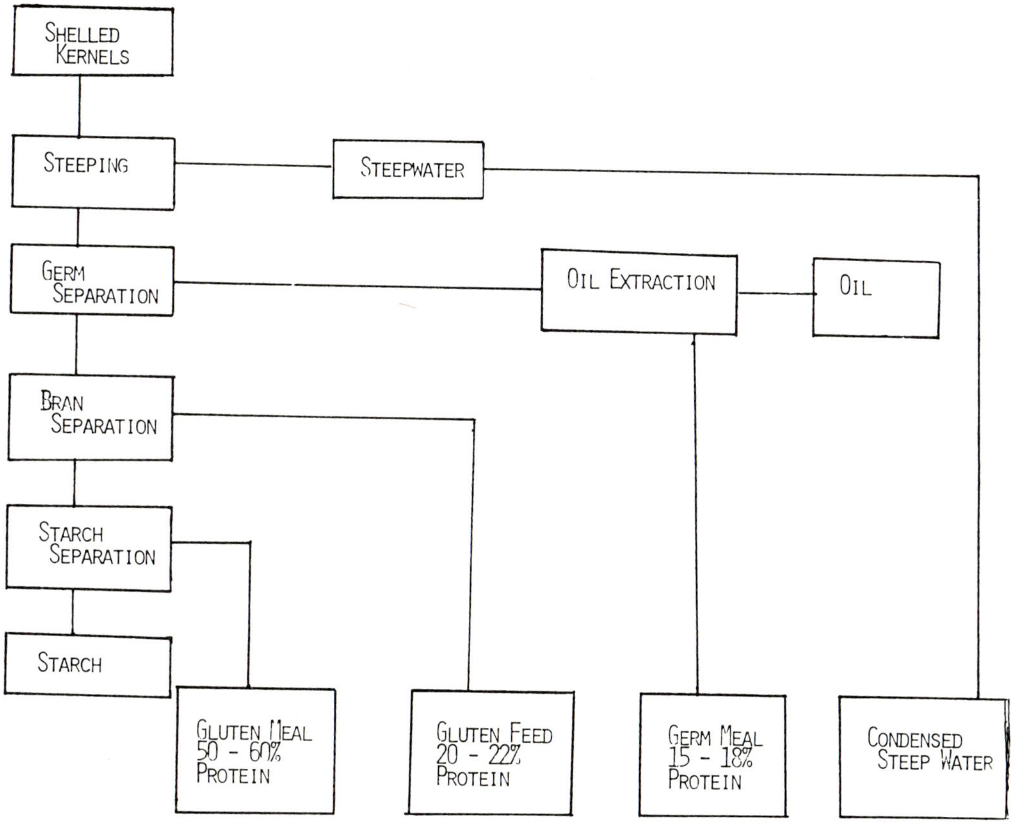

FIGURE 1

Schematic Flow of Maize Wet Milling

TABLE 1. Maize Crop in Major Producing Countries (in thousands of metric tons)

COUNTRY	1971	1972	1973	1974	1975	1976	1977	1978
Argentina	9,930	5,860	9,000	9,900	7,700	5,855	8,800	8,300
Brazil	14,130	14,891	14,109	16,284	16,354	17,885	18,800	17,500
France	7,581	8,840	8,177	10,671	8,699	8,194	5,368	8,140
South Africa	8,600	9,483	4,160	11,105	9,140	7,312	9,592	8,500
Thailand	1,938	2,300	1,315	2,350	2,450	3,000	2,700	1,750
U.S.	105,472	143,422	141,734	144,043	119,421	147,252	157,895	161,730
World Total	262,768	299,483	292,955	318,057	287,376	321,813	336,502	339,236

Source: Corn Annual 1978

The wet gluten is a bright yellow filter cake having approximately 70% moisture. Of the 15-18% fat, 20-30% is in the form of free fatty acids. The feed gluten has a strong odor and flavor and a tendency to develop rancidity due to the high content of unsaturated fatty acids.

To upgrade the gluten meal for human use it is necessary to extract the oil, pigments and typical flavor. Leaving any of these would either reduce the potential applications or shorten the shelf life. One other factor which has to be considered is the high moisture which precludes typical hydrocarbon extraction unless prior drying has taken place. The process which we developed (3) in order to prepare a maize protein concentrate (MPC) is shown schematically in Figure 2. It consists in the extraction at 70°C of wet gluten with ethyl acetate-ethanol-water azeotropic mixture or with water saturated ethyl acetate. After solvent separation, a hot water wash removes traces of solvent, oligosaccharides and soluble nitrogen.

The MPC is a bland, light colored material having approximately 70% protein and under 0.6% fat (Table 3). The carbohydrate fraction is essentially all cellulose and hemicellulose. There are no oligosaccharides present. (4)

The usefulness of any protein ingredient is determined by its functional and nutritional properties. MPC contains protein that is low in solubility through most of the pH range. Figure 3 shows the nitrogen solubility profile for MPC as well as the gluten meal and maize flour from which it was ultimately derived. It can be seen that the already low solubility of the native protein is further reduced by processing to a concentrate.

Water binding capacity, compared to that of soya flour and soya concentrate is given in Table 4. The values for MPC expressed as the ratio of bound water to dry solids remain fairly constant while those of the soya flour and soya concentrate vary considerably. This is probably due to decreasing solubility of the soya proteins with heat. The denatured proteins then bind water rather than dissolve. Oil binding of the dry material at room temperature was also determined in comparison to soya concentrate under the same conditions (Table 5). The maize protein concentrate bound a significantly greater amount of oil than did the soya concentrate. This was not due to a difference in bulk density as has been reported for other proteins, as both materials had the same density.

The essential amino acids profile of MPC is compared to those of soya concentrate in Table VI. The complimentary levels of lysine and sulfur amino acids are obvious. An optimal mixture of the soya and maize concentrates is a w/w

Fig. 2

Schematic Extraction of Gluten Meal

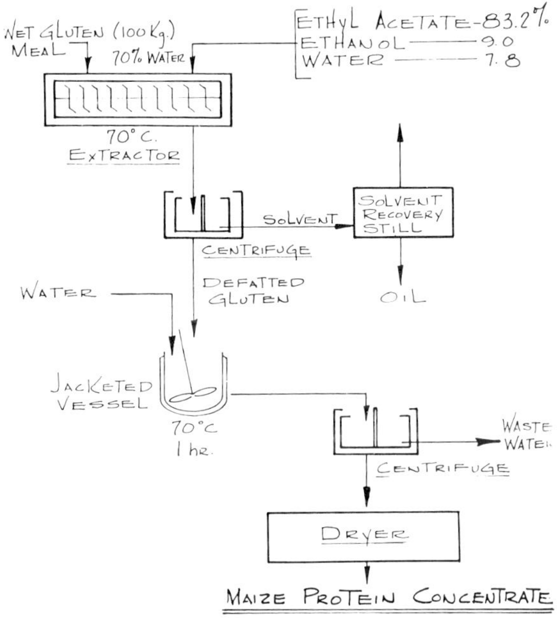

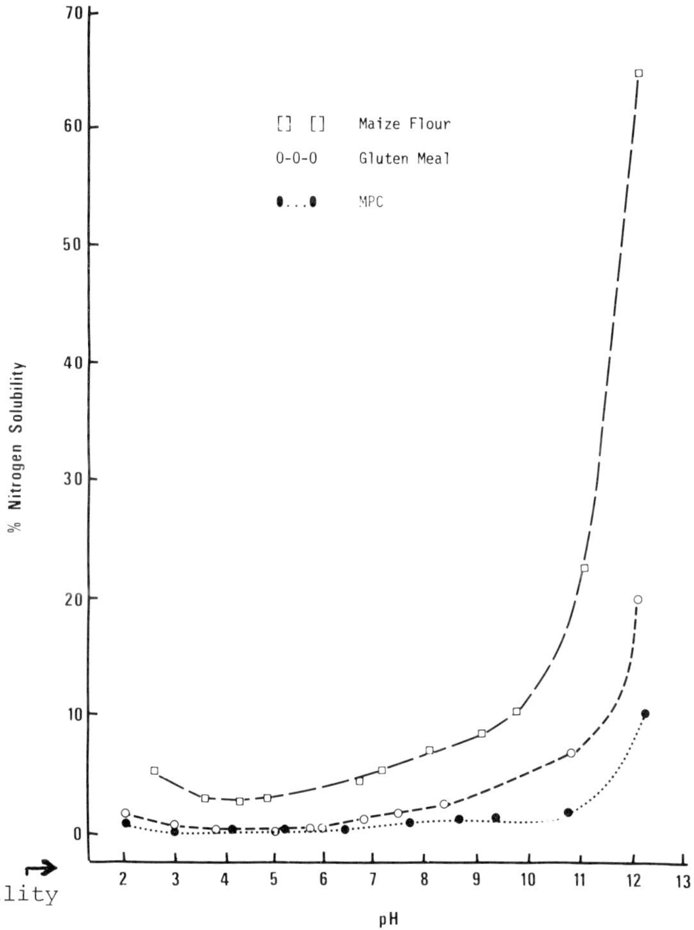

Fig. 3. Nitrogen Solubility

TABLE 2. Composition of Gluten Meal

Water	50-70%
Protein	50-56%
Fat	15-18%
Carbohydrates	20-25%
Fiber	3.5%
Ash	1-2%

TABLE 3. Proximate Analysis of MPC*

	% Composition
Protein	67 - 72
Fat	<0.6
Carbohydrate	18 - 27
Fiber	∿ 5
Ash	∿ 3
Moisture	(5-7)

TABLE 4. Water Binding By MPC Soya Flour and Soya Concentrate

Sample	Ratio of Bound Water to Dry Solids (w/w) Under Various Incubation Conditions		
	Room Temp.	82°C	121°C
MPC	2.2	2.3	2.2
Soya Flour	0.7	1.5	2.4
Soya Concentrate	1.9	2.3	2.8

* Dry Substance Basis

TABLE 5. Oil Binding by MPC and Soya Concentrate

Sample	Ratio of Bound Oil to Solids (5% H_2O)
MPC	1.2
Soya Concentrate	1.0

TABLE 6. Essential Amino Acids of MPC and a Typical Soya Concentrate

Essential Amino Acid	Content (G/16 G Nitrogen)	
	Maize Protein Concentrate	Soya Concentrate
Lysine	1.7	6.3
Methionine + Cystine	5.4	2.8
Threonine	3.6	4.3
Leucine	15.4	7.9
Isoleucine	3.6	4.6
Phenylalanine + Tyrosine	11.1	8.9
Valine	4.7	4.8
Tryptophan	0.7	1.5

ratio of 80/20 respectively. At this ratio both lysine and methionine plus cystine content are optimized within 10% of the provisional FAO recommended amino acids profile (Table 7).

The nutritional effect of combining MPC with soya materials was determined as the PER of rats. Table 8 shows the results. In each case, the mixtures were significantly superior to the respective soya preparation alone. Improvements were on the order of 20-30%. Concurrent with some of the PER studies apparent digestibility was determined for the proteins fed. Values for maize soya mixtures as given by Table 9, were only slightly lower than those for casein or soya isolate.

Our studies of food applications for MPC were based on its functional as well as particular nutritional properties. The oil binding capacity was tested for stabilization of peanut butter observing the effect on oil separation of unstabilized peanut butter. Increasing levels of MPC decreased oil separation and levels above 5% withstood centrifugation better than a commercial peanut butter containing hydrogenated fats as a stabilizer. No off-flavors were apparent at the 10% use level. An added benefit is the complimentary nature of the essential amino acids of the two proteins.

The reduced water retention capacity of MPC was tested as a means to reduce the sogginess of breakfast cereals. A prototype breakfast flake was prepared by drum drying a precooked corn slurry using various MPC substitution levels. Rehydration values were not changed during the first five minutes in cold milk but were reduced after 7.5 minutes for 10% and 15% substitution levels. No adverse flavor changes were evident at the tested levels of MPC.

An application of particular importance is the use of MPC as a companion ingredient for soya in extruded meat extenders and meat analogs. Here advantage is taken of the complimentary nature of the essential amino acids compositions in optimal blends. Soya flour and MPC mixtures were extruded into small and large size granules. After hydration the granules were used as a 30% replacement of ground beef in hamburgers, meat loaf, spaghetti sauce and chile. In all instances, organoleptic evaluations were favorable in comparison to the controls and in some applications it was found that the usual soya flavor associated with vegetable meat extenders was much reduced (Table 10).

The data presented seems to indicate that maize protein concentrate may in the future have a place among the major vegetable proteins used by food manufacturers. An ample supply of gluten meal, a by-product of maize refining,

TABLE 7. Essential Amino Acids Composition of An Optimal Mixture of MPC and Soya Concentrate Compared to FAO Recommendations

Essential Amino Acid	Content (G/16 G Nitrogen)	
	MPC/Soya Concentrate 20:80	FAO
Lysine	5.4	5.5
Methionine + Crystine	3.3	3.5
Threonine	4.2	4.0
Leucine	9.4	7.0
Isoleucine	4.4	4.0
Phenylalanine + Tyrosine	9.3	6.0
Valine	4.8	5.0
Tryptophan	1.4	1.0

TABLE 8. The Effect of MPC on PER of Rats Fed Soya Preparations

Protein Source	Corrected PER
Soy Flour	1.5
Soy Flour/MPC, 75:25	1.8
Soy Concentrate	1.8
Soy Concentrate/MPC, 80:20	2.2
Soy Isolate	<0.5
Soy Isolate/MPC 60:40	1.7

TABLE 9. Apparent Digestibilities of Proteins Used for PER Determinations

Protein Source	Apparent Digestibility Week 2	Week 4
ANRC Casein	.93	.93
Soya Concentrate	.91	.90
Soya Concentrate/MPC 80:20	.89	.88
Soya Isolate	.94	.97
Soya Isolate/MPC 60:40	.90	.90

TABLE 10. Explored Applications of MPC

Application	Characteristics
1. Extrusion of soya products.	Complimentary essential amino acids. Reduction in soya flavor - Improved nutritional value. Extension of meat.
2. Use in meat analogs.	Partial replacer of casein.
3. Breakfast cereals.	Reduction of sogginess. Increased protein content without additional starch.
4. Peanut Butter.	Prevention of oil separation.

already exists and can be upgraded by a relatively simple technology to become acceptable for food use as a Maize Protein Concentrate. The functional properties of the MPC indicate applications in diverse fields. No toxic or anti-nutritional factors have been reported in maize while the color, taste and odor complement the value of MPC as a desirable new protein ingredient.

REFERENCES

1. Encyclopedia Britannica, 1941. Volume 14, pg. 696.
2. Katz, S.H., Hediger, M.L., and Valleroy, L.A. (1974) Science 184, 765.
3. Phillips, R.D. U.S. Pat. 4,024,120.
4. Phillips, R.D. and Sternberg, M. (1979) J. Food Sci. in press.

UTILIZATION OF MAIZE IN BREWING

H. Hug
H. Pfenninger

Swiss Breweries Research Station
Zürich, Switzerland

In this survey the various maize products used in breweries and their preparation are described. The requirements made by the breweries with regard to the characteristics of maize grits are critically reviewed. On the basis of extensive tests and experiments, the much discussed influence of a high maize fat content on the beer quality is questioned. The lecture concludes with practical information on the processing of maize products in the brewhouse.

I. INTRODUCTION

With the exception of those countries which brew their beers according to the "Reinheitsgebot", maize has already been used as a component of the grist for a long time in most beer producing countries. The answer to the question as to whether a partial surrogation of malt by maize is worthwhile, depends on many factors. Not only economic reasons predominate here. The technological and qualitative aspects are equally significant. It must be said straight away that the addition of maize alters the beer type to a greater or lesser extent according to the grist pro portion and the maize product. This need not be in a negative sense, however. Beers with maize components in the grist are of a desirable lighter colour and more elegant than all malt beers. They have lower polyphenol and anthocyanogen contents. By using maize, the technologist has an easy method of compensating for the high protein con-

ISBN: 0-12-370960-1

tent of the malt available The economic advantages resulting from the use of maize cannot be summed up in generally valid figures. This will always be a question of the difference to the actual prices of other adjuncts, as well as of malt. Since the prices are often also influenced by measures of economic policy, clear advantages do not always result from the use of non-malted cereals.

This lecture deals with the various maize products used in the brewery, the role of the maize fat content in beer production, as well as several technological aspects of maize processing in the brewhouse.

II. MAIZEPRODUCTS IN THE BREWERY

Maize can be used in the brewery in different forms. Whereas in Europe brewers grits are used, in North America and Canada corn syrups are widespread. In isolated cases cornflakes and refined corn starch are also employed.

A. Preparation and Properties of Brewers Grits

1. Degermination Systems. The crushing and especially the degreasing of the maize grains requires specific installations and equipment. A lot of technical progress has been made in this field and it is very advanced. Basically there are two processing systems available at present (4), viz. the dry impact degermination and the wet or semi wet degermination. Apart from the technology, the two processes differ mainly in the size of the required investments as well as in the yield and quality of the finished products. Thus factors which are determining the price of maize grits.

In the case of the dry impact degermination system, the actual separation of the germs and endosperms is carried out with vertical impact degerminators. The mixture of endosperm pieces, hulls and germs is sorted into 2-3 fractions. The hulls are separated with aspirators. The most important step apart from the degermination, is the sorting of the pure germs with the gravity table. The endosperm pieces are reduced to grits or powder with rollers, plansifters and grits cleaning machines, whereby the number of passages determines the quality of the granulation

The wet or semi wet degermination is a newer process. It was developed out of the large demand for very low-fat products on the one hand and the recovery of maize oil on the other hand. The difference here lies in the specific preparation of the maize grains which consists of wetting them with cold water, allowing them to stand and in treating the grain surface with water and steam in a second step. These measures make it possible to subsequently remove the fatty hulls completely. The actual degermination with the vertical impact degerminator is then the third step. Further processing and sorting is the same as for the dry method.

A comparison of the two methods shows that both have their advantages and disadvantages. In the wet degermination system the higher investment costs are practically compensated by the larger yield of brewers grits with a guaranteed lipid content of less than 1 %. Nevertheless for this system the price of the end product is higher, due to the extra expenses required for preparing the maize grains and the finally drying the grits to reduce the water content from 16-18 % down to 12 %. The dry impact degermination system produces brewers grits with a higher lipid content (1,3-1,5 % by weight of dry substance), however. Further degreasing would be technically possible with this process but is not practical for brewers grits because of the low yields obtained in this case.

2. The Requirements to Brewers Grits.

a. The demands regarding the granulation. The requirements which must be met by the brewers grits differ largely. The demands regarding granulation are often exaggerated. In the case of maize grits only the very large and the very fine particles play a large role. If too many large particles are present in the maize grits, the extract yield is lower. A correction by better thermal breakdown is only partly possible, since coarse granules remain even when the mash is cooked for a long time and are found practically unchanged in the spent grains. On the other hand if too many fine grain fractions and a lot of flour is present, this can lead to run-off difficulties in the lautering process. Thus when preparing brewers grits, a compromise must be reached between a maximum yield on the one hand and a minimum lautering time on the other hand. The brewhouse equipment and facilities must also be taken into consideration. According to our opinion the main emphasis must still be placed on obtaining an optimum quality for the granulation of the malt. From the quantitative aspect already, the influence of the maize granulation can only be slight. From this, as well as from numerous practical observations, we conclude that the demands made on the granulation of the maize should not be too great. Limiting the number of coarse and fine particles is nevertheless important. In practice, the granulation, with the approximate values indicated in TAB. I, have proved to be successful.

b. The demands regarding the lipid content. With regard to the lipid content of the maize grits the requirements also vary in many places. The well known tenet that maize products should be degreased as far as possible for use in brewing, still applies (2). We will come back to this point later.

c. The demands regarding the water content. According to our experience, the water content usually required (< 13 %) is still too large in some cases. At this level - the lowest for the growth of micro-organisms - enzymatic reactions can still take place and the fat decomposition due to lipase activity can continue. 13 % water also appear dangerous since the flow properties already change in this range. This can lead to difficulties in air lifts and on emptying silos (sticking, forming bridges and lumps). If one would want to eliminate all these disadvantages, the water content of the maize grits would have to be adjusted to values usually employed for malt.

TABLE I. Optimal Granulation of Brewers Grits

Percentage	Particle Size[a]
0	> 2
< 5	> 1,27
20-30	1,01-1,27
50-60	0,546-1,01
15-25	0,152-0,546
	< 0,152

[a]Units are in mm

B. Preparation and Properties of Corn Syrups

The preparation of the maize grains for the production of corn syrups is similar to that of the wet milling method. The isolation of almost pure corn starch requires even more expenditure than for the production of maize grits. Not only the germs and hulls are separated in this case, but all the fats and proteins are extracted. The isolated starch is then hydrolyzed using either the acid process or the acid-enzyme process (3). The latter involves an enzymatic decomposition with amylglucosidases after the acid hydrolysis, which produces highly fermentable syrups. The glucose to maltose ratio can be varied according to the measure of the acid hydrolysis and the enzyme specificity, within certain limits. If the desired degree of conversion has been reached, neutralization with sodium carbonate is carried out and the pH adjusted to a value of 4-5. This also facilitates the subsequent elimination of undesirable substances (lipids and substances causing colour and turbidity). The density is adjusted by subsequent evaporation.

The corn syrups produced usually contain 80 % extract, 40 % glucose, 25 % maltose and 9 % maltotriose, which corresponds to an attenuation limit of 65-70 %. They do not contain any starch at all and only traces of lipids.

The use of corn syrups led to irregular and incomplete fermentation of such worts in many breweries. As was shown by Pfisterer et al (4) this observation is due to a carbohydrate spectrum which differs from that of the pure malt wort in that it contains up to 300 % more glucose. According to their investigations, not all yeast strains are in a position to adapt themselves equally well to this substrate. Several show typical catabolite-repression phenomena, in that the synthesis of several enzymes is repressed by the high glucose concentration. This also affected the maltose uptake system, for example, which resulted in the fact that it was then only assimilated when the glucose had been almost completely absorbed. These disadvantages could be eliminated if the corn syrups are prepared with β-amylases instead of amyloglucosidases. It was possible to obtain carbohydrate spectrums which were very similar to those found in all malt worts.

C. Preparation and Properties of Cornflakes

In order to prepare cornflakes, the coarse maize grits which has been degreased as a result of degermination, is cooked, flaked with steam heated rollers and then dried with hot air. During this procedure a complete gelatinization of the maize starch occurs. The most important characteristics of cornflakes are: 11-12 % water, 0,5-0,8 % fat and an apparent density of 200-500 kg/m^3.

D. Preparation and Properties of Refined Corn Starch

This is a pure starch product, which is obtained in the same way as for the corn syrup preparation, but is not hydrolyzed. This starch powder is extremely fine and approx. 96 % passes through a 200 mesh sieve.

III. THE SIGNIFICANCE OF THE MAIZE FAT CONTENT IN BEER PRODUCTION

As already mentioned, most breweries want degreased maize products (5). We believe this is because the influence of the fat content of the raw materials on the beer quality is overestimated. The demand for a fat content of less than 1 % in maize products is even less understandable since a fat content of over 2 % is present in malt and this is usually tolerated without complaint.

The significance of the lipids in beer production has not yet been investigated sufficiently at present. Their influence on various properties of the beer, such as the taste and off-flavour (7,8), foam stability (9,10), gushing (11), and the yeast viability (12), has been described by various authors but has never been studied in connection with an increased fat content in maize products.

In order to test this influence on the beer quality, we carried out numerous brewing experiments at the Swiss Breweries Research Station in 1976. The tests were carried out on the pilot plant of our institute as well as on an industrial scale, using maize with various fat contents (< 1, ∿ 1,5, > 4 %). (13). All the maize, which originated from France, was milled by Bühler-Miag Ltd., Uzwil. For fat contents below 1 %

the semi wet degermination method was used and for grits with approx. 1,5 % the dry impact degermination method. Batches which were not degreased were also milled using rollers but all the fractions were combined to a single end product.

A. Experimental Results

From our big number of experiments carried out only a few results will be mentioned here. The results summarized in TAB. II are for a series of tests carried out with 9 brews containing 40 % maize and different fat contents, as compared to 3 brews of malt only. The figures in the table are the average obtained from 3 brews.

If the maize brews are considered first, there are no differences between these worts and beers in the analytical date presented. The uniform foam stability of all the maize brews must be mentioned in particular. The influence of a high maize fat content on the foam has obviously been overrated up to now.

TABLE II. Analytical Data of Worts and Beers from the Pilot Plant Trials

Grist	Malt %	60	60	60	100
	Maize %	40	40	40	-
Maize Fat Content	%	0,9	1,5	4,3	-
Worts					
Extract	% w/w	12,1	12,2	12,1	12,1
Colour	EBC	8,5	8,5	8,5	8,5
pH		5,60	5,60	5,60	5,50
Total Nitrogen[1]	mg/l	688	708	749	1064
Free Amino Nitrogen	mg/l	150	156	165	221
Beers					
Original Gravity	% w/w	12,2	12,2	12,2	12,1
Alcohol	% w/w	4,22	4,19	4,16	4,11
Degree of Fermentation	%	81,9	82,0	81,4	81,2
Attenuation Limit	%	84,1	84,6	84,6	82,5
Foam Stability[2]	Σ	119	117	118	122

[1]Determined with TNBS according to Analytica-EBC III,
[2]Determined according to Ross and Clark

As can also be seen from TAB. II, the use of maize leads to a lighter colour, higher pH than for the pure malt products. There is also a lower total nitrogen and free amino nitrogen content but a higher attenuation limit.

As can be seen from the results in TAB. III and IV, these findings were also confirmed in large scale experiments.

In this case the adjunct rate was 30 % maize with a lipid content indicated on the top of the tables. Here also the maize worts were of a lighter colour, higher pH and lower total nitrogen content than the all malt brew. The level of the free amino nitrogen, which is normally too low when maize is used, was corrected by the addition of a bacterial proteinase. It is important to note that the polyphenol content in maize worts and beers is significant lower than in the all malt ones which is an advantage in view of a better colloidal stability of the beers. Here there was also no negative effect on the foam stability.

TABLE III. Analytica Data of the Worts from the industrial Trials with 30 % Maize in the Grist compared with an all Malt Brew

Brew		Maize	Maize	Maize	Malt
Lipid Content	%	0,7	1,7	4,3	-
Extract	% w/w	11,3	11,3	11,3	11,3
Colour	EBC	8,5	8,5	9,5	11,0
pH		5,6	5,6	5,6	5,5
Total Nitrogen	mg/l	895	895	941	1194
Free Amino Nitrogen[1]	mg/l	205*	205*	215*	214
Bitterness[2]	BU	38	38	40	38
Polyphenols[3]	mg/l	191	176	193	272
Anthocyanogen[4]	mg/l	40	40	37	49

[1] Determined with TNBS according to Analytica-EBC III
[2] Determined according to Analytica-EBC III
[3] Determined with CMC according to Analytica-EBC III
[4] Determined according to Harris and Ricketts
* Addition of bacterial protease

TABLE IV. Analytical Data of the Beers from the industrial Trials with 30 % Maize in the Grist compared with an all Malt Brew

Brew		Maize	Maize	Maize	Malt
Lipid Content	%	0,7	1,7	4,3	-
Original Gravity	% w/w	11,0	11,2	11,3	11,3
Alcohol	% w/w	3,81	3,86	3,88	3,86
Degree of Fermentation	%	83,2	83,2	82,2	81,0
Attenuation Limit	%	85,9	85,8	85,8	83,8
Difference	%	2,7	2,6	3,6	2,8
Colour	EBC	7,0	7,0	7,5	9,0
pH		4,4	4,4	4,5	4,5
Carbon Dioxide[1]	% w/w	0,43	0,47	0,46	0,48
Foam Stability[2]	Σ	126	124	124	126
Reducing Potential[3]		51	50	50	57
Total Nitrogen	mg/l	647	624	699	875
Free Amino Nitrogen[4]	mg/l	100	103	125	155
Bitterness[5]	BU	26	26	27	26
Polyphenols[6]	mg/l	144	125	117	173
Anthocyanogen[7]	mg/l	22	19	21	34
Diacetyl[8]	mg/l	0,04	0,03	0,02	0,02
2,3-Pentandion[8]	mg/l	0,03	0,02	0,02	0,02
Viscosity (20° C)	Pa·s	1530	1540	1550	1610

[1] Determined according to Blom and Lund
[2] Determined according to Ross and Clark
[3] Determined according to Steiner
[4] Determined with TNBS according to Analytica-EBC III
[5] Determined according to Analytica-EBC III
[6] Determined with CMC according to Analytica-EBC III
[7] Determined according to Harris and Ricketts
[8] Determined gaschromatographically

In these tests a lot of importance was placed on the sensory assessment, of course. When beers from the pilot plant (28 tasters) or from the industrial scale (78 tasters) were tasted, no significant difference was found at all between the beers with maize having fat contents of 0,7 and 4,3 %. Even after being kept at 20° C for 6 weeks there was no difference, which shows that both beers have an equally good flavour stability. These findings indicate that a high maize fat content also has no effect on the beer flavour. This applies only in the case of fresh maize grits, however, as will be shown later. According to our tests and also to the literature (14), the main part of the lipids (∿ 95 %) in the raw material remains in the spent grains according to the mash and lautering conditions. Another important part is eliminated with the hot trub. Only about 0,1 to 0,7 % of the total fatty acids in the raw material are still present in the wort. Of this small remainder, more than 80 % are eliminated with the cold trub and during fermentation. Since the lipids (mainly the unsaturated fatty acids) are assimilated by the yeast during fermentation, a higher lipid content in the maize has hardly a negative effect on the beer quality. According to studies by Grill et al (15) the kettle worts, prepared with 20 % non-degreased maize, contained 30 % less free fatty acids than worts made of malt only, with which they were compared.

It is also well known, however, that undesirable changes in the lipids can take place when the maize grits are stored (5). The one reaction which is possible, is the hydrolytic decomposition of glycerides (by lipases) which liberates fatty acids. The other reaction is defined as an autooxidation and leads to the formation of aldehydes and ketones from the unsaturated fatty acids. Usually both reactions occur simultaneously. The result is a rancid odour, which should also affect the finished beer.

Apart from the rather subjective assessment of the odour of maize batches, we determined the so-called acid value (AV) by the titration with sodium hydroxide and the peroxide number (PN) which give an indication of the changes taking place in lipids (3). To test the influence of the storage time and temperature, we periodically studied the above mentioned quality criteria with regard to maize samples (whole grains as well as grits with different fat contents). The samples were stored at 15 and 35° C for 24 weeks in sealed milk cans. In order to assess the odour we always used a freshly milled and a clearly spoilt sample of maize as a reference (AV = 128, PN = 3,2).

As can be seen from the values in TAB. V, all the maize grits samples stored at 15° C show a moderate increase in the acid value, while those at 35° C have a large increase in this value. The peroxide numbers also increase considerably in some cases. Their relatively low concentration indicates that the autooxidation, which is a chain reaction, was in the so-called initial phase during the entire experimental period and that it thus did not play a decisive part in the decomposition of the fat. In the whole kernels practically no changes in the lipids were observed. The odour tests showed that all the grits samples were dubious after 4 weeks and impure and spoilt

TABLE V. Changes of Lipid Quality Criteria in different Maize Products during Storage at 15 and 35° C

Product		Grits		Grits		Grits		whole kernels	
Lipid Content		< 1,0		∿ 1,5		> 4,0		> 4,0	
Water		14,1		11,0		10,7		12,0	
Storage weeks	Temp. °C	AV[1]	PN[2]	AV[1]	PN[2]	AV[1]	PN[2]	AV[1]	PN[2]
0	-	-	-	-	-	-	-	18	0,7
4	15	25	2,2	39	1,3	31	0,9	20	0,7
4	35	28	2,2	43	1,3	38	0,9	21	0,7
10	15	33	2,1	41	1,3	41	1,0	23	0,7
10	35	44	2,4	57	1,4	54	1,1	25	0,7
19	15	40	3,1	51	1,4	53	1,0	23	1,0
19	35	59	3,3	80	1,9	78	1,2	23	1,1
24	15	43	3,3	52	1,6	54	1,1	20	1,1
24	35	61	3,5	87	2,0	86	1,4	21	1,2
Δ total[3]	15	25	2,6	34	1,1	32	0,4	2	0,4
Δ total[3]	35	43	2,8	69	1,3	68	0,7	3	0,5

[1] AV = Acid Value
[2] PN = Peroxide Number
[3] Δ total = Total Increase during 24 weeks

after approx. 6 weeks. It must be pointed out, however, that this off-flavour was no longer detected in beers made from these raw materials, since it was eliminated when the mash and wort was cooked.

On the basis of these results, we recommend for reasons of safety, that maize grits which are not degreased should be processed at the brewery when they are as fresh as possible. This is best achieved if whole maize grains instead of grits are stored and if milling is carried out immediately before mashing in (as for malt). Our preliminary tests with a Condux Hammer Mill (16) (2 mm round hole sieve, 300 rpm) gave satisfactory results from the very beginning. The cost savings resulting from this procedure are abvious and any disadvantages of higher maize fat contents are thus eliminated.

IV. THE PROCESSING OF VARIOUS MAIZE PRODUCTS IN THE BREWERY

The maize products used in the brewery can be divided into two groups: the mash tun or cooker adjuncts and the kettle adjuncts. All starch type products belong to the first category (grits, cornflakes and refined corn starch), while corn syrup belongs to the second one, since the starch has already been converted to fermentable sugars.

A. Brewers Grits

This is the maize product which is used the most frequently in the breweries, mainly for reasons of price. They can be used as up to 50 % of the grist but usually the upper limit is 30-40 % because of the poorer lautering properties. But as we found in extreme tests using 60 % maize, the lautering time can be reduced up to 15 % when the adjunct mash is heated only to 85° C and gelatinized as this temperature for 30 min. instead of cooking it at 100° C. The yield only decreases by 0,5 % on an average. This slight loss in extract is more than compensated by the saving in energy. In this case the yield was only unsatisfactory (60 %) in tests where the mash was heated to 75° C. In cases where the lautering time on using maize is the time limiting factor, corresponding tests are certainly worth recommending. Further practical information on the processing of maize grits in the brewery can be found in our former publication (17).

B. Corn Syrups

These products were used more and more in the past few years. The price difference between corn syrups and brewers grits was 6 $/lb in 1975 and now it is only 2 $/lb. The main advantage of corn syrups is that they can be added directly to the kettle, so that they are chosen in case of bottlenecks in the brewhouse and for high gravity brewing. Their utilisation rate in the brewery can amount to 60 %, whereby a thorough mixing with the conventional wort must be ensured.

C. Cornflakes

The use of corn flakes in the brewery is only found in isolated cases in Great Britan, North and South America. Since the starch is already gelatinized during the preparation of the actual cornflakes, they can also be processed with infusion methods.

D. Refined Corn Starch

This product has also not been used extensively up to now. In this case it is essential to earth the transport pipes and tanks to avoid explosions resulting from static electricity sparks. The processing of refined corn starch is the same as for grits, whereby no lautering problems should arise, according to Coors (6).

ACKNOWLEDGMENTS

We would like to thank the Foundation for Scientific Research of the Swiss Breweries Association for financial support of this work and for the permission to publish this paper.

REFERENCES

1. Wyss, E., *Bühler-Diagramm Nr. 58* (7.1974), Gebrüder Bühler AG, Maschinenfabrik, Uzwil, Switzerland.
2. Bradee, L. H., *MBAA Techn. Quart. 11*, 57 (1974).
3. Swain, E. F., *MBAA Techn. Quart. 13*, 108 (1976).
4. Pfisterer, E. A., Garrison, I. F., and Mckee, R. A., *MBAA Techn. Quart. 15*, 59 (1978).
5. Bradee, L. H., in "The Practical Brewer" (H. M. Broderick ed.) p. 41, MBAA, Madison, Wisc. (1977).
6. Coors, J., *MBAA Techn. Quart. 13*, 117 (1976).
7. Äyräpää, T., and Lindström, I., in "Proc. Europ. Brew. Conv.", 14th Congr., Salzburg 1973, p. 271, Elsevier, Amsterdam (1974).
8. Krauss, G., Forch, M., and Holstein, H., *Mschr. Brauerei 28*, 229 (10.1975).
9. Zürcher, Ch., *Mschr. Brauerei 24*, 276 (10.1971).
10. Klopper, W. J., Tuning, B., and Vermeire, H. A., Proc. Europ. Brew. Conv., 15th Congr. Nice 1975, p. 559, Elsevier, Amsterdam (1975).

11. Carrington, R., Collet, R. C., Dunkin, I. R., and Halek G., *J. Inst. Brew. 78*, 243 (1972).
12. Thompson, C. C., and Ralph, D. J., Proc. Europ. Brew. Conv., 12th Congr. Madrid 1967, p. 177, Elsevier, Amsterdam (1968).
13. Hug, H., and Pfenninger, H., *Schweiz. Brauerei-Rdsch. 87*, 105 (5.1976).
14. Krauss, G., Zürcher, Ch., and Holstein H., *Mschr. Brauerei 25*, 113 (6.1972).
15. Grill, W., and Püspök, J., Proc. Europ. Brew. Conv. 16th Congr. Salzburg 1977, p. 195, Elsevier, Amsterdam (1977).
16. Pfenninger, H., *Schweiz. Brauerei-Rdsch. 87*, 13 (2.1976).
17. Hug, H., Knöpfel, H. P. and Pfenninger, H., *Schweiz. Brauerei-Rdsch. 87*, 135 (6.1976).

THE MORPHOLOGY AND PHYSIOLOGY OF MALTING BARLEYS

G.H. Palmer

Department of Brewing and Biological Sciences
Heriot-Watt University
Edinburgh, Scotland

INTRODUCTION

Barley-malt is the primary raw material of many foods, alcoholic drinks and beverages. The main objective of the malting process is to encourage the development of flavour compounds and to promote enzymatic hydrolysis (modification) of the protein and carbohydrate reserves of the germinated grain. The conversion of barley to malt in industry is therefore brought about by the same physiological mechanisms which control the growth and development of the grain in the field.

Early physiological studies of germinating barley grains were carried out by Horace Brown and Haberlandt in 1890. The classic studies of these investigators suggested that enzymatic hydrolysis and solubilization of the endosperm food reserves were under the dual control of the growing embryo and the aleurone cells. From a practical viewpoint the work of Brown and Haberlandt established that the brewing quality of malt was determined by the physiology of the malting grain.

Nearly half a century later, notable brewing chemists such as Hopkins and Krause (1937) arrived at the same conclusion. They stated that malt quality was determined in the malthouse by "(a) morphological changes (growth of the acrospire and rootlets); (b) histochemical changes (disappearance of the cell walls of the endosperm, and the associated softening of the corns); (c) various metabolic changes such as degradation of proteins, starch etc. to simple substances; and (d) the formation and liberation of enzymes, to the activities of which all other changes are due, either directly or indirectly". Although this view of the malting process is very general in outline, it's basic concepts closely correspond with present-day knowledge of the science and technology of barley germination (Palmer & Bathgate, 1976 and Bewley & Black, 1978).

ISBN: 0-12-370960-1

Although the biological properties which determine the malting potential of barleys are not fully understood, there has been significant progress in the understanding of the physiological processes which control the germination and growth of the grain. The progression from scientific discovery to technological development is the foundation on which the economic production of high quality food products, such as malt, is based. In this paper, important relationships between grain structure, function and malting performance will be discussed.

Germination

Rapid and even germination are important properties of malting barleys. From a morphological viewpoint, germination is completed when the coleorhiza (chit) penetrates the husk (see Figs 1, 2). The subsequent growth of the young root and shoot (acrospire) is regarded as seedling growth (Fig 8a). In malting practice, no morphological distinction is made between germination and subsequent seedling growth. The entire growth phase is called "germination". In this paper the biological distinction between germination and seedling growth will be followed.

In the absence of externally applied malting aids such as gibberellic acid (see below), enzymatic breakdown of the starchy endosperm will not occur if the grain fails to germinate. Germination failure may be caused by natural metabolic lesions which, for a time, ensure that the grain will not germinate efficiently. In other cases, germination failure may arise because the harvested grains were injudiciously dried and stored. The former reflects an innate but temporary condition of dormancy, whilst the latter amounts to an induced state of dormancy from which the grain may not recover, thus rendering it useless for malting. Some degree of innate dormancy can be over-come by storing dried barley (12%) at 20-25°C for a limited period or by abrading the grain, i.e. scarification of the husk/pericarp (Pollock, 1962; Baxter et al. 1974 and see below).

Innate Dormancy

Dormancy can be a serious problem to some European maltsters who wish to malt barley soon after harvest. This industrial requirement often conflicts with the natural tendency of some barleys to remain dormant for a limited period (2 to 3 months). From a practical or a scientific point of view very little is known about the causes of dormancy in barley. Water-sensitivity (i.e. failure to grow in an excess of water) was first described for barley by Essery et al. in 1954. It was observed that grains which failed to germinate efficiently in large volumes of water could be made to do so if the steeping (soaking) periods were interrupted by exposing the grain to air or oxygen. Gaber & Roberts (1969) and Crabb & Kirsop (1969) confirmed the earlier work of Urion & Chapon (1955) that water-sensitive and normal grains had similar rates of respiration. Gaber & Roberts suggested that high relative humidity and poor illumination during ripening promoted water-sensitivity. Although these workers agreed that water-sensitive grains had a higher than normal requirement for oxygen, Gaber & Roberts (1969) concluded that high levels of micro-organisms in the husk of water-sensitive grains deprived the embryo of oxygen. In contrast, Crabb & Kirsop (1970) suggested that high levels of oxygen were required to overcome metabolic inhibitors which were present in the embryo. Although these putative inhibitors were not identified, abscisic acid declines to minimal levels during barley maturation (Slominsky et al. 1979) and small residues of this growth (germination) inhibitor may, in certain circumstances persist in the mature grain and cause dormancy.

Recent studies of dormant freshly harvested barleys by Roberts (1973); Hendricks & Taylorson (1974) and Roberts & Smith (1978) provide new and interesting evidence that the metabolism of pentose sugars in the embryo may play an important part in the dormancy-breaking process. These authors suggested that the pentose phosphate (PP) pathway is inoperative in dormant freshly harvested barleys. It is well known that when the PP pathway is operating, Carbon 1 of carbon-labelled glucose is released as carbon dioxide. In contrast, Carbon 6 is metabolized to carbon dioxide when the usual aerobic Embden-Meyerhoff-Parnas (EMP) pathway is active. Roberts (1973) and Roberts & Smith (1978) have suggested that dormant barleys with low C6/C1 ratios (0.04-0.20) tend to germinate, whilst those with higher ratios failed to do so.

From a practical viewpoint, the use of the C6/C1 ratio to predict the germinative potential of malting barleys would be limited because barleys with similar intermediate ratios can have different levels of germination (see Roberts & Smith, 1978). Notwithstanding, the possible role of the PP pathway in barley dormancy provides an interesting link with earlier studies of the germinating grain by Pollock et al. (1955) and Pollock and Kirsop (1956). It has always been puzzling why inhibitors of aerobic respiration such as carbon monoxide (Pollock et al. 1955), hydrogen sulphide (Pollock &

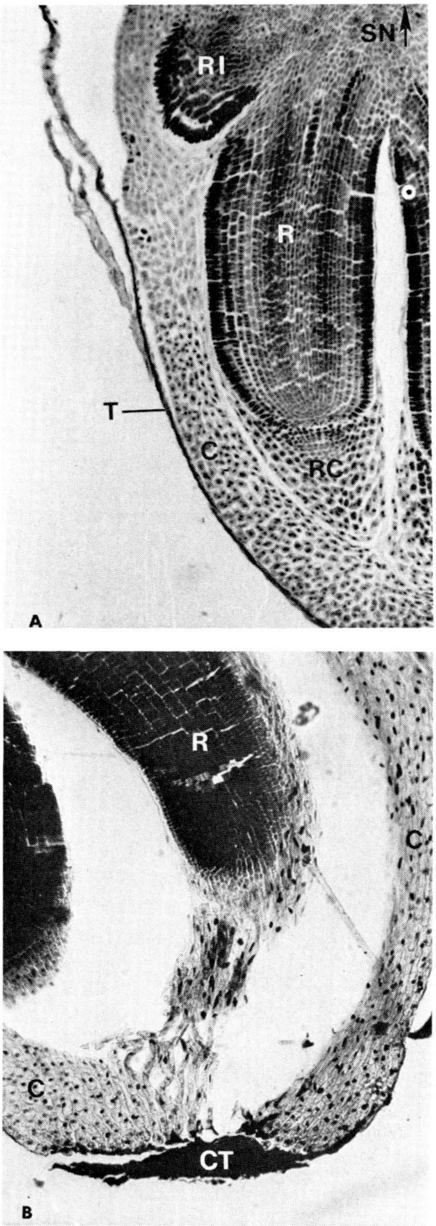

FIGURE 1. Light micrographs showing stages in the differential elongative growth of the coleorhiza and root during germination. Fig. 1 - a = 0hr; b = 10hr; c = 17hr; d = 20hr.

SN = Scutellar Node; RI = Root Initial; R = Young Root; T = Testa; C = Coleorhiza; RC = Root Cap; CT = Corky Tip.

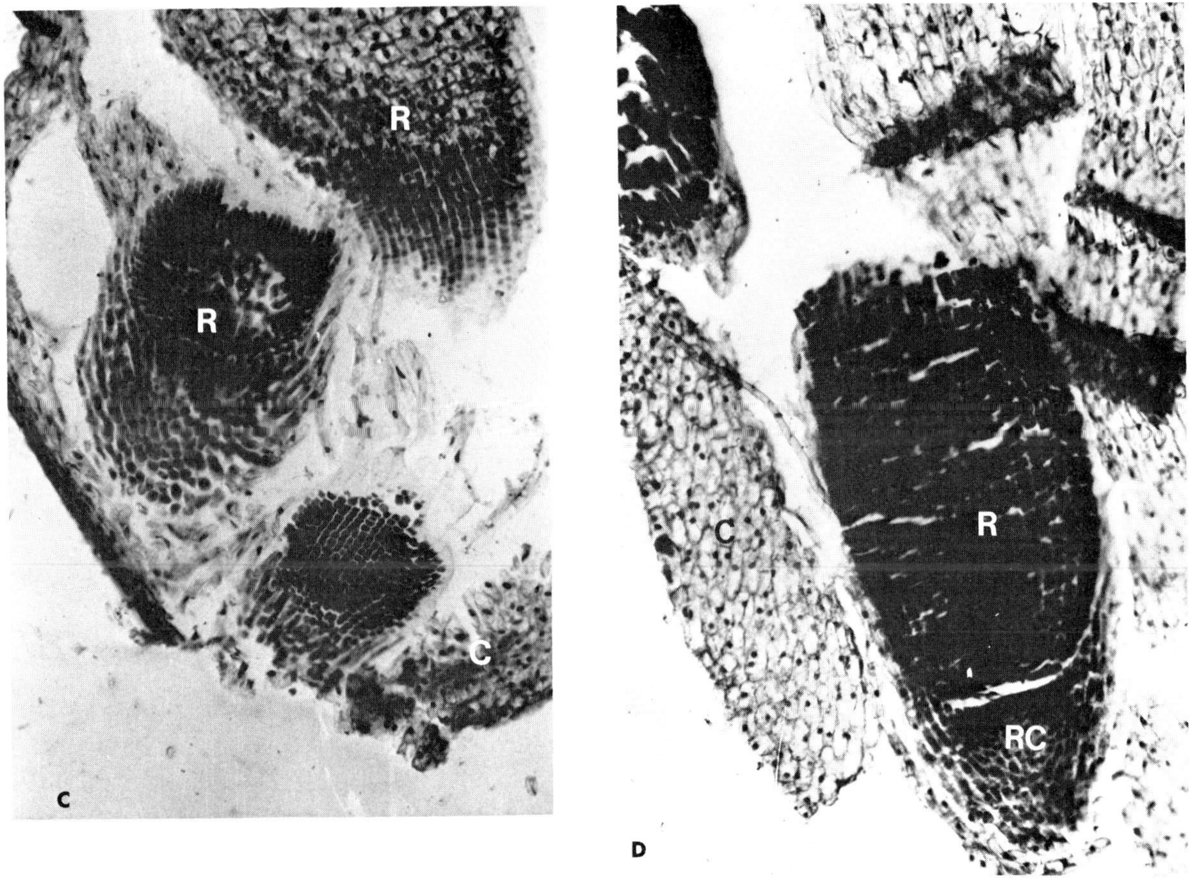

C

D

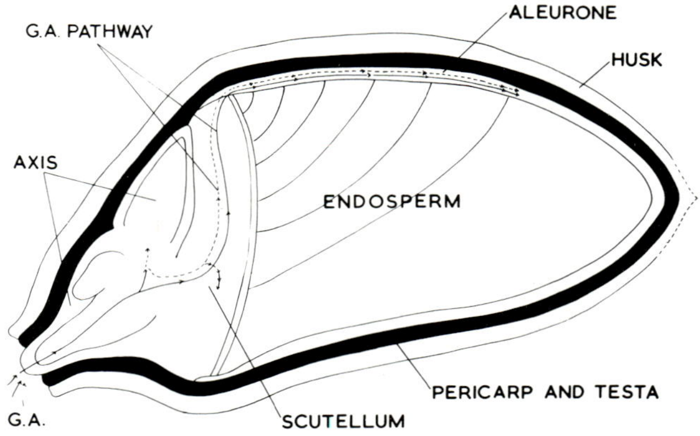

FIGURE 2. Diagrammatic representation of a longitudinal section of germinated barley. The subsequent asymmetric distribution of gibberellic acid and mainly one-sided pattern of enzymic modification of the endosperm are illustrated. GA = gibberellic acid.

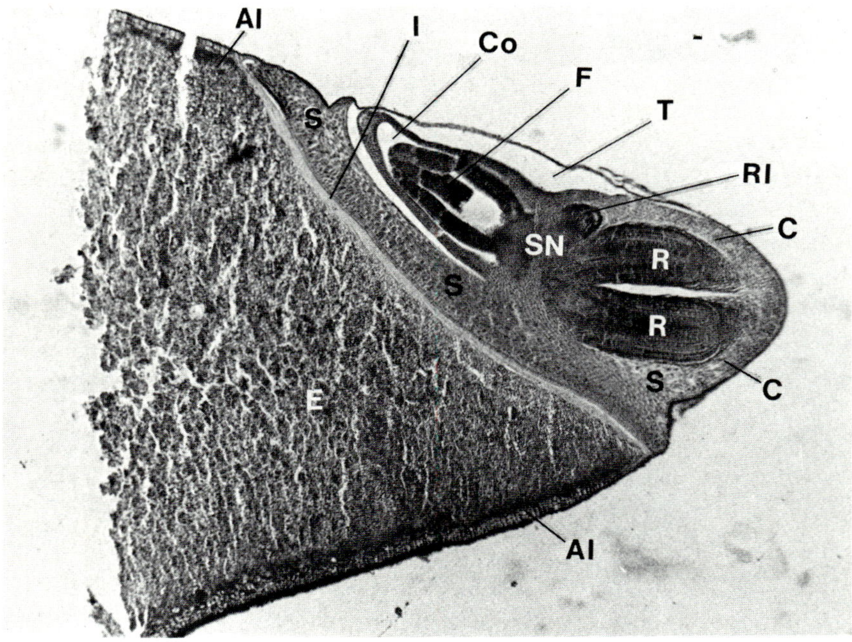

FIGURE 3. Light micrograph of longitudinal section of Ohr embryo and proximal endosperm.

Al = Aleurone layer; S = Scutellum; I = Intermediate layer of empty endosperm cells; Co = Coleoptile; F = Foliar shoot; T = Testa; RI = Root Initial; R = Young Root; C = Coleorhiza; E = Endosperm.

Kirsop, 1956), Cyanide (Fischnich et al. 1962) and malonate (Oganara & Hayashi, 1964) improved the germination levels of dormant barleys. Roberts (1973) and Roberts & Smith (1978) have proposed that these inhibitors and other promoters of barley germination such as nitrite, nitrate, gibberellic acid, imbibition below 15°C or warm storage at 25°C all reduced the C6/C1 ratio of dormant barleys.

The PP pathway metabolism of Carbon 1 of glucose to carbon dioxide requires NADP (Nicotinamide adenine dinucleotide phosphate). This compound is a cofactor of the enzymes, glucose-6-phosphate dehydrogenase and 6-phosphogluconate dehydrogenase, which play major roles in the PP pathway. NADPH and 5 and 4-carbon compounds are produced. The NADPH may function to promote reductive synthetic reactions (Roberts & Smith, 1978). The PP pathway may be an effective means of providing essential metabolites for a limited loss of carbon as CO_2, before germination begins. There is no evidence for this or the alternative possibility that pentose sugars may be required to initiate germination. However, the cells of the coleorhiza elongate rapidly during germination (see below) and it is interesting to note that Sutcliffe & Sexton (1974) have reported that the PP pathway is very active in elongating cells of roots.

Although the PP pathway may, to some unknown degree, help to overcome dormancy in barley, it should not be overlooked that non-dormant grains respire aerobically and the operation of the PP pathway does not appear to be obligatory for their germination (see Roberts & Smith, 1978). Therefore, the PP pathway may normally function during after-ripening to reduce dormancy and prepare the embryo for aerobic respiration during germination.

Structure and Function of Barley Embryos

A special characteristic of mature cereal (wheat) embryos is that they contain, before hydration, the ribosome apparatus which initiate early protein synthesis during germination (Brooker et al. 1978). Barley embryos are also morphologically well developed. Each embryo is essentially a whole plant in microcosm (see Figs 2, 3). The young embryo is protected in the grain by the testa, the pericarp and thickened areas of the husk (see Fig 4). As shown in Figs 1, 2, 3 the small delicate roots (about five per embryo) are enclosed by the coleorhiza (chit). During germination the coleorhiza elongates faster than the roots (Fig 1b). After breaking through the husk, the coleorhiza ceases to elongate and germination is completed. The corkiness of the apical

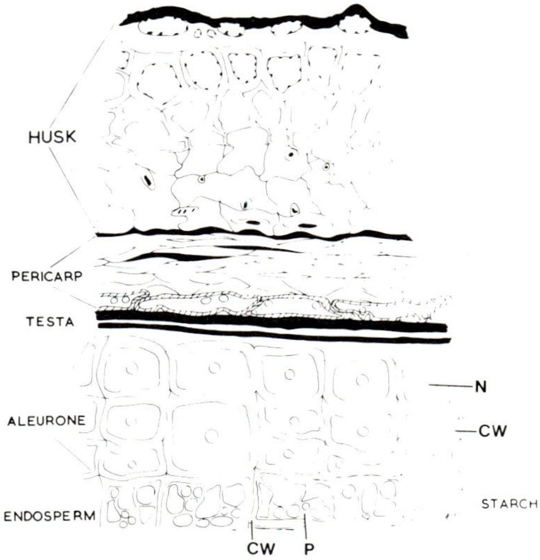

FIGURE 4. Diagram of a transverse section of outer layers of ungerminated barley.

N = Nucleus; CW = Cell Wall; P = Storage Protein area.

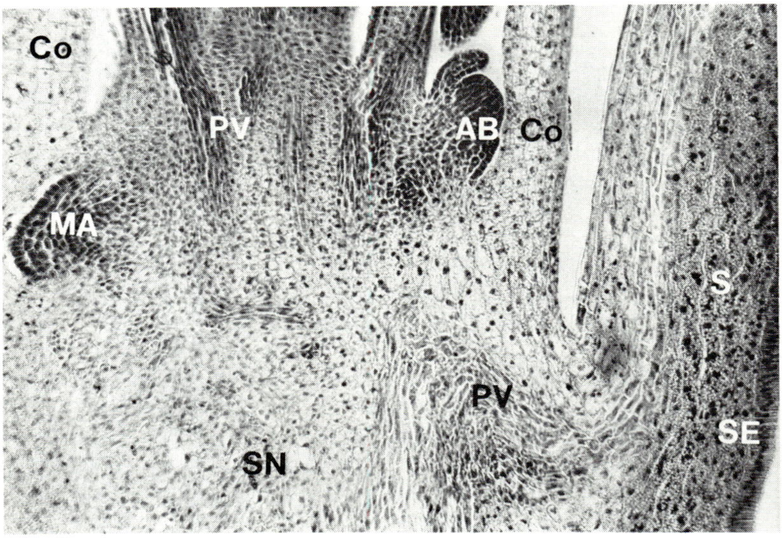

FIGURE 5. Light micrograph of longitudinal section of the 20hr germinated embryo of barley. The base of the foliar shoot, the scutellar node and adjoining scutellum are shown. Note axillary bud.

Co = Coleoptile; MA = Meristematic Area; PV = Differentiating vascular tissue; SN = Scutellar Node; S = Scutellum; SE = Scutellar Epithelial cells.

tip of the coleorhiza increases (Fig 1b). There is an acceleration of the growth rate of the largest root (Fig 1c) which then emerge near the corky apex of the elongated coleorhiza (Fig 1d). In barley, the metabolic processes which produce germination must, to a great part, reside in the short-lived coleorhiza. It would therefore be interesting to know whether germination-promoting mechanisms such as the PP pathway (Roberts 1973 and Roberts & Smith 1978) or the activation of messenger RNA and the polymerization of pre-formed ribosomes (Brooker et al. 1978) operate in the coleorhiza. Generally, there is a serious lack of evidence linking proposed biochemical mechanisms of germination with the growth behaviour of the different tissues of the embryo. Improved knowledge of this relationship should help the malting technologist to effect better control of the germination process.

A similar differential growth rate occurs between the coleoptile and foliar shoot (acrospire) during post-germination development (Figs 2 and 3). The acrospire grows slowly under the pericarp, eventually emerging towards the rear end of the grain if the malting period is excessive (see Palmer & Bathgate, 1976).

Structural and Physiological Changes in the Embryo

Although the dissection of plant tissue can cause misinterpretation of physiological date, it is still the best method of investigating the correlative processes which control growth and development. In general, it probably produces far fewer difficulties of interpretation than the biochemical techniques of homogenization.

Early growth of the young roots of barley embryos is mainly brought about by cell elongation. Table I shows that root growth was only marginally affected when cell division was stopped in the presence of maleic hydrazine (+ gibberellic acid). Similarly, root growth was not significantly increased when cell division in the root was doubled by gibberellic acid treatment. Table I also shows that CCC (2-Chloroethyl Trimethyl Ammonium Chloride), a compound which inhibits the early (pre-Kaurene cyclization) stages of gibberellin synthesis (Robinson & West, 1970), did not reduce the incidence of cell division in barley roots. This suggests that Kaurene-derived gibberellins may not promote cell division in barley roots. Regarding those factors which may be involved in root growth of the embryo - Palmer (1967) found that whereas indole acetic acid (IAA) at 10^{-4}M inhibited root elongation, IAA at 10^{-10}M with gibberellic acid effected

TABLE I. Relative Contribution of Cell Division and Cell Elongation to Root Growth and Germination in Barley.

Treatment	Mitotic Index of Root	Root Number	Root Length (cm)
H_2O	80	6.0	1.3
GA_3 10^{-4}M	120*	6.0	1.5
CCC 10^{-3}M	80	6.0	1.26
CCC+ + GA_3 10^{-4}M	120	6.0	1.43
MH 10^{-3}M	0	5.7	0.90
MH+ + GA 10^{-4}M	2	5.7	1.26

* All other root cells at prophase
\+ Also at 10^{-3}M
CCC = 2-Chloroethyl Trimethyl Ammonium Chloride
MH = Maleic Hydrazine
Grains were germinated at 25°C and examined at 36 h.
(From Palmer, 1967)

greater root growth than did either compounds alone. Whilst galactose did not promote significant root growth, sucrose, glucose and fructose did. The young roots and coleorhiza of barley contain a high level of invertase and sucrose metabolism is higher in the roots than in the other parts of the embryo (Palmer, 1967 and 1969a). The correct balance between growth hormones and sugars in the embryo may therefore be more critical for germination than the presence or absence of any one kind of compound.

Brown & Morris (1890), MacLeod & Palmer (1966, 1967 and 1969) and Palmer (1967, 1972a, 1974a and 1979) have described many of the morphological changes which occur in the germinating embryo of barley. The scutellum and its 1-cell layer of epithelial cells (Figs 2, 5, 8a) is biologically equivalent to the food-storing cotyledons of dicotyledonous plants. The scutellum contains large reserves of sucrose, raffinose, protein and lipids, and is very active during germination. Fig 5 shows the central region of the scutellum and axis, 20 h after the grain was placed to germinate. The epithelial cells are in the process of elongation. They elongate from 30 to 70μ during the first 72 h of embryo growth. These cells were thought to secrete most (Brown & Morris, 1890) or substantial quantities (Briggs, 1972) of hydrolytic enzymes into the endosperm during the seedling growth phase of malting. This view is no longer tenable for barley (see Palmer, 1974a, 1975a, 1979). The hydrolytic enzymes of the malting grain are now known (see below) to come from the aleurone cells which enclose the starchy endosperm (Fig 8a) and are present in the scutella of excised embryos (Fig 9).

Fig 5 also shows that the cells of the body of the scutellum are more granular in appearance at 20 h than they are in the dry grain. The pro-vascular strand has already undergone lignification and links the shoot, buds and roots of the growing axis with the food reserves of the scutellum. The metabolism of some of the food reserves of the scutellum is illustrated in Figs 6, 7. Fig 6 is an electron micrograph of the 12 h scutellum. It can be seen that even though the embryo is only half-way through the germination process starch synthesis is well advanced and enzymatic hydrolysis of protein and lipid reserves is extensive. The scutellum of the ungerminated grain contains numerous protein and lipid bodies but no starch. At 24 h the protein bodies (Fig 7) have disappeared, the contents of the lipid bodies are reduced and starch synthesis is clearly the dominant event in these cells of the scutellum. A similar sequence of metabolic events also occur in the epithelial cells (Nieuwdrop & Buys, 1964).

It is noteworthy that starch synthesis in the scutellum started before the initiation of enzymatic hydrolysis (modification) of the carbohydrate reserves of the endosperm - i.e. 18-20 h at 25°C (Palmer, 1967, 1969; MacLeod & Palmer, 1966). As regards starch synthesis in the scutellum, Palmer (1967) observed that the isocitritase enzyme is very active in the embryo, and especially the scutellum, during the early stages of germination. It is therefore likely that starch synthesis in the scutellum is promoted by the glyoxylate pathway. Both lipids and the normal sucrose reserve of the scutellum may be, to different degrees, converted into starch. The reasons for this rapid synthesis of starch in the scutellum is not clear. However, it may help to regulate the osmotic balance of the scutellum.

The presumptive transporting tissue of the axis and scutellum contains only pro-vascular cells at 0 h. At 24 h the vascular transporting system is fully developed and exists as two sets of strands in the dorsal half of the scutellum (see Fig 8a, 8b and MacLeod & Palmer, 1966). The heavy reticulate thickening of lignin material which occurs in the vascular cells of the scutellum is shown in Fig 8b. The constituents of this insoluble polymer are likely to have been derived from simple phenolic materials. The vascular system (5-6 cells per strand, Palmer, 1979) is used for transporting exogenous and endogenous gibberellins to the aleurone cells of the endosperm (MacLeod & Palmer, 1966, 1969;

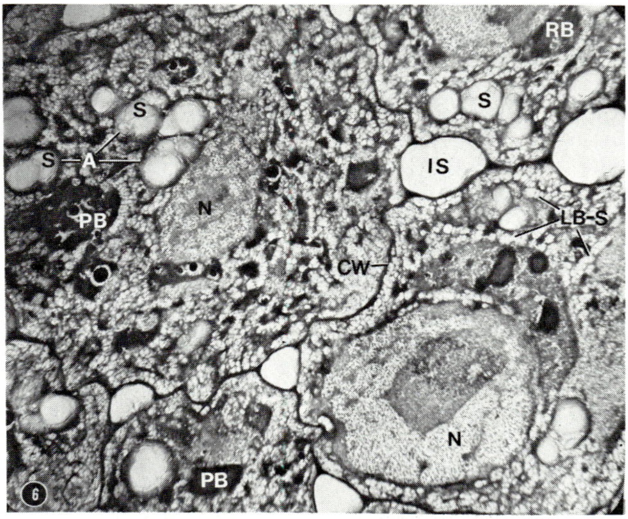

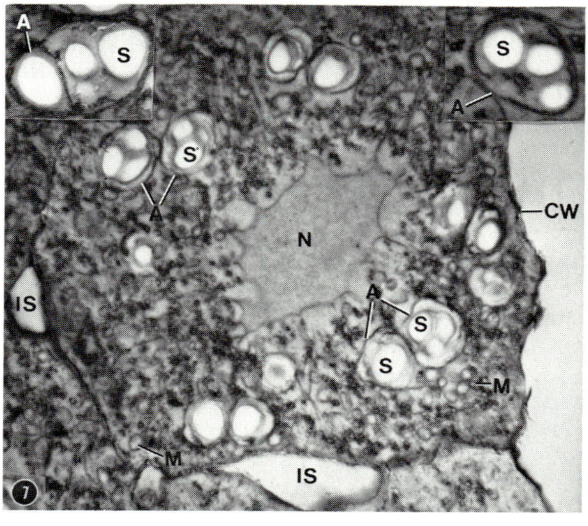

FIGURE 6. Electron micrograph of the scutellum of germinating barley at 12hr. These cells contained protein bodies and much lipid reserves but no starch-containing amyloplasts at 0hr.

N = Nucleus; A = Amyloplasts with developing starch granules; S = Developing starch granules; PB = Protein body; IS = Intercellular space; CW = Cell Wall; LB-S = Lipid Bodies/ Spherosomes.

FIGURE 7. Electron micrograph of the scutellum of 24hr germinated barleys showing more fully developed amyloplasts. Insets show amyloplasts at 48hr.

Palmer, 1972). This system is also used for the reverse transfer of soluble endosperm sugars and proteins to the growing axis. As in other plants, the lignification process in barley embryo is likely to be under the control of auxins which were secreted from the coleoptile tip.

As far as malting is concerned, the most important consequence of the germination phase of development is the production of gibberellic acid. This natural plant hormone passes from the germinated embryo to the adjoining aleurone layer of the endosperm. Within 6 h of entering the aleurone cells it induces the production of hydrolytic enzymes which initiate the asymmetric mobilization of the food reserves of the endosperm (see Figs 2, 8, 8a, reviews by Palmer, 1974; Palmer & Bathgate, 1976 and Yomo 1958; Paleg, 1960; Yomo & Iinuma, 1963; Varner & Ram-Chandra, 1964; MacLeod & Miller, 1964; MacLeod & Palmer, 1966; Cohen & Paleg, 1967; Radley, 1967, 1969 and Groat & Briggs, 1969). With regard to the physiological relationship between the embryo and the endosperm, Cohen & Paleg (1967) reported that the growing barley embryo secreted gibberellic acid (GA_3) into the endosperm during the first two days of seedling growth. Later work by Groat & Briggs (1969) showed that the GA_3 content of the embryo declined after a similar period of growth. However, although the ability of the embryo to produce and secrete gibberellins was intimated by Haberlandt in 1890 and demonstrated by Yomo in 1958, the site of gibberellin-production in the embryo was not known. Using a variety of dissection steps MacLeod & Palmer (1966, 1967) and Palmer (1967, 1979) showed that α-amylase production and enzymic hydrolysis (modification) of the endosperm reserves did not occur if the axis was removed 2 h after the grain was placed to germinate. Enzyme production in the endosperm was more linked to the metabolism of the young roots and scutella node than to the activities of the shoot or the scutellum. Indeed, when the GA-requiring foliar shoot was removed (Fig 3) the enzyme-producing activity of the adjoining aleurone layer of the endosperm was increased over that of the intact control. The gibberellins of the growing embryo are therefore divided between the growth of the foliar shoot and the enzyme-producing activities of the grain. This balance may be an important aspect of the different malting potential of barleys.

Although MacLeod & Palmer concluded that the growing axis was the primary site of gibberellin production, Radley (1968, 1969) concluded that the scutellum was the tissue-site of gibberellin production. It would seem that when the axis is excised at 2 h, the fully excised scutellum can not only produce gibberellins, it can also develop α-amylase in its

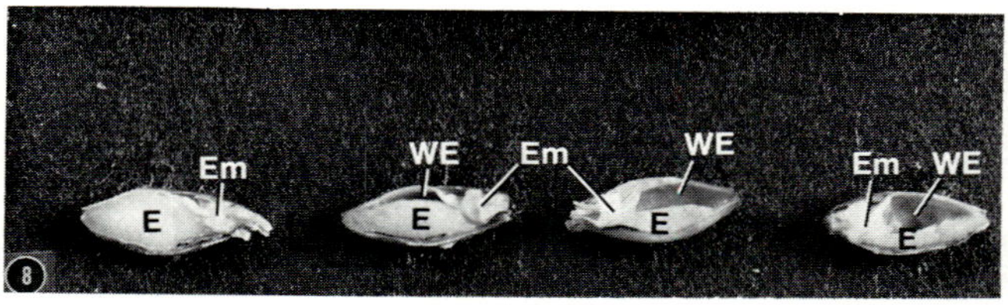

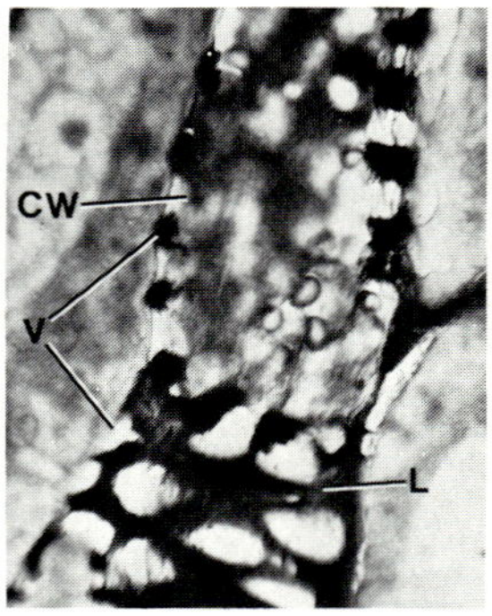

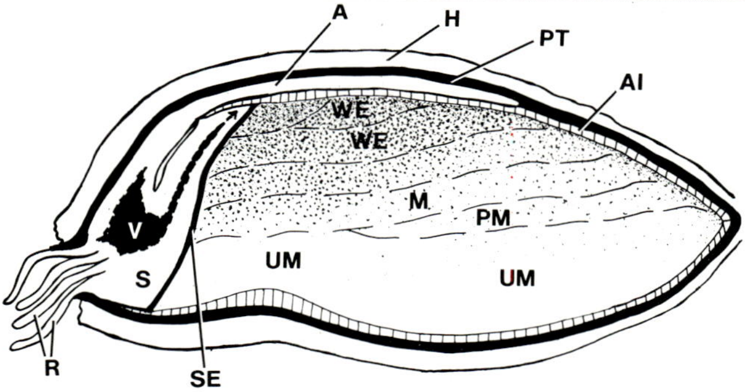

A = Amyloplasts with developing starch granules; S = Starch; IS = Intercellular Space; M = Mitochondria; CW = Cell Wall; N = Nucleus.

FIGURE 8. Light micrograph of washed out longitudinal half of malted (grown) barley grains. Note, roots have been removed and the progressive asymmetric pattern of endosperm degradation (modification) can be seen.

Em = Embryo; E = Under-modified Endosperm; WE = Washed out (Modified) Endosperm.

FIGURE 8a. Diagram of the longitudinal half-section of malted (grown) barley. The different degrees of enzymic modification present in a malted grain are illustrated.

A = Acrospire; H = Husk; PT = Pericarp-Testa; Al = Aleurone; V = Vascular System; R = Roots; WE = Well-modified Endosperm; M = Modified Endosperm; PM = Partly-Modified Endosperm; UM = Under-Modified Endosperm; S = Scutellum; SE = Scutellar Epithelial cell layer.

FIGURE 8b. Electron micrograph showing lignification of a singular vascular cell of the scutellum of germinated barley. V = Vascular element; CW = Carbohydrate Cell Wall; L = Lignin.

peripheral ring of aleurone cells (Palmer, 1967 and MacLeod & Palmer, 1966). The reason why excised scutella develop gibberellins is not clear because if the scutellum, devoid of its axis, is left on its endosperm, it fails to develop gibberellins (Radley, 1969) and α-amylase (MacLeod & Palmer, 1966). Radley proposed that endosperm sugars may act to inhibit gibberellin development in the scutellum if the axis is absent. This suggestion is at variance with the fact that the sugar content of the endosperm is marginally low in the absence of axis growth (Aisien, unpublished). Indeed, most of the sugars of the ungerminated grain is located in the embryo - i.e. the scutellum and the axis (Harris, 1962; Palmer, 1969). Failure to appreciate the complexity of the correlative physiology of the grain has led various authors (Jones & Stoddart, 1978 and Briggs, 1978) to conclude that the sugar content of the endosperm controls gibberellin production in the embryo or α-amylase production in aleurone layers of the intact grain. There is no evidence to suggest that this is the case in the intact grain.

Radley found that the gibberellin of excised scutella was GA_1 whilst that of the intact grain has been variously reported by Yomo & Iinuma (1963), Cohen & Paleg (1967) and Groat & Briggs (1969) to be GA_3. In addition, active synthesis of gibberellic acid in the embryo Groat & Briggs (1969) coincides with the most active phase of sucrose synthesis in the intact scutellum (Palmer, 1969). Both the excised axis and scutella will develop various morphological abnormalities when cultured in vitro. For example, Radley found that excised axes did not develop gibberellins. This is not surprising because 2 h excised axes will show no semblance of normal growth and development. In marked contrast, Cohen & Paleg (1967) showed that intact axes contained, as suggested by MacLeod & Palmer (1966, 1969), more GA_3 than the adjoining scutellum. In all the abovementioned studies of the normal physiology of the embryo, it is clear that completely excised tissues of the embryo can behave abnormally. Thus although excised scutella can produce GA_1 there is no evidence to suggest that the intact scutellum is a major site of early gibberellin production. Also, added sucrose may alter the normal metabolism of the barley scutellum. Indeed, the author has found that sucrose can inhibit GA-induced growth of the foliar shoot. In contrast, the auxin-induced growth of the coleoptile is stimulated by sucrose. In passing, full excision of the intact embryo inhibits the normal elongative growth of the scutella epithelial cells (Palmer, 1967). This is not usually taken into account by many who work on excised embryos and scutella (Briggs, 1972, 1978).

Although, the role of gibberellic acid in the malting grain is fairly well understood, that of the auxin hormones is still not clear. As stated earlier, auxins are likely to be involved in the development of the coleoptile and vascular system of the embryo. At low concentrations they may also be involved in the early elongative growth of the young roots. Additionally, whilst the production of hydrolytic enzymes in the aleurone layer is mainly stimulated by gibberellic acid, low concentrations of auxins (10^{-10}M) may synergize with GA_3 in the aluerone of the malting grain to produce optimal levels of hydrolytic enzymes (MacLeod & Palmer, 1969 and Verbeek-Wyndaele, 1973).

Structural and Physiological Changes in the Endosperm

Gibberellic acid will induce the synthesis and development in the aleurone layer of many hydrolytic enzymes such as α-amylase, endo-protease, limit dextrinase, pentosanases, β-1,3,1,4-glucanases, cellobiase and laminaribiase (MacLeod & Miller, 1962; MacLeod et al. 1964; Briggs, 1964; Varner & Ram-Chandra, 1964; Jacobsen & Varner, 1967 and see reviews by Palmer & Bathgate, 1976 and Bewley & Black, 1978). Higgins et al. (1976) have recently confirmed earlier work of Varner & Ram-Chandra that GA_3 initiates the production of a m-RNA. It is Poly(A) rich and, with the translating cell free system of wheat embryos, will initiate α-amylase production in vitro in 5-6 hours. The production of other important hydrolytic enzymes in the aleurone such as endo-β-1,3-glucanase and phytase is not altogether dependent on the presence of gibberellic acid (Jones, 1971; Palmer, 1973; Preece et al. 1960 and Eastwood & Laidman, 1971). Hydrolytic enzymes such as β-amylase and carboxypeptidase are found in the starchy endosperm and are activated, possibly by endoproteases or reducing compounds during malting (Laberge et al. 1967; Hejgaard 1978; Mikola & Kolehmainen, 1972 and Baxter, 1978).

Gibberellic acid causes a rapid reorganisation and subsequent enzymatic disorganisation of the sub-cellular structure of the aleurone tissue (Paleg & Hyde, 1964; Jones, 1969 and Van der Eb & Nieuwdrop, 1967). α-Amylase production is very active during the reorganisation phase of development, and is linked with the development of polysomes and rough endosplasmic reticulum (Jones, 1969; Evins & Varner, 1971 and Locy & Kende, 1978). The aleurone cells contain lipid and protein carbohydrate bodies (the aleurone grain). The latter probably provides the amino acids from which many of the hydrolytic enzymes are produced (Jacobsen & Varner, 1967; Melcher & Varner, 1971 and Jacobsen et al. 1971). Production of these hydrolytic enzymes requires oxygen (MacLeod & Miller, 1962). Potassium bromate is a protease inhibitor and is used to reduce root growth and proteolysis during malting (Macey & Stowell 1961 and Palmer & Bathgate 1976). Reduced proteolysis in the aleurone will also cause a reduced production of α-amylase (Melcher & Varner, 1971).

Although the synthesis of enzymes such as α-amylase, endo-β-glucanase and endo-protease is of primary importance in malting, effective degradation of the starchy endosperm will only occur if these and other hydrolytic enzymes are rapidly released from the aleurone layer. Although complex theories regarding the active release across the plasmolemma of enzymes from the aleurone have been proposed (Varner & Meuse 1972 and Locy & Kende, 1978), some release may simply occur because the entire membrane system of the aleurone cytoplasm autolyses after enzyme synthesis. In this context of morphological change, Taiz & Jones (1970) have reported that endo-β-1,3-glucanase is primarily involved in the degradation of the aleurone cell wall and the release of hydrolytic enzymes into the endosperm. This conclusion proved to be incorrect when it was discovered that the aleurone cell walls contain 85% arabinoxylan and 8% cellulose (McNeil et al. 1975). It is therefore noteworthy that MacLeod & Miller (1962) reported some time ago that gibberellic acid stimulated the production of xylanase. Palmer (1973) also reported that xylose and pentosan oligosaccharides were released from the aleurone cells in significant quantities if they were treated with gibberellic acid. Taiz and Honigman (1976) have confirmed these earlier reports and concluded that the release of hydrolytic enzymes from the aleurone layer is controlled by the activity of the endo-β-1,4-xylanase enzymes.

The quantity of enzymes which is produced in the germinated grain is not only dependent on the gibberellin-producing potential of the embryo (Horgan, 1970). The physiological properties of the aleurone layer also determine the levels of enzymes which are produced in the grain. Thus, in equivalent experimental conditions the isolated aleurone layers of the six-rowed barley Olli produced more α-amylase than did two-rowed barleys such as Proctor, Zephyr or Golden Promise (Palmer, 1978). In addition, α-amylase required less gibberellic acid for optimal production than did endo-β-1,3/1,4-glucanase (Palmer, 1973).

Activating the aleurone layer to produce enzymes is not the only property of the grain which can influence the malting performances of barleys. There is a large body of evidence which suggests that the structural properties of the starchy endosperm can affect the rate at which its own reserves are hydrolysed during malting. The three major reserves of the endosperm are starch (55-65%), proteins (12-8%) and cell walls

(8–10%), Enzymatic action in the endosperm during malting facilitates the release of starch and protein extract during brewing. A primary objective of malting research is to acquire a better understanding of the relationships between the enzymes of the endosperm and the substrates which they degrade during malting (see Harris, 1962; Palmer & Bathgate, 1976).

Fincher (1975) and Ballance & Manners (1978) have shown that the cell walls of the endosperm (Figs 10, 11, 12) contain 70–75% β-glucan, 20–25% arabinoxylan, 3–5% protein and about 2.0% mannan. The β-glucans are therefore the primary constituents of the walls of the storage cells of the endosperm (Figs 10, 11). Chemically, the β-glucans contain β-D-glucose residues, 70% of which are β(-1,4-)linked and 30% are β(-1,3-) linked. β-Glucans which are extracted in water at 65°C from milled barley contain the same ratio of 1,4/1,3-links as β-glucans extracted at 40°C. However, the former have longer 1,3-linked units,are more viscous and more susceptible to the action of the endo-β-1,3-glucanase enzyme complex (Bathgate et al. 1974). The arabinoxylan polymers contains β(-1,4-)linked residues by D-xylose. Side chains consisting of α-arabinofuranose residues at the 2 and/or 3 positions of the xylan backbone are usually present. Generally, the chemical nature of these materials and the fact that they are major constituents of the cell walls which entrap the starch and protein reserves has led many researches to conclude that β-1,4/1,3-glucanases, the β-1,3-glucanases and the pentosanases (xylanases) must play a primary role in extract development during malting (Preece & Mackenzie, 1952; Meredith & Anderson, 1955; Preece & MacDougall, 1958; MacLeod et al. 1964; Luchsinger et al. 1965; Manners & Marshall, 1969; Bathgate & Dalgleish, 1974 and see reviews by Palmer & Bathgate, 1976). Although significant degradation of β-glucan occurs during malting (Forrest & Wainwright, 1977) only limited hydrolysis of the arabinoxylan polymers takes place in the endosperm (Preece & Hobkirk, 1955; Preece & MacDougall, 1958 and Preece, 1964).

From a structural viewpoint MacLeod and McCorquodale (1958) and Palmer (1971) have reported that the middle lamella which joins the endosperm cells is mainly protenaceous (see Fig. 12). Palmer (1975) has also reported that both the protenaceous lamella and the arabinoxylans are readily extracted from the walls of the endosperm cells by diluted solutions of sodium hydroxide. The remaining cell walls (see Fig. 13) contained mainly β-glucan which dissolved more slowly. Palmer therefore suggested that although β-glucans are rapidly degraded when released from the cell walls of the malting grain, proteolytic and pentosanase enzymes may have

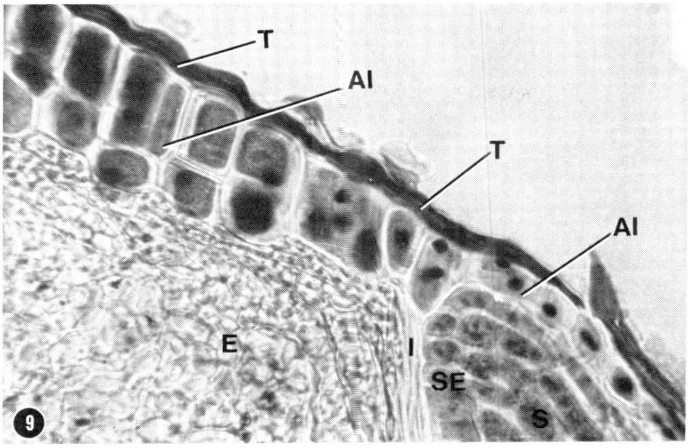

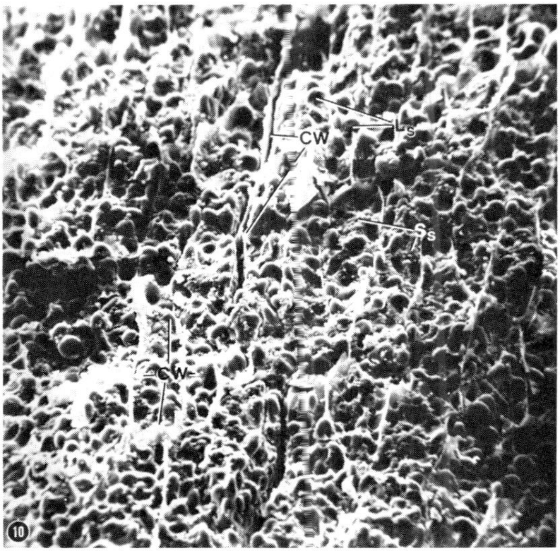

FIGURE 9. Light micrograph showing testa and the tissue relationship between aleurone layer, the endosperm and the scutellum of the embryo of barley.

T = Testa (2 fat-staining layers); Al = Aleurone layer; I = Intermediate layer; S = Scutellum; SE = Scutellar Epithelial cells; E = Starchy Endosperm.

FIGURE 10. Scanning electron micrograph of the starchy endosperm of Ohr barley. Note cellular compartmentation of endosperm reserves into discrete sacs.

CW = Cell Wall; L_S = Large Starch Granules; S_S = Small Starch Granules + Storage Protein Matrix.

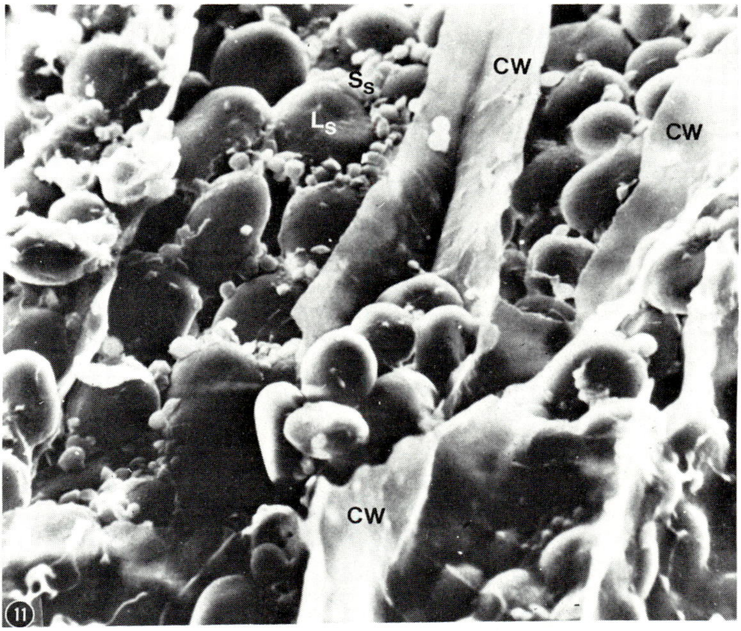

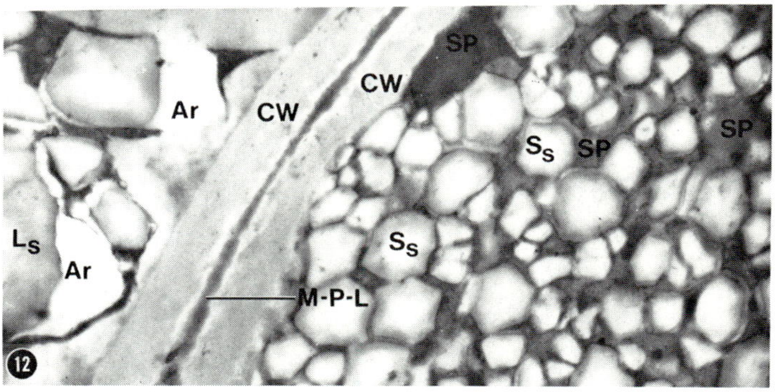

FIGURE 11. Scanning electron micrograph of the starchy endosperm of barley. Note the β-glucan/pentosan cell walls and the structural relationship between the large and small starch granules.

S_s = Small Starch Granules; L_s = Large Starch Granules; CW = Cell Wall. Note that not all large granules are associated with small granules.

FIGURE 12. Transmission electron micrograph of the starchy endosperm of barley. Note close association between the small starch granules and cell wall with electron dense protein materials.

CW = Cell Wall; SP = Storage Proteins; S_s = Small Starch Granules; Ar = Artifact; L_s = Large Starch Granules.

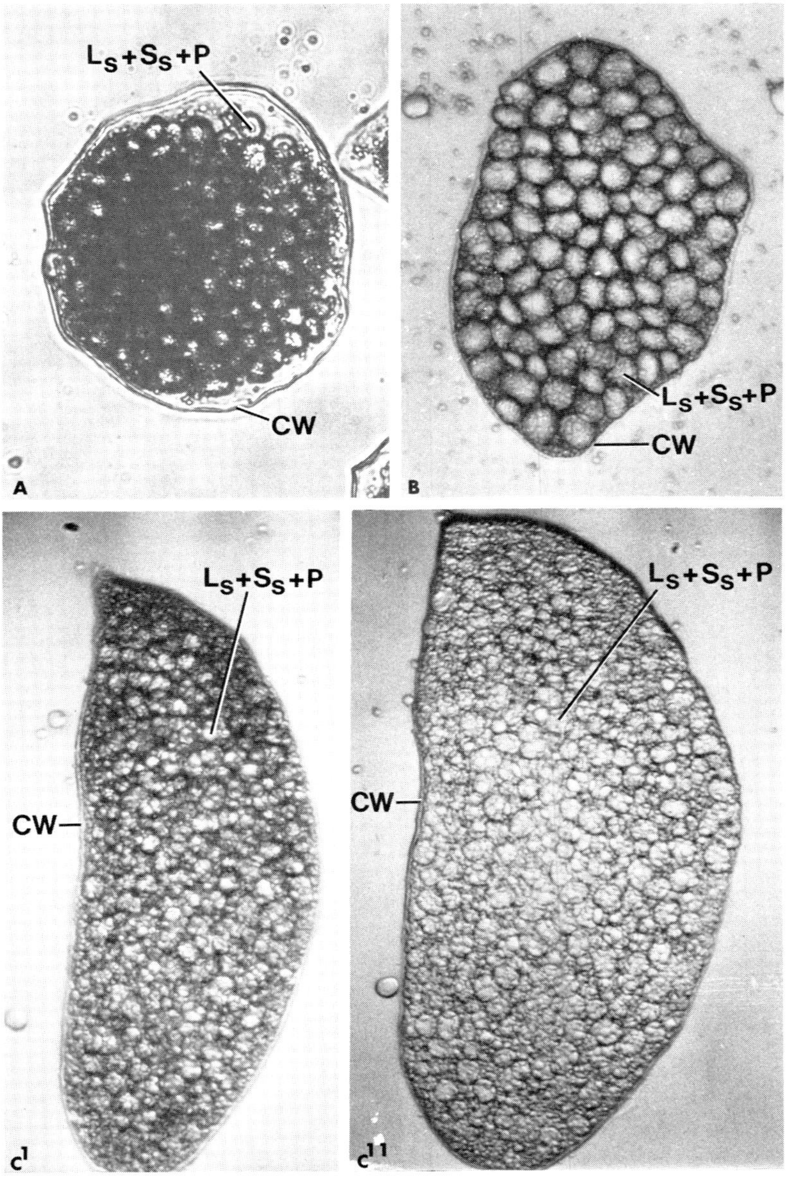

FIGURE 13. Showing the water mounted, alkali separated starchy endosperm cells of Julia (a), Proctor (b) and Senta (c). Compare the more rigid structure of the Julia cell with the more elastic properties of those of Proctor and Senta (C^1 to C^{11}) especially.

L_S + S_S + P = Large and Small Starch granules + Storage Protein Matrix; CW = Cell Wall.

primary roles in the release of starch and protein extract during malting. Although Burger (1973) and Baxter (1979) have reported that peptide hydrolases and carboxypeptidases may help to release β-glucans during malting, it should not be overlooked that enzymatic modification of the endosperm begins when the pentosanases, endo-proteases and β-glucanases are released from the aleurone layer.

Baxter et al. (1978) have suggested that endo-proteases play a major role in the rate at which different barleys released their extract during malting. Although these endo-proteases (Hordeinases) are likely to play an important role in extract development, earlier work by Palmer (1975) indicated that even when the fast-malting Proctor or the slow-malting Julia barleys are induced by gibberellic acid to produce equivalent levels of β-glucanases, α-amylases and proteases, Proctor always showed the faster malting rate. The work of Baxter et al. 1978 also showed that endo-proteases levels had a greater influence on the malting rate of Julia than on the malting rate of Proctor. Palmer (1975) suggested that differences in the structural properties of the endosperm were the main causes of the different malting behaviours (see below).

Figs 14, 10, 11 show the basic structure of the starchy endosperm of barley (see Palmer, 1972 and Pomeranz 1974). The large (25μ) and small (5μ) starch granules are embedded in a matrix of reserve proteins (glutelin and hordein). About 35–40% of the reserve proteins is solubilized during malting. In contrast only 5–10% of the starch is degraded. Although the release of starch from the confines of the endosperm cells is more important to the brewer than starch degradation in the malting grain, it is worth noting that proteolytic enzymes exposed the starch granules to extensive or limited amylolytic attack during malting. Thus, Palmer (1972) and Bathgate & Palmer (1972; 1973) showed that deproteinated starch granules were more susceptible to attack by α-amylase and that proteolytic and amylolytic attack on the starches in the malting grain made them more susceptible to further hydrolysis during brewing. Recent work by Slack et al. (1979) suggests that the carboxypeptidase enzymes may also play a part in preparing the starch granules for subsequent amylolytic attack. Fig 15 shows a well-modified area of the endosperm of a malted barley grain. It is clear that the protein matrix and small starch granules are extensively degraded. The limited surface pitting of the large starch granules is mainly brought about by α-amylase (Manners & Bathgate, 1969). This pitting conceals more extensive internal attack of the granules - see Insets A and B. The latter is stained with iodine solution. Such internal digestion of malt-starch may, in rapid brewing cycles, facilitate starch conversion to sugars and oligosaccharides

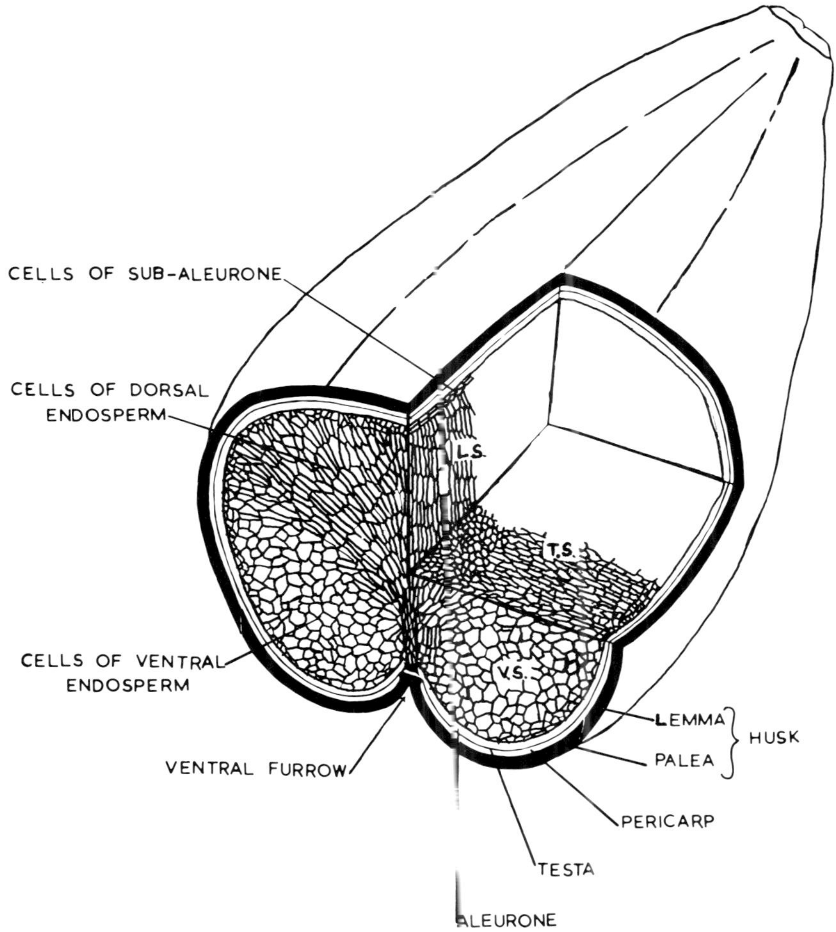

FIGURE 14. The general cellular plan of the starchy endosperm of barley is illustrated.
L.S. = Longitudinal Section; T.S. = Tranverse Section; V.S. = Vertical Section.

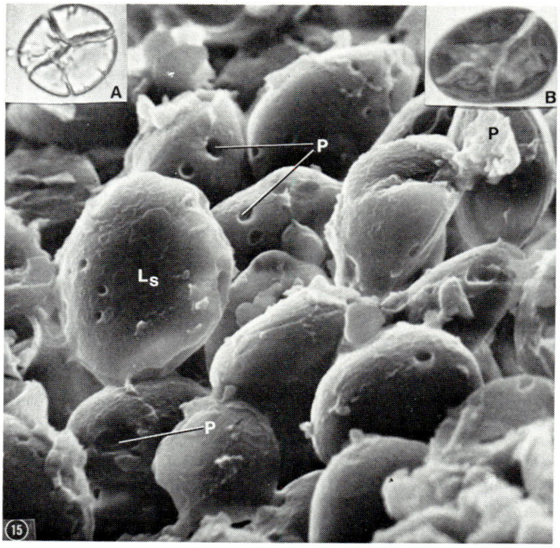

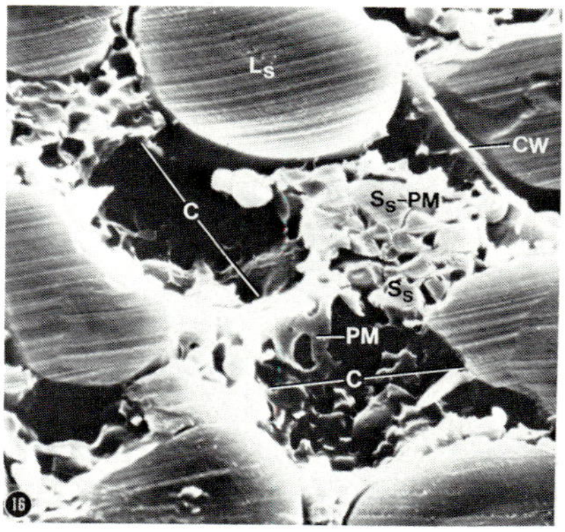

FIGURE 15. Scanning electron micrograph of the enzymically well-modified area of the endosperm of malted barley. Note absence of cell wall, and small starch granule-protein matrix material. Compare with Figs. 11, 16, 17, 18 of barley. P = Pits caused by amylolytic enzymes; L_S = Large Starch Granules.

Inset A = Light micrograph of large starch granule of malt showing internal attack by amylolytic enzymes.

Inset B = Same as Inset A but stained with diluted Iodine.

FIGURE 16. Scanning electron micrograph of the mealy endosperm of Proctor. Note crumb-like open texture of the small starch granule - storage protein matrix area.

L_S = Large Starch Granule; S_S + PM = Small Starch Granule + Protein Matrix; C = Cavity of large starch granule. CW = Cell Wall.

(see Greenwood & Thompson, 1961; Manners & Bathgate, 1969; Enevoldsen & Schmidt, 1973 and Palmer & Bathgate, 1978). The association of "starch units" in the amyloplasts of the developing endosperm of barley (Williams & Duffus, 1978) is similar to that of the amyloplasts of the scutellum (Figs 6, 7). The inset A of Fig 15 suggests that early amylolytic attack on the internal areas of endosperm starch occurs most readily between the existing "units" of the granule (see Palmer, 1972).

This brief survey of the action of the major hydrolytic enzymes which effect protein solubilization and starch release reveals little of the structural properties of the endosperm which limit enzyme action and malting rate. Thus, although increased hydration and the use of gibberellic acid will accelerate the malting rate of slow-malting barleys, the response of better quality barleys is always quicker (Narziss & Friedrich, 1970; Palmer, 1974, 1975 and Baxter et al. 1975). Numerous studies have been carried out to identify those endosperm properties which may limit the rate of malting. Greenberg (1974), Iruegas et al. (1973) and Bendelow (1975) have suggested that slow-malting barleys tend to have high viscosities and large quantities of β-glucan (Anderson et al. 1978). Morgan (1977) has recently reported that unspecified changes in the viscosity curves of barleys indicate malting quality. Curiously enough most of these studies of the relationship between β-glucan and malting quality developed from the earlier studies of Preece & Mackenzie (1962), Preece & Aitken (1954), Preece et al. (1954) and Bourne & Pierce (1970). However, Preece concluded from detailed work that the β-glucan content of a barley was not related to its malting quality. He also suggested that the viscosities of the β-glucans which are extracted from milled barleys were likely to give a variable and misleading indication of malting quality (Preece, 1964). Chemical studies of the cell walls by Fincher (1975) of the fast-malting Maris Otter and the slow-malting Julia showed that the entire β-glucan content of the walls of both barleys contained β-1,4 to 1,3-links in a ratio of 2.5/1. Palmer (1975) could find no difference in the β-glucan levels of Proctor and Julia barleys. However, the isolated cell walls of Julia released β-glucan more slowly. The released β-glucans of Julia were more susceptible than were those of Proctor to the action of β-1,3-glucanase enzymes. Since very little is known about the pentosan or the alkali-soluble (glutelin-like) lamella protein of the native cell wall, the overall influence of the native cell wall on extract release in malting barleys is still not known. Recent studies of the endosperm properties which may control malting rate have been mainly concerned with the general structure of the endosperm and with the polypeptide patterns of the hordein food reserves (see below).

Fig 14 shows the cellular arrangement of the starchy endosperm cells of a European two-rowed barley. Palmer & Harvey (1977) reported that the cells of the dorsal endosperm are smaller and more elongated than the cells of the basal check areas. The cellular organization in the fast-malting Proctor and the slow-malting Julia was found to be similar. Fig 10 shows a longitudinal section of the endosperm of a malting barley. The close packing of the starch-containing cells of the endosperm is clearly illustrated.

Mealiness and steeliness of barley endosperms have, for a long time, been regarded as useful indications of malting quality. Mealy grains of a barley sample usually contain less protein than corresponding steely grains. The view that high nitrogen barleys tend to malt slower than low nitrogen barleys is not always correct. Indeed, good malting barleys usually malt faster than poor quality barleys even with a nitrogen difference of 1-2%. Detailed microscopic examination of Proctor and Julia barleys by Palmer (1975) and Palmer & Harvey (1977) has revealed important differences in the structure of the endosperms of these two malting barleys. Fig 16 shows the structure of the intracellular starch-protein matrix of the mealy endosperm of Proctor. The small starch granule-protein matrix is crumb-like. The corresponding mealy grains of Julia had a similar structure but the overall appearance of the matrix was more rigid. Fig 17 shows the starch-protein matrix of the steely endosperm of Proctor. Again, the corresponding grains of Julia were of equivalent structure but appeared to be more compact (see Palmer & Harvey 1977). With regard to the percentage of mealy and steely grains which were present in the separate bulk samples of Proctor and Julia (both at 1.7% nitrogen): the Julia sample contained 5% of <u>very</u> steely grains (2.1% nitrogen), 5% mealy grain (1.5% nitrogen) and 90% of steely grains (1.7% nitrogen). In contrast, the Proctor bulk sample contained 5% steely grains (2.0% nitrogen) and 20% mealy grains (1.5% nitrogen). The remaining 75% of Proctor grains were neither mealy nor steely in structure and contained 1.75% nitrogen (Fig 18). The structures of the mealy (Fig 16), steely (Fig 17) and intermediate type grains (Fig 18) indicated that, in the Proctor bulk sample, increased steeliness was associated with increased nitrogen. However, the relationships between grain nitrogen and mealiness and steeliness in the slow-malting Julia was less clear. It is noteworthy that the nitrogen content of the 90% steely fraction of Julia grains was similar to that of the 75% fraction of Proctor grains which were <u>less</u> steely in structure. It is therefore clear that equivalent nitrogen

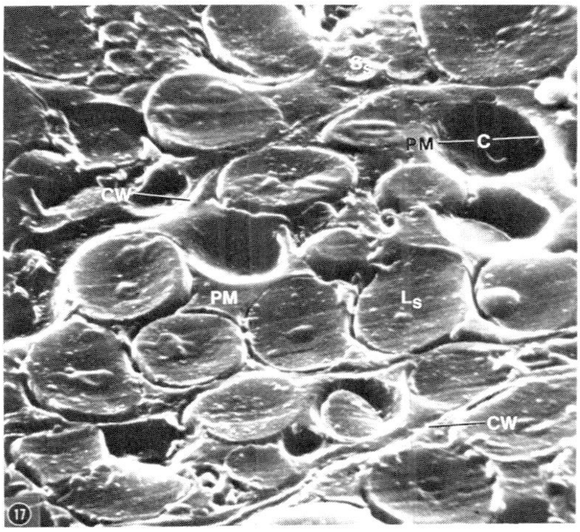

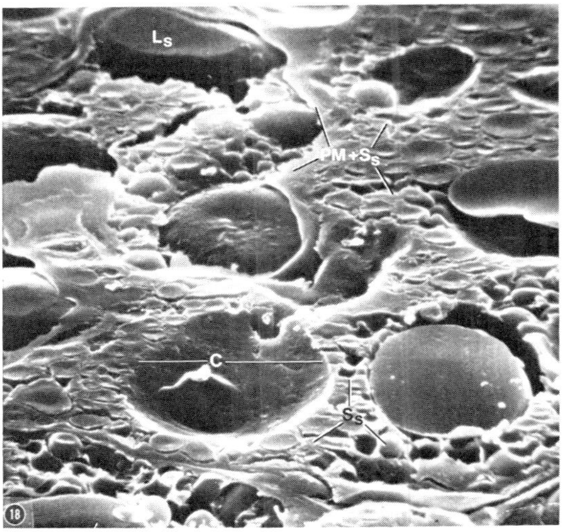

FIGURE 17. Scanning electron micrograph of the starchy endosperm of the steely endosperm of Proctor. Compare with Fig. 16 and note that here the small starch granule-protein matrix area is more compact giving a steely appearance.

L_S = Large Starch Granule; S_S = Small Starch Granule; PM = Protein matrix; C = Cavity of large starch granule; CW = Cell Wall.

FIGURE 18. Scanning electron micrograph of the starchy endosperm of Proctor showing reduced steely structure. Compare with the more mealy (Fig. 16) and the steely (Fig. 17) endosperms of Proctor.

L_S = Large Starch Granules; PM + S_S = Protein Matrix and Small Starch Granules.

values of barleys do not reflect equivalent states of endosperm morphology. This observation may, in part, explain why high nitrogen levels in Proctor and Maris Otter are less inhibiting to malting rate than equivalent nitrogen levels in Julia and Zephyr. Generally, the structural (cytoplasmic) framework on which the reserve proteins is deposited in the developing endosperm may be a more important indicator of malting quality than traditional concepts of mealiness, steeliness and nitrogen levels. The development of steeliness may be more restricted in Proctor than in Julia. The true relationship between nitrogen content, steeliness and malting performance is still not understood. Structurally, the situation is very complex and is unlikely to be resolved by the general extraction and characterization of the polypeptides of unseparated bulk samples of barley grains (see Scriban & Strobbel, 1979 and below). Preliminary microscopic studies of some of the separated grain types of Proctor, Julia and Maris Trojan barleys show that a superficial appearance of mealiness or steeliness is not an adequate guide to the subcellular structure of the endosperm. For example, the endosperm of Maris Trojan has a mealy appearance but the matrix protein and the small starch granules are tightly packed (see Fig 19). Thus low nitrogen levels in Maris Trojan is not linked with the crumb-like mealiness which is found in Proctor or Julia. In other cereals such as sorghum, the central mealy area of the grain can contain as much nitrogen as the peripheral area which is extremely steely (see Fig 20 and Palmer, 1972). In sorghum, steeliness is caused by the tight packing of the starch granules. It is interesting to note that while this kind of packing is not typical of the endosperm of modern malting barleys, it can be found in the endosperm of the wild <u>Hordeum spontaneum</u>.

Studies of the milling behaviour barleys (Palmer, 1975 and Allison et al. 1976) have suggested respectively that "good" barleys tend to release large quantities of free large starch granules or fragment readily during milling. The release of starch granules in the sedimentation test (Palmer, 1975) is brought about during the limited milling period. If milling is excessive (Baker & Amr, 1978) the relative rates of sedimentation, illustrated in Fig 21 are not obtained when similar weights of milled flour are made to sediment in cold 70% ethanol. Although the fast malting barleys such as Proctor, Maris Otter, Imber and Wing have high sedimentation scores (cloudy suspension) whilst slow-malting barleys have correspondingly low sedimentation scores, this clear difference in milling behaviour possibly indicates previously discussed structural differences of the endosperms of Proctor and Julia which are likely to accelerate or retard malting rate. Since

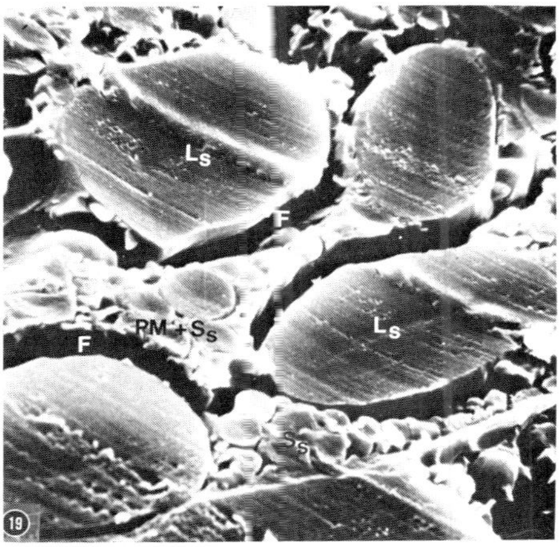

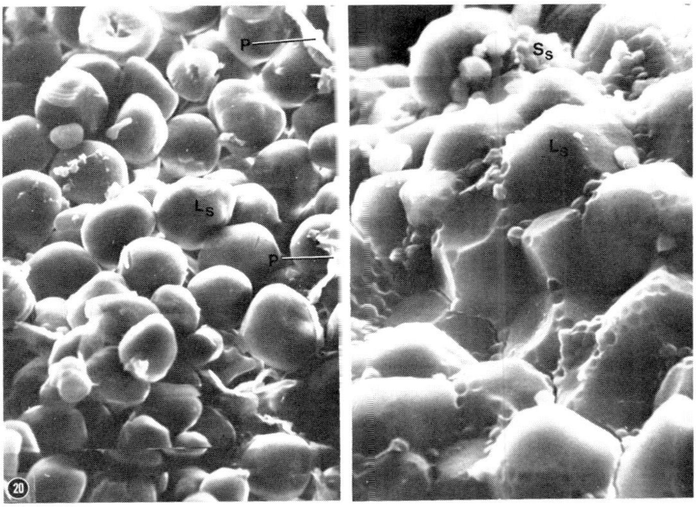

FIGURE 19. Scanning electron micrograph of the starchy endosperm of Maris Trojan barley. Although of meal appearance, contains compact steely-like matrix which causes fissures.
L_S = Large Starch Granule; S_S = Small Starch Granule; PM + S_S = Protein Matrix + Small Starch Granule; CW = Cell Wall; F = Fissures.

FIGURE 20. Scanning electron micrograph of the mealy and steely areas of the endosperm of a sorgum grain. Note that in this grain, in contrast to barley (Fig. 16 and Fig. 17), mealiness and steeliness are more strongly influenced by packing of the large starch granules rather than by the nature of the small starch granule-protein matrix materials.
P = Protein; L_S = Large Starch Granule; S_S = Small Starch Granule.

the sedimentation test assesses the structural nature of the cell wall and proteins of the endosperm, it cannot be used as an overall guide to malting quality which is dependent on other properties of the grain and the conditions of malting (see above and Palmer, 1975; Palmer & Harvey, 1977 and Ewertson, 1977). The same reservations can be drawn about the milling energy test of Allison et al. (1976) where again barleys such as Proctor, Wing and Maris Otter give a lower milling energy than slower malting barleys such as Julia, Maris Mink and Lofa Abed.

The possible influence of the hordein reserves on malting quality is not understood. Hordeins are regarded as the reserve protein fraction that is soluble in hot 70% ethanol or 55% propan-2-ol + 2% 2-mercaptoethanol. Brandt & Ingversen (1978) have reported that the hordein content of the endosperm is determined by an 11S messenger RNA. The newly synthesised hordein is transferred to protein bodies which evenutally disappear in the mature grain (Munck & Wettstein, 1976). Although the exact locations of hordein and glutelin in the endosperm are not known, Nielsen et al. (1970) have suggested that the glutelin protein is closely associated with the cell wall and other carbohydrate materials of the endosperm. The hordein material, as stated above, is closely associated with the small starch granules. About 60% of the hordein reserve of the endosperm is degraded during malting (Pollock & Kirsop, 1959 and Pollock, 1962).

Recent studies by Shewery et al. (1978) have shown that the hordein polypeptides of different barley varieties will separate electrophoretically into three major groups: "A", "B" and "C". The barleys tested were placed into different groups on the basis of their "B" and "C" polypeptide patterns because the "A" pattern was common to all varieties. It is noteworthy that when the sedimentation scores of these barleys (see Palmer, 1975a and 1975b) are compared with their polypeptide patterns (see Shewery et al. 1978), a highly significant relationship is observed. For example, almost all the varieties which Shewery et al. placed into the Proctor and Wing groups had high sedimentation scores (e.g. Imber, Gerkra, Berac, Golden Promise and Armelle) whilst those of the Julia, Maris Mink and Lofa Abed groups had low sedimentation scores (see Fig 21). Although the relationship between the glutelin protein and the sedimentation scores of barleys is not known, it is possible that the structural properties of the hordein proteins and those of the glutelin-carbohydrate materials of the cell walls may combine to limit the physical (by milling) and the enzymatic (by malting) release of starch extract. There is little direct evidence to suggest that the role of hordein in milling and malting is more important than that of

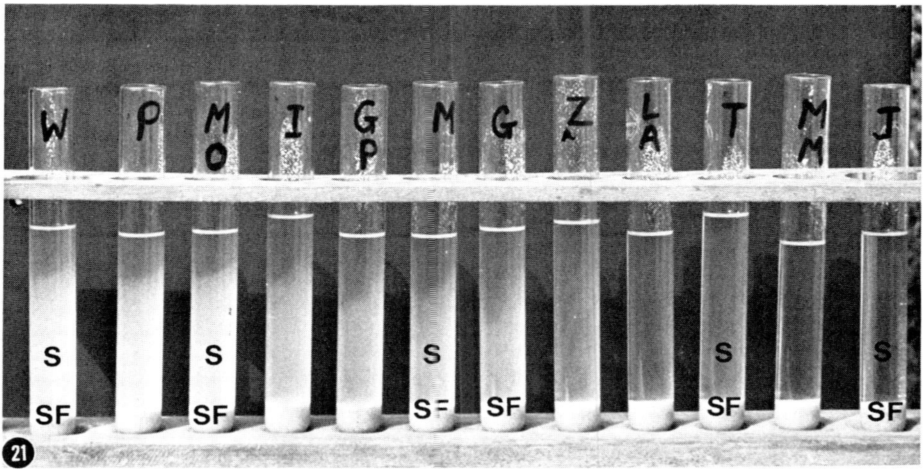

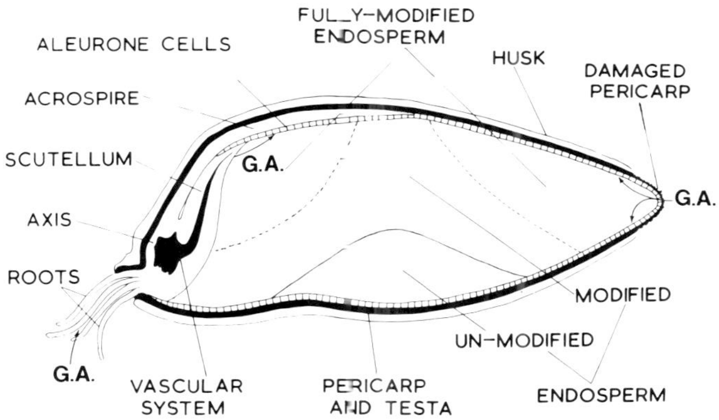

FIGURE 21. Light micrograph showing that the milled flours (250μ fraction) of European barleys varieties have different sedimentation behaviours in cold 70% ethanol. Sedimentation time = 5 minutes.

S = Sedimenting fraction; SF = Sedimented fraction. W = Wing; P = Proctor; M.O. = Maris Otter, I = Imber; G.P. = Golden Promise; M = Mazurka; G = Gerkra; Z = Zephyr; L.A. = Lofa Abed; T = Tern; M.M. Maris Mink; J = Julia.

FIGURE 22. Illustrates the two ended pattern of enzymic modification of the endosperm of abraded grains which causes such grains, when treated with gibberellic acid, to malt faster than control (unabraded) grains (Fig. 2). Abrasion damage to the side of the grain also accelerates malting but is less efficient.

G.A. = Gibberellic Acid.

the cell wall. Indeed, there is evidence that although Julia modifies more slowly than Proctor, localized protein breakdown can occur in Julia during malting. This suggests that the limitations which hordein places on starch extract development may be secondary to those of the native cell walls of the endosperm.

Various features of the morphology of the endosperm cells of Julia, Proctor and Senta are shown in Fig 13. As stated above (Palmer, 1975) diluted (0.4%) sodium hydroxide will dissolve the lamella protein, pentosan and some β-glucan materials from the walls of the endosperm cells (Fig 10, 11, 12). The cells then separate and the remaining cell wall contains mainly β-glucan. Fig 13a shows that the cell walls and the reserve-protein and starches of the separated Julia cell are fairly rigid. Those of Proctor (Fig 13b) shows full expansion. Further expansion tends to rupture the cell walls of this variety (Palmer, 1975). In marked contrast, the separated cells of Senta (Fig 13c) expand rapidly after separation. The relationships between cell expansion in Senta and malting performance are not clear. However, the rigid structure of the Julia cell, in contrast to the more flexible one of Proctor, confirm the foregoing observations that, structurally, the endosperm of Julia is more compact and "plastic" than that of Proctor. This plasticity may reflect a higher degree of polymerization of cell wall and protein materials.

Although the embryo, the aleurone and the starchy endosperm play different parts in the malting process, optimal activity between the separate tissues is the basis of efficient malting. Thus, high out-of-steep moistures tend to increase proteolysis, root growth and reduce malting time. However, reduced production of some amylolytic enzymes may also result. Figs 2, 8, 8a also show that since the normal pattern of enzymic modification of the endosperm is asymmetric, excessive application of gibberellic acid can cause over-production of enzymes at the embryo end of the endosperm. The benefits of increased enzyme production and extract development is often reduced by excessive proteolysis and colour-forming potential. Uneven modification of the endosperm may also cause poorly hydrolysed proteins, β-glucan and pentosans to pass into the finished beer (Enari & Loisa, 1974; Jacobsen et al. 1974; Bellmer, 1975; Gersten, 1970; Patel, 1978 and Schur & Pfenninger, 1978).

Overall consideration of the one-sided pattern of enzymatic modification of the endosperm (Figs 2, 8, 8a) leads to the conclusion that malting conditions which limit aleurone activity will increase the tendency of the grain to malt unevenly. Palmer (1969, 1971, 1974, 1978 and 1979a) has shown that significant improvements in the malting rate can be achieved if barley grains are abraded. Abrasive treatment causes limited damage to the pericarp layers over the distal (non-embryo) half of the grain which then absorbs gibberellic acid directly. The embryo and the usually slow-malting distal half of the malting grain (Fig 8a) now produce enzymes simultaneously. Although the total level of hydrolytic enzymes of an abraded grain is only increased by 15-20% that of the distal half is doubled. The consequence of this improved distribution of enzymes in the endosperm is that abraded grains malt faster than non-abraded grains. Generally malting time is reduced and extract yield (both nitrogen and carbohydrate) is increased (Fig 22 and Palmer, 1974).

Physiologically, gibberellic acid (endogenous and exogenous) is the primary factor in malting technology (Sandegren & Beling, 1959 and see above). The dramatic accelerating effect of the abrasion process on endosperm hydrolysis (modification) suggests that the physiological basis of accelerated malting of all barley varieties rests primarily in increased aleurone activity - not merely in terms of enzyme level but in terms of enzyme distribution. The great body of physiological work which has accumulated since Hopkins and Krause (1937) has narrowed down the important objectives of malting science and technology. Very few barleys combine high grain yield with good malting quality. It has been far easier to select for yield than for quality factors such as fast malting rate and high enzymatic potential.

In general, the inheritance of malting quality is poorly understood and is difficult to achieve (Bell, 1962; Sparrow, 1970). Thus, the future use of slow-malting barleys of high yield must be associated with improved developments in malting technology. Although small exogenous doses of gibberellic acid and high out-of-steep moistures can improve the malting rate of difficult barleys, localized over-modification can occur and problems of extract loss, poor solubilization of starch and protein materials and reduced filtration and stability of the beer may result (Bellmer, 1975). Despite the technical limitations of abrading machines (Palmer, 1974;1979) the process is being used to significant commercial advantage, especially by brewing companies which have their own maltings. The improved enzymatic activity of the whole aleurone layer of abraded grains has minimized the processing difference between slow- and fast-malting barleys (Palmer & Barrett, 1973; Northam & Button 1973; Halford et al. 1976; Schlimme, 1977 and Hyde & Brooks, 1978). A recently published paper (Smith & Briggs, 1979) has claimed from limited studies on punctured and infected dehusked grains that distal end modification and α-amylase production do not occur in abraded grains. However, this view was not supported by their references and experimenal results. Final consideration of the physiological principles on which the grain is constructed leads to the conclusion that new malting techniques which promote general, rather than localized, aleurone performance are likely to be of greatest value to the maltster of the future (Macey, 1974; Palmer 1974).

Addendum: Using a most promising immunofluorescent technique. Gibbons, G.C., (2nd International Symposium on Pre-harvest Sprouting Damage in Cereals, Cambridge, 1979) has reported that the scutellum or the epithelial cells of barley can secrete large quantities of one species α-amylase into the endosperm of the germinated grain. Unfortunately, the α-amylase contribution of the contaminating aleurone cells of the scutellum was not determined. Furthermore, whilst the immunofluorescent response of the endosperm was distinct that of the scutellum (or aleurone layer) was surprisingly poor. Although the scutellum was given an important enzyme secreting role during the early stages of seedling growth, the evidence presented did not rule out the possibility that significant quantities of α-amylase could have been secreted down from the aleurone cells across the surface of the epitheial cells rather than our from the epithelial cells. In fact, as far as enzymic degradation of the endosperm is concerned, Gibbon's Figs 5, 6 and 7 have confirmed that the pattern of enzymic hydrolysis of the endosperm is asymmetric. As suggested above by the author, the distribution of α-amylase in the endosperm of malting barleys is not a reliable guide to the general pattern of endosperm hydrolysis. Indeed, a better understanding of the importance of the aleurone layer in the mobilization (modification) of the endosperm of barley may occur when this immunofluorescent technique is applied to more important hydrolytic enzymes such as the endo-β-glucanases and the endo-proteases.

Although the general physiology of cereal embryos is poorly understood, the presence of α-amylase, in some cereal grains, may reflect sprouting (Manchylo, B.A., La Croix, L.J. and Kruger, J.E., also in 2nd Sprouting Symposium, Cambridge, 1979). However the physiological importance of this enzyme in the malting grain is more related to its site of production and limited action rather than to its general localization in complex organs such as the scutellum and the endosperm.

REFERENCES

Aisien, A. (1979). Personal Communication.

Allison, M.J., Cowe, I., and McHale, R.J. (1976). J. Inst. Brew. 84, 166.

Anderson, M.A., Cooke, J.A., and Stone, B.A. (1978). J. Inst. Brew. 84, 233.

Baker, C.W., and Amr, A.S. (1978). Amer. Soc. Brew. Chem. 36, 87.

Ballance, G.M., and Manners, D.J. (1978). Carbohydrate Research 61, 107.
Bathgate, G.N., and Dalgleish, C.E. (1974). Amer. Soc. Brew. Chem. Proc. 32,
Bathgate, G.N., Palmer, G.H., and Wilson, G. (1974). J. Inst. Brew. 80, 278.
Baxter, E.D., Booer, C., and Palmer, G.H. (1974). J. Inst. Brew. 80, 549.
Baxter, E.D. (1978). J. Inst. Brew. 84, 271.
Baxter, E.D., Booer, C.D., and Wainwright, T. (1978). J. Inst. Brew. 84, 30.
Bell, G.D.H. (1962). Soc. Chem and Ind. Monograph, No. 16, 15.
Bellmer, H.G. (1975). E.B.C. Barley and Malting Symposium II Zeist, 41.
Bendelow, V.M. (1975). J. Inst. Brew. 81, 127.
Bewley, J.D., and Black, M. (1978). Physiology and Biochemistry of Seeds, Vol. 1, Springer-Verlag, Berlin, N.Y.
Bourne, D.T., and Pierce, J.S. (1970). I. Inst. Brew. 76, 328.
Brandt, A., and Ingversen, J. (1978). Carlsberg Research Communs. 32, 451.
Briggs, D.E. (1964). J. Inst. Brew. 70, 14.
Briggs, D.E. (1972). Plants (Berlin) 108, 351.
Briggs, D.E. (1978). Barley (Chapman & Hall), 209.
Brooker, J.D., Chung, C.P., and Marcus, A. (1977), in The Physiology and Biochemistry of Seed Dormancy and Germination, North-Holland, N.Y., Oxford, 348.
Brown, H.T., and Morris, G.H. (1890). J. Chem. Soc. 57, 458.
Burger, W.C. (1973). Plant Physiol. 51, 1015.
Cohen, D., and Paleg, L.G. (1967). Plant Physiol. 42, 1288.
Crabb, D., and Kirsop, B.H. (1969). J. Inst. Brew. 75, 254.
Crabb, D., and Kirsop, B.H. (1970). J. Inst. Brew. 76, 158.
Eastwood, D., and Laidman, D.L. (1971). Phytochem. 10, 1459.
Van der Eb, A.A., and Nieuwdrop, P.J. (1967). Acta Botan. Neer. 15, 690.
Enari, T-M., and Loisa, M. (1974). Monatsschr. Brau. 27, 254.
Enevoldsen, B.S., and Schmidt, F. (1974). J. Inst. Brew. 80, 52.
Essery, R.E., Kirsop, B.H., and Pollock, J.R.A. (1954). J. Inst. Brew. 60, 473.
Evins, W.H., and Varner, J.E. (1972). Plant Physiol. 49, 348.
Ewertson, G. (1977). Agri. Hortique Genetica XXXV, 1.
Fincher, G.B. (1975). J. Inst. Brew. 81, 116.
Fischnich, O., Thielbein, M., and Grahl, A. (1962). Saatgut-Wirtsch. 14, 12 and 39.
Forrest, I.S., and Wainwright, T. (1977). Proc. Eur. Brew. Conv. (Amsterdam), 401.
Gaber, S.D., and Roberts, E.H. (1969). J. Inst. Brew. 75, 299 and 303.
Gjertsen, P. (1970). Brew. Dig. 45(9), 68.

Greenberg, D.C. (1974). J. Inst. Brew. 80, 435.
Greenwood, C.T., and Thompson, J. (1959). J. Inst. Brew. 65: 346.
Groat, J.I., and Briggs, D.E. (1969). Phytochem. 8, 1615.
Haberlandt, G. (1890). In: Brown and Morris (1890).
Halford, M.H., Cordiner, B.G., and Clarke, B.J. (1976). Proc. Conv. Inst. Brew. Aust.-N.Z.-sect., 45.
Hejgaard, J. (1978). J. Inst. Brew. 84, 43.
Harris, G. (1962). In Barley and Malt (ed. Cook, A.) Academic Press (N.Y., London), 492.
Hendricks, S.B., and Taylorson, R.B. (1972). Nature, London, 193, 247.
Higgins, T.J.V., Zwar, J.A., and Jacobsen, J.V. (1976). Nature, London, 197, 1313.
Hopkins, R.H., and Krause, B. (1947). Biochemistry Applied to Malting and Brewing, Allen & Unwin, London.
Horgan, R. (1970). Proc. 11th Conv. Inst. Brew. Aust.-N.Z.-sect. Hobart, 69.
Hyde, W.R. and Brookes, P.A.(1978). J. Inst. Brew.84,167.
Iruegas, A., Pyler, R.E., Banasik, O.J., and Deckhard, E. (1973). Amer. Soc. Brew. Chem. Proc. 152.
Jacobsen, J.V., and Varner, J.E. (1967). Plant Physiol. 42, 1596.
Jacobsen, J.V., Scandalios, J.G., and Varner, J.E. (1970). Plant Physiol. 45, 367.
Jacobsen, J.V., Lie, S., and Chapon, L. (1974). Tech. Quart. Master Brew. Ass. Amer. 11, 225.
Jones, R.L. (1969). Plant Physiol. 85, 359.
Jones, R.L. (1971). Plant Physiol. 47, 412.
Jones, R.L., and Stoddart, J.L. (1977). In The Physiology and Biochemistry of Seed Dormancy and Germination, North-Holland, N.Y., Oxford, 77.
LaBerge, D.E., Clayton, J.W., and Meredith, W.O.S. (1967). Amer. Soc. Brew. Chem. Proc, 18.
Locy, R., and Kende, Hans. (1978). Planta 143, 89.
MacLeod, A.M., and McCorquodale, H. (1958). J. Inst. Brew. 64, 162.
MacLeod, A.M., and Miller, A.S. (1962). J. Inst. Brew. 68, 322.
MacLeod, A.M., Duffus, J.H., and Johnston, C.S. (1964). J. Inst. Brew. 70, 521.
MacLeod, A.M., and Palmer, G.H. (1966). J. Inst. Brew. 72, 580.
MacLeod, A.M., and Palmer, G.H. (1967). Nature 216, 1342.
MacLeod, A.M., and Palmer, G.H. (1969). New Phytologist 68, 295.
Macey, A. (1974). Proc. Eur. Brew. Conv. Wort Symp. (Zeist), 339.
Macey, A., and Stowell, K.C. (1961). J. Inst. Brew. 67, 396.
McNeil, M., Albersheim, P., Taiz, L., and Jones, R.L. (1975). Plant Physiol. 55, 64.
Manners, D.J., and Bathgate, G.N. (1969). J. Inst. Brew. 75,

169.
Manners, D.J., and Marshall, J.J. (1969). J. Inst. Brew. 75, 550.
Melcher, V., and Varner, J. (1971). J. Inst. Brew. 77, 456.
Mikola, J., and Kolehmainen, L. (1972). Planta (Berlin) 104, 167.
Morgan, A.G. (1977). J. Inst. Brew. 83, 231.
Munck, L., and Von Wettstein, D. (1976). Proc. Nat. Acad. Sciences (Washingston D.C.), 71.
Narziss, L., and Friedrich, G. (1970). Brauwelt, 114, 851, 886.
Nieuwdrop, P.J., and Buys, M.C. (1964). Acta Botanica Neerlandica 12, 295.
Northam, P., and Button, A.H. (1973). Proc. Eur. Brew. Conv. Salzburg, 99.
Ogowara, K., and Hayashi, J. (1964). Ber. Ohara Inst. Landwirtsch. Biol. Okayama Univ. 12, 153.
Paleg, L.G., and Hyde, B. (1964). Plant Physiol. 39, 673.
Palmer, G.H. (1967). Ph.D. Thesis, Edinburgh University, Scotland.
Palmer, G.H. (1969). J. Inst. Brew. 75, 536.
Palmer, G.H. (1969a). J. Inst. Brew. 75, 505.
Palmer, G.H. (1971). Proc. Eur. Brew. Conv. (Estoril), 59.
Palmer, G.H. (1972a). J. Inst. Brew. 78, 326.
Palmer, G.H. (1972). J. Inst. Brew. 78, 470.
Palmer, G.H. (1973). J. Inst. Brew. 79, 513.
Palmer, G.H. (1974a). J. Inst. Brew. 80, 13.
Palmer, G.H. (1974). Brew. Dig. 49(2), 40.
Palmer, G.H. (1975a). J. Inst. Brew. 81, 408.
Palmer, G.H. (1975b). Amer. Soc. Brew. Chem. Proc. 174.
Palmer, G.H. (1975c). J. Inst. Brew. 81, 71.
Palmer, G.H., and Bathgate, G.N. (1976). In Advances in Cereal Science and Technology, American Association of Cereal Chemists (Pomeranz, Y. ed.) St. Paul, Minnesota, 237.
Palmer, G.H., and Harvey, A. (1977). J. Inst. Brew 83, 295.
Palmer, G.H. (1978). Malting Symposium (Guinness-Dublin), 7.
Palmer, G.H. (1978). Tech. Quart. Master Brew. Ass. Amer. 15(3), 157.
Palmer, G.H. (1979). Brewers Digest, June, in the press.
Palmer, G.H. (1979a). Brewers Guardian, September, in the press.
Patel, M. (1978). M.Sc. Thesis, Heriot-Watt University, Edinburgh.
Pollock, J.R.A., Kirsop, B.H., and Essery, R.E. (1955). J. Inst. Brew., London 61, 301.
Pollock, J.R.A., and Kirsop, B.H. (1956). J. Inst. Brew., London 62, 323.
Pollock, J.R.A. (1962). In Barley & Malt (ed. Cook, A.) Academic Press (N.Y., London), 303.
Pomeranz, Y. (1974). Cereal Chem. 51, 545.
Preece, I.A., and Mackenzie, K.G. (1952). J. Inst. Brew. 64, 489.

Preece, I.A., and Aitken, R.A. (1953). J. Inst. Brew. 59, 453.
Preece, I.A., and Hobkirk, R. (1955). J. Inst. Brew. 61, 393.
Preece, I.A., and MacDougall, M. (1958). J. Inst. Brew. 58, 353.
Preece, I.A., Garg, N.K., and Hoggan, J. (1960). J. Inst. Brew. 66, 331.
Preece, I.A. (1964). Tech. Quart. Master Brew. Ass. Amer. 1, 127.
Radley, M. (1967). Planta (Berlin) 75, 164.
Radley, M. (1969). Planta (Berlin) 86, 218.
Roberts, E.H. (1973). In Seed Ecology, Butterworths, London, 189.
Roberts, E.H., and Smith, R.D. (1977). In The Physiology and Biochemistry of Seed Dormancy and Germination, North-Holland, N.Y., Oxford, Amsterdam, 386.
Robinson, D.R., and West, C.A. (1970). Biochemistry 9, 80.
Sandegren, E., and Beling, H. (1959). Proc. Eur. Brew. Conv, 13.
Schlimme, G. (1977). Brauwelt 117, 1894.
Schur, F., and Pfenninger, H. (1978). Brauerei Rundschau 89(2), 17.
Scriban, R., and Strobbel, B. (1979). Tech. Quart. Master Brew. Ass. Amer. 16(1), 28.
Shewry, P.R., Ellis, J.R.S., Pratt, H.M., and Miflin, B.J. (1978). J. Sci. Food Agric. 29, 1.
Slack, P.T., Baxter, E.D., and Wainwright, T. (1979). J. Inst. Brew. 85, 112.
Slominsky, B., Rejowski, A., and Nowak, J. (1979). Physiol. Plant. 45, 167.
Smith, M.T., and Briggs, D.E. (1979). J. Inst. Brew. 85, 167.
Sparrow, D.H.B. (1970). Proc. 2nd Iternat. Barley Genetic Symposium, Pullman, Washington, 559.
Sutcliffe, J.R., and Sexton, R. (1974). In Structure and Function of Primary Root Tissue, Veda, Bratislava, 203.
Taiz, L., and Jones, R.L. (1970). Planta (Berlin) 92, 73.
Taiz, L., and Honigman, W.A. (1976). Plant Physiol. 58, 380.
Urion, E., and Chapon, L. (1955). Proc. Eur. Brew. Conv. (Baden-Baden), 172.
Varner, J.E., and Ram-Chandra, G. (1964). Proc. Nat. Acad. Sci. U.S.A. 52, 100.
Varner, J.E., and Mense, R.M. (1972). Plant Physiol. 49, 187.
Verbeek-Wyndaele, R. (1973). Proc. Eur. Brew. Conv. (Salzburg), 75.
Williams, J.M., and Duffus, C.M. (1978). J. Inst. Brew 84, 47.
Yomo, H. (1958). Hakko Kyokaishi 16, 444.
Yomo, H., and Iinuma, H. (1963). Proc. Eur. Brew. Conv. (Brussels), 520.

THE MOLECULAR BIOLOGY OF BARLEY STORAGE PROTEIN SYNTHESIS

Verena Cameron-Mills
Anders Brandt
John Ingversen

Department of Physiology
Carlsberg Laboratory
Copenhagen, Denmark

I. INTRODUCTION

Barley is the world's fourth most important cereal after wheat, rice and maize and is also the most widely cultivated. In Western countries barley is primarily used as an animal feed, but in Eastern countries large quantities, particularly the naked grain varieties, are consumed in human food and drink. Although grain may comprise some 75% of a human diet and 80%-85% of a pig diet, the poor nutritional quality of barley protein necessitates a dietary supplement of high quality protein. Supported by world-wide pressure to increase and improve the production of plant proteins, considerable effort has been directed towards an understanding of the biochemistry and genetics of cereal grain development.

Selection of cereal varieties of improved protein quality, based on a screening programme for elevated grain lysine levels, has led to the recognition of the maize high-lysine lines Opaque-2 (17) and Floury-2 (19). The first high-lysine line to be recognised in barley was Hiproly, screened from the world's barley collection (18). Subsequently mutant no. 1508 has been isolated after mutagen treatment, and its high-lysine character attributed to a single recessive gene on barley chromosome 7 (9,14). A survey of the wild (Hordeum spontaneum) and cultivated (Hordeum vulgare) barley varieties has revealed considerable polymorphism in the genes determining the storage protein composition of the grain (10).

It is now rapidly becoming clear that a real understanding of the genetic regulation of cereal protein synthesis and composition will depend upon a study of these processes at the molecular level. Considerable progress has been made in the characterization and in vitro synthesis of both maize and barley storage proteins (3,5,7,15). In each case the endoplasmic reticulum of the endosperm cells has been recognised as the

ISBN: 0-12-370960-1

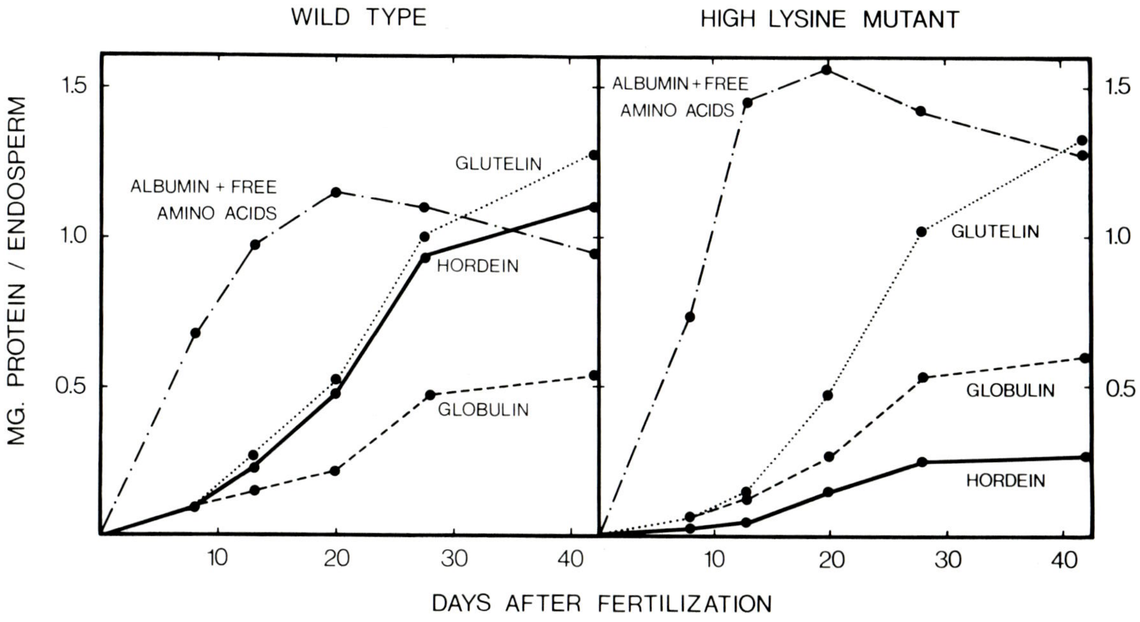

FIGURE 1. Synthesis of albumin + free amino acids, globulin, hordein and glutelin during endosperm development in Bomi barley and the high-lysine mutant no. 1508 (from 3).

the site of storage protein synthesis and the mRNA molecules coding for these proteins have been purified and translated in vitro (5,15). DNA cloning techniques have been developed to prepare double-stranded cDNA, complementary to mRNA coding for maize and barley storage proteins (4,24,32) and the cloned cDNA is now being used as a probe to study the location and organization of the storage protein structural genes. Our current knowledge of the molecular biology and genetics of barley storage proteins will now be described in greater detail.

A. The Structure and Composition of Barley Storage Proteins

The cereal endosperm proteins are classically termed albumin, globulin, prolamin and glutelin according to their solubility in water, dilute salt solutions, dilute alcohol and acid or alkali, respectively (21). The biosynthesis of the proteins contained in the classical Osborne fractions during the grain-filling period, from fertilization to maturity, are shown for Hordeum vulgare L. cv. Bomi (wild type) in figure 1. The albumin fraction, which also includes the free amino acids is synthesized early in grain development, whereas hordein (barley prolamin), glutelin and globulin are synthesized from 10 days after fertilization until maturity. The hordein fraction, whose synthesis is at a maximum between 10 and 27 days after fertilization, is a major component of the endosperm storage proteins, which are primarily deposited in protein bodies (13). Storage proteins, which are predominant in the mature grain, are characterized by a high glutamine and proline content and low levels of the nutritionally essential amino acids lysine, methionine and tryptophan. Mutant no. 1508 derived from Bomi, owes its high-lysine character to depressed hordein synthesis [Fig. 1] (3,14).

Figure 2 shows a transmission electron micrograph of a thin section cut through a wild type endosperm cell in the period of active storage protein synthesis. The protein bodies develop inside a vacuolar membrane and consist of two components, one homogeneous and one granular. The developing endosperm cell also contains starch grains which evolve within amyloplasts, and an extensive rough endoplasmic reticulum. At higher magnification (Fig. 3) the two component structure of the protein bodies from wild type barley is revealed in detail, the homogeneous component being predominant. Isolation of these protein bodies and their characterization by SDS-polyacrylamide gel electrophoresis have identified the homogeneous component as hordein (13). The depressed hordein content of the mutant endosperm correlates with the very small amounts of the homogeneous component detected in the mutant (Fig. 4).

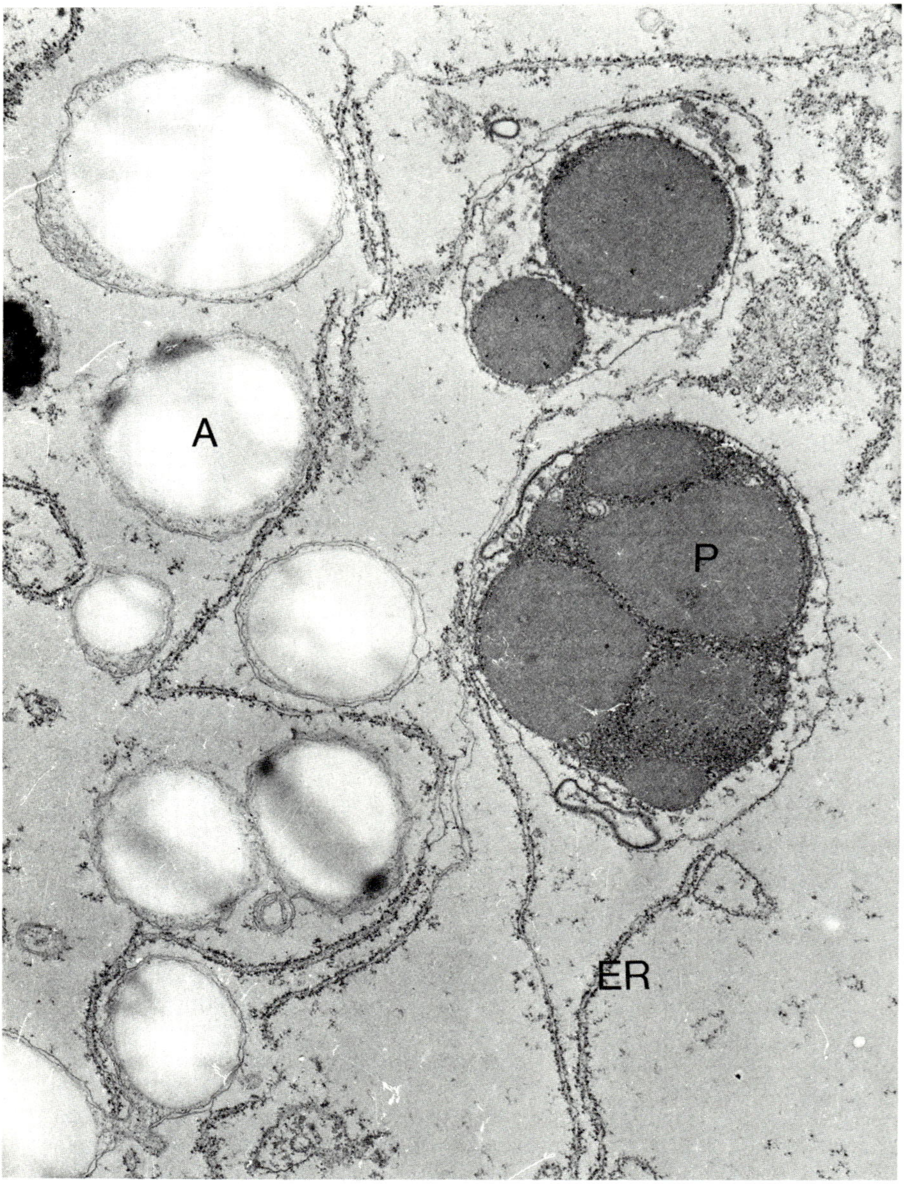

FIGURE 2. Electron micrograph of a section through an endosperm cell of developing Bomi barley endosperm, 3 weeks after fertilization. The protein bodies (P) are formed inside vacuoles and starch grains are deposited in amyloplasts (A). ER = profiles of the ribosome-studded endoplasmic reticulum. Magnification: x 18000 (from 33).

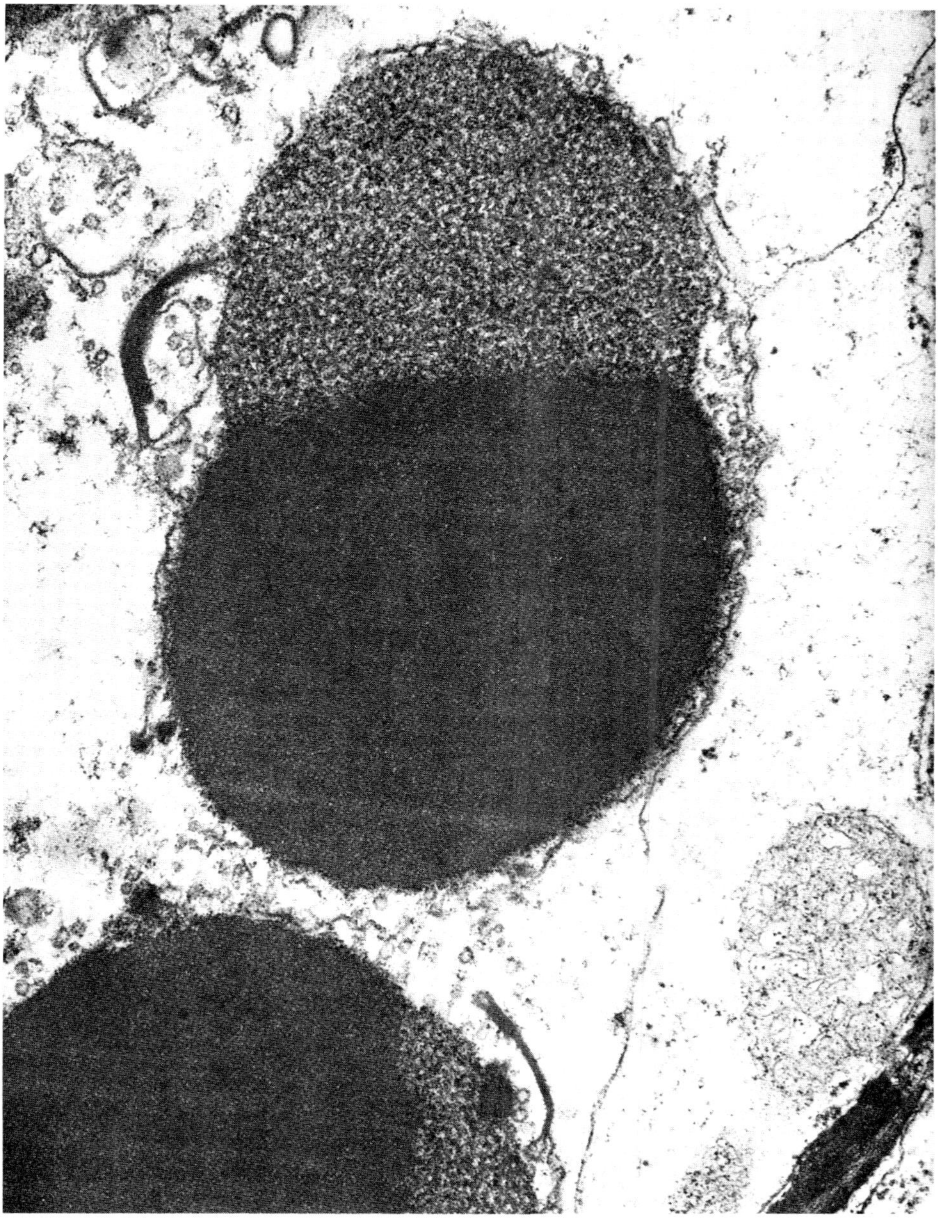

FIGURE 3. Electron micrograph of a protein body from a 3 week old Bomi barley endosperm. The protein body consists primarily of an homogeneous component combined with a granular component. Magnification: x 34000 (from 33).

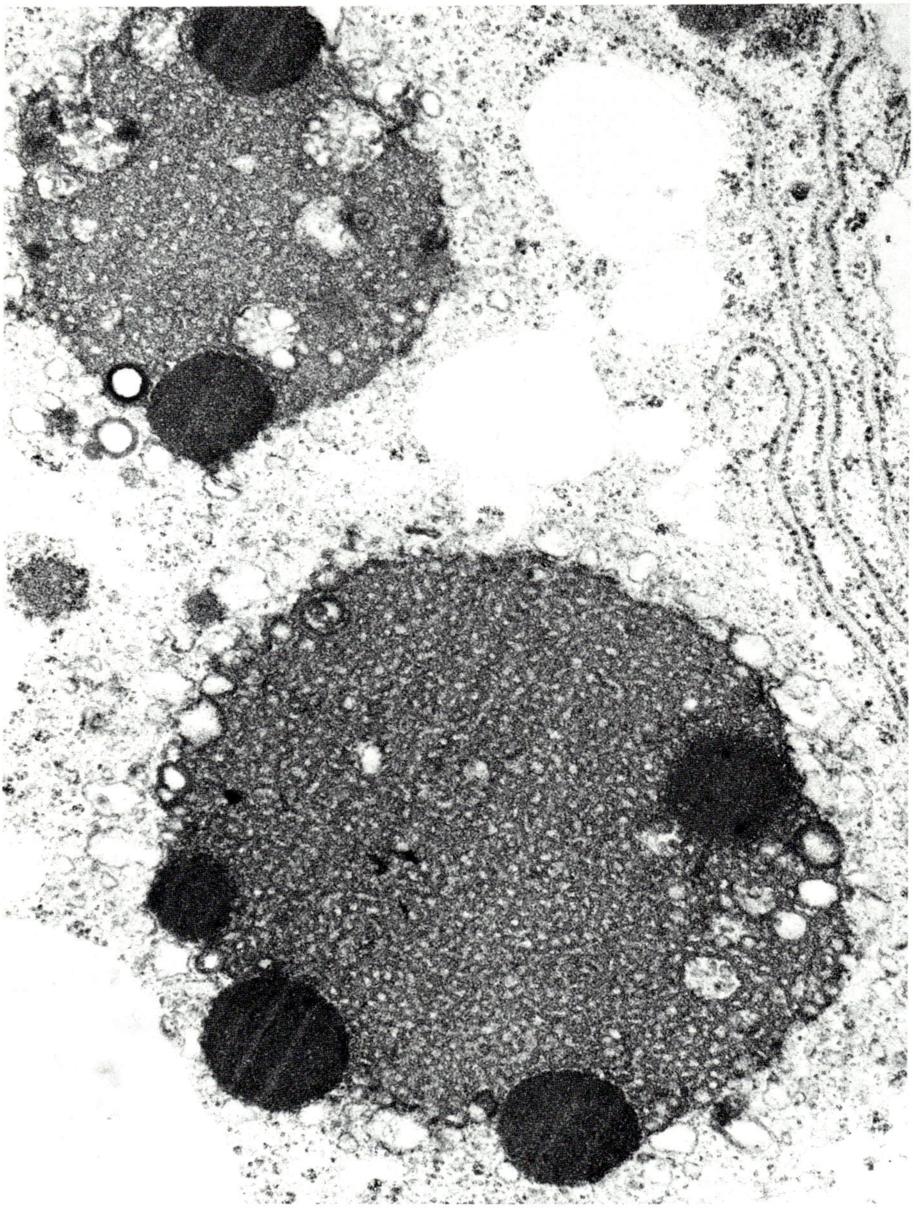

FIGURE 4. As figure 3 but for the high-lysine mutant no. 1508. The protein bodies of the mutant endosperm consist mainly of the granular component. The homogeneous component is only present in small amounts. Magnification: x 34000 (from 33)

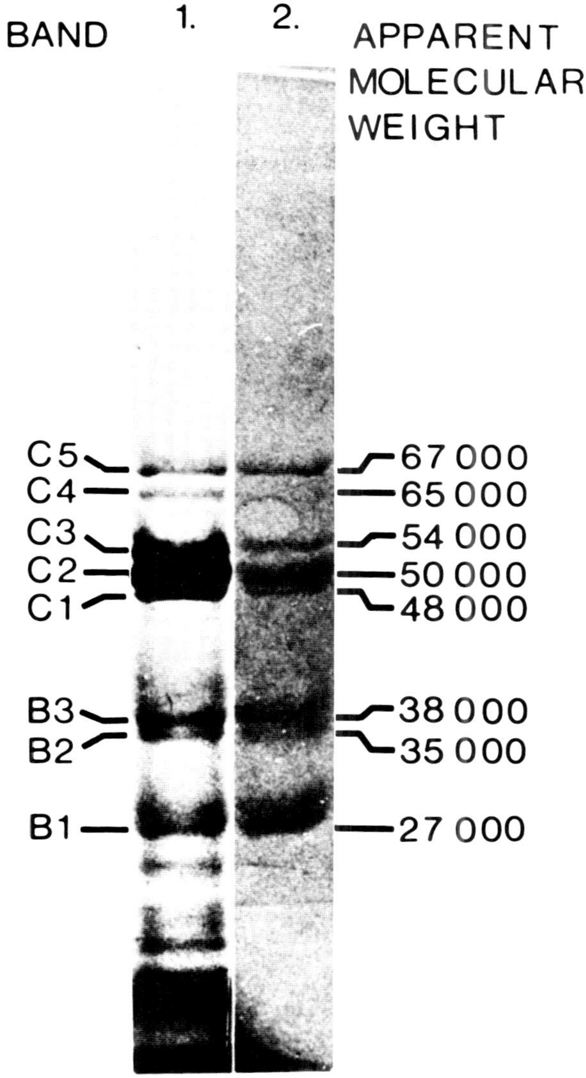

FIGURE 5. Electrophoretic separation of hordein polypeptides. Track 1 is an autoradiogram of hordein polypeptides synthesized in vitro with ^{35}S-labelled methionine. Track 2 is an autoradiogram of in vivo synthesized hordein polypeptides labelled with ^{14}C-amino acids (from 12).

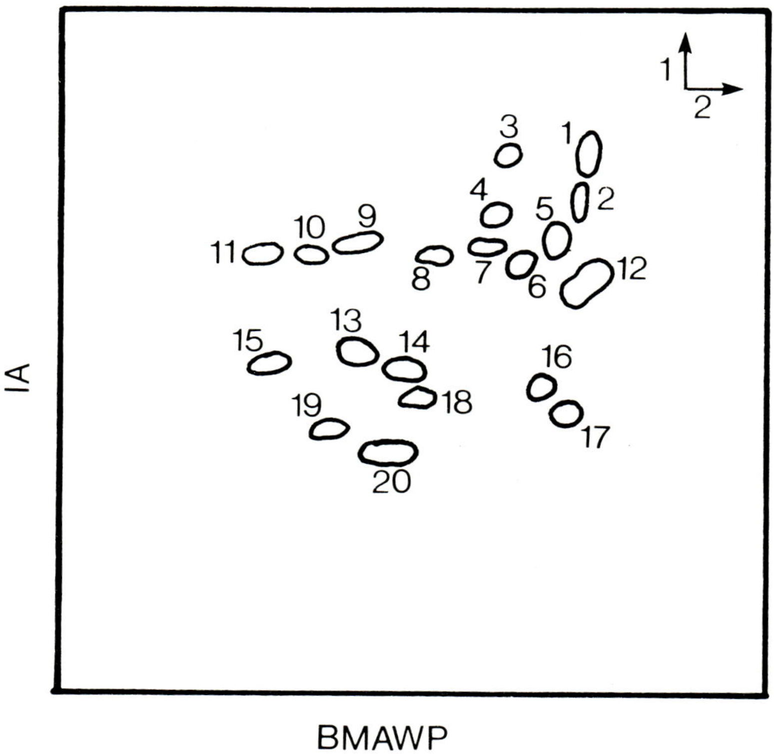

FIGURE 6. Two-dimensional chromatographic separation of ^{35}S-methionine-containing peptides which were released by chymotrypsin treatment of acrylamide gel slices containing in vitro synthesized hordein polypeptides. BMAWP butanol:methanol:acetic acid:water:pyridine (45:30:6:24:20, v/v). IA isopropanol:25% ammonia (65:35 v/v). (from 12).

Distribution of ^{35}S-methionine containing peptides on the chymotryptic peptide maps of individual hordein polypeptides. The nomenclature for polypeptides refers to figure 5 and that for peptides to figure 6.

Polypeptide band	Peptide number 1	2	3	4	5	6	7	8	9	10	11	12	13	14	15	16	17	18	19	20
B1	+++	+	+	-	-	+++	-	+	-	-	+	+++	+	++	+++	-	-	-	+	-
B2	+++	-	-	-	-	+++	-	-	-	-	+	+	++	+	+++	-	-	-	-	-
B3	+++	-	+	-	-	+++	-	-	+	-	+	++	+++	+	+++	-	-	-	-	-
C1	+++	-	-	+	+++	+++	-	-	+	+	+	-	+++	++	+++	+++	+++	+	-	++
C2	+++	-	-	-	+++	+++	++	+	++	+	+	-	+++	+	+++	++	+++	+	-	+
C3	+++	++	-	+	+++	++	-	-	+	+	+	-	+++	+	+++	++	++	+	-	+
C4	+++	-	-	-	+	+++	-	-	-	-	+	+	+	++	+++	+	-	-	-	-
C5	+++	-	-	-	+++	+++	-	-	+	-	+	-	++	+	+++	+	+	-	-	+

+++ = strong autoradiographic image

++ = moderate autoradiographic image

+ = weak autoradiographic image

- = no autoradiographic image

When mature barley seeds are ground and extracted with 55% isopropanol, hordein is solubilized in the supernatant. The hordein can be separated into at least eight molecular species by SDS-polyacrylamide gel electrophoresis. Figure 5 shows the separation of ^{35}S-methionine- and ^{14}C-labelled hordein polypeptides. Each of the eight ^{35}S-methionine-labelled polypeptides has been excised from the gel and subjected to chymotryptic digestion. Two-dimensional chromatographic separation of the chymotryptic peptides resulted in peptide maps with a total of 20 recognizable peptides (Fig. 6). The distribution of the radioactive peptides amongst the chymotryptic products from the individual hordein polypeptides is shown in table 1. The eight in vitro synthesized hordein polypeptides have six methionine containing peptides in common (peptides 1,6,11,13,14,15). Peptides 5, 16, 17 and 20 were only present in the hordein polypeptides with the higher molecular weights of 48000 to 67000 daltons. The similarity between the hordein polypeptides (Table 1) suggests homology in their primary structure and a common genetic origin.

When hordein from different varieties of Hordeum vulgare is analysed by SDS-polyacrylamide gel electrophoresis, substantial polymorphism is observed (10,28). The variation in electrophoretic banding pattern in one group of hordein polypeptides was shown to be controlled by co-dominant alleles at one locus on chromosome 5 (20). In a recent study (29) it has been reported that the polymorphism in a second group of hordein polypeptides is directed by co-dominant alleles at another locus, linked to the first described hordein locus (20). The designations <u>Hor-1</u> and <u>Hor-2</u> have been suggested for the two linked loci specifying the hordein-1 polypeptides (molecular weights from 48000 to 67000 daltons) and the hordein-2 polypeptides (molecular weights from 27000 to 38000 daltons) (10). Since the different alleles of <u>Hor-1</u> and <u>Hor-2</u> loci specify a large number of polypeptides varying in both charge and molecular weight, they must include several genes. However, intralocus recombination at either loci has not been detected. A fine structure analysis of the hordein loci is now possible employing cloned double-stranded cDNA sequences which code for hordein polypeptides. A cDNA molecule containing the coding sequence for two of the hordein-2 polypeptides has recently been isolated (4).

B. A DNA Cloning Technology for the Study of Barley Storage Proteins

As a first step towards analysis of the hordein structural genes, hordein polypeptides have been synthesized in vitro with barley endosperm polyribosomes (5). A cell-free protein

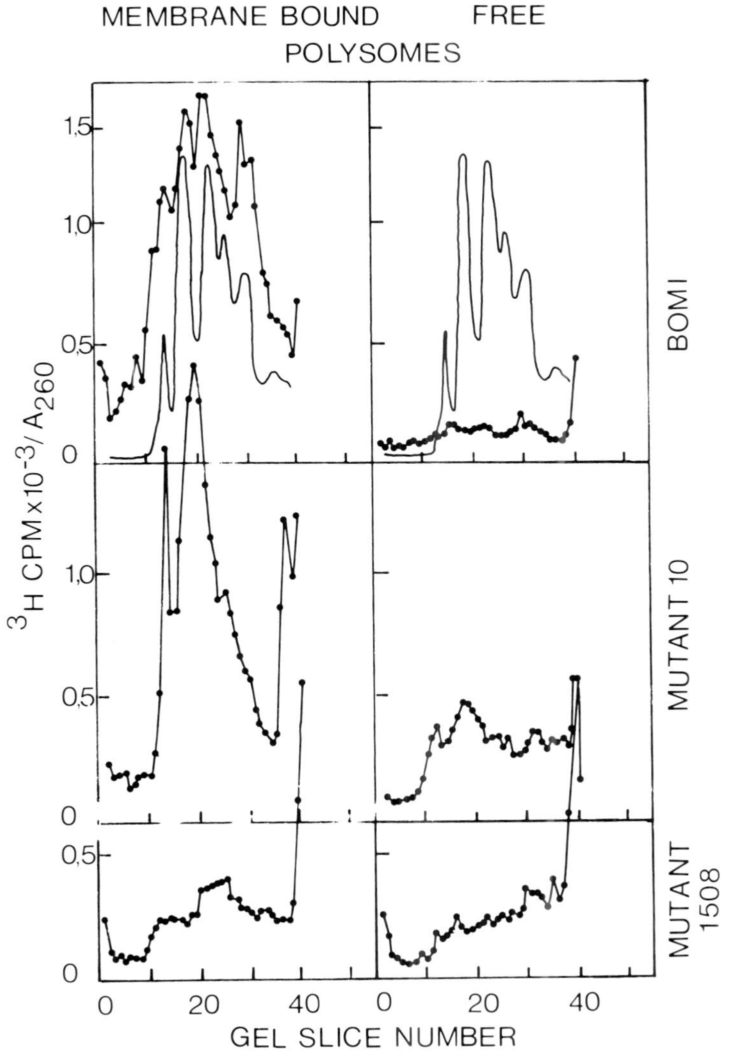

FIGURE 7. The in vitro synthesis of hordein polypeptides by membrane-bound and free cytoplasmic polyribosomes from the endosperms of Bomi barley (wild type), the hordein-rich mutant 10 and the hordein-deficient mutant no.1508. Solid curves in the uppermost two panels are the electrophoretically separated hordein polypeptides. The curves with the dots show the amount of radioactive amino acids incorporated in the different electrophoretically separated fractions (from 5).

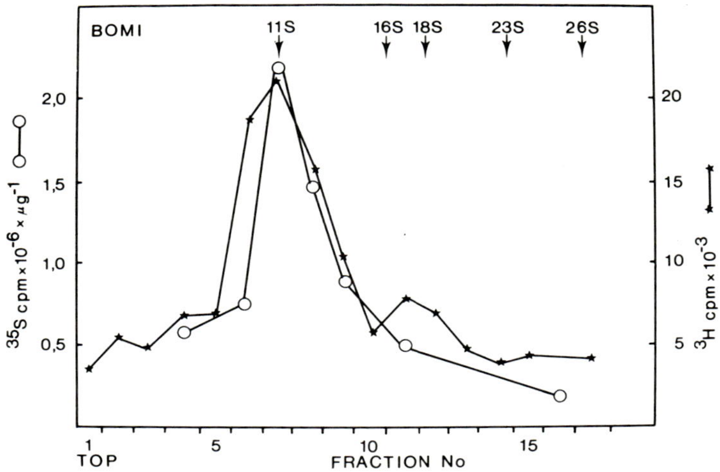

FIGURE 8. Isolation of 11S mRNA from Bomi barley endosperms by sucrose gradient sedimentation. RNA retained on Poly(U)-Sepharose 4B columns was eluted and then separated on 5-20% sucrose gradients. After centrifugation, fractions were collected and the RNA precipitated with ethanol and then dissolved in H_2O. An aliquot of this RNA was hybridized to ^{3}H-polyuridylic acid and the RNAase resistant radioactivity determined. Another portion of the RNA was used for analysis of template activity in the wheat-germ, cell-free protein synthesizing system, using ^{35}S-methionine as label (from 6).

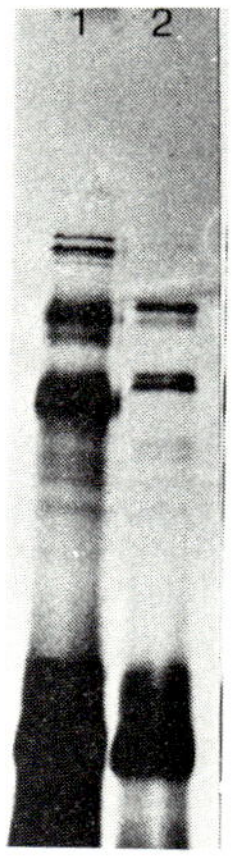

FIGURE 9. Autoradiograms of the translation products of mRNA isolated from Bomi barley endosperms. ^{35}S-Methionine-labelled polypeptides synthesized in vitro by initially membrane-bound polyribosomes (track 1) and 11S polyadenylated RNA (track 2) extracted with 55% isopropanol, were analysed by electrophoresis and autoradiography (from 6).

synthesizing system is employed containing ribosomal subunits and transfer RNA, with all the additional factors essential for protein synthesis provided by a wheat-germ extract or a rabbit reticulocyte lysate. The newly synthesized hordein polypeptides are detected by the inclusion of ^{35}S-methionine in the cell-free system and autoradiography of the translation products separated on SDS-polyacrylamide gels. Hordein polypeptides are selectively extracted from the translation products prior to electrophoresis using 55% isopropanol.

The capacity of the two polyribosome populations in the endosperm, namely those bound to the endoplasmic reticulum and those free in the cytoplasm, to direct hordein synthesis is shown in figure 7. The banding position of the native hordein polypeptides in the gel may be compared with the presence of corresponding radioactively-labelled polypeptides amongst the isopropanol-soluble translation products. It is clear that template activity for hordein resides almost exclusively in the membrane-bound polyribosome population. Mutations leading to hordein-enriched (mutant 10) or hordein-deficient seeds can be directly correlated with a change in hordein polypeptide template activity in the membrane-bound polyribosomes of these mutants (figure 7). Using in vivo ^{14}C-labelled hordein, a closer comparison can be made between the polyribosomal translation products [^{35}S-methionine-labelled] and native hordein (Fig. 5). Polyribosomes from Bomi endosperm synthesize all eight of the major hordein polypeptides detected in native hordein.

The mRNA fraction from membrane-bound polyribosomes, coding for hordein polypeptides, has been isolated and purified (6) using poly U-sepharose 4B affinity chromatography and sucrose density centrifugation. The fractionation of polyadenylated RNA from Bomi endosperm into different molecular species during sucrose gradient sedimentation is shown in figure 8. The 18S and 26S ribosomal RNA fractions are separated from 11S polyadenylated RNA in which the bulk of the hordein template activity is located. The purified 11S mRNA has an eight-fold increase in the hordein template activity over that of total polyadenylated RNA. Although a polyadenylated RNA fraction has also been purified from mutant no. 1508 polyribosomes, its template activity is much lower. Figure 9 shows that template activity for all the major hordein polypeptides synthesized in vitro by Bomi membrane-bound polyribosomes is retained in the purified 11S mRNA. Although the molecular weights of hordein polypeptides synthesized in vitro by polyribosomes are similar to those of native hordein, the hordein translation products of purified mRNA are all approximately 2000 daltons larger. The significance of larger primary translation products and their processing into the native proteins will be discussed in the final section of this chapter.

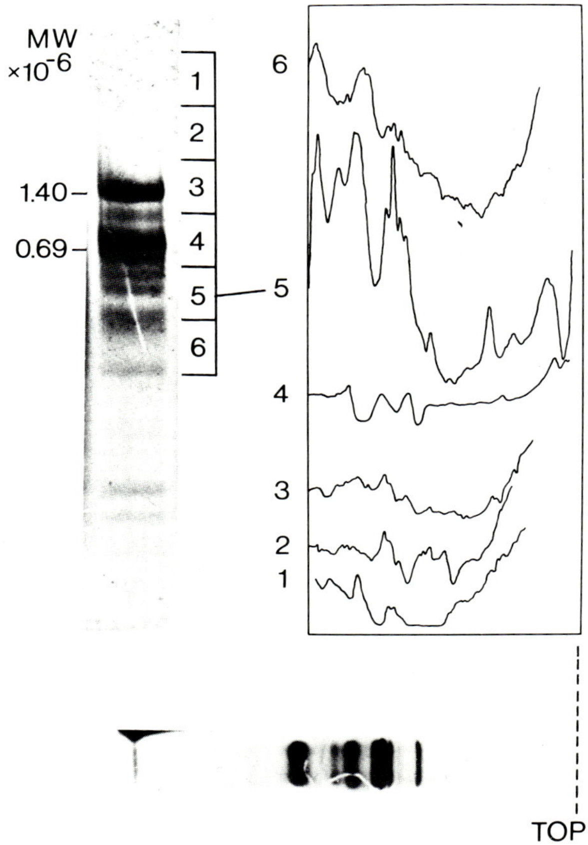

FIGURE 10. Densitometric tracings of autoradiograms from in vitro ^{35}S-methionine-labelled polypeptides made with RNA fractionated by preparative gel electrophoresis. RNA isolated from initially membrane-bound polyribosomes from Bomi endosperms was separated on a composite agarose-polyacrylamide gel (left). The RNA was extracted from gel sections 1 to 6 and purified by oligo(dT)-cellulose affinity chromatography. The polyadenylated RNA was used as template in the wheat-germ, cell-free protein synthesizing system. In vivo ^{14}C-labelled hordein marker polypeptides are presented at the bottom. Significant template activity for hordein is restricted to fraction 5 (from 6).

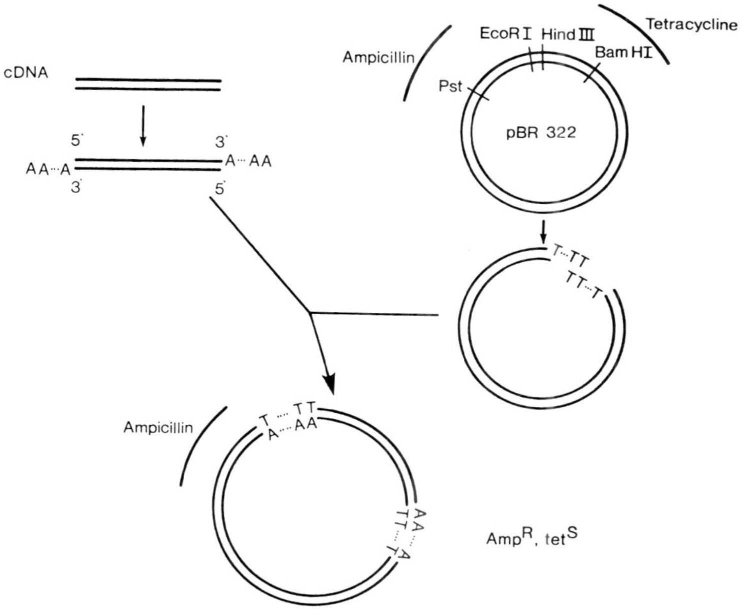

FIGURE 11. Diagrammatic representation of the construction of a hybrid bacterial plasmid containing hordein genes using the dA.dT tailing method. The plasmid pBR322 was linearized with the restriction enzyme HindIII and the 3-OH' ends of the plasmid and the cDNA were tailed with thymidine and deoxy-adenosine, respectively. The two molecules were annealed and used to transform E.coli HB101 cells.

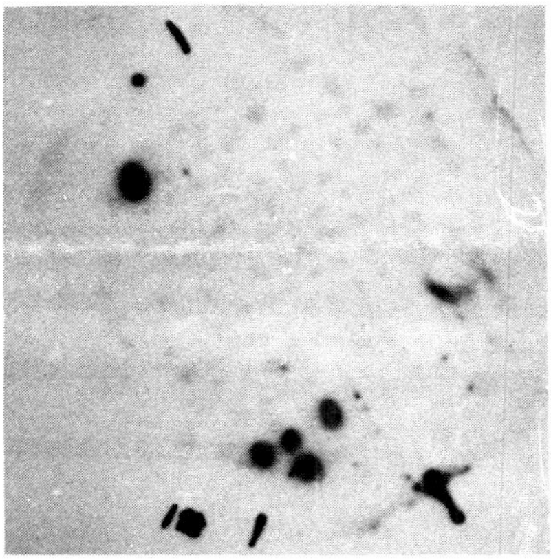

FIGURE 12. Detection of E.coli clones containing recombinant plasmids. Colonies were grown on Millipore filters, lysed and hybridized to ^{32}P-labelled cDNA. The autoradiogram shows 5 colonies containing DNA hybridizable to the cDNA (from 4).

The size of the mRNA molecules which code for hordein polypeptides has been determined. Figure 10 shows the fractionation of Bomi polyribosomal RNA by preparative gel electrophoresis and the isopropanol-soluble translation products of the RNA extracted from sections one to six of the gel. The densitometric tracings of the electrophoretically separated translation products reveal that template activity for hordein polypeptides is localized in section 5. Since section 5 contained 2/3 of the total template activity recovered from the gel sections, translatable mRNA for hordein must be highly abundant in the developing endosperm. Molecular weights for the hordein mRNA of 0.45, 0.36 and 0.30 megadaltons have been derived by a number of independent estimations. These mRNA sizes are compatible with the predicted coding capacity required for the translation of the hordein precursor polypeptides.

Although polyadenylated 11S RNA has also been purified in significant amounts from mutant no. 1508 endosperm, its template activity is low and limited to a single low molecular weight polypeptide. This deficiency of template activity in the mutant may either stem from an absence of mRNA coding for hordein or the presence of untranslatable hordein mRNA.

The second stage in these studies has involved the production of cDNA probes complementary to hordein mRNA, which can be used to localize the hordein structural genes (4). Single-stranded cDNA has been synthesized from a template of purified hordein mRNA, employing avian myoblastosis virus reverse transcriptase. The second strand of the cDNA molecule is synthesized with E.coli DNA polymerase I, giving a double-stranded cDNA molecule with an average size of 1000 base pairs. This double-stranded cDNA is cloned by insertion into the bacterial plasmid vector pBR322. The main features of this splicing technique are depicted in figure 11. The circular plasmid is first linearized with the restriction endonuclease HindIII and the two 3-OH' ends of the molecule are then tailed by the addition of thymidine residues with calf thymus terminal transferase. The 3-OH' ends of the double-stranded cDNA are similarly extended with deoxyadenosine residues. The tailed plasmid is then annealed with the tailed double-stranded cDNA and the hybrid plasmids cloned by transformation of E.coli HB101.

Since the hybrid plasmids carry a resistance gene for ampicillin, but have lost their tetracycline resistance, those transformed E.coli cells carrying hybrid plasmids could be isolated using antibiotic selection growth media. 314 colonies resistant to ampicillin but sensitive to tetracycline were obtained. In order to select those colonies containing hybrid plasmids with large cDNA inserts, the DNA in the lysed cells of each colony is hybridized to ^{32}P-labelled cDNA in the pre-

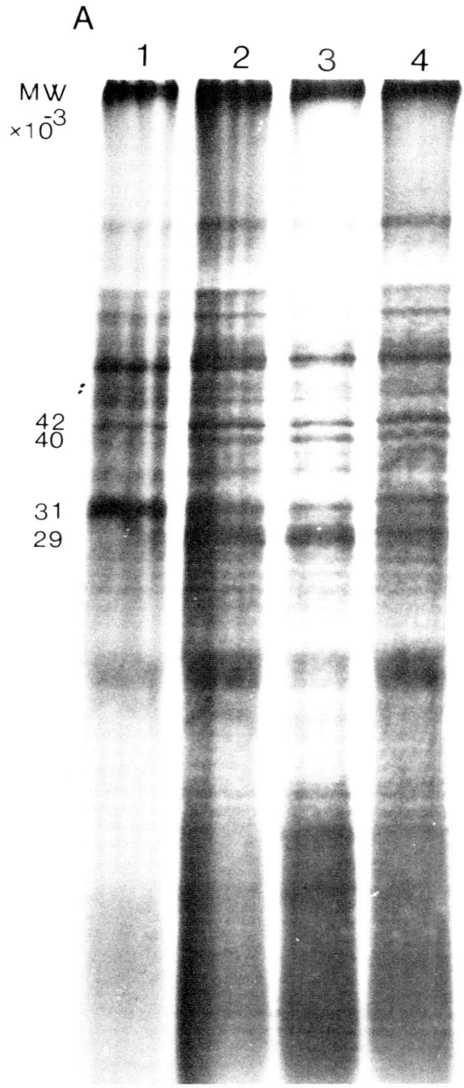

FIGURE 13. Hybrid-arrested translation. Autoradiograms of cell-free translation products using polyadenylated mRNA isolated from Bomi barley endosperms and translated in the rabbit reticulocyte lysate system. ^{35}S-Methionine-labelled polypeptides translated from mRNA after its hybridization to the cDNA containing plasmid in track 1. The mRNA hybridized to cDNA was melted by heat prior to translation in track 2. ^{35}S-Methionine-labelled polypeptides from mRNA mixed with pBR322 under conditions allowing hybridization in track 3. The mRNA hybridized to pBR322 was melted prior to translation in track 4 (from 4).

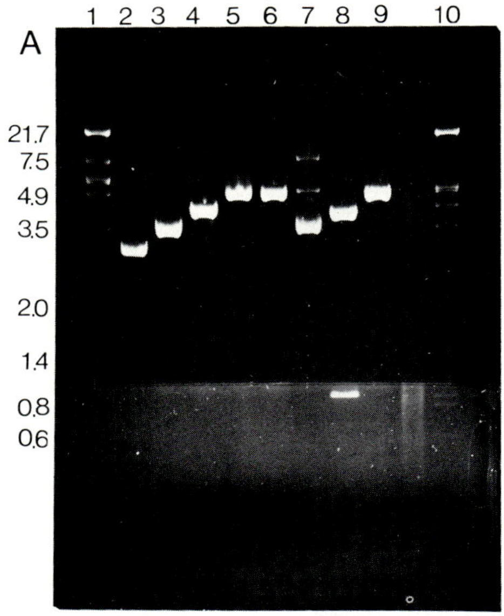

FIGURE 14. Restriction endonuclease analysis of the recombinant plasmid. Single restriction endonuclease digestion patterns of fragments separated on a 0.8% (upper 2/3) and 1.5% (lower 1/3) agarose gel. Size marker (kb), λ DNA digested with EcoRI in tracks 1,10. Native pBR322 (major band is closed circular DNA) in track 2. Native recombinant plasmid (major band is closed circular DNA) in track 3. pBR322 digested with Eco RI in track 4. Recombinant plasmid digested with Eco RI (track 5), HindIII(track 6), SalI(track 7), PstI (track 8), BamHI (track 9), (from 4).

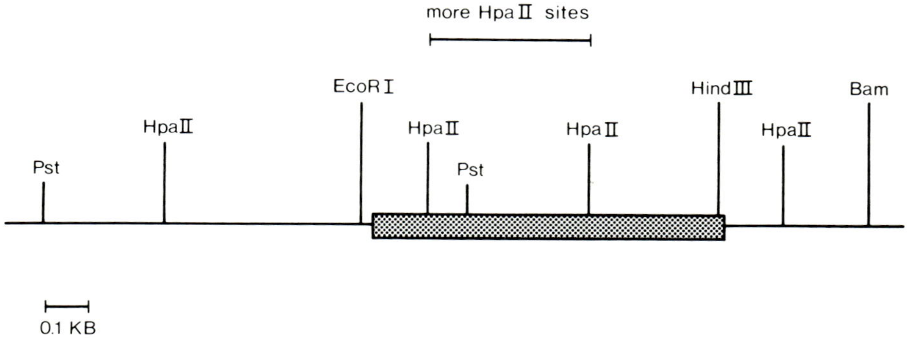

FIGURE 15. Restriction endonuclease map of the hordein specific insert. Cross marked heavy line depicts the insert and the thin line pBR322 DNA. The cleavage sites for the various restriction endonucleases are indicated above the line (from 4).

sence of sheared E.coli DNA (Fig. 12). In this manner 12 colonies were selected and the size of their hybrid plasmids determined by agarose gel electrophoresis. The cDNA inserts in most of the plasmids were less than 100 base pairs in length, except one which was found to be larger.

The ability of this largest insert to hybridize with a hordein mRNA molecule was determined by hybrid-arrested translation. Hordein mRNA was hybridized with an excess of the hybrid plasmid, which had been linearized with the restriction endonuclease EcoR1. The hybridized mRNA was divided into two samples of which one half was heated to denature any mRNA-DNA hybrids. Comparison of the template activity in the two samples reveals that hybridization to the recombinant plasmid has arrested translation of the 29 kilodalton hordein polypeptide completely, and that of the 40 kilodalton polypeptide partially (Fig. 13, track 1). Synthesis of both of these polypeptides is restored by prior melting of the DNA-RNA hybrids (Fig. 13, track 2). In parallel experiments, hybridization with linearized native plasmid pBR322 failed to arrest translation of any of the hordein polypeptides (Fig. 13, tracks 3,4). The DNA sequence of the large insert in the hybrid plasmid must therefore be complementary to two mRNA molecules coding for two hordein polypeptides in the hordein-2 group.

The technique of restriction enzyme mapping, using a series of endonucleases which cleave specific nucleotide sequences, has evolved for the fine mapping of DNA. This technique, when applied to the hybrid plasmid, may be used to determine the size of the inserted sequence and the position of the restriction sites within the insert. Further, the removal of the inserted sequence intact from the cloned plasmid and its subsequent cleavage into mapped fragments, is a pre-requisite for nucleotide sequence analysis.

The closed circular plasmid pBR322 has a single restriction site for each of the endonucleases EcoRI, PstI, BamHI and HindIII. Digestion with any one of these enzymes yields a linear molecule with an apparent molecular weight of 4.36 kilobase pairs in agarose gels (Fig. 14, track 4). Digestion of the hybrid plasmid with EcoRI, HindIII or BamHI similarly yields a single linear molecule with a higher molecular weight of 5 kilobase pairs, from which the size of the insert can be derived (Fig. 14, tracks 5,6,9). Since the HindIII site on the native plasmid is destroyed during the recombination event, one must conclude that the restriction site for HindIII in the hybrid plasmid lies within the inserted sequence.

By compairing the number and size of the fragments produced by endonuclease digestion of the native and hybrid plasmids, one may deduce the location of the restriction sites within the plasmid (Fig. 15). Thus, the presence of two fragments of 4.3 and 0.8 kilobase pairs in the PstI digest of the hybrid plasmid, defines a PstI restriction site within the insert (Fig. 14, track 8). Digestion of the hybrid plasmid with the endonuclease HpaII, yields several fragments from which two restriction sites within the insert, in addition to the two in the native plasmid, have been deduced (results not shown). Two of the larger fragments obtained (620 and 440 base pairs) correspond to the fragments lying between the HpaII restriction sites within and outside the inserted sequence. However, a 350-400 base pairs fragment corresponding to the sequence between the two insert restriction sites has not been detected. This implies that additional HpaII sites may lie within this insert sequence, yielding fragments too small to be discerned by this technique. A preponderance of HpaII sites within the insert, whose recognition sequence is GGCC, is compatible with the coding sequence for hordein, since the high percentage of proline (GGC codon) and certain other amino acids in hordein would increase the incidence of the HpaII recognition sequence.

Localization of the structural genes for hordein polypeptides among the barley chromosomes may now be attempted using the cloned cDNA as a probe. The fine structure of the Hor-2 locus on chromosome 5, which codes for the hordein-2 group, may now be explored and the number and distribution of its component genes determined. These investigations, combined with a further understanding of the genetic regulation of the hordein genes, will provide an important basis for future breeding strategies aimed at producing high-yielding, high-lysine barley varieties.

C. The Synthesis and Deposition of Barley Storage Proteins

A better understanding of the genetics and biochemistry of the storage proteins in developing barley grain may also be gained from an elucidation of the molecular biology of their synthesis and deposition. Hordein, in common with most storage proteins in eukaryotic cells, is deposited in membrane-bound vesicles (33). The transport of these proteins into membrane vesicles, in addition to those proteins destined to become internal components of cell organelles or a part of the cell membranes, has presented scientists with a considerable physico-chemical problem. How can proteins, several thousand daltons in size, be moved through an eukaryotic cell which is compartmented by numerous membranes, whose semi-permeable properties selectively restrict the movement of ions and low molecular weight molecules ?

One important step towards solving this problem lay in the discovery that most membrane, storage and secretory proteins are synthesized on membrane-bound rather than cytoplasmic

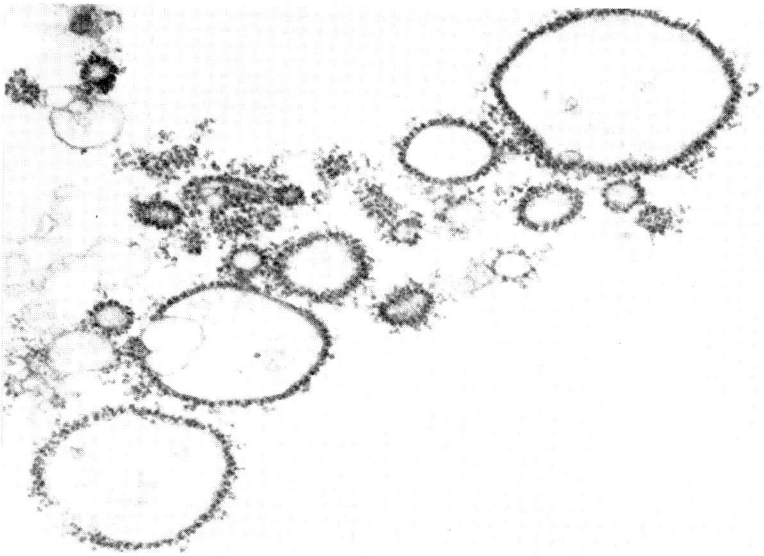

FIGURE 16. Rough microsomes from Bomi barley endosperm. Rough microsomes were prepared from 20 day old Bomi endosperms by discontinuous sucrose gradient centrifugation of the membrane fraction derived from the homogenized endosperm. The isolated material was fixed, stained, embedded and sectioned for electron microscopy. Magnification: x 60,000.

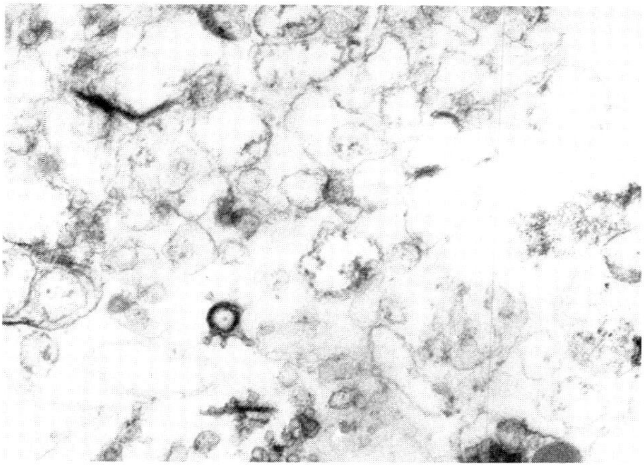

FIGURE 17. Stripped microsomes from Bomi barley endosperm. Stripped microsomes were obtained from the rough microsomes by a 30 min treatment with 30mM-EDTA and subsequently purified on a discontinuous sucrose gradient. The isolated material was examined by electron microscopy. Magnification: x 60,000.

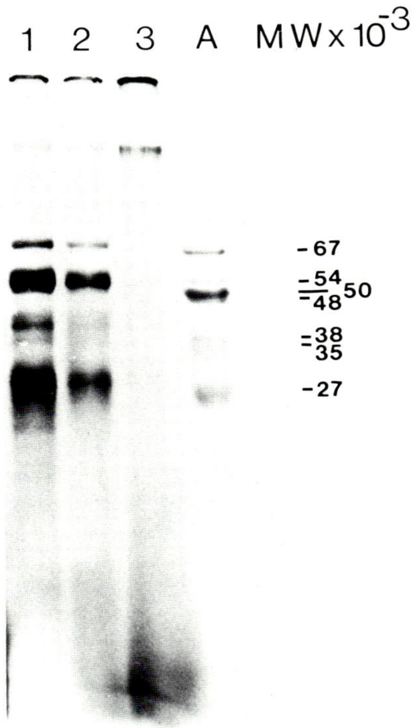

FIGURE 18. SDS-Polyacrylamide gel electrophoresis of 55% isopropanol-soluble translation products of rough microsomes from Bomi barley endosperm inaccessible to enzymic proteolysis. Cell-free protein synthesis was performed with rough microsomes prepared from 20 day old Bomi endosperm. The 55% isopropanol-soluble translation products remaining after a 30 min treatment with chymotrypsin at $0^{o}C$, were analysed by electrophoresis and autoradiography. The autoradiogram shows the translation products of rough microsomes in track 1; and after chymotrypsin treatment (0.4 $mg.ml^{-1}$) in track 2; and after chymotrypsin treatment with detergent (1% Triton X-100) in track 3. Track A contains ^{14}C-labelled hordein marker (from 8).

polyribosomes (31). Furthermore, the transport of such proteins into or across a membrane is predominantly a co-translational event (1,2). Thus the growing nascent chain could move across the membrane prior to folding and assumption of its secondary and tertiary structure. It has also been found that larger precursor polypeptides are the primary translation products of many mRNA molecules coding for secreted proteins (23,25,27). The additional sequence of some 15-25 residues at the N-terminus is thought to trigger ribosome membrane binding shortly after the initiation of translation (2). Although the movement of a nascent polypeptide comprising both polar and charged residues across the membrane lipid bilayer is not yet fully explained (11), the presequence with its preponderance of hydrophobic residues, may be important for both membrane binding and the initiation of membrane transport (23, 26). The presequence or 'signal' is later cleaved from the nascent polypeptide during synthesis by a membrane-bound peptidase (2).

The 'signal hypothesis', first propounded by Blobel and Dobberstein (1,2) to explain the transport of secreted proteins, has found considerable support in studies of protein secretion in animal (16,22,26,27,30) and bacterial cells (25). It has now been found (8) that a plant storage protein, namely hordein, is synthesized on the endoplasmic reticulum and released into the lumen for subsequent transport to and packageing in membrane-bound vacuoles in a similar manner. The synthesis and transport of hordein polypeptides has been demonstrated in vitro in a model system employing intact membrane vesicles (microsomes). During homogenization of barley endosperms, fragmentation of the endoplasmic reticulum yields rough microsomes which can be purified by sucrose density centrifugation (Fig. 16). The numerous polyribosomes bound to the outer microsomal surface are detached by EDTA treatment, allowing the isolation of stripped microsomes (Fig. 17).

Rough microsomes from Bomi endosperms synthesize all the major native hordein polypeptides during in vitro cell-free protein synthesis (Fig. 18, track 1). When these rough microsomes are treated with the proteolytic enzyme chymotrypsin after the cessation of protein synthesis, the majority of the synthesized hordein is resistant to degradation (Fig. 18, track 2). However, when a membrane-solubilizing detergent is included during proteolytic treatment the translation products are exposed to hydrolysis (Fig. 18, track 3). These experiments demonstrate that hordein polypeptides synthesized on microsomes are co-translationally transported across the membrane and released into the microsomal lumen where they are inaccessible to proteolytic attack.

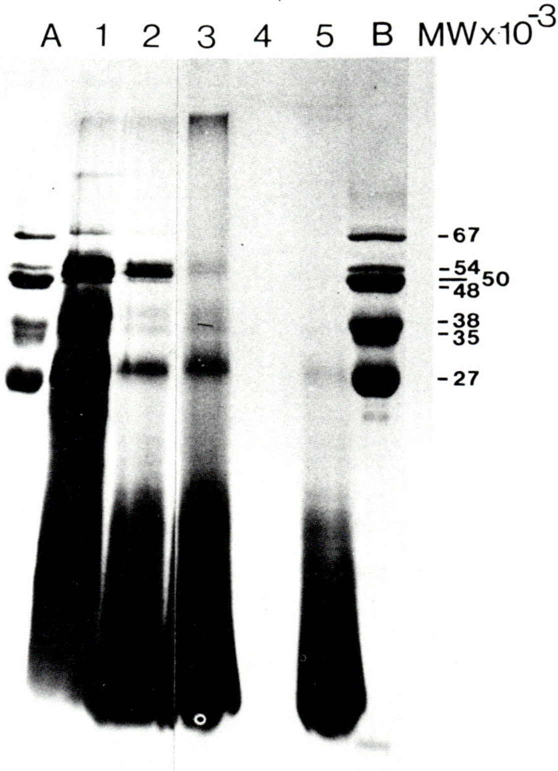

FIGURE 19. SDS-Polyacrylamide gel electrophoresis of 55% isopropanol-soluble translation products of rough microsomes reconstituted from polyribosomes and stripped microsomes from Bomi barley endosperm. Cell-free protein synthesis was performed with initially membrane-bound polyribosomes and stripped microsomes from Bomi endosperm. The 55% isopropanol-soluble translation products remaining after a 30 min treatment with chymotrypsin at $0^{o}C$, were analysed by electrophoresis and autoradiography. The autoradiogram shows translation products after the following treatments. Tracks A and B contain native ^{14}C-labelled hordein marker (from 8)

Track	Template	Co-translation additions	Chymotrypsin ($0.4 mg.ml^{-1}$)	Triton X-100 (%)
1	Polyribosomes	0	0	0
2	Polyribosomes	Stripped microsomes	+	0
3	Polyribosomes	Stripped microsomes	+	+
4	0	Stripped microsomes	0	0
5	Polyribosomes	0	+	0

It has previously been mentioned that translation of hordein mRNA molecules yields larger precursor hordein polypeptides. The ability of these precursor polypeptides to trigger ribosome-membrane binding and co-translational transport has been demonstrated in vitro with detached polyribosomes and stripped microsomes. The hordein translation products of polyribosomes are totally susceptible to proteolysis (Fig. 19, tracks 1,5). However, when stripped microsomes are present during protein synthesis, the hordein polypeptides are protected during the subsequent chymotrypsin treatment. The co-translational transport of the polyribosomal translation products into the lumen of the stripped microsomes is further evidenced by their accessibility to proteolysis after membrane solubilization (Fig. 19, track 3). The stripped microsomes themselves have lost their template activity for hordein polypeptides (Fig. 19, track 4).

It is concluded that the in vitro synthesis of hordein polypeptides on rough endoplasmic reticulum involves the translation of precursor polypeptides from hordein mRNA, which trigger ribosome-membrane binding and initiate co-translational transport of the polypeptides. The additional peptide sequence is cleaved from the precursor hordein polypeptides during their discharge into the microsomal lumen. We have now to ascertain the precise route and mechanism whereby the hordein polypeptides, after vectorial discharge into the lumen of the endoplasmic reticulum, are transported to and packaged in membrane-bound vacuoles, to become protein bodies.

REFERENCES

1. Blobel, G., and Dobberstein, B., J. Cell Biol. 67, 835 (1975).
2. Blobel, G., and Dobberstein, B., Ibid. 67 852 (1975).
3. Brandt, A., Cereal Chem. 53, 890 (1976).
4. Brandt, A., Carlsberg Res. Commun. 44, in press (1979).
5. Brandt, A., and Ingversen, J., Ibid. 41, 311 (1976).
6. Brandt, A., and Ingversen, J., Ibid. 43, 451 (1978).
7. Burr, B., Burr, F.A., Rubenstein, I., and Simon, M.N., Proc. Natl. Acad. Sci. USA 75, 696 (1978).
8. Cameron-Mills, V., and Ingversen, J., Carlsberg Res. Commun. 43, 471 (1978).
9. Doll, H., Hereditas 74, 293 (1973).
10. Doll, H., and Brown, A.H.D., Can. J. Genet. Cytol., in press.
11. Heijne, G. von, and Blomberg, C., Eur. J. Biochem. 97, 175 (1979).

12. Holder, A.A., and Ingversen, J., Carlsberg Res. Commun. 43, 177 (1978).
13. Ingversen, J., Hereditas 81, 69 (1975).
14. Ingversen, J., Køie, B., and Doll, H., Experientia 29, 11151 (1973).
15. Larkins, B.A., and Hurkman, W.J., Plant Physiol. 62, 256 (1978).
16. Maurer, R.A., and McKean, D.J., J. Biol. Chem. 253, 6315 (1978).
17. Mertz, E.J., Bates, L.S., and Nelson, O.E., Science 145, 279 (1964).
18. Munck, L., Karlsson, K.E., Hagberg, A., and Eggum, B.O., Ibid. 168, 985 (1970).
19. Nelson, O.E., Mertz, E.T., and Bates, L.S., Ibid. 150, 1469 (1965).
20. Oram, R.N., Doll, H., and Køie, B., Hereditas 80, 53 (1975).
21. Osborne, T.B., J. Am. Chem. Soc. 17, 539 (1895).
22. Palade, G., Science 189, 347 (1975).
23. Palmiter, R.D., Gagnon, J., Ericsson, L.H., and Walsh, K. A., J. Biol. Chem. 252, 6386 (1977).
24. Park, W.D., Lewis, E.D., and Rubenstein, I., Abstract p. 108 from FEBS Advanced Course 'Genome Organisation and Expression in Plants' Edinburgh Scotland 11-21 July, 1979.
25. Randall, L.L., Hardy, S.J.S., and Josefsson, L-G., Proc. Natl. Acad. Sci. USA 75, 1209 (1978).
26. Schechter, I., and Burstein, Y., Ibid. 73, 3273 (1976).
27. Scheele, G., Dobberstein, B., and Blobel, G., Eur. J. Biochem. 82, 593 (1978).
28. Shewry, P.R., Ellis, J.K.S., Pratt, H.M., Finch, R.A., and Miflin, B.J., J. Sci. Food Agric. 29, 433 (1978).
29. Shewry, P.R., Pratt, H.M., Finch, R.A. and Miflin, B.J., Heredity 40, 463 (1978).
30. Shields, D., and Blobel, G., Proc. Natl. Acad. Sci. USA 74, 2059 (1977).
31. Shore, G.C., and Harris, R., J. Cell Biol. 74, 315 (1977).
32. Weinard, V., Brüschke, C., and Feix, G., Nucleic Acid Res. 6, 2707 (1979).
33. Wettstein, D. von., Proc. Eur. Brewery Conv. Congr. Berlin, in press (1979).

USE OF PROANTHOCYANIDIN-FREE BARLEY IN BEER BREWING

Kenneth Erdal
B. Ahrenst-Larsen

Department of Brewing Chemistry

B. Jende-Strid

Department of Physiology

Carlsberg Research Center
Copenhagen (DK)

I. INTRODUCTION

Beer brewed according to traditional methods will become cloudy after a few months' storage unless preventive measures are taken. It is known that the main constituents of beer haze are proteins and polyphenols. The polyphenols most prominent in this respect are the proanthocyanidins (by brewers called anthocyanogens) of which the amount in beer for 80-100% derive from barley, depending on whether natural hops or tannin-free hop extract has been used.

In beer the molecular weight and the reactivity of the polyphenols may increase by oxidative condensations, and at a certain level colloidal complexes can be formed with the beer proteins. Already during the cold storage period at the brewery colloidal haze is formed and removed during the clarification of the beer, however, formation of colloidal haze may continue after filtration.

The rate of haze formation is dependent on the amount and the state of reactivity of the two main components mentioned but also dependent on temperature, presence of metal ions and oxygen. If no preventive measures are taken visible haze in beer kept at room temperature will develop within weeks.

ISBN: 0-12-370960-1

TABLE I. Examples of Stabilisation Techniques

	Proteins	Polyphenols
Modification by	Proteolytic enzymes	(Anthocyanogenases in green malt)
Removal by	Silica gel	PVPP
Avoidance by		Genetic blocking of biosynthesis

The problem could of course be neglected either by selling the beer before the haze develops or by making hazy beer accepted by the consumers; apart from the change in visual appearance the occurrence of haze is not deleterious to beer quality. However, the brilliant look of a glass of beer is generally considered as a quality characteristic and therefore to day all breweries are using stabilising agents to ensure beer brilliance.

Stabilising treatments may be aimed at one or both of the haze constituents protein and polyphenol with the purpose of either removal or modification. Examples of techniques are shown in Table I.

Modification of proteins by the use of proteolytic enzymes, for instance of the papain type, is probably the most widely used technique to day. The enzyme degrades beer proteins and thereby reduces the possibility of reaction between proteins and polyphenols.

Proteins can also be partly removed from the beer by adsorption on the surface of insoluble adsorbents like silica gels, preparations of silicic acid. The silica gels are added to the beer either in storage tank prior to final filtration or used as an addition to the filter, whereby the beer is treated simultaneously with the filtration process. Some proanthocyanidins will also be removed from the beer by this treatment.

Enzymatic modification of the polyphenols has not yet been used at a technical scale but it has been reported (1) that in brews made with green malt, anthocyanogens are decomposed during mashing due to anthocyanogenases from the unkilned malt. In practice this method is complicated by certain problems, for instance grinding of the green malt and the taste of the finished beer.

Polyvinylpolypyrrolidone (PVPP, Polyclar AT), an insoluble synthetic protein analog, adsorb polyphenols effectively. In this procedure PVPP is added to the beer prior to final filtration. After removal of part of the reactive polyphenols from the beer, the high molecular weight proteins have less opportunity to form colloidal haze.

Alternatively, the presence of polyphenols can be avoided by blocking their biosynthesis in the barley grain with the aid of suitable mutations and beer with very good haze stability has been obtained in brewing experiments with the first such mutant tried (2). This last method has the advantage of avoiding special operations and extra costs for additives and dosing. It also does not involve beer losses and deleterious effects on beer quality as is inherent with some of the other techniques. Such an alternative technique of haze control without the use of additives has furthermore the advantage of meeting the demands in countries where the use of stabilisers may be subject to legislative limitations or public disapproval.

Barley cultivars contain proanthocyanidins and catechins with relatively small quantitative variations among varieties (3,4,5). These compounds are considered to be the most potent of the beer haze polyphenols. By screening induced mutants affecting anthocyanin pigmentation for the presence of proanthocyanidins and catechins, the mutant anthocyanin-free 13 (*ant*13-13) was identified as lacking both anthocyanins, catechins and proanthocyanidins in the vegetative parts of the plants as well as in the grain, whereas many other anthocyanin-free mutants contained normal amounts of catechins and proanthocyanidins (5).

From the biosynthetic pathway of the flavonoids (Figure 1) it is apparent that the mutant *ant*13-13 is blocked prior to the branch point leading to anthocyanidins and catechins, respectively. Since precursor flavonoids can be identified in the mutant, the gene *ant*13 most likely controls a step close to the branch point, i.e. between the dihydroflavonol and the flavenol intermediate.

The mutant *ant*13-13 has been shown to be constantly free of proanthocyanidins over several years (Table II). It has been propagated and used for further malting and brewing experiments in pilot scale and in full scale. The results of these experiments will be reported in the following.

BIOSYNTHESIS OF PROANTHOCYANIDINS

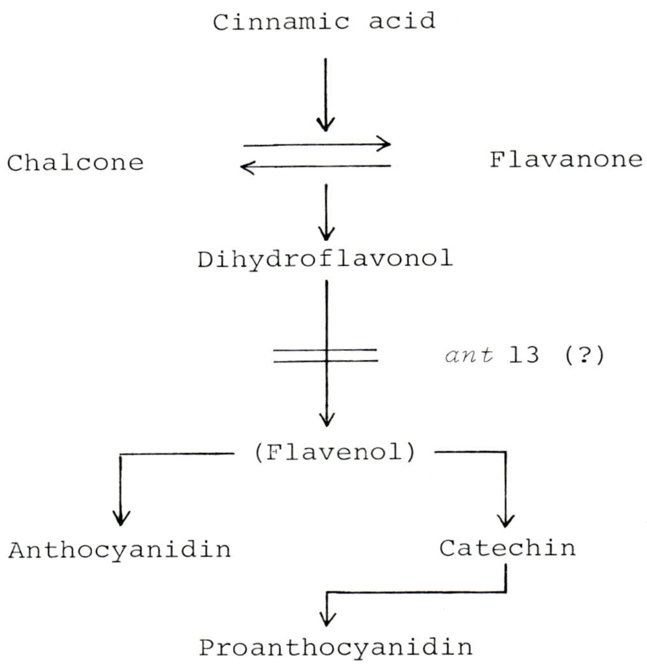

FIGURE 1. Scheme for proanthocyanidin biosynthesis

TABLE II. Proanthocyanidin in barley

	Proanthocyanidin mg/100 g
ant-13, 1969	2
ant-13, 1973	3
ant-13, 1974	3
ant-13, 1975	2
ant-13, 1976	0
ant-13, 1977	0
Foma (and other barleys)	150-200

II. MATERIALS AND METHODS

A. Barley and Malt

The ant-13 barley used for pilot malting and brewing was from the 1975 and 1976 harvest whereas the ant-13 barley used in the full scale trial was from the 1977 harvest. All samples were obtained from the experimental farms of Carlsberg Plant Breeding. The reference malts from Foma barley were produced in pilot scale whereas the malts from Mala and Nordal were regular malts.

B. Pilot Malting

Portions of 14 kg of barley were steeped and germinated in a rotating drum of stainless steel net. Steeping was carried out over 2 days with alternating periods of several hours in water and in air at 14-15°C until a moisture content of 41-43% was reached. During germination for 5-6 days at 12°C, the moisture content was maintained at 41-43%. The kilning was carried out in an oven. During the first few hours the temperature of the malt increased to 50°C and was kept at this temperature up to 10 hours, whereafter the temperature was raised over a period of 3 hours to a maximum of 78°C. Curing was carried out at this temperature for an additional 3 hours. The cured malt was replaced in the drum and the rootlets were removed by drum rotation.

C. Pilot Brewing

1. All-malt Brews. 10 kg of ground malt was mashed in the pilot brewery with 42 l of deionized water by infusion mashing. The following standard mashing procedure was empolyed: 30 min at 52°C, heating from 52°C to 63°C in 20 min, 60 min at 63°C, heating from 63°C to 78°C in 40 min, 10 min at 78°C. A lauter tub was used for the wort filtration at 73°C, and four spargings with 6+6+10+15 l of deionized water were performed. Total filtration time for 61-62 l of wort amounted to about 160 min. After addition of 8.7 g tannin-free hop extract from Horst Company, the wort was boiled for 90 min and cooled to 5°C by standing overnight in a wort holding vessel cooled with ice water. On the following day the cold wort was separated from the hot and cold break, transferred to two 30 l fermentation vessels and pitched with yeast, Saccharomyces carlsbergensis strain U, from which

possible adhering polyphenols had been removed by washing the cells six times with water. The amount of pitching yeast was 90 g of centrifuged yeast per 28 kg of wort. The fermentation vessels were placed in a water thermostat at 10°C. After fermentation for one week, the beer was racked into two 25 l storage tanks with CO_2-pressure and placed in a refrigerated cabinet. During 6 weeks of storage, the beer temperature was decreased gradually from 5°C to -1°C. The beer was filtered through 20 x 20 cm Seitz K-7 sheets and the CO_2 content adjusted to 0.5%. Bottling and crowning were carried out manually, the bottles being knocked to reduce the air content in the headspace. The bottled beer was pasteurized at 62°C for 20 min.

2. Adjunct Brews. In the adjunct cooker 1.2 kg of malt and 3.5 kg of maize grits were mixed with 12 l of water and kept 30 min at 52°C, heated from 52°C to 100°C in 40 min and boiled for 20 min. In the mash tun 5.3 kg of malt was mixed with 20 l of water at 35°C 30 min before the end of boiling in the adjunct cooker. After boiling, the adjunct mash was transferred to the mash tun and the main mash adjusted to 63°C. After 30 min at 63°C the temperature was raised to and kept at 69°C for 40 min, followed by heating to 78°C over 10 min where the temperature was held constant for 20 min. The lautering and further procedures were performed as described for all-malt brews.

D. Full Scale Experiment

Malting of 7 tons of ant-13 barley was performed as a production malting on a reduced scale at the Carlsberg Breweries.

Brewing was carried out following a standard procedure for lager brewing at the Carlsberg Breweries using 35% of maize grits. No special precautions against the possible contributions of polyphenols from hops and yeast were taken.

E. Analyses

Proanthocyanidins and polyphenols were assayed as described by Harris & Ricketts (6) and Jerumanis (7). Additional testing for haze stability was carried out according to Chapon (8). HMW-nitrogen was determined by the amidoblack adsorption technique by Klopper (9). Wort and beer analyses were performed according to Analytica-EBC except for head retention which was determined after Blom (10).

III. ALL-MALT PILOT BREWS

Portions of ant-13 barley and of its mother variety Foma barley were pilot malted and used for brewing a 10% Plato all-malt lager in our pilot brewery. To avoid possible contributions to proanthocyanidins and polyphenols from hops and yeast, wort boiling was performed with a tannin-free hop extract and pitching yeast was washed until free of proanthocyanidins.

Analyses of the beers (Table III) show that proanthocyanidins are almost absent in the ant-13 beers and the polyphenol values are very low. The corresponding values for Foma beers are at levels normally observed.

The haze values measured with three different tests are low for the ant-13 beers. Such low values for beers produced from Foma, or other barleys, can only be obtained after treatment with chemical stabilisers.

TABLE III. Analyses of All-malt Beer Produced by Pilot Brewing from ant-13 and Foma Malt

Malt	ant-13		Foma	
Harvest	1975	1976	1975	1976
Original extract, % Plato	9.7	10.5	10.1	10.0
Attenuation, real, %	58	58	61	60
Bitterness, B.U./EBU	20	21	20	17
Nitrogen, mg/100 g	115	102	105	66
pH	4.6	4.7	4.7	4.6
CO_2, % w/w	0.51	0.51	0.51	0.50
Colour, EBC units	7.5	4.5	7.5	5.0
Head retention, sec.	114	100	108	111
Air in headspace, ml/bottle	1.6	0.5	1.3	0.5
Proanthocyanidins, mg/l	0	4	29	36
Polyphenols mg/l	7	15	86	117
Chill-haze, 5 d./60°C *, EBC	1.9	2.2	12	12
Chill-haze, 60 d./20°C**, EBC	-	1.6	-	4.8
Haze after 40 min at -8°C (Chapon), EBC	3.5	2.8	>12	>12

* Total haze measured after 5 days at 60°C and 1 day at 0°

** Total haze measured after 60 days at 20°C and 2 days at 0°.

The traditional beer analyses demonstrate that the mutant beers are in all respect of good quality. Except for total nitrogen, where Foma 1976 was low, very similar analytical results are found for the four beers.

IV. ADJUNCT BREWS

Additional pilot brewings were performed with ant-13 and with Nordal malt using 35% maize grits as adjunct. The variety Nordal may be taken as a typical Danish malting barley.

The beer analyses (Table IV) reflect the dilution effect of the maize grits when compared with the corresponding analyses of all-malt beer given in Table III. The results obtained are within the normal range of pilot beers.

It is well known that one of the factors controlling haze development is oxygen. In order to examine the sensitivity of these beers to oxygen, bottling was performed so that some of the final pasteurized beer bottles contained 1 ml and others 4 ml of air in headspace.

For both beers it is seen that more haze is developed in the bottles with increased air content, but the Nordal beer

TABLE IV. Analyses of Beer Produced by Pilot Brewing from ant-13 and Nordal Malt with 35% Maize Grits

Malt	ant-13		Nordal	
Original extract, % Plato	10.5		11.0	
Attenuation, real, %	59		65	
Bitterness, B.U./EBU	24		17	
Total nitrogen, mg/100 g	56		40	
HMV-nitrogen, mg/100 g	9		10	
pH	4.3		4.3	
CO_2, % w/w	0.48		0.46	
Colour, EBC units	4.5		3.6	
Head retention, sec.	90		82	
Proanthocyanidins, mg/l	0		22	
Polyphenols, mg/l	13		61	
Air in headspace, ml/bottle	1	4	1	4
Chill-haze, 5 d./60°C, EBC	1.0	1.8	4.9	8.2
Haze after 40 min at -8°C (Chapon), EBC	3.0	3.1	>12	>12

is far the most sensitive. The chill-haze values for both beers are lower than observed for the all-malt beers, so the expected improvements as a result of the dilution effect of the adjunct were obtained. The ant-13 beer shows an excellent haze stability, which is even better than the stability of commercial beers stabilised in a conventional manner.

V. MALT BLENDING

From a theoretical as well as from a practical point of view it was considered interesting to investigate the properties of beer, where the ant-13 malt was partly substituted by ordinary malt.

A series of 5 pilot brews was then made with 35% maize grits as adjunct. The malts in these experiments were mixtures of the mutant malt and an ordinary malt from the barley variety Mala. The contents of ant-13 malt were 0, 25, 50 75 and 100% of the total amount of malt.

The beer analyses are given in Table V. Malt-derived characteristics such as attenuation and nitrogen reflect the content of mutant malt in the malt mixture.

With increasing amount of ant-13 malt an almost proportional reduction of the proanthocyanidin and polyphenol content is found. The values for the brew from Mala malt (0% ant-13) may be taken as typical for this type of beer produced from normal malt and adjunct.

The chill-haze values obtained by two methods follow the same trend as found for proanthocyanidins and polyphenols. This malt blending experiment shows that it is feasible to use mixtures of the mutant malt and ordinary malt without any adverse effect upon the quality of the resultant beer. Mixing of the mutant malt with ordinary malt results in beer with a reduced content of anthocyanogen and a corresponding improvement of the chill-haze stability. In this series beer produced with equal parts of ant-13 malt and ordinary malt obtained a chill-haze stability corresponding to the stability that can be obtained by the addition of 0.75 g per hl of Collupulin (a commercial chill-proofing enzyme preparation) to a normally produced reference beer.

An important question is whether the flavour of beer produced from the proanthocyanidin-free barley deviates from the control beer. This is not the case neither for fresh beers nor for beers aged for one week at 37°C or 13 months old beers (Table VI). The absence of proanthocyanidins in the mutant beer did not affect the flavour, as these beers were classified by United Breweries' taste panels to be within

TABLE V. Analyses of Beer Produced by Pilot Brewing from 35% Maize Grits and Mixtures of ant-13 and Mala Malt

ant-13 malt, % of total malt	0	25	50	75	100	Ref.*
Original extract, % Plato	10.7	10.4	10.3	10.7	10.3	10.6
Attenuation, real, %	63	64	61	62	59	67
Bitterness, B.U./EBU	20	21	22	23	18	22
Nitrogen, mg/100 g	32	36	42	48	50	36
pH	4.2	4.3	4.3	4.3	4.4	4.1
CO_2, % w/w	0.50	0.48	0.49	0.49	0.49	0.48
Colour, EBC units	3.0	3.0	4.0	4.0	8.0	5.5
Head retention, sec.	88	91	95	89	86	82
Air in headspace, ml/bottle	0.5	0.5	0.5	0.5	0.5	0.5
Proanthocyanidins, mg/l	22.8	16.4	12.6	7.2	2.3	23.1
Polyphenols, mg/l	61	45	32	21	8	67
Chill-haze, 5 d./60°C, EBC	5.1	3.6	2.5	1.7	0.7	2.1
Chill-haze, 60 d./20°C, EBC	3.3	1.7	1.1	1.0	0.5	1.5

* Reference production beer, pilot filtered and stabilised with 0.75 g Collupulin per hl.

TABLE VI. Taste tests (mean panel score)

Adjunct beer	Foma	Mala	Mala/ ant-13 1+1	ant-13
Beer, 1-2 weeks after bottling	2.0	2.1	2.3	1.7
Beer, aged 7 days at 37°C	2.2	2.0	1.8	2.2
Beer, aged 13 months at 10°C	1.8	2.0	1.9	1.9

score: 3 normal
2 with faults within acceptable range
1 with faults that can not be accepted
0 with grave faults.

the normal range for pilot beers. This finding apparently conflicts with the prevailing opinion that polyphenols are the most important single group of beer constituents affecting flavour, colour, stability and quality (11). Stabilisation has been believed to be a balancing act in preferential, partial removal of proanthocyanidins without infringing noticeably on the flavour or colour of the beer. This opinion was based on results derived from experiments where normal beer was deprived of up to 95% of the proanthocyanidins by means of adsorbtion on excessive amounts - from 2 to 10 kg per hl - of nylon. Such procedures probably result in adsorbtion of other important beer components as well.

VI. BEER BLENDING

In 1977 the harvest of the ant-13 barley grown at our experimental farms was increased to 7 tons and for the first time it was possible to carry out malting and brewing on full scale. The last series of experiments to be reported deals with the characteristics of blendings of this beer with normal beer (Table VII).

The filtered mutant beer (100%) was blended with filtered unstabilised production beer (0%) and three mixtures containing 25%, 50% and 75% mutant beer were obtained. With the mutant beer characteristics reported (Table IV and V) in mind it is not surprising that the attenuation decreases and nitrogen and colour increases with increasing ant-13 beer content.

TABLE VII. Analyses of Beer Obtained by Blending of ant-13 Adjunct Beer Produced in Full Scale and Unstabilised Production Beer.

ant-13 beer, % Production beer, %	0 100	25 75	50 50	75 25	100 0
Original extract, % Plato	10.4	10.5	10.5	10.5	10.8
Attenuation, real, %	69	66	64	62	59
Bitterness, B.U./EBU	23	22	23	24	25
Nitrogen, mg/100 g	40	44	50	54	56
pH	4.2	4.2	4.3	4.4	4.4
CO_2, % w/w	0.52	0.47	0.44	0.43	0.43
Colour, EBC units	6.5	8.0	8.5	10.0	11.5
Head retention, sec.	91	91	92	95	96
Air in headspace, ml/bottle	0.5	0.5	0.7	0.9	1.1
Proanthocyanidins, mg/l	23.9	20.2	15.7	10.6	6.5
Polyphenols, mg/l	65	54	42	32	23
Chill-haze, 5 d./60°C, EBC	8.4	5.3	5.9	4.2	3.6
Chill-haze, 60 d./20°C, EBC	4.1	2.2	2.1	1.6	1.0

The content of proanthocyanidin and polyphenol in these beers is reduced proportional to the percentage of ant-13 beer in the blendings. The values for the pure full scale mutant beer are low, although somewhat higher than found in the pilot beers. This may be due to some intermingling (up to 1%) with varieties containing proanthocyanidins during the large scale field propagation of the mutant as well as to contributions from hops and yeast which in the full scale brew was used according to normal production practice.

Except for a single odd value the trend of the values for chill-haze is clear: Chill-haze stability is controlled by the proportion of mutant beer. The stability of the pure mutant beer is extremely good thus making it possible to produce beer of desired chill-haze stability by blending.

VII. POLYPHENOLIC HAZE PRECURSORS IN BARLEY

The fact that we now have at our disposal beer virtually without proanthocyanidins at any process step may present certain advantages to the brewing chemist. Thus it will now be possible in a more direct way to evaluate the haze forming potential of specific proanthocyanidins by addition of such compounds, e.g. to the finished proanthocyanidin-free beer. An example will be given in the following.

Proanthocyanidins from barley have been extracted with 60% ethanol and further fractionated on Sephadex LH-20 as shown in Figure 2 (12).

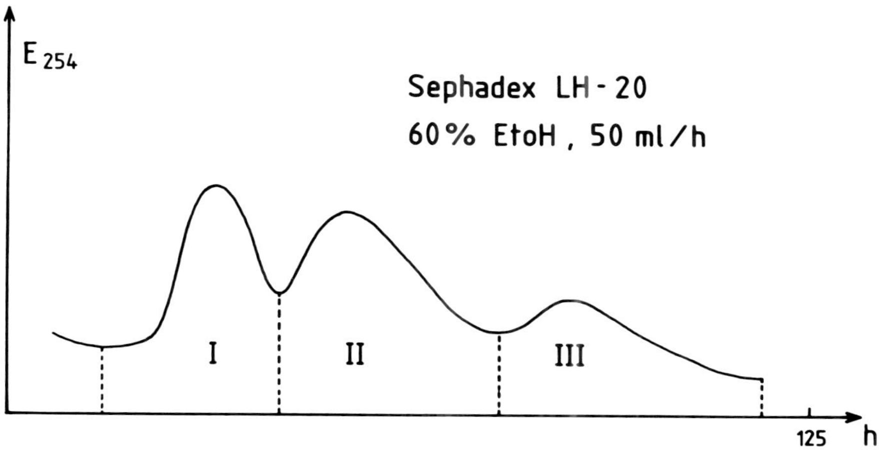

FIGURE 2. Separation of barley proanthocyanidins on Sephadex LH-20.

TABLE VIII. Effect on Chill-haze of Addition of a Proanthocyanidin Fraction from Barley to ant-13 Beer

	Chill-haze	
	Initial EBC	After 5 days at 60°C EBC
ant-13	1.2	1.0
ant-13 + 20 ppm I	1.2	5.3

Fraction I was collected and shown by HPLC to consist mainly of two compounds, identified by chemical methods and by NMR as the two dimeric proanthocyanidins prodelphinidin and procyanidin (B-3).

Fraction II and III have not yet been identified but consist of higher oligomers of (+)catechin and/or (+)gallocatechin.

The fractions were freeze-dried and 20 ppm added to ant-13 beer at bottling. Chill-haze was measured before and after 5 days storage at 60°C. In Table VIII the results from the addition of the well characterized fraction I are given. The initial chill-haze is not affected, but a marked increase in chill-haze after the forced ageing test is observed. This value is at a level we normally observe for unstabilised pilot beers from normal malt.

VIII CONCLUSION

It is feasible to brew beer with an excellent chill-haze stability without the use of stabilising agents from proanthocyanidin-free barley and at the same time maintain the high quality of the beer.

The elimination of stabilising additives in brewing as a regular practice requires the production of agronomically acceptable high yielding proanthocyanidin-free barley varieties. Extensive breeding programs have been initiated to produce such high yielding varieties (13). By the induction and isolation of 51 new proanthocyanidin-free mutants in the spring barley varieties Nordal, Alf, Triumph, Gula and Ark Royal it has been shown that genetic blocking of proanthocyanidin synthesis can be introduced into any high yielding malting barley variety. Recombinants from crosses between the mutant ant-13 and different mildew resistant varieties have resulted in lines with a yield fully comparable to present day commercial cultivars.

REFERENCES

(1) Hudson, J.R., Proceedings of the European Brewery Convention Congress, Brussels, 1963, 422.

(2) Wettstein, D. von, Jende-Strid, B., Ahrenst-Larsen, B. & Sørensen, J.A., Carlsberg Research Communications *42*, 341 (1977).

(3) McFarlane, W.D. & Sword, P.T., the Journal of the Institute of Brewing, *68*, 344 (1962).

(4) Pollock, J.R.A., Pool, A.A. & Reynolds, T., the Journal of the Institute of Brewing, *66*, 389 (1960).

(5) Jende-Strid, B., Carlsberg Research Communications, *43*, 265 (1978).

(6) Harris, G. & Ricketts, R.W., Proceedings of the European Brewery Convention Congress, Rome, 1959, 290.

(7) Jerumanis, J., Bulletin de l'Association Royale des Anciens Etudiants en Brasserie de l'Université de Louvain, *64*, 223 (1968).

(8) Chapon, L. & Chemardin, M., Proceedings of the European Brewery Convention Congress, Madrid, 1967, 389.

(9) Klopper, W.J., Brauwissenschaft, *21*, 373 (1968).

(10) Blom, J., Proceedings of the European Brewery Convention Congress, Copenhagen, 1957, 51.

(11) Dadic, M. & van Gheluwe, G.E.A., Technical Quarterly of the Master Brewers Association of America, *10*, 69 (1973).

(12) Outtrup, H., personal communication.

(13) Wettstein, D. von, Jende-Strid, B., Ahrenst-Larsen, B. & Erdal, K., Technical Quarterly of the Master Brewers Association of America (in press).

RECENT PROGRESS IN SOY SAUCE PRODUCTION IN JAPAN

Gengo Ogawa

Central Research Laboratory
Nisshin Flour Milling Co. Ltd.
Saitama, Japan

Akio Fujita

Central Research Laboratory
Nisshin Flour Milling Co. Ltd.
Saitama, Japan

I. INTRODUCTION

Soy sauce has occupied a place as the most familier condiment or seasoning in an oriental kitchen and at a dinner table. It is not too much to say that soy sauce is a so-called all purpose seasoning and used for most of dishes and foods in Japan.

The natinal annual consumption of soy sauce for 1977 is about 1,200,000 Kl which was produced by as many as 3,500 companies. About ten years ago, one of the biggest soy sauce producers began to export to outside of Japan. Recently this producer established a soy sauce plant in U.S.A.. Since then, soy sauce has been becoming familier with occidental people as a sauce of barbecues, steaks, hamburgers and virtually any dish.

Soy sauce is undoubtedly the oldest processed condimint consumed by human beings. Soy sauce is historically originated in brine in which meat or fish was soaked for the purpose of preservation. Soy sauce was introduced with Buddihism to Japan from China in the six century. According to records of the Imperial Court 'hishio' (the Japanese historical name

ISBN: 0-12-370960-1

of soy sauce) was given to the Emperors subjects and military officials as a part of annual wage. The word 'shoyu' (the today's Japanese name for soy sauce) was used for the first time in ancient records in 1552, when soy sauce production seems to begin. That time, however, its process was operated only on a small scale due to lack of the knowledge to control the fermentation. Generally speaking the modern industrial soy sauce production was established after the Second World War, especially for past ten years in Japan.

A typical modern soy sauce factory has the following equipment; a large continuous autoclave for soybean cooking and oven for wheat roasting which can process several tons per hour; automated semi-continuous mold cultivation rooms accommodating 15-20 tons of 'koji'; completely controlled fermenting tanks containing a hundred ton of 'moromi' mash; giant continuous pressing apparatus; and finally auto-packing system. Pipes and conveyers are running all over the place. Workers only have to take care of the important spots. The conventional physical workers with sweat on brow have been eliminated from the soy sauce production.

Such wonderful alteration in soy sauce industry was not gained without numerous scientific and engineering advances and efforts. In these advances and efforts I will introduce following four revolutionary investigations from the standpoint of industrial aspects.

(a) Optimum cooking condition for soybean.
(b) Microbial control in 'moromi' fermentation.
(c) The solution cf the reasons why the pressing process is so difficult.
(d) Development two types of puffed granules by an extruder as processed materials.

II. TRADITIONAL SOY SAUCE PROCESS

Soy sauce could be manufactured by two basic methods, fermentation method and chemical one. Only the former will be discussed because most of soy sauce is produced by fermentation method from the standpoint of its superior quality and wholesomeness. Figure 1 shows the process of soy sauce production.

Soy sauce is a sort of seasoning made from soybean and wheat with brine. Soybean is mainly a source of nirtogen and its protein is finally decomposed to amino acids and soluble peptides by proteinase generated and activated by mold cultivation. After the Second World War defatted soybean flakes or meals have been used for political and economical reasons

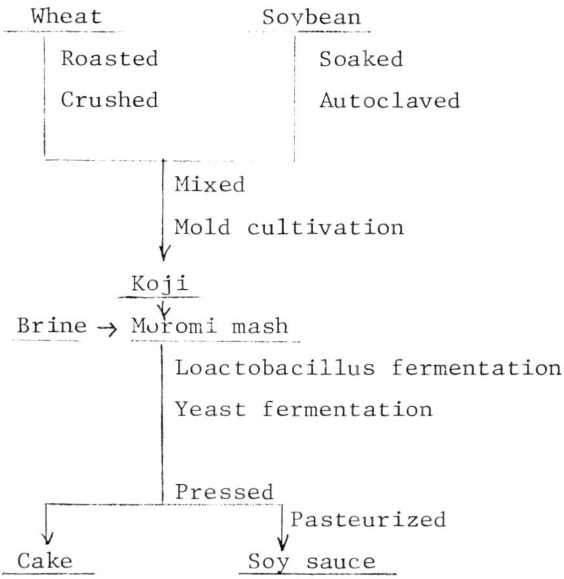

FIGURE 1. The process of soy sauce production.

instead of whole soybean. (Hereinafter soybean means defatted soybean flakes or meals unless otherwise stated.) Wheat is used as a carbohydrate source for effective fermentation, but plays another role in contributing to the good taste and flavor. Some desirable organic acids, alcohols and esters are synthesized from wheat starch. In general wheat is roasted with sand, separated and crushed before mixed with cooked soybean for adjusting the moisture content of the resulting mixture.

Molds are cultivated on the mixture for three days at 30 °C for the purpose of producing various enzymes; proteinase, peptidase, glutaminase, amylase and so on. Among molds, *Aspergillus soyae* or *Aspergillus oryzae* is generally used. One part of the matured 'koji', is dispersed in two parts of 20% brine. This suspension is so-called 'moromi' in Japanese. 'Moromi' is allowed to stand for six months in an opened concrete vessel or a closed fermenting tank. Chemical and biochemical changes occured in 'moromi' will be discussed later. Soy sauce is expelled from the matured 'moromi' and pasteurized before packed. In general, five tons of soy sauce can be produced with each ton of soybean, wheat and salt and three tons of water.

III. OPTIMUM COOKING CONDITIONS FOR SOYBEAN

The ratio of solubilized nitrogen into 20% brine to the total nitrogen used to be 65% before the Second World War. Recently this ratio has increased up to 85-90% because of selection of molds, improvements in mold cultivation and especially consideration of cooking condition for soybean. Cooking condition for soybean plays important role in not only digestibility but also somewhat turbidity of the final product when it is diluted or heated for cooking. This turbidity is caused by insolubilization of undenaturated protein which is dispersed in concentrated salt solution. This turbidity spoils the commercial value of the product.

Many investigations on these mechanisms and countermeasures were done by Mogi and his colleagues. Their excellent works will be outlined in the following context (1, 2, 3).

Figure 2 shows the degree of heat-denaturation of protein in soybean at various moisture contents under various steaming pressures (temperatures) and periods. With increasing steaming pressure or moisture content of soybean, soy protein tends to be denatured more easily.

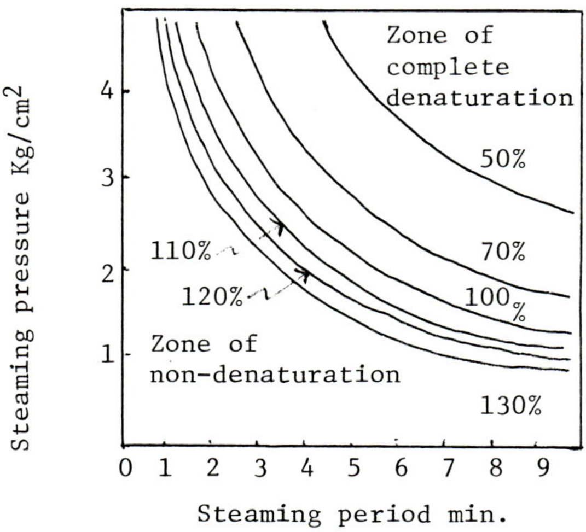

FIGURE 2. Heat-denaturation curve for protein in soybean at various moisture contents.

On the other hand, over-denatured soy protein loses an accessibility for enzyme reaction. Consequently yield of soluble nitrogen and other soluble compounds will be down and soy sauce productivity will be poor. Figure 3 suggests a suitable condition for steaming pressure and period for heat-treatment of soy protein to produce completely unturbid soy sauce with high yield.

The zone of under oblique lines is optimum condition. The cooking period becomes shorter with increasing temperature. However considering yield of digestibility of 'moromi', the high temperature and short period cooking is desirable. Therefore recently cooking of soybean is operated under the more severe conditions of upper-left zone in Figure 3, to compare with that of 0.9 Kg/cm^2 and 30-60 minutes employed till several years ago. Of course operating technic became difficult and it would not overcome until sophisticated devices were developed such as a continuous autoclave connected with a quickly cooling conveyer.

Table 1 shows various cooking conditions of soybean and results of digestibility. Generally No. 2 condition is used industrially in Japan. An industrial process is shown in Figure 4.

Cooked and excluded soybean from autoclave can be cooled instantaneously because it is in part puffed just after cooking.

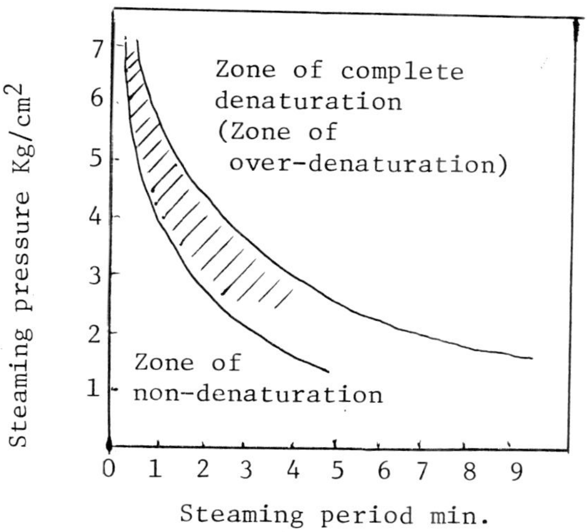

FIGURE 3. The relationship between steaming pressure and cooking period in heat-denaturation of soy protein.

TABLE 1. Various cooking conditions of soybean and results of digestibility

No.	Steaming Pressure Kg/cm^2	(Temperature) °C	Steaming Period min.	Digestibility %	Turbidity
1	0.9	(117)	45	86.13	(-)
2	2.0	(133)	5	91.60	(-)
3	4.0	(152)	2	93.74	(-)
4	7.0	(170)	1/4	95.10	(-)

However new problems arise as follows; an uniformity of water sprinkling over soybean in the pre-process is required, bacterial infection is easier than batch system, growth of yeast and lactobacillus become slow and weak in the 'moromi' process because of high nitrogen solubility and digestibility and so on. So defatted soybean flakes have been changing to small granule of meals. Hygienic consideration and microbial control become more severe and accurate recently.

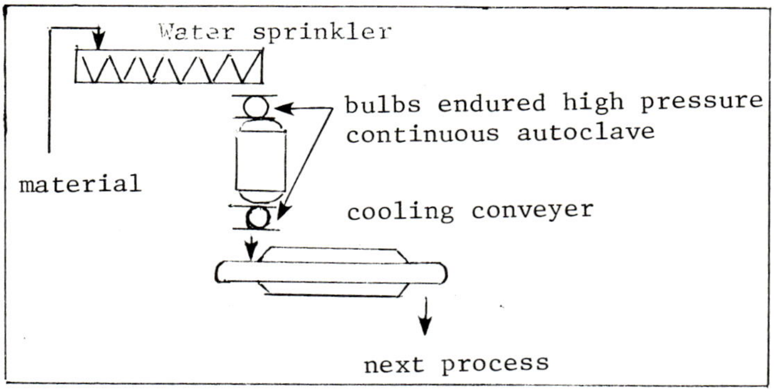

FIGURE 4. A flowsheet of modern cooking process of soybean under conditions of high pressure and short period.

IV. MICROBIAL CONTROL IN 'MOROMI' FERMENTATION

Before the Second World War it took a year or even more to mature 'moromi' mash. Nowadays it can be performed in six months. The quality of final product, soy sauce expelled from it, depends on the season when 'moromi' is prepared. 'Moromi' prepared in summer is rich in lactic, acetic and pyruvic acid and low in alcohol and glutamate, compared with one in winter. The latter has generally considered to be better than the former with respect to taste and flavor. But recently the quality of soy sauce is considerably uniform throughout a year, and is not effected by producing seasons any more.

This progress is due to numerous biochemical and engineering researchs to control overall process for soy sauce production. Among these works, one of the most important feature was an addition or an injection of the specific yeast and bacteria to 'moromi' in the early stage of the 'moromi' process. Only these microorganisms were originally isolated from 'moromi', idetified and cultured. By the addition of the microorganisms, the fermentation of the 'moromi' system was activated in desirable way and easily controlled. This technique was applied to practical 'moromi' production by Kimura, one of the pioneers in this field about 25 years ago. He successfully organized an association to promote this in soy sauce industry (4). For this purpose the activated 'koji' molds was developed by blending with lyophilized yeast and lactobacillus. Nowadays *Saccharomyces soyae* or *rouxii* and spesies of *Torulopsis* as yeast and *Pediococcus soyae* or *halophilus* as lactobacillus may be added directly in the 'moromi' process.

The 'moromi' process consists of three stages and a suitable artificial control of fermentation should be carried out at each of them. Figure 5 shows temperature control and injection period of microorganisms at each stage. Soy sauce 'koji' is transfered to a fermenting tank with about 20% brine cooled nearly at 5°C. It is recomended to hold it at low temperature about 15°C for 3-4 weeks. In this first stage aeration of this suspension should be moderate for proper lactic acid fermentation by the injected lactobacillus. The temperature should be rised to 30°C gradually. When pH value of 'moromi' gets down to 5.2 in 30-35 days, yeast will be injected for subsequent alcohol fermentation. This is the second stage. An occasional vigorous aeration should be provided for the good growth of the yeast and for preventing the growth of undesirable anaerobic bacteria. After 4-5 months from starting the 'moromi' process, species of

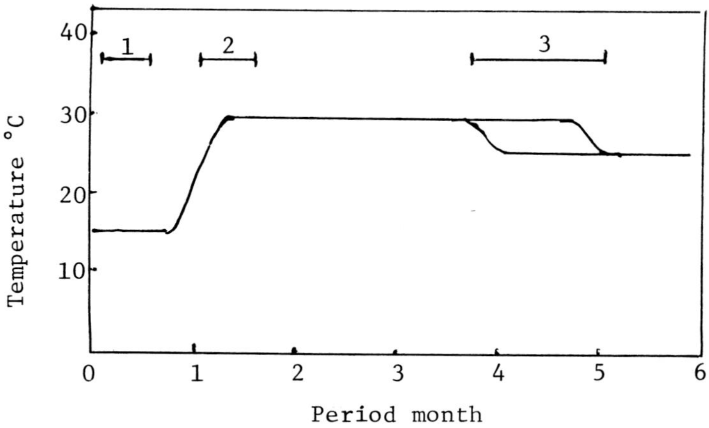

Period of microorganisms injection
1. Pediococcus 2. Saccharomyces 3. Torulopsis

Figure 5. Temperature control and period of microorganisms injection in the 'moromi' process.

Torulopsis is injected to the system for 4-ethyl-guaiacol production which is one of the most desirable flavor of aged soy sauce. This is the third stage. The temperature should be kept at 25-28°C for a month or so to obtain the efficient amount of the compound. Naturally the hydrolysis of protein and carbohydrate is occured throughout the 'moromi' process especially first 3-4 months.

Table 2 shows composition of a typical commercial soy sauce which is guaranteed by Japanese Agricultural Standards (5).

TABLE 2. The composition of a typical commercial soy sauce.

Salt %	17.44
Total nitrogen %	1.58
Formol nitrogen %	0.86
Direct reducing sugar %	2.69
Alcohol %	2.21
Acidity I ml.	12.61
Acidity II ml.	11.34
pH	4.90
Solid(except. salt) %	18.10

V. THE SOLUTION OF THE DIFFICULT PRESSING PROCESS

With regard to pressing process in soy sauce production, scientific investigations and technical improvements have only started recently because of its complexity. The pressing process seems to expel and separate solution, soy sauce, from insoluble substances, but in fact it is to press out or squeeze liquid from water swelling mass or a sort of gel. Therefore extraordinary high pressure by a large equipment is demanded and many workers are required for the operation.

Recently Kikuchi and his co-workers (6) achieved a break-through to clarify the mechanism of pressing process. They reported that whole soybean contains 30% carbohydrate, and a half of it is a polysaccharide which consists of arabino-galactan, soy-pectin and celullose in the ratio of 5:3:2. They proposed the structural model for cell wall of soybean as shown in Figure 6.

It is postulated that in the process for soy sauce production soy-pectin is a main component making the pressing difficult. In other words, a small part of soy-pectin of the cell wall becomes soluble by cooking soybean and enzymatic decomposition of arabinogalactan in mold cultivation. But most parts of soy-pectins easily swells or gelatinizes in brine because it contains much galacturonic acid. The activ-ities of pectinase generated in mold cultivation is so weak

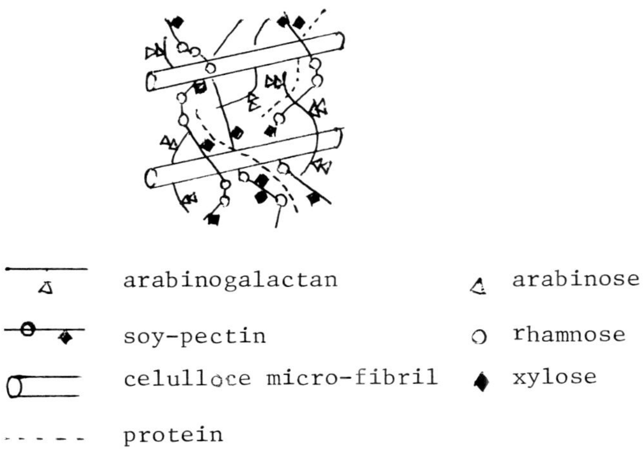

FIGURE 6. The stractural model for cell wall of soybean.

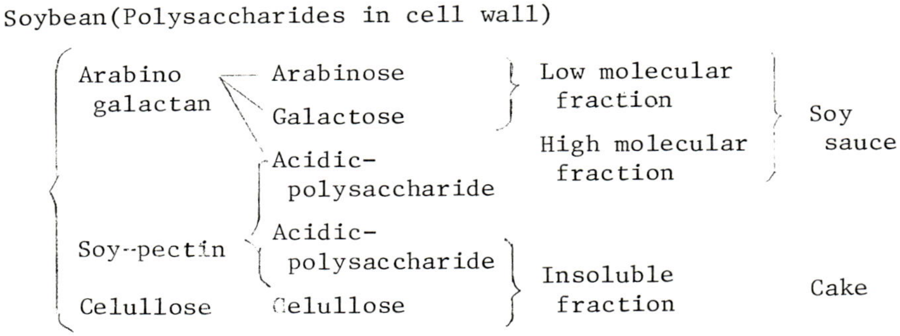

FIGURE 7. Behaviors of polysaccharides in soybean in the 'moromi' process.

that soy-pectin is considered to be gel in early stage of the 'moromi' process. The gel makes the press difficult. Figure 7 summarizes the mechanism mentioned above.

The simple and practical method for solubilizing soy-pectin is to activate pectinase generated in mold cultivation or to add any pectinase to 'moromi'. Figure 8 shows an experimental results in the effect of pectinase added to 'moromi' on soy sauce yield. Undoubtedly these techniques will be applied in the soy sauce industry in the near future.

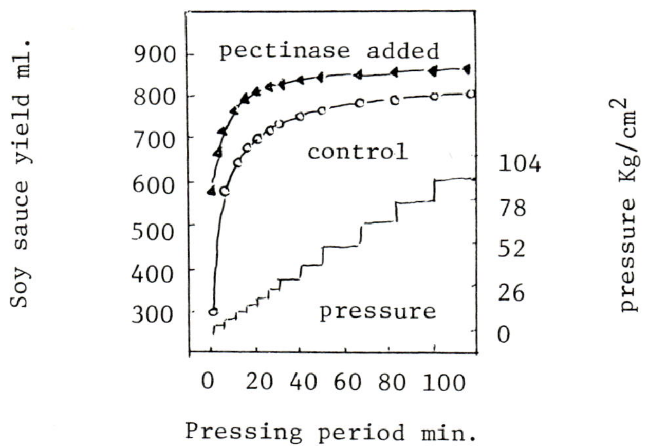

Figure 8. The effect of pectinase added in the pressing test.

VI. DEVELOPMENT OF TWO TYPES OF PUFFED GRANULES BY AN EXTRUDER AS PROCESSED MATERIALS

Reduction and rationalization of steps in soy sauce production is a dream for researchers and engineers in this field for a long time. Fujita, Kishi and their colleagues, Nisshin Flour Milling Co. Ltd., are working on the subject by applying technology of flour milling for several years.

Starchy granules were produced from wheat flour and bran by steaming on a sort of commercial pelleting machine instead of roasting and crushing wheat kernel. Characteristics of the starchy granules are as follows;

(a) The traditional process of roasting and crushing of wheat kernel can be eliminated in the soy sauce factory.

(b) The starchy granules can be attacked by enzyme much easier than traditional roasted wheat.

(c) Composition of the starchy materials can be controlled by selecting and blending various flours and any other streams of wheat milling. For example, starchy granules containing high wheat protein results in good tasting soy sauce because wheat protein contains high glutamate content in the amino acid balance. For another example, by eliminating wheat bran it is possible to produce extremely light brown soy sauce.

In applying the developed starchy granules, the effects of size and size distribution, gelatinization of starchy granules by puffing with an extruder instead of the pelleting machine and the conditions of mold cultivation on the quality of soy sauce were investigated (7, 8, 9, 10).

A few years ago, development of the processed material for soy sauce production entered in the second phase. Processed puffed granules were tried containing both soybean and wheat developed by using an extruder.

In employing an extruder some problems arised;

(a) What is the optimum condition for thermal denaturation of soy protein without causing turbidity in diluted soy sauce?

(b) How to supply water to the processed granules?

(c) How to prevent infection and other hygienic troubles?

It has been considered that optimum thermal denaturation of soy protein for soy sauce production could not be obtained as far as an extruder was used (11). However it was successfully overcome by adjusting granular size of soybean, water content, feeding quantity, total surface area of die's holls, degree of steaming and so on (12).

Figure 9 and Figure 10 show procedures for producing the puffed granules containing both soybean and wheat, and new

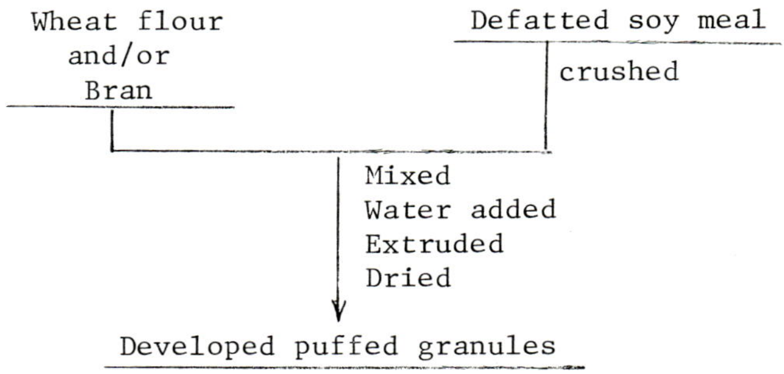

FIGURE 9. Procedures for developed puffed granules.

flow-sheet of soy sauce production respectively. A new continuous agitating mixer was developed to supply water to the developed puffed granules. Using these granules, about a half of facilities and operation for soy sauce production, will be reduced; 2-3 times more enzymatic activities are obtained in mold cultivation because granules are more homogeneous in water and protein and carbohydrate distribution than the traditional 'koji' mixture. These granules are so porous in stracture that the mold can grow fast and steady. Hydolyzation of protein and carbohydrate proceeds easily and quickly in the 'moromi' process because of good affinity of substrates for enzymes and high enzymatic activities. As a result, the period for the 'moromi' process can be shorten from six to 3-4

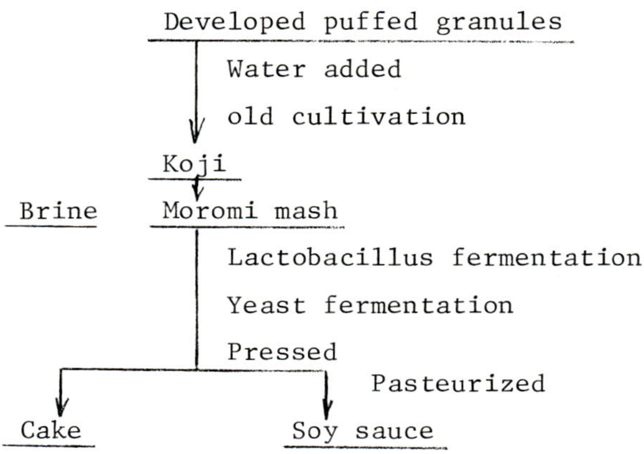

FIGURE 10. The new process of soy sauce production applying developed puffed granules.

TABLE 3. Analytical data for materials.

	Defatted soy meal	Roasted wheat	Developed puffed granules
Moisture %	9.1	3.9	11.3
Ash %	5.52	1.59	3.60
Total nitrogen %	8.21	2.56	5.33
Total sugar %	20.3	73.7	44.8
Specific gravity g/cm^3	0.70	0.42	0.43
Degree of gelatinization	-	32.5	95.5

TABLE 4. Analytical data for soy sauce 'Koji'.

	Control (by traditional materials)	Tested (by developed puffed granules)
Moisture %	28.6	30.5
pH	6.40	6.21
α-amylase activity u./g	1800	2200
Gluco-amylase activity u./g.	40.5	48.5
Proteinase activity Tyr.μg./g.		
in pH 7.0	320	720
in pH 9.0	205	656

TABLE 5. Analytical data for unrefined soy sauce.

	Control (by traditional materials)	Tested (by developed puffed granules)
Salt %	17.95	17.82
Total nitrogen %	1.78	1.99
Formol nitrogen %	0.80	1.02
Total sugar %	3.85	3.31
Direct reducing sugar %	3.63	3.25
pH	4.79	4.82
Acidity I ml.	12.5	12.2
Acidity II ml.	13.8	13.0
Alcohol %	1.80	2.75
Digestibility %	85.4	94.5

months. By selecting suitable raw material, operation of filter is expected to be easy and the quantity of residue of soy sauce to be reduced. Tables 3, 4, and 5, show experimental data by Fujita and his co-workers (12) using the developed puffed granules and the traditional materials. A few field tests are almost finished successfully, before a new plant for this product will be planned.

Recent progress in soy sauce production in Japan involving four revolutionary investigations and developments have been reviewed from the industrial point of view. More detailed contents have been omitted regretfully by limit space.

There are and will be many investigations and developments on soy sauce industry especially in Japan. Nowadays Japanese traditional soy sauce becomes popular and familiarity gradually increases in many countries as a prepared condiment. It will be expected hopefully that excellent scientists and engineers in the world might be interested in and devote themselves to the development in this field.

REFERENCES

1. Mogi, T. *J. of Jap. Soy Sauce Rech. 2*, 177 (1976).
2. Jap. Patent No. 929910 (1978).
3. Yashuda, A. *Chomi Kagaku. 20* (7), 20 (1973).
4. Nagase, I. *j. of Ferm. Assoc. Jap. 28*, 381 (1970).
5. *J. of Jap. Soy Sauce Rech. Inst. 5*, 98 (1979).
6. Kikuchi, T. *Ibid, 5*, 71 (1979).
7. Fujita, A. and Kishi, K. *Ibid, 1*, 13 (1975).
8. Fujita, A. and Kishi, K. *Ibid, 2*, 16 (1976).
9. Fujita, A. et al. *Ibid, 3*, 200 (1977).
10. Fujita, A. et al. *Ibid, 4*, 18 (1978).
11. Yokozuka, T. *Ferm. and Ind. 36*, 946 (1978).
12. Fujita, A. Presented at annual meeting of "*Jap. Soc of Food Sci. and Tech.*"(1979).

Most of the peaks have been identified as derivatives of amino acids, but the 3 peaks 16, 17 and 18 in Figure 1 have been positively identified as 3 dipeptides, namely the derivatives of valylvaline, isoleucylvaline and isoleucylisoleucine. The mass spectra of valylvaline and isoleucylvaline can be seen in Figures 2 and 3. The fragmentation patterns of these N-acetyl, n-propyl derivatives are similar to those reported for N-acetyl peptides (Heyns and Grützmacher, 1963), ethyl esters of amino acids (Biemann *et al.*, 1961) and methyl and ethyl esters of amino acids (Andersson *et al.*, 1961).

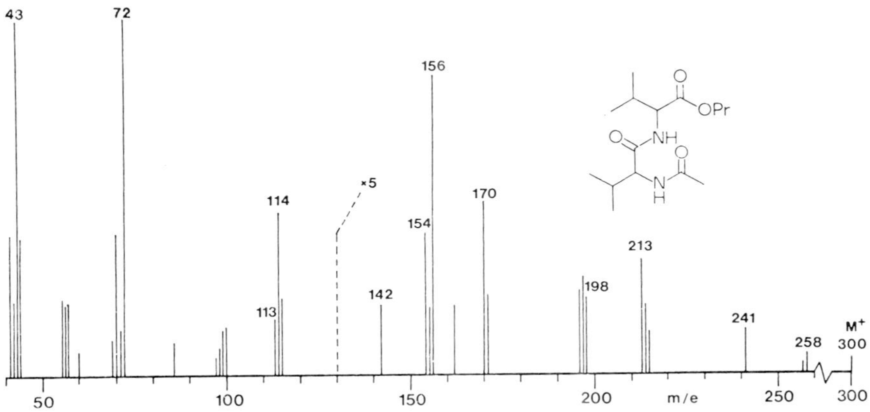

FIGURE 2. Mass spectrum of N-acetyl, n-propyl valylvaline.

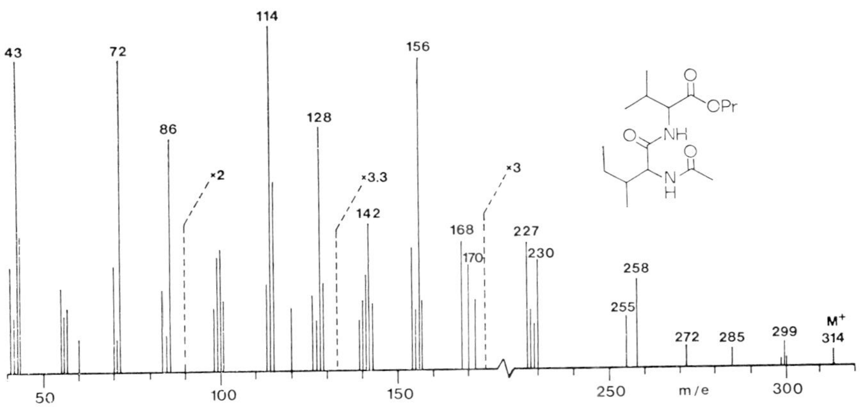

FIGURE 3. Mass spectrum of N-acetyl, n-propyl isoleucyl-valine.

Thermal cyclization of dipeptides is known to occur to a degree where most of the observed ion current is from the cyclic product (Svec and Junk, 1964). In the spectrum of N-acetyl, valyl-valine propyl ester the peak m/e = 198 is the molecular ion for cyclo (val-val). The m/e = 156 ion is the cyclo (gly-val) ion. A detailed discussion of the fragmentation of cyclic dipeptides has been previously reported (Eriksen and Fagerson, 1976).

It appears doubtful that any linear peptides would be present in the hydrolyzates, but it is known that linear dipeptides can cyclize to diketopiperazines under acidic conditions (Abderhalden and Komm, 1924) (Figure 4). This type of compound is thus formed by intramolecular condensation of a dipeptide which is an intermediate in the acid hydrolysis of protein. In the hydrolysis process peptide linkages consisting of amino acids having bulky side chains are inferred to be split with difficulty because of steric hindrance, and, therefore, comparatively stable dipeptides are formed and the anhydrides are produced by the action of the acid. All the cyclic dipeptides may not be formed in the hydrolysis process. Dansi _et al_. (1970) who isolated cyclic dipeptides from corn steep liquor were able to isolate some directly from corn as well as from other cereals (rye, sorghum, wheat and oat).

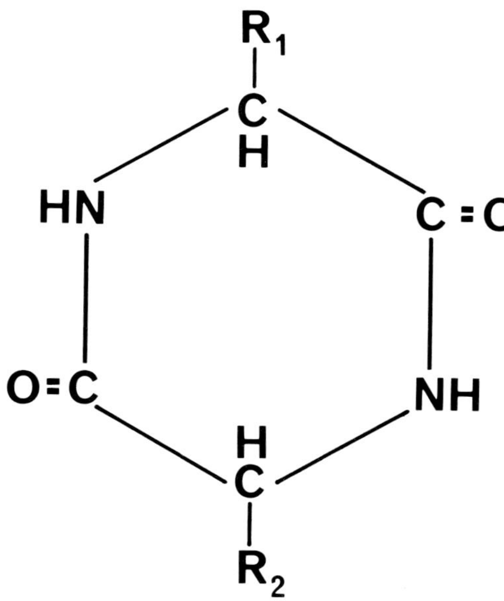

FIGURE 4. A diketopiperazine. R_1 and R_2 are amino acid side chains.

In the derivatization procedure of the samples for GC analysis, propylation precedes acetylation. Propylation opens the ring structure of the DKP's leaving the amino groups acetylatable.

Acetylation prior to propylation would not make any difference with respect to the derivatization of amino acids and linear dipeptides, but in the case of cyclic dipeptides no acetylation would occur. In a chromatogram run after a reverse derivatization procedure the 3 dipeptide peaks are missing indicating the cyclic nature of these compounds.

Aqueous Extract

Compounds identified in the aqueous extract are listed in Table 2 and a gas chromatogram of the derivatized extract

TABLE II. Compounds Identified in Fraction X of Soy hydrolyzate

Peak number	Name of compound	GLC	MS	Comment
1	Ac-alanine-OPr	+	+	P
2	Ac-valine-OPr	+	+	P
3	Ac-2-amino-butyric acid-OPr	+	+	P
4	Ac-glycine-OPr	+	+	P
5	Ac-isoleucine-OPr	+	+	P
6	Ac-leucine-OPr	+	+	P
7	Ac-2-amino-4-(2-furyl)-crotonic acid-OPr		+	T
8	Ac-isoleucine-OBu	+	+	P
9	Ac-proline-OPr	+	+	P
10	Ac-threonine-OPr	+	+	P
11	Ac-phenylethylamine	+	+	P
12	Ac-serine-OPr	+	+	P
14	Ac-methionine-OPr	+	+	P
15	Ac-phenylalanine-OPr	+	+	P
16	Ac-glutamic acid-diOPr	+	+	P
17	Ac-phenylalanine-OBu	+	+	P
20	Ac-valylvaline-OPr	+	+	P
21	Ac-isoleucylvaline-OPr	+	+	P
22	Ac-isoleucylisoleucine-OPr	+	+	P
25	O,N-diAc-tyrosine-OPr	+	+	P
26	O,N-diAc-tyrosine-OBu	+	+	P

P indicates positive identification
T indicates tentative identification only

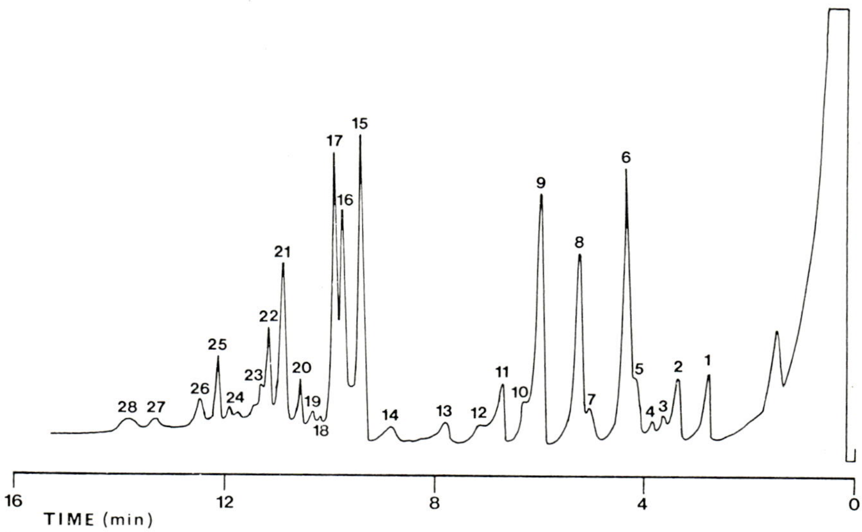

FIGURE 5. Gas chromatogram of derivatized Fraction X. See Figure 1 for GLC conditions.

is shown in Figure 5. As in Figure 1 most of these peaks have been identified as derivatives of amino acids. There are, however, several peaks of interest. The peaks numbered 8, 17 and 26 have been identified as the butyl esters of isoleucine, phenylalanine and tyrosine, respectively. This can only by from partial butylation during the butanol extraction. This was confirmed by running a gas chromatogram of only acetylated sample, which produced identical peaks.

Peak 11 has been positively identified as phenylethylamine. This compound has only been isolated from a few foods, bitter chocolate, some samples of mature cheeses and certain heavy red wines (Chaytor *et al.*, 1975), where its presence is known to initiate migraine attacks in dietary migraine sufferers at a level of only 3 mg (Sandler *et al.*, 1974). Phenylethylamine most probably arises as the decarboxylation product of phenylalanine, and is found in many families throughout the plant kingdom (Smith, 1977).

It is interesting that glutamic acid (peak 16) is present in the aqueous extract following butanol extraction, since dicarboxylic acids are not extracted with butanol (Dakin, 1918). This would indicate that some of the glutamic acid in the hydrolyzate must be present in a different form. Greenstein and Winitz (1961) pointed out that glutamic acid originally present in the highly acid hydrolyzate may be partly converted to pyrrolidonecarboxylic acid (PCA) in the later stages of the manufacturing process of HVP when the product is neutralized with salt formation and the mixture concentrated. Derivatization of PCA for GLC analysis would open the pyrrolidone ring and PCA would be detected as glutamic acid (Adams, 1974).

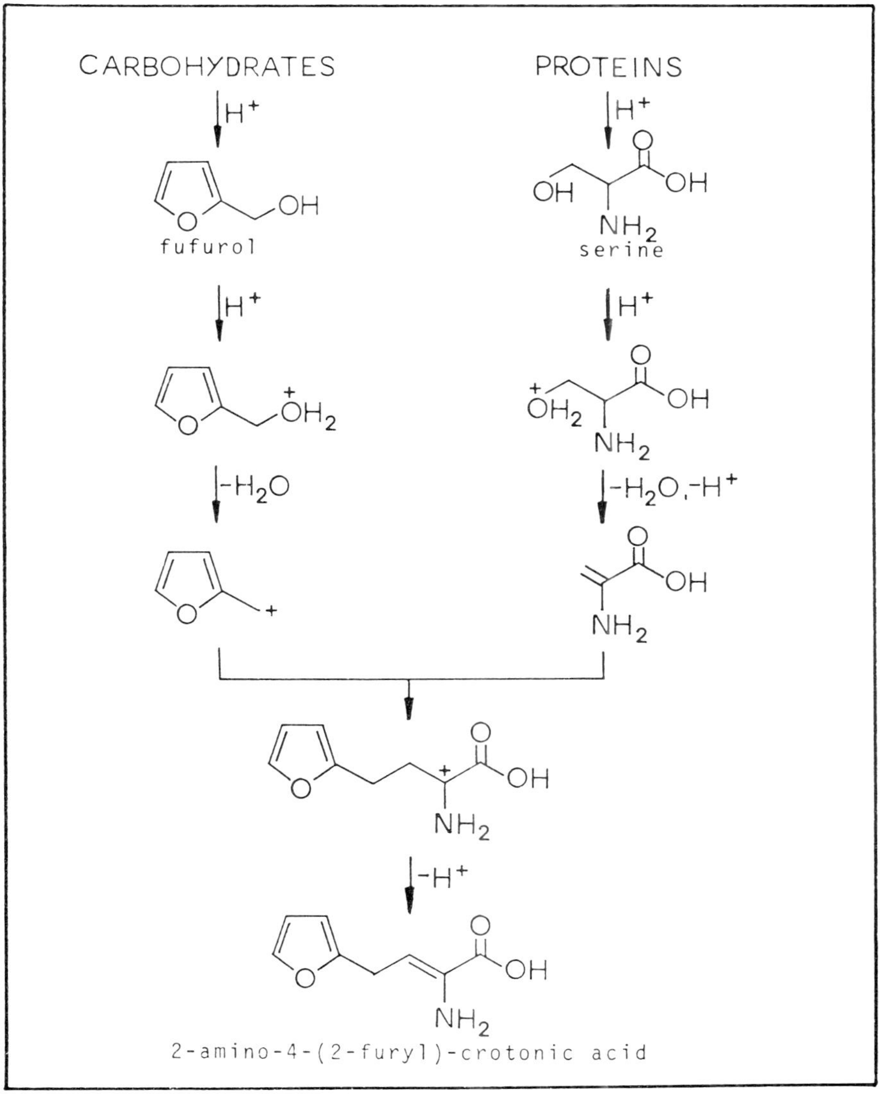

FIGURE 6. Proposed reaction scheme for the formation of 2-amino-4-(2-furyl)-crotonic acid.

Peak 7 in Figure 5 has been tentatively identified as 2-amino-4-(2-furyl)-crotonic acid on the basis of mass spectrometry. Figure 6 outlines a proposed mechanism for the formation of the compound from furfurol, a carbohydrate breakdown product, and serine during the hydrolysis process. Furfurol was claimed by Markh and Vinnikova (1973) to be one of the most important components responsible for the typical aroma of soy hydrolyzates. All the steps are basic organic reactions occuring under acidic conditions.

Ethanolic Precipitate

The main objective of using Dakins extraction method (Dakin, 1918) was to extract peptide anhydrides (diketopiperazines) present in the hydrolyzate. The diketopiperazines (DKPs) are soluble in n-butanol but only sparingly soluble in ethanol. So a precipitate from the ethanol extraction would most likely be DKPs.

The cyclic nature of the precipitate was confirmed with infrared spectroscopy. The spectrum obtained, although more detailed, is in aggreement with the spectrum published by Obata and Mitzutani (1959) of cyclo(ile-val) which they had isolated from acid hydrolyzed soybean meal. This conclusion was further substantiated by NMR-analysis (see Figure 7) where the observed spectrum is in agreement with the suggested structure. The NMR-results are summarized in Table 3.

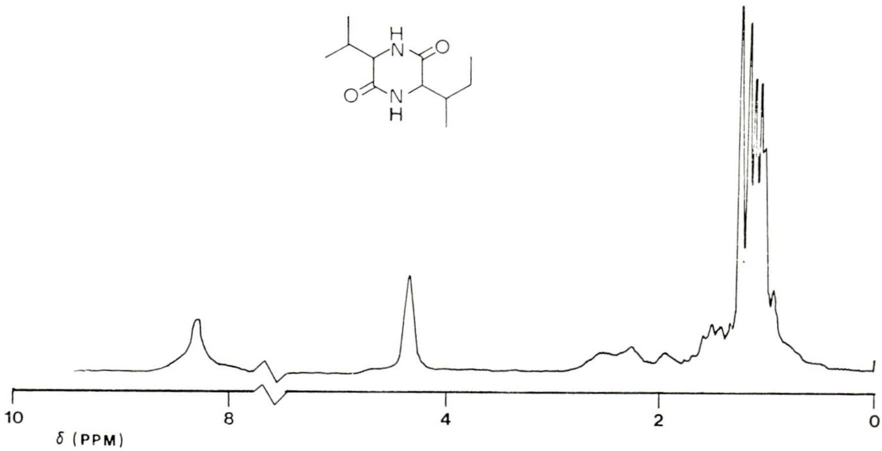

FIGURE 7. NMR spectrum of precipitate from ethanol extraction identified as cyclo (ile-val).

TABLE III. Summation of NMR Results of Precipitate from Ethanol Extraction

Type of H	Chemical shift δ	Coupling constant J (Hz)	No. of H
a	1.02 (t)	6.5	6
b	1.08 (d)	7.0	
c	1.21 (d)	7.0	6
d	1.5		2
e	2.25		1
f	2.55		1
g	4.32		2
h	8.30		2

d: doublet
t: triplet

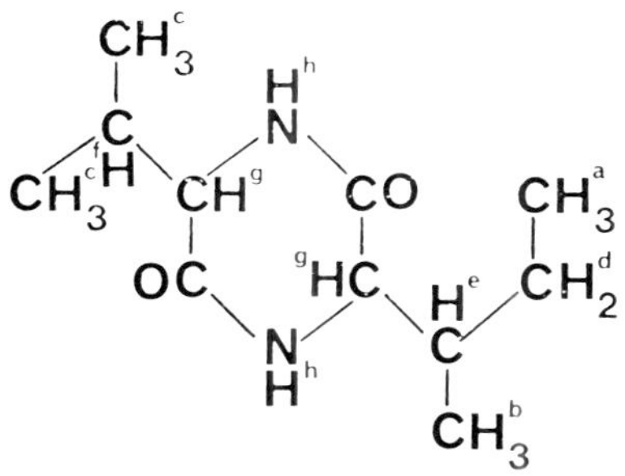

Gas chromatography of the precipitate after recrystallization from 50 % acetic acid showed 3 peaks. Hydrolysis for 70 hours with 6N HCl at 110 °C revealed only valine and isoleucine as the hydrolysis products. The fact that only 2 amino acids occur as a result of the acid hydrolysis and 3 compounds are present before hydrolysis indicate the possibility of the compounds previously mentioned, namely cyclo (val-val), cyclo (ile-val) and cyclo (ile-ile). The above conclusions are not in disagreement with the NMR and IR analysis where cyclo (ile-val) was the only possibility. If cyclo (val-val) and cyclo (ile-ile) were present in equimolar amounts, which is the case, the NMR and IR results would be identical.

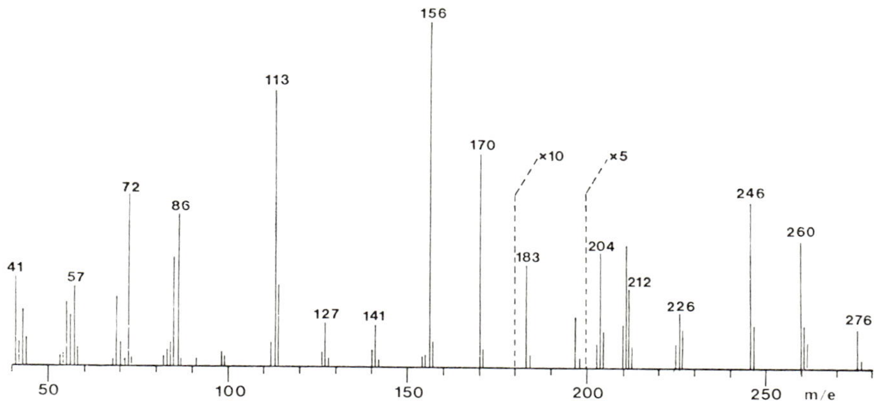

FIGURE 8. Mass spectrum of ethanolic precipitate obtained at 50 eV.

A mass spectrum of the crude precipitate in the solid state was obtained with the results shown in Figure 8. The explanation of the major fragmentation ions has previously been reported (Eriksen and Fagerson, 1974). The ions m/e = 198, 212, and 226 are the molecular ions of cyclo (val-val), cyclo (ile-val) and cyclo (ile-ile), respectively. GLC-analysis of the crude precipitate after hydrolysis revealed traces of phenylalanine and leucine. This is in agreement with the observed ions m/e = 246 and 260 in the mass spectogram which are the molecular ions of cyclo (val-phe) and cyclo (leu-phe), respectively. Both these ions eliminate their side chains by cyclic elimination, leaving the ion m/e = 204. A small peak at m/e = 91 supports the presence of the two compounds. The presence of cyclo(leu-tyr) is suggested by the molecular ion at m/e = 276.

Figure 9 summarizes the identified cyclic dipeptides in HVP. It has been pointed out by Matoba and Hata (1972) that several dipeptides become more bitter, when their amino and carboxyl groups are blocked or they are converted to diketopiperazines. A mixture of the 3 top compounds listed in Figure 9 were tested for bitternes in the solid state and were found to have no taste. Hence, their contribution to the flavor of HVP can be neglected.

CYCLO (VALYL - VALYL)
CYCLO (ISOLEUCYL - VALYL)
CYCLO (ISOLEUCYL - ISOLEUCYL)
CYCLO (PHENYLALANYL - VALYL)
CYCLO (PHENYLALANYL - LEUCYL)
CYCLO (TYROSYL - LEUCYL) (T)

FIGURE 9. Diketopiperazines identified in HVP (T = tentative identification)

The presence of diketopiperazines in HVP is, however, interesting, since the ban in USA on the use of the sweetener "Aspartame" which is a dipeptide methyl ester (L-aspartyl-L-phenylalanine methyl ester) is due to the fact that this compound in aqueous solutions slowly cyclizes to its DKP derivative.

REFERENCES

Abderhalden, E. and Komm, E. (1924). *Z. Physiol. Chem. 139*, 147-168.
Adams, R.F. (1974). *J. Chromatogr. 95*, 189-212.
Andersson, C.-O., Ryhage, R. and Stenhagen, E. (1961) *Arkiv för Kemi 19*, 417-434.
Biemann, K., Seibl, J. and Gapp, F. (1961). *J. Am. Chem. Soc. 83*, 3795-3804.
Chaytor, J.P., Crathorne, B., and Saxby, M.J. (1975). *J. Sci. Food Agric. 26*, 593-598.
Dakin, H.D. (1918). *Biochem. J. 12*, 290-317.
Dansi, A., Pozzo, A.D., Mariotti, V., Bonferoni, B. and Piccioni, M. (1970). *Die Stärke 22*, 305-309.
Eriksen, S. and Fagerson, I.S. (1976). *J. Agr. Food Chem. 24*, 1242-1243.

Greenstein, J.P. and Winitz, M. (1961). *In* "Chemistry of the Amino Acids, Vol. 3", pp. 1929-1954. John Wiley and Sons, Inc., New York, N.Y.

Heyns, K. and Grützmacher, H.F. (1963). *Ann. 669*, 189-201.

Manley, C.H. and Fagerson, I.S. (1971). *Flav. Ind. 2.*, 686-690.

Markh, A.T. and Vinnikova, L.G. (1973). *Prikl. Biokhim. Mikrobiol. 9*, 913-917.

Matoba, T. and Hata, T. (1972). *Agr. Biol. Chem. 36*, 1423-1431.

McLafferty, F.W. and Venkataraghavan, R. (1979). *J. Chrom. Sci. 17*, 24-29.

Obata, Y. and Mitzutani, J. (1959). *Bull. Agr. Chem. Soc. Japan 23*, 121-125.

Sandler, M., Youdim, M.B.H. and Hanington, E. (1974). *Nature 250*, 335-337.

Smith, T.A. (1977). *Phytochem. 16*, 9-18.

Svec, H.J. and Junk, G.A. (1964). *J. Am. Chem. Soc. 86*, 2278-2282.

RICE: RECENT PROGRESS IN CHEMISTRY AND NUTRITION[1]

Bienvenido O. Juliano

Chemistry Department
The International Rice Research Institute[2]
Los Baños, Laguna, Philippines

I. INTRODUCTION

Chemists at the International Rice Research Institute (IRRI) are involved in its Genetic Evaluation and Utilization (GEU) program on the grain and nutritional quality of rice. Although grain properties can be improved through postharvest treatment and processing, IRRI's basic philosophy is to incorporate or build in as many of these desirable characteristics into a rice variety, whenever possible, without additional cost to the consumer. In the grain quality aspect of the GEU program, chemists are obtaining a better understanding of varietal differences in heritable grain properties related to grain quality and have developed simple, sensitive, and rapid indicators of eating quality applicable to the breeding program. In the GEU-nutritional quality aspect, the studies so far have been on the effect of increased protein content on the nutritional properties of the grain. However, we plan to study shortly the nutritional properties of other grain constituents.

II. GRAIN COMPOSITION AND STRUCTURE

The major constituents of milled rice are starch and protein (Table 1). The minor constituents are fat (lipids),

---

[1]The protein work was supported in part by National Institutes of Health (U.S.A.) contract NO1-AM-7-0726, April 1967 to December 1979.

[2]Mail address: P.O. Box 933, Manila, Philippines.

ISBN: 0-12-370960-1

TABLE 1. Mean Composition of Waxy and Nonwaxy Milled Rice[a]

Constituents	Waxy rice	Nonwaxy rice
Starch	~90	~90
Amylose	1-2	12-33
Amylopectin	88-89	57-78
Protein (N x 5.95)	8	8
Fat (lipids)	1	1
Nonstarch	0.9	0.4-0.6
Starch	0.1	0.4-0.6
Crude Fiber	~0.5	~0.5
Dietary fiber	~1	~1
Crude ash	~0.5	~0.5

[a]Weight percentage, wet basis.

crude fiber (dietary fiber), and crude ash, and vitamins. The structure of the developing and mature grain was recently studied in detail (Hoshikawa, 1967, 1968; Bechtel and Pomeranz, 1977, 1978a and b). In addition to the spherical protein bodies, crystalline-type protein bodies, which are more susceptible to proteolysis than the spherical protein bodies (Bechtel and Pomeranz, 1978b; Tanaka *et al.*, 1978), have been observed in the subaleurone layer (Harris and Juliano, 1977; Bechtel and Pomeranz, 1978b). The three types of protein bodies differ in origin (Bechtel and Juliano, 1979).

Staff in the Rice Quality Laboratory of the IRRI Plant Breeding Department routinely measure in thousands of milled-rice samples the following starch properties:

a) Amylose content, by iodine colorimetry (Perez and Juliano, 1978).

Waxy	0- 2.9% milled rice dry wt
Very low	3.0- 9.9% milled rice dry wt
Low	10.0-19.9% milled rice dry wt
Intermediate	20.0-24.9% milled rice dry wt
High	25.0-33.0% milled rice dry wt

b) Gelatinization temperature (gel. temp.), by the alkali digestibility test of Little *et al.* (1958). The degree of disintegration of milled rice previously soaked 23 hours in 1.7% KOH at 30°C is a measure of gel. temp.

Low gel. temp. (<70°C) - high alkali spreading value of 6-7

Intermediate gel. temp. (70–74°C) - intermediate alkali spreading value of 4–5
Intermediate-high gel. temp. - intermediate-low alkali spreading value of 3
High gel. temp. (>74°C) - low alkali spreading value of 1–2

c) Gel consistency measured by the length of the gel of 100 mg rice flour in 2 ml 0.2 *N* KOH in a 13- x 100-mm test tube held in a horizontal position for 1 hour at room temperature (Cagampang *et al.*, 1973).

Soft gel consistency - 61–100 mm
Medium gel consistency - 41–60 mm
Hard gel consistency - 26–40 mm

Starch occurs in the endosperm as 3- to 9-μm compound granules. Protein occurs as 0.5- to 4-μm protein bodies that are distinct from the aleurone granules in that they have no globoid inclusions (phytate bodies) (Tanaka *et al.*, 1973).

Because rice is consumed as whole grain, its grain structure is important. Grain properties affect the equilibrium moisture content, hardness, tendency to crack, resistance to attack by insect pests, and breakage during milling (Juliano, 1979). Crystallinity, as indexed by residual lintnerized starch granules, tends to be higher in starches with high gel. temp. and high amylose content (Maniñgat and Juliano, 1979a). This difference in starch granule structure may explain the negative relationship between equilibrium water content of steeped rice grain and amylose content (Antonio and Juliano, 1973). Among high-amylose rices, low-gel.-temp. grains tended also to have higher equilibrium water content (IRRI, unpublished data; Bhattacharya, 1979). Differences in equilibrium moisture content of rough rice above 75% r.h. (Juliano, 1964) may be similarly explained. We are checking the importance of these starch granule properties to resistance of brown rice to rice weevil (*Sitophilus oryzae* L. and *S. zeamais*), red flour beetle (*Tribolium castaneum*), and fungi *Aspergillus flavus* and *A. parasiticus*.

III. COOKING AND EATING QUALITIES

Volume expansion and water absorption during cooking are positively correlated with amylose content of milled rice. In Asia rice is usually cooked in the optimum amount of water it can absorb, but in Western countries it is usually cooked in excess water. Cooking time is directly affected by gel.

temp. and protein content (Juliano, 1979). The amount of solids in cooking gruel is negatively correlated with amylose content. However, high-amylose rices with hard gel consistency have starch with lower hot water solubility, particularly amylopectin, than those with soft gel (Juliano, 1979).

The negative correlation of glossiness and color of cooked rice and amylose content (Juliano *et al.*, 1965) may be explained by the relative expansion and bulk density of the cooked grain. Waxy rices are glossy but cream-colored. High-amylose rices are dull but white in appearance. Stickiness and tenderness of cooked rice, properties more attributable to amylopectin than to amylose, are negatively correlated with amylose content (Juliano *et al.*, 1965).

Although amylose content is the major determinant of texture of cooked rice, rices of similar amylose content, and even waxy rices, show varietal differences. We use two additional indexes, gel consistency and gel. temp., to differentiate among varieties of similar amylose content.

The gel consistency test was specifically developed for high-amylose rice (Cagampang *et al.*, 1973). Among the early IRRI varieties, IR5 is considered of better quality than IR8; in India, the IR8-type rices were considered inferior to traditional rices (Bhattacharya, 1979). The early semidwarf rices obtained from their dwarf parents their high amylose content, low gel. temp., and high Amylograph setback viscosity (hard gel consistency). Other high-amylose varieties such as IR5 have intermediate gel. temp. and soft gel (Table II). Although cooked IR5 rice is softer than cooked IR8 rice, the taste panel scores did not differ significantly. Instron, Ltd. measured hardness and cohesiveness of the two samples and the relative values obtained were expected from our previous observations.

In Sri Lanka, the preferred small-grained "samba" varieties have the same amylose content and gel. temp. as the common varieties, but they tend to have a soft gel.

Less data on intermediate-amylose rices are available. In Indonesia low-gel.-temp. "bulu" or javanica varieties are preferred to intermediate-gel.-temp. indica varieties, and a soft gel is preferred to a hard gel (Perez and Juliano, 1979).

In the Philippines the variety C4-63G is preferred to BPI-121-407. The two varieties differ mainly in gel. temp. (Table III). C4-63G cooked rice is softer by about 1 kg, as shown by the Instron OTMS cell. A consumer panel last January considered IR4570-74-2 (derived from BPI-121-407) at par with C4-63G although it had a harder cooked rice. The IR4570 line has a longer grain than the four other rices tested. Thus factors other than texture, such as grain size and shape, percentage head or whole-grain rice, and color, influence the choice of consumers.

TABLE II. Properties of Raw and Cooked IR5 and IR8 Milled Rice. IRRI, Institute of Human Ecology, Univ. of the Philippines, Los Baños, and Instron, Ltd., England, 1975

Property	IR5[a]	IR8[a]
Gel. temp.	interm.	low
Amylose content (% dry basis)	28.4	27.9
Gel consistency (mm)	98	36
Mean taste panel scores[b]		
Tenderness of cooked rice	4.4	3.8
Cohesiveness of cooked rice	4.6	4.1
Cooked-rice Instron hardness[c] (kg)	140	160
Cooked-rice Instron cohesiveness[d] (g)	286	277

[a]1974 dry-season crop.
[b]Mean of duplicate assessments of warm cooked rice by a taste panel of six of the Institute of Human Ecology, Univ. of the Philippines, Los Baños. Scores based on a scale of 1-9 with the higher number indicating a greater intensity of the character (Juliano *et al.*, 1965).
[c]Ottawa Texture Measuring System (OTMS) cell, 100 mm/min.
[d]Compression anvil and compression table, 50 mm/min.

Among low-amylose rices, the first Korean varieties from an indica x japonica crossing program, represented by Tongil, demonstrate the importance of these secondary factors. Tongil has an analysis identical to that of japonica rices such as Jinheung but it has a more flaky (poorer quality) cooked rice (Table IV). Amylograph of 10% paste and gel consistency of 100 mg/2 ml 0.2 *N* KOH revealed no difference although Instron hardness data did. When 120 mg rice flour was used in the gel consistency test, differences were noted (Table IV). The results from Amylograph consistency (viscosity cooled to 50°C minus final viscosity at 94°C) of 12% pastes (Perez and Juliano, 1979) were similar. When we analyzed in 1978 the newer varieties from the Korean program, Milyang 23, Suweon 264, and all the others had physicochemical properties similar to those of japonica rices. Egyptian japonica and indica low-amylose rices also showed the same trend in cooked-rice hardness.

Waxy rices with low gel. temp. (about 65-68°C) had the softest textured cooked processed-rice products (Table V) (Perez *et al.*, 1979). Panpet 63 and RD4, rices with high gel. temp., have a harder cooked rice, combined with a harder neutral

TABLE III. Properties and Preference Scores of Five Intermediate-Amylose Total Milled Rices. Institute of Human Ecology, Univ. of the Philippines, Los Baños and IRRI, 1979

Property	IR4570-74-2	C4-63G (check)	IR9129-209-2	C4-63G	BPI-121-407
Gel. temp.	low	interm.	interm.	interm.	low
Amylose content (% d.b.)	24.7	23.7	21.6	24.6	23.9
Gel consistency (mm)	84	87	82	94	70
Kett Whiteness reading	40.8	43.1	37.4	40.8	38.0
Cooked-rice Instron					
Hardness (kg)	6.4	4.8	5.6	5.7	6.6
Stickiness (g.cm)	79	87	105	89	99
Mean Preference score[a]					
Cooked rice	0.25a	0.09ab	0.09ab	-0.14bc	-0.29c
Raw rice[b]	0.02ab	0.18a	-0.26b	0.08ab	-0.02ab

[a]Mean of 39 consumer panelists who rated the five samples. 1st = 1.16; 2nd = 0.50; 3rd = 0.0; 4th = -0.50; and 5th = -1.16. Means in the same line followed by a common leter are not significantly different at the 5% level. Mean acceptability scores were all high and ranged from 82-96%.

[b]Brokens ranged from 25-73 wt % of total milled rice.

gel and higher Amylograph peak viscosity. We use 200 mg flour/1 ml 0.3 *N* KOH, neutralized with 1 ml 0.3 *N* acetic acid to determine the neutral gel consistency. Among waxy rices with low gel. temp., those with soft gel are preferred. IR29 is inferior in texture to Malagkit Sungsong but new IRRI selections are being developed.

The rice cakes prepared from Japanese waxy rices with very low gel. temp. (56-63°C) are inferior to those from rices with 65-68°C gel. temp. (Palmiano and Juliano, 1972), although the rices have similar gel consistencies (Table V). The preferred varieties, similar to the low-gel.-temp. waxy rices from the Philippines and Thailand, have higher Amylograph peak viscosities.

The chemical aspects of rice grain quality were reviewed in an international workshop held October 1978 at IRRI (IRRI, 1979). As a follow-up, chemists are doing ring tests on five grain quality tests to identify sources of variation and standardize the method; gel consistency (1 method), alkali test (2 methods), amylose (2 methods), instrument methods for cooked-rice texture, and degree of milling (2 methods). IRRI is

TABLE IV. Physicochemical Properties of Japonica and Indica Low-Gelatinization Temperature, Low-Amylose Milled Rices from Korea and Egypt.[a] IRRI, 1978

Sample source and name	Grain type[a]	Amylose content (% d.b.)	Gel consistency (mm) 5% gel	Gel consistency (mm) 6% gel	Cooked-rice Instron hardness (kg)
Korea (1977-1978 crops)					
Jinheung	j	19.1	89	76	5.7
Akibare	j	17.5	89	76	4.6
Tongil	i x j	19.9	86	54	7.4
Milyang 23	i x j	19.5	92	72	4.7
Suweon 264	i x j	18.6	92	86	4.4
Egypt (1977 crop)					
Giza 171	j	18.4	86	79	5.1
Reiho	j	19.6	98	86	5.6
IR1626-203	i	19.6	90	62	6.4
IR1615-246	i	20.0	89	74	6.6

[a]j = japonica; i = indica.

coordinating the first four groups and S. Barber of Instituto Agroquimica y Tecnologia de Alimentos, Valencia, Spain, is chairing the test on degree of milling. The groups compare no more than two methods on five or six samples. In the case of the instrument methods, however, several instruments are compared on 10 samples representing all rice types. The first set of samples for the first four tests should be completed in 1979.

Samples aged at least 4 months after harvest are used for milling and cooking and eating tests, since during storage the grain undergoes such physicochemical changes as increased hardness, greater total and head-rice recovery, greater volume expansion and water absorption on cooking, less dissolved solids, and flakier cooked rice (Villareal *et al.*, 1976).

IV. PROCESSED-RICE PRODUCTS

Parboiled rice is the major processed-rice product. A large proportion of the rice in India and Sri Lanka is par-

TABLE V. Physicochemical Properties of Waxy Milled Rices from Three Countries (Perez *et al.*, 1979)

Sample source and name	Processed-rice quality	Final gel. temp. (°C)	Neutral gel consistency (mm)	Amylograph peak visc. (Brabender units)	Cooked-rice Instron hardness (kg)
Philippines					
Malagkit Sungsong	good	62.5	76	440	3.5
IR29	fair	63	52	560	5.6
UPLB-Ri-1	--	64	62	400	3.3
Panpet 63	poor	77	35	870	5.9
Thailand					
RD6	good	63.5	74	450	5.9
Niaw San Pahtawng	good	64	94	570	4.9
RD4	poor	73.5	46	820	6.6
Japan					
Norin No. 1	poor	58.5	86	350	6.9
Nakatamochi	poor	63	90	350	5.9
Koganemochi	good	67.5	86	560	5.6
Hatsunemochi	good	68.5	81	595	6.0

boiled. Parboiling consists of steeping rough rice in either cold or warm water, steaming it to gelatinize the starch without much swelling, and slowly drying the processed grain. Parboiling makes the grain translucent and harder, and, therefore, less prone to breaking during milling. The bran of parboiled rice contains more oil than that of nonparboiled rice because it is more selectively removed during milling. Parboiled milled rice contains more thiamin and riboflavin because even though heat during parboiling degrades vitamins, it also causes water-soluble vitamins to diffuse into the endosperm (Padua and Juliano, 1974). Recently we noted that parboiled milled rice has a softer gel than raw milled rice. The effect of parboiling seems more evident in nonwaxy rices (Raghavendra Rao and Juliano, 1970) but many properties of parboiled rice are still not fully understood. For example parboiling makes protein less extractable and less digestible (Eggum *et al.*, 1977).

We noted that equilibrium moisture content determines the degree of parboiling of rough rice, which, in turn, determines volume expansion during puffing (Antonio and Juliano, 1973).

The texture required for yeast-leavened bread and fermented rice cakes depends mainly on the type of boiled rice preferred, which is that produced by intermediate-amylose rice in the Philippines (Perdon and Juliano, 1975b) and low-amylose rice in California (Nishita *et al.*, 1976). Various waxy rice products have also been studied (Palmiano and Juliano, 1972; Antonio and Juliano, 1974; Antonio *et al.*, 1975).

The use of overmilled white-core rice for rice wine or "sake" production takes advantage of the soft, loose packing of cell contents in the opaque area of nonwaxy rice and the lower protein content of the core of the endosperm.

V. RELATIONSHIPS AMONG STARCH PROPERTIES

Although amylose content and gel. temp. are independent properties of starch, the popular varieties grown in the world do not contain them in all combinations. By screening part of the world rice collection for these properties, we have been able to identify these rare types: intermediate gel. temp. waxy and very low- and low-amylose rices and high-gel.-temp. intermediate- and high-amylose rices. We are currently studying the properties of these varieties to obtain a better insight into their interrelations with eating quality.

Among waxy rices, the molecular size of amylopectin, as indexed by intrinsic viscosity and sedimentation constant, tended to show a U-shaped correlation with gel. temp. with minimum values at 65-68°C (Perdon and Juliano, 1975a). Low-gel.-temp. rices has softer and less viscous starch gels (160 mg/2 ml 0.2 *N* KOH) than intermediate- and high-gel.-temp. rices (Perdon and Juliano, 1975a; Perez *et al.*, 1979). A study of seven nonwaxy starches indicated that amylopectin contributes more to gel viscosity and consistency than amylose (Juliano and Perdon, 1975) (Table VI). Gel viscosity of starch correlated positively with intrinsic viscosity of amylopectin only (Juliano and Perdon, 1975a). Since the degree of branching of rice amylopectin corresponds to a narrow range of mean chain length of 23-28 glucose units by periodate oxidation, our results suggest that the molecular size of amylopectin is the major factor determining the gel viscosity of starches of both waxy and nonwaxy rice. The ratio of A- to B-branches of nonwaxy rice amylopectin is similar to that of wheat amylopectin (Marshall and Whelan, 1974).

We are currently checking the sedimentation constants of amylopectin and amylose of high-amylose varieties differing in gel consistency. In hard-gel types, amylopectin, and, to a lesser extent, amylose, are less soluble in boiling water

TABLE VI. Range and Mean Intrinsic Viscosity and Gel Viscosity in Potassium Hydroxide of Starch Fractions of Seven Nonwaxy Rice Starches (Juliano and Perdon, 1975)

Property and starch fraction	Range	Mean
Intrinsic viscosity (ml/g)		
Amylopectin	159-197	177
Amylose	136-230	170
Gel viscosity[a] (8% in 0.2 *N* KOH, cps)		
Starch	140-686	423
Amylopectin	354-605	487
Amylose	26- 73	40
Reconstituted starch	148-240	210

[a]Wells-Brookfield microviscometer model RVT.

(Maniñgat and Juliano, 1978; Bhattacharya *et al.*, 1978). Gel filtration even on Sepharose 2B-CL has little value in molecular weight estimation of amylopectins with $\overline{DP}$ of $>10^6$. Rice starch offers an interesting material for starch chemists because its properties vary widely although it consists only of small granules.

VI. RICE PROTEINS AND NUTRITIONAL VALUE

Rice, the principal source of protein in the diets of tropical Asia, contributes from 30 to 80% of the dietary protein. In terms of the classical solubility fractions, rice proteins consist of 5% albumin (water-soluble protein), 10% globulin (salt-soluble protein), <5% prolamin (alcohol-soluble protein), and >80% glutelin (alkali-soluble protein). Any increase in protein that is due to genetics or environment is found only in the glutelin and prolamin of milled rice (Cagampang *et al.*, 1966).

The major proteins of milled rice have been purified and characterized recently - glutelin subunits (Juliano and Boulter, 1976; Villareal and Juliano, 1978); prolamin (Mandac and Juliano, 1978), and α-globulin (Perdon and Juliano, 1978). We are completing our characterization of the second major milled-rice globulin and the two major albumins (Cagampang *et al.*, 1976).

Isoelectric focusing of proteins of rice grain soluble in 3 M urea-β-mercaptoethanol has been suggested as being useful for varietal identification (du Cros *et al.*, 1979). Cooperative work with C.W. Wrigley of CSIRO Wheat Research Unit, North Ryde, Australia, on electrophoregrams of milled-rice protein fractions suggests that the protein differentials are albumins and globulins, not prolamins.

A slight decrease in lysine content of protein may accompany an increase of up to 10% in protein content of milled rice (Juliano *et al.*, 1973). The decrease in lysine is ascribed mainly to the increase in prolamin content, which is poor in lysine (Mandac and Juliano, 1978). Glutelin and milled-rice protein have similar aminograms. The low prolamin content of rice proteins suggests that rice already has the high-lysine gene that increases lysine content of cereal protein by suppressing prolamin synthesis (Nelson, 1976).

Cooperative feeding trials with microorganisms, *in-vitro* digestibility, and with growing rats confirmed the results of chemical or amino-acid scoring that the utilizable protein of milled rice increases with an increase in protein content (Juliano, 1978). Studies of nitrogen balance in adults and children -- in which low-protein rice was replaced by an equal amount of high-protein rice -- confirmed the nutritional utility of the higher-protein rice since apparent nitrogen retention did not deteriorate with a higher intake of rice protein (Juliano, 1978). In a one-year cooperative work with S. M. Pereira of the Christian Medical College and Hospital, Vellore, Tamil Nadu, India, preschool children in an orphanage were fed the South-Indian diet using either local (6% protein) or IR2183-58-3 (9% protein) milled rice. No significant difference in height and weight was noted, probably because the children in the control diet had adequate calories and protein.

Nitrogen balance in Peruvian and Filipino children was studied using similar protocols, to determine the effect of staple food on the children's dietary protein utilization (MacLean *et al.*, 1978; Roxas *et al.*, 1979). The results showed similar values for the two rice diets in both Peru and the Philippines (Table VII), suggesting that regional nutritional evaluation laboratories can achieve comparable results provided their protocols are similar. The results also verify that protein content is the major determinant of the nutritional value of rice as there is relatively less variation in the lysine content of its protein.

The true digestibility of rice protein in man is about 85% compared with 90% for wheat flour (Viteri *et al.*, 1971), which verifies earlier results of Graham (1977) on the poorer digestibility in Peruvian children of rice protein relative to wheat-flour protein. In rats, digestibility of both raw and boiled

TABLE VII. Nitrogen Balance Data of Low-Protein and High-Protein Milled Rices in Preschool Children (MacLean *et al.*, 1978; Roxas *et al.*, 1979)

Property	Peruvian children		Filipino children	
Protein content of rice (% N x 6.25)	7.2	11.0	7.2	11.0
Age of children (years)	1.0 -	1.5	1.2 -	2.0
Children (no.)	8	8	8	8
N intake (mg/kg body wt)	240	240	250	250
Apparent N digestibility (% of intake)	66.6	64.9	66.2	60.0
Apparent N retention (% of intake)	28.6	23.0	26.9	23.4
Apparent N retention of casein I control (% of intake)	35.2	35.2	30.8[a]	30.8[a]
Lysine content of diet (g/16 g N)	3.9	3.4	3.9	3.4

[a]Based on the mean for first four children.

wheat flour is 90%, whereas cooking reduces the digestibility of milled rice from 100 to 89% (Eggum *et al.*, 1977) (Table VIII).

Tanaka *et al.* (1975) reported the presence in the feces of adults on a rice diet of protein particles, which they identified as rice protein bodies. Y. Tanaka worked with us on this problem in 1977. We characterized the fecal protein particles and pepsin-treated cooked-rice protein bodies to represent the core 10-15% of the large spherical protein bodies (Tanaka *et al.*, 1978). The undigested protein particles were richer in lipids and had much lower nutritional value (lower lysine content) than whole protein bodies (Table IX). Their high cysteine content may explain their insolubility in cooked rice, despite the low molecular weight (mol. wt.) of the polypeptide subunit of <5000. Cooking probably enhances protein-protein interaction.

We are isolating and purifying this polypeptide, which does not adsorb protein stains. SDS-Sephadex G-75 gel filtration indicated that this polypeptide accounts for 10% of the total UV absorption of milled-rice protein bodies. Its amino acid composition is different from that of rice proteins we have analyzed (Table X).

TABLE VIII. Mean Nutritional Properties of Three Raw and Freeze-Dried Cooked Milled Rices (Eggum *et al.*, 1977)

Property	Raw	Cooked
Crude protein[a] (% N x 6.25)	8.93	9.04
Lysine (g/16 g N)	3.60	3.51
Amino acid score[b] (%)	65.5	63.8
Nitrogen balance in growing rats		
True digestibility (%)	99.7	88.6
Biological value (%)	67.7	78.2
Net protein utilization (%)	67.5	69.2
Utilizable protein[a] (%)	6.02	6.24

[a]Wet basis. [b]Based on 5.5 g lysine/16 g N as 100%.

TABLE IX. Properties of Whole and Pepsin-Treated (Undigested) Protein Bodies of Cooked IR480-5-9 Milled Rice. IRRI, 1978

Property	Whole	Pepsin-treated
Weight recovery[a] (%)	100.0	34.5
Protein content[a] (% N x 5.95)	79.1	62.4 (27.2%)[b]
Lipid content[a] (%)	9.51	22.0 (80.1%)[b]
Lysine (g/16.8 g N)	4.0	1.3
Cysteine (g/16.8 g N)	2.6	4.6
Methionine (g/16.8 g N)	3.1	4.8
Neutral lipid:glycolipid: phospholipid ratio	92/5/3	92/5/3
Fatty acid composition of neutral lipids		
Palmitic (wt % of total)	32	34
Oleic (wt % of total)	24	24
Linoleic (wt % of total)	42	40

[a]Wet basis. [b]Values in parentheses are recoveries.

The poorer quality of the protein rendered less digestible by cooking explains the data on rats in Table VIII. Despite the poorer digestibility of cooked rice (and also cooked parboiled rice), net protein utilization of raw and cooked rice are similar since the biological value of cooked rice is better. This relationship between structure, composition, and nutritive

TABLE X. Content of Selected Amino Acids (g/16 g N) and Molecular Weight of Milled Rice Proteins (Perdon and Juliano, 1978; Mandac and Juliano, 1978; Villareal and Juliano, 1978; IRRI, unpublished data)

Property	α-Globulin	Other globulins	Prolamin	Glutelin subunit	Pepsin-treated cooked-rice protein bodies
Lysine	1.1	1.9	0.1	1.9	1.2
Histidine	0.4	2.7	1.0	2.0	1.8
Arginine	15.2	12.8	5.8	7.4	5.5
Serine	8.8	3.8	5.2	6.1	5.6
Cysteine	3.0	5.5	0.2	0.5	4.4
Valine	4.0	5.8	6.6	6.2	5.9
Methionine	4.0	3.9	0.1	2.4	4.6
Isoleucine	1.8	2.4	4.6	4.1	4.7
Leucine	6.6	7.3	13.0	8.6	11.5
Phenylalanine	3.4	3.0	6.3	5.1	6.4
Tryptophan	0.6	--	0.5	--	0.5
Mol. wt. x 10^{-3}	18	>20	17	16	<5

value of protein bodies may be unique only to milled rice as the endosperm of other cereals contains both protein bodies and matrix proteins. Wheat has no distinct endosperm protein bodies in the mature grain.

The nonstarch lipids that were recovered from pepsin-treated protein bodies represented 80% of the total and had similar properties to those of the total lipids of whole protein bodies (Table IX). However, various amylases tend to give different ratios of neutral lipids:glycolipids:phospholipids in destarched milled rice.

A comparison of the nitrogen balance data for wheat and rice in children (Table XI) again demonstrates the danger of judging nutritional value from digestibility alone (MacLean *et al.*, 1976; IRRI, 1976). Since the better quality rice proteins are not rendered less digestible by cooking, a rice-milk diet had the same apparent retention in children as a wheat-milk diet (Table XI). All these studies indicate that protein quality is more reliably predicted from apparent retention than from apparent digestibility. Also digestibility cannot be predicted from amino acid composition of the protein alone.

A current interest in rice products is focused on rice-based weaning foods. Bulk is the major problem in using cooked

TABLE XI. Nitrogen Balance Data of Low-Protein and High-Protein Wheat Flour and of Cereal-Milk (50:50 N source) Diets in Preschool Children (MacLean *et al.*, 1976; IRRI, 1976)

Property	Wheat flour		Wheat-milk	Rice-milk
Nationality of subjects	Peruvian		Filipino[a]	
Protein content of cereal (% N x 6.25)	11.4	16.1	12.0	7.4
Age of children (years)	0.7 -	2.3	1.2 -	2.0
Children (no.)	4	4	3	3
N intake (mg/kg body wt)	240	240	200	200
Apparent N digestibility (% of intake)	86.0	82.4	87.1	77.2
Apparent N retention (% of intake)	31.3	26.4	26.5	30.3
Apparent N retention of casein I control (% of intake)	39.5	39.5	--	--
Lysine content of diet (g/16 g N)	2.0	2.0	5.5	5.9

[a]Cooperative work with the Food and Nutrition Research Institute, Manila.

rice in weaning food. Bulk can be reduced by homogenizing the rice into a milk with the addition of oil. In the Philippines, waxy rice has not been popularly used in weaning diets because of diarrhea problem. In north and northeast Thailand and in Laos, however, waxy rice is the staple food. Studies in Thailand indicate that malnutrition is more prevalent in the waxy rice area than in the south (Gershoff *et al.*, 1975). The explanation is that the mother determines the amount of rice her baby consumes. Thus, inspite of the greater bulk density of cooked waxy rice, a north Thai baby does not actually consume more rice than a south Thai baby. In the Philippines, eating cooked waxy rice instead of an equivalent volume of cooked nonwaxy rice results in overeating. In Peruvian children waxy and nonwaxy rice starch are equaly highly digestible (MacLean *et al.*, 1978).

There is some interest in obtaining protein concentrates from rice bran (Betschart *et al.*, 1977; Barber and Marquieira, 1977). However, these concentrates should not be considered for supplementing rice-based weaning diets, because rice bran protein is also deficient in lysine. Legume and fish proteins are preferable.

Although the presence of hemagglutinin (Bog-Hansen *et al.*, 1976; Benedito de Barber and Barber, 1978; Takahashi *et al.*, 1973) and protease inhibitors (Tashiro and Maki, 1978; Barber *et al.*, 1978; Kanaya *et al.*, 1976) in rice grain has been reported, they are mainly confined to the bran fraction. That explains why the nutritional value of rice protein is readily predicted from chemical analysis.

VII. MINOR CONSTITUENTS OF MILLED RICE

The minor constituents that probably relate also to varietal differences in grain quality are lipids, nonstarch polysaccharides, and minerals (Table I). Rice is also an important source of B vitamins. High- and low-protein rice have the same content of these minor constituents but differ only in starch content (Resurreccion *et al.*, 1979).

Recent studies on lipids of milled rice indicate almost equal amounts of nonstarch and starch lipids in nonwaxy rices, with intermediate-amylose rice having the largest amount of starch lipids (Choudhury, 1979). Waxy rice has the highest level of nonstarch lipids, which may be important in fat rancidity during storage (Villareal *et al.*, 1976). Nonstarch lipids are associated with protein bodies and are mainly triglycerides. Starch lipids are mainly lysophosphatidyl choline, lysophosphatidyl ethanolamine, and free fatty acids (Choudhury, 1979; Maniñgat and Juliano, 1979b).

Defatting of starch with ethanol and butanol affected its gel viscosity and consistency (Maniñgat and Juliano, 1979b). Brown rice has a hard gel, whereas defatted brown or milled rice always has a soft gel. Starch lipids of tropical milled rice, particularly the glycolipids, have aroused the interest of the group of Y. Fujino, Obihiro University, Japan, in view of the presence of oligoglycosyl diglycerides in rice (Fujino and Miyazawa, 1979).

Cell walls contribute to endosperm integrity during cooking. Varieties differ in the direction of expansion. Most varieties expand girthwise, but some premium varieties, such as Basmati, Sadri, and D25-4, elongate when their presoaked grains are cooked (Juliano, 1972). This property is not related to any unique starch granule property, although the varieties that have it tend to have low gel. temp., intermediate amylose content, and medium gel consistency (Perez *et al.*, 1979). Also because of the current interest in dietary fiber in relation to health and Zn and Fe availability, we initiated some research on properties of cell-wall polysaccharides of milled rice. B. A. Stone of La Trobe University, Bundoora, Australia, is working on the isolation of intact cell walls using a 70% ethanol medium. Several papers on milled-rice nonstarch polysaccharides (Mod *et al.*, 1978; Shibuya and Iwasaki, 1978; Shibuya and Misaki, 1978; Anderson *et al.*, 1978) confirmed the low content of water-soluble nonstarch polysaccharides in milled rice (Cartaño and Juliano, 1970).

Proteoglycans of rice bran have been characterized and *O*-α-L-arabinofuranosyl hydroxyproline has been identified among the degradation products (Yamagishi *et al.*, 1976).

The mineral content of milled rice is low. In some countries, salt is added during rice cooking, which takes advantage of the known effect of salts on the gelatinization characteristics of rice starch (Ghosh, 1959). Screening of the germplasm bank and elite lines for tolerance to problem soils showed that some traditional varieties and even IR42 have tolerance for adverse soil conditions, such as salinity, alkalinity, and Zn deficiency. We have started to study the effect of problem soils on the mineral content of milled rice of IR42 and other tolerant varieties. If mineral content is affected, we plan to study any difference in cooking properties and actual Fe and Zn levels of milled rice. The availability of Zn and Fe in Asian diets is less than 3% (Hallberg *et al.*, 1978).

The talc coating of polished rice, it was reported recently, is not directly related to the high incidence of Japanese stomach cancer (Stemmermann and Kolonel, 1978).

Published data on the volatile components of scented rice are scanty. However, the volatile components characteristic of cooked rice were studied in detail recently (Yajima *et al.*, 1978).

REFERENCES

Anderson, M. A., Cook, J. A., and Stone, B. A. (1978). *J. Inst. Brew.* 84, 233.

Antonio, A. A., and Juliano, B. O. (1973). *J. Food Sci.* 38, 915.

Antonio, A. A., and Juliano, B. O. (1974). *Philippine Agriculturist* 58, 17.

Antonio, A. A., del Mundo, A. M., and Juliano, B. O. (1975). *Philippine Agriculturist* 58, 351.

Barber, S., and Marquieira, A. (1977). *Rev. Agroquim. Tecnol. Aliment.* 17, 209.

Barber, S., Benedito de Barber, C., Flores, M. J., and Montes, J. J. (1978). *Rev. Agroquim. Tecnol. Aliment.* 18, 80.

Bechtel, D. B., and Juliano, B. O. (1979). *Ann. Bot.* 43 (in press).
Bechtel, D. B., and Pomeranz, Y. (1977). *Am. J. Bot.* 64, 966.
Bechtel, D. B., and Pomeranz, Y. (1978a). *Am. J. Bot.* 65, 75.
Bechtel, D. B., and Pomeranz, Y. (1978b). *Am. J. Bot.* 65, 684.
Benedito de Barber, C., and Barber S. (1978). *Rev. Agroquim. Tecnol. Aliment.* 18, 89.
Betschart, A. A., Fong, R. Y., and Saunders, R. M. (1977). *J. Food Sci.* 43, 1088.
Bhattacharya, K. R. (1979). *In* "Workshop on Chemical Aspects of Rice Grain Quality," p. 135. Intern. Rice Res. Inst., Los Baños, Philippines.
Bhattacharya, K. R., Sowbhagya, C. M., and Indudhara Swamy, Y. M. (1978). *J. Sci. Food Agr.* 29, 359.
Bog-Hansen, T. C., Ory, R. L., and Mod, R. L. (1976). *In* Proc. 16th Technical Working Group, p. 80. Texas A & M Univ., College Station, Texas.
Cagampang, G. B., Cruz, L. J., Espiritu, S. G., Santiago, R. G., and Juliano, B. O. (1966). *Cereal Chem.* 43, 145.
Cagampang, G. B., Perez, C. M., and Juliano, B. O. (1973). *J. Sci. Food Agr.* 24, 1589.
Cagampang, G. B., Perdon, A. A., and Juliano, B. O. (1976). *Phytochem.* 15, 1425.
Cartaño, A. V., and Juliano, B. O. (1970). *J. Agr. Food Chem.* 18, 40.
Choudhury, N. H. (1979). Ph.D. dissertation, Univ. Philippines, Los Baños. 98 pp.
du Cros, D. L., Wrigley, C. W., and Blakeney, A. B. (1979). *Riso* 27 (in press).
Eggum, B. O., Resurreccion, A. P., and Juliano, B. O. (1977). *Nutr. Rept. Intern.* 16, 649.
Fujino, Y., and Miyazawa, T. (1979). *Biochim. Biophys. Acta* 572, 442.
Gershoff, S. N., McGandy, R. B., Suttapreyasri, D., Nondasuta, A., Pisolyabutra, U., and Tantiwongse, P. (1975). *Am. J. Clin. Nutr.* 28, 170.
Ghosh, B. P. (1959). *Ann. Biochem. Exptl. Med.* 19, 83.
Graham, G. G. (1977). *In* "Nutritional Evaluation of Cereal Mutants," p. 1. Intern. Atomic Energy Agency, Vienna.
Hallberg, L., Björn-Rasmussen, E., Rossander, L., and Suwanik, R. (1978). *Am. J. Clin. Nutr.* 31, 1403.
Harris, N., and Juliano, B. O. (1977). *Ann. Bot.* 41, 1.
Hoshikawa, K. (1967). *Proc. Crop Sci. Soc. Japan* 36, 151.
Hoshikawa, K. (1968). *Proc. Crop Sci. Soc. Japan* 37, 207.
International Rice Research Institute (1976). *In* "Annual Report for 1975," p. 111. Los Baños, Philippines.

International Rice Research Institute (1979). "Chemical Aspects of Rice Grain Quality." Los Baños, Philippines.
Juliano, B. O. (1964). *Cereal Chem.* 41, 191.
Juliano, B. O. (1972). *In* "Rice Breeding," p. 389. Intern. Rice Res. Inst., Los Baños, Philippines.
Juliano, B. O. (1978). *Food Chem.* 3, 251.
Juliano, B. O. (1979). *In* "Chemical Aspects of Rice Grain Quality," p. 69. Intern. Rice Res. Inst., Los Baños, Philippines.
Juliano, B. O., and Boulter, D. (1976). *Phytochem.* 15, 1601.
Juliano, B. O., and Perdon, A. A. (1975). *Stärke* 27, 115.
Juliano, B. O., Oñate, L. U., and del Mundo, A. M. (1965). *Food Technol.* 19, 1006.
Juliano, B. O., Oñate, L. U., and del Mundo, A. M. (1972). *Philippine Agriculturist* 56, 44.
Juliano, B. O., Antonio, A. A., and Esmama, B. V. (1973). *J. Sci. Food Agr.* 24, 295.
Juliano, B. O., Perdon, A. A., Perez, C. M., and Cagampang, G. B. (1974). *In* "Proceedings 4th International Congress of Food Science and Technology," vol. 1, p. 120. Agroquim. Tecnol. Aliment., Valencia, Spain.
Kanaya, K., Yamamoto, K., and Mitsuda, H. (1976). *J. Japan. Soc. Food Nutr.* 29, 341.
Little, R. R., Hilder, G. B., and Dawson, E. H. (1958). *Cereal Chem.* 35, 111.
MacLean, Jr., W. C., Lopez de Romaña, G., and Graham, G. G. (1976). *J. Nutr.* 106, 362.
MacLean, Jr., W. C., Klein, G. L., Lopez de Romaña, G., Massa, E., and Graham, G. G. (1978). *J. Nutr.* 108, 1740.
Mandac, B. E., and Juliano, B. O. (1978). *Phytochem.* 17, 611.
Maniñgat, C. C., and Juliano, B. O. (1978). *Stärke* 30, 127.
Maniñgat, C. C., and Juliano, B. O. (1979a). *Stärke* 31, 5.
Maniñgat, C. C., and Juliano, B. O. (1979b). *Stärke* 31 (in press).
Marshall, J. J., and Whelan, W. J. (1974). *Arch. Biochem. Biophys.* 161, 234.
Mod, R. R., Conkerton, E. J., Ory, R. L., and Normand, F. L. (1978). *J. Agr. Food Chem.* 26, 1031.
Nelson, O. E. (1976). *In* "Genetic Improvement of Seed Protein," p. 383. Natl. Acad. Sci., Washington, D.C.
Nishita, K., Roberts, R. L., and Bean, M. M. (1976). *Cereal Chem.* 53, 626.
Padua, A. B., and Juliano, B. O. (1974). *J. Sci. Food Agr.* 25, 697.
Palmiano, E. P., and Juliano, B. O. (1972). *Agr. Biol. Chem.* 36, 157.
Perdon, A. A., and Juliano, B. O. (1975a). *Stärke* 27, 69.
Perdon, A. A., and Juliano, B. O. (1975b). *Stärke* 27, 196.

Perdon, A. A., and Juliano, B. O. (1978). *Phytochem.* 17, 351.
Perez, C. M., and Juliano, B. O. (1978). *Stärke* 30, 424.
Perez, C. M., and Juliano, B. O. (1979). *Food Chem.* 4, 185.
Perez, C. M., Pascual, C. G., and Juliano, B. O. (1979). *Food Chem.* 4, 179.
Raghavendra Rao, S. N., and Juliano, B. O. (1970). *J. Agr. Food Chem.* 18, 289.
Resurreccion, A. P., Juliano, B. O., and Tanaka, Y. (1979). *J. Sci. Food Agr.* 30, 475.
Roxas, B. V., Intengan, C. Ll., and Juliano, B. O. (1979). *J. Nutr.* 109, 832.
Shibuya, N., and Iwasaki, T. (1978). *Agr. Biol. Chem.* 42, 2259.
Shibuya, N., and Misaki, A. (1978). *Agr. Biol. Chem.* 42, 2267.
Stemmermann, G. N., and Kolonel, L. N. (1978). *Am. J. Clin. Nutr.* 31, 2017.
Takahashi, N., Yamada, N., Iwamoto, K., Shimabayashi, Y., and Izutsu, K. (1973). *Agr. Biol. Chem.* 37, 29.
Tanaka, K., Yoshida, T., Asada, K., and Kasai, Z. (1973). *Arch. Biochem. Biophys.* 155, 136.
Tanaka, Y., Hayashida, S., and Hongo, M. (1975). *Agr. Biol. Chem.* 39, 515.
Tanaka, Y., Resurreccion, A. P., Juliano, B. O., and Bechtel, D. B. (1978). *Agr. Biol. Chem.* 42, 2015.
Tashiro, M., and Maki, Z. (1978). *Agr. Biol. Chem.* 42, 1119.
Villareal, R. M., and Juliano, B. O. (1978). *Phytochem.* 17, 177.
Villareal, R. M., Resurreccion, A. P., Suzuki, L. P., and Juliano, B. O. (1976). *Stärke* 28, 88.
Viteri, R., Bressani, R., and Arroyave, G. (1971). Expert Committee FAO/ESN document FAO/WHO/PR.78b (iii); cited in *PAG Bull.* 5(3), 30 (1975).
Yajima, I., Yanai, T., Nakamura, M., Sakakibara, H., and Habu, T. (1978). *Agr. Biol. Chem.* 42, 1229.
Yamagishi, T., Matsuda, K., and Watanabe, T. (1976). *Carbohydr. Res.* 50, 63.

ON THE PRESENCE AND DISTRIBUTION OF SOME BIOACTIVE AGENTS IN OAT VARIETIES

Radomir Lásztity
Éva Berndorfer-Kraszner
Magda Huszár

Department of Biochemistry and Food Technology
Technical University of Budapest
Budapest, Hungary

Oat /Avena sativa L./ has an important role in both animal fodder and human food. Its particular role can be attributed to the specific chemical composition of the plant, more particularly of the grain, which is different from that of other cereals.

We have investigated in detail the chemical composition of 13 varieties /9 European, 3 Scotch and 1 American/ and the quantitative relationships of some bioactive agents in them.

The protein content of oat varieties studied ranges between 9.8 and 13.8 %. Among the amino acids lysine is present in remarkably higher concentrations than in other cereals /3.89 - 5.58 %/. The amount of crude fat was found to vary between 3.9 and 6.9 %. Among fatty acids linoleic acid was present in the highest concentration /38.1-46.9%/.

The oat varieties studied contained 1.53-4.73 mg/100 g tocopherols. Among them the percentage amount of α-T-3 was the largest, and the presence of α- and γ-tocopherylquinones could also be detected. The tocopherol/fat ratio was found to vary between 0.23 and 0.79 /mg/g fat/.

Among the prostetic vitamins vitamin B_1 was present in substantial quantities /590-900 μg of B_1/100 g dry substance/. Vitamin B_6 was found to

ISBN: 0-12-370960-1

be widely varying in quantity, between 110 and 500 µg/100 g dry substance. There is a significant positive correlation between protein content and the amount of vitamin B_6.

I. INTRODUCTION

Although oat does not belong to the cereals grown in the largest quantities all over the world /see Fig.1/, its various possibilities of use motivate a deeper investigation /1-3/. In addition to feeding animals oat may be a constituent of baby food as well as of breakfast, roborative and diabetic foods. Several patents are available to its application in semi-finished or ready-made meals, primarily flakes. The production of protein concentrates, applicable in several fields, has also

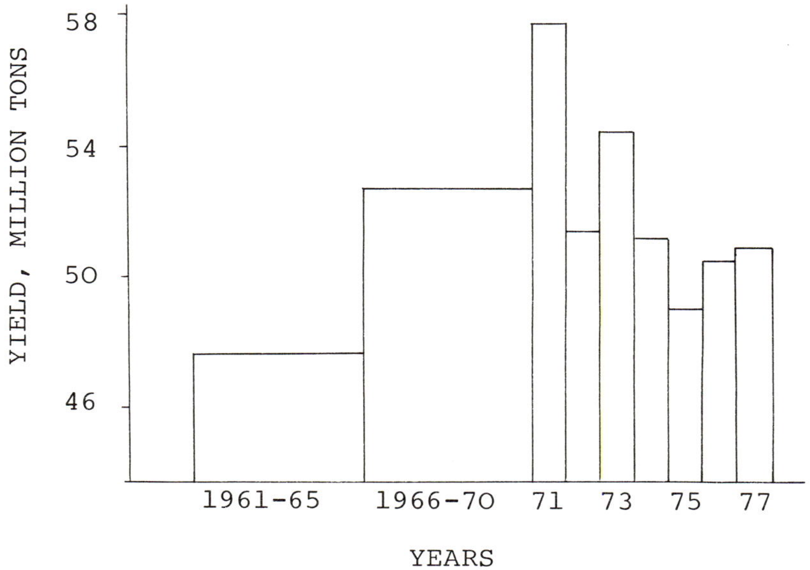

Fig. 1. World's oat yields in million tons from 1961 till 1977 /1-3/

been elaborated /4-6/. It is of food technological importance that the fat content of oat has the quality of edible oils, and is comparable in stability with soybean oil /7/. Among the industrial applications the production of oat pentosanes can be mentioned, which are used for fermentation purposes in the production of furfurol, and to produce acetic acid as a by-product /8/. Furthermore oat flour is used in cosmetic preparations, and experiments are in progress on its application in pharmaceutical industry.

When discussing the composition of oat, its most valuable components, the value of proteins should be mentioned first.

The primary limiting amino acid of cereals is lysine. It is therefore of utmost importance that among the cereals the lysine content of oat protein is the highest /3.2-5.5 %/, closely approaching the ideal composition recommended by FAO/WHO, which is 5.2 % lysine/100 g protein /9/. The secondary and tertiary limiting amino acids are threonine-which, according to chemical analysis is not absent, but biological studies suggest that it is present in inaccessible forms /10/, - sulphur containing amino acids and tryptophan /11/.

The ratio of protein fractions and amino acids vary with the physiological state of the plant, and are different in different tissues /12/.

Oat contains more lipids than other cereals. Fat is accumulated in the germ to the greatest concentration /13/, but the fat content of endosperm and aleuronic layer is also significant /14-15/.

From the aspects of application it is important to determine the essential fatty acid composition of fats. According to the present state of our knowledge linoleic acid and arachidonic acid can be considered essential,of which only the former occurs in plant tissues. The amount of linoleic acid in oat fat is between 33.2 and 46.2 % according to various authors /16-18/.

The importance of fat is further increased by the bioactive agents dissolved in it. It has been known for a long time that oat has favourable effect on the physiological condition of young and breeding animals. Among other factors /favourable ratio of nutrients, vaniline content of the grain etc. /19// the main reason for this effect may be the presence of tocopherols.

Tocopherols have in vivo antioxidative and enzymatic effects explaining the active role of tocopherols in the electron transport system of living organisms.

One of the groups of bioquinones namely tocopheryl-quinones, the reduction products of which are tocopherols, also occur in relatively large amounts in oat. Their presence accounts for the antioxidant properties of oat.

Among the prostetic vitamins mainly vitamin B_1 in oat grains is emphasized by Mitrofanov and Mitrofanova /20/. Smith /21/ recommending oat preparations as food points out the high thiamine, riboflavine, niacine and minerals /Ca, Mg, S, Fe/ content of oat.

II. MATERIALS AND METHODS

13 oat varieties /9 European, 3 Scotch and 1 American/ were investigated. The investigations were extended to the study of macro composition, amino acid and fatty acid composition. Furthermore, the total tocopherol content, within it the percent distribution of various fractions, and the amounts of the prostetic vitamins vitamins B_1, B_2 and B_6 - equally important in animal feeding and human nutrition - have been studied.

The oat varieties investigated are listed in Table I.

The samples studied were grown in the 1978 season. After sampling the grains were ground in a laboratory grinder and all analyses were carried out on the complete grist. Three parallel measurements performed in the analysis, the data given are the averages of parallel determinations.

TABLE I. Oat Varieties Investigated

Number	Oat Varieties	Origin	Remarks
1	Leanda	Netherlands	Grown in Hungary to the greatest extent
2	Condor	Netherlands	Rich in yield
3	Natal	Netherlands	Rich in yield
4	American	USA	Imported for feeds
5	Czechoslovakian-1	Czechoslovakia	
6	Czechoslovakian-2	Czechoslovakia	
7	Czechoslovakian-3	Czechoslovakia	
8	Czechoslovakian-4	Czechoslovakia	
9	Early from Szeged	Hungary	In 1978 acknowledged variety, rich in yield
10	Naked from Táplán	Hungary	Acknowledged naked variety for food
11	Astor	Scotland	
12	Karin	Scotland	
13	Maris Tabaret	Scotland	

Experimental Methods

Dry substance, crude fiber, crude protein and crude fat contents were determined according to the Hungarian standards /22-23/.

Amino-acid composition was determined in the Agrotechnical Research Institute of the Hungarian Academy of Sciences /Martonvásár/ with an automatic amino acid analyser /JEOL-ILC-5 AH, Japan/, according to the conventional Moore-Stein technique /24/. Sulphur containing amino acids were pretreated by oxidative protecting hydrolysis before analysis.

Fatty acid composition was determined gas chromatographically after hydrolysis and conversion into fatty acid methyl esters /25/.

The total amount of tocopherols was determined according to Tsen /26/, the individual fractions were identified and determined by the method of Berndorfer - Kraszner /27/.

Vitamins B_1, B_2 and B_6 were determined fluorimetrically /28-30/.

III. RESULTS AND DISCUSSION

The macro composition of oat varieties studied is shown in Table II. The results indicate that there is no significant difference in the crude fiber content. Protein content varied between 9.8 and 13.8 %, in a global agreement with literature data. Of the samples investigated the protein content of Early from Szeged, Natal, American and Condor varieties is high.

Crude fat is 5.2 % in average. This is in the range of 3.1 - 11.6 % obtained by Brown and Craddock /31/. For Canadian varieties a range of 5.57 - 8.84% was found by Sahasrabudhe /32/.

TABLE II. Macro Composition of Oat Varieties

No.	Oat Var.	Fiber	Crude Protein	Fat
		% / D.S.		
1	Leanda	10,6	12,1	5,3
2	Condor	11,3	13,0	4,7
3	Natal	11,8	13,3	4,6
4	American	10,5	13,1	5,2
5	CS-1	10,1	11,9	4,5
6	CS-2	10,9	11,0	4,5
7	CS-3	11,0	9,8	4,8
8	CS-4	9,1	11,1	5,0
9	Early Sz.	9,4	13,8	4,3
10	Naked T.[a]	1,5	19,6	7,1
11	Astor	10,1	11,8	6,8
12	Karin	10,4	11,1	6,2
13	M. Tabaret	10,1	11,5	6,9
	Average	10,4	12,0	5,2

[a]Not included in the average

The amino acid composition of the oat varieties studied are given in Table III.

On average lysine content of 4.51 % was found. The lysine content of Czechoslovakian-4 and Czechoslovakian-3 varieties was higher than average /5.58 and 5.05 %/. The values found approach the ideal composition recommended by FAO/WHO /5.2 % lysine in 100 g protein/. Fig. 2 illustrates the trends of lysine and threonine content of oat varieties.

The fatty acid compositions of the fat of oat varieties are shown in Table IV, listing the concentrations of palmitic, stearic, oleic, linoleic and linolenic acids. Fatty acids 20:2 appearing in certain varieties, and fatty acids 22:0 and 22:1 present only in traces are given under the heading "others". It can be seen from the table that linoleic acid is present in the highest amount - which is very important - but remarkable amounts of oleic and palmitic acids can also be detected. The linolenic acid content of the Czechoslovakian-2 variety is especially high.

TABLE III. Amino Acid Composition of Oat Varieties[a] part A

No.	Oat Var.	Crude protein % / D.S.	ALA	ARG	ASP	CYS	GLU	GLY	HIS	ILE
1	Leanda	12,1	5,18	7,42	8,92	1,94	21,04	5,26	2,26	3,91
2	Condor	13,0	5,14	7,34	9,14	1,80	21,11	5,22	2,25	3,91
3	Natal	13,3	5,02	7,62	8,96	1,89	20,79	5,20	2,29	4,02
4	American	13,1	5,15	7,05	8,36	1,48	22,05	5,25	2,12	4,07
5	CS-1	11,9	5,29	6,55	8,49	1,13	20,67	5,43	2,73	3,79
6	CS-2	11,1	5,52	6,62	8,59	2,36	20,20	5,64	2,58	3,83
7	CS-3	9,8	5,13	6,96	8,51	2,16	20,76	5,48	2,79	3,75
8	CS-4	11,1	5,01	7,00	8,60	0,98	20,35	5,04	3,18	3,67
9	Early Sz.	13,8	4,85	7,49	8,30	2,49	21,43	5,14	2,19	3,95
10	Naked T.[b]	19,6	4,87	7,35	7,83	3,38	22,45	4,95	2,21	3,80
11	Astor	11,8	5,60	6,53	8,11	1,83	20,44	6,10	2,04	3,89
12	Karin	11,1	4,34	6,73	9,47	0,74	22,27	5,51	2,61	4,04
13	M. Tabaret	11,5	5,23	6,40	8,71	1,94	21,52	5,45	2,49	4,02
	Average	12,0	5,12	6,98	8,68	1,73	21,06	5,39	2,46	3,90

[a]Units are in amino acid g / 16g N

[b]Not included in the average

TABLE III. Amino Acid Composition of Oat Varieties[a] part B

No.	Oat Var.	LEU	LYS	MET	PHE	PRO	SER	THR	TYR	VAL	NH_3
1	Leanda	7,67	4,49	1,87	5,07	5,28	5,09	3,67	3,34	5,29	2,30
2	Condor	7,65	4,39	1,39	5,07	5,41	5,27	3,87	3,33	5,26	2,47
3	Natal	7,64	4,40	1,69	5,10	5,49	5,03	3,67	3,59	5,20	2,40
4	American	7,79	4,07	1,28	5,42	6,36	4,76	3,42	3,76	5,46	2,18
5	CS-1	7,34	4,90	1,73	4,98	6,98	4,89	3,56	3,62	5,36	2,58
6	CS-2	7,28	4,80	1,82	4,44	6,10	5,07	3,67	3,87	5,25	2,34
7	CS-3	7,27	5,05	1,46	4,63	7,70	4,75	3,47	2,91	4,80	2,42
8	CS-4	7,32	5,58	1,63	4,88	6,87	4,83	3,46	3,70	4,97	2,95
9	Early Sz.[b]	7,69	4,14	1,63	5,20	5,77	5,02	3,55	3,69	5,13	2,33
10	Naked T.[b]	7,66	4,06	1,02	5,11	5,50	4,97	3,57	3,84	5,14	2,35
11	Astor	7,29	3,89	2,51	4,78	5,04	6,75	3,90	3,75	5,48	2,06
12	Karin	7,83	4,02	1,57	4,75	5,40	5,21	3,82	4,02	5,31	2,37
13	M. Tabaret	7,85	4,44	1,17	4,83	6,01	4,96	3,63	3,77	5,36	2,21
	Average	7,55	4,51	1,65	4,93	6,03	5,14	3,64	3,61	5,24	2,38

[a]Units are in amino acid g / 16g N

[b]Not included in the average

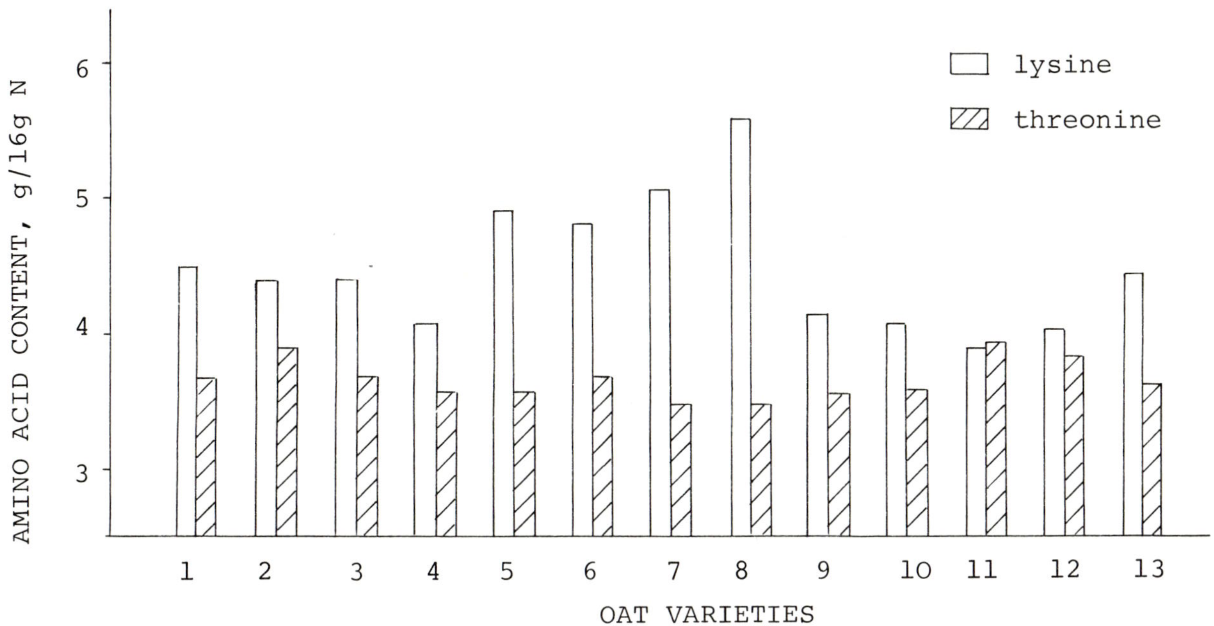

Fig. 2. The lysine and threonine content of oat varieties

The vitamin B_6 content shows an interesting trend, it varies between 110 and 500 ug/100 g dry substance, and there appears to be a correlation between its amount and the protein content /Table VII/.

Linear regression analysis gives a correlation coefficient of +0.9755.

TABLE VII. Correlation Between Vitamin B_6 and Crude Protein Contant

No.	Oat Var.	Crude protein	Vitamin B_6
		% / D.S.	µg/100g D.S.
1	Leanda	12,1	3,90
2	Condor	13,0	450
3	Natal	13,3	460
4	American	13,1	440
5	CS-1	11,9	300
6	CS-2	11,0	220
7	CS-3	9,8	110
8	CS-4	11,1	210
9	Early Sz.	13,8	500
10	Naked T.[a]	19,6	520
11	Astor	11,8	250
12	Karin	11,1	250
13	M. Tabaret	11,5	250
Average		12,0	320

[a]Not included in the average

REFERENCES

1. "Nemzetközi Statisztikai Évkönyv", Központi Statisztikai Hivatal /"Intern. Stat. Yearbook", Centr. Stat. Office/, Budapest /1974/.
2. Production Yearbook, Vol. 28, FAO, ROMA /1974/.
3. "Nemzetközi Statisztikai Zsebkönyv", Központi Statisztikai Hivatal /"Intern. Stat. Pocket-book." Centr. Stat. Office/ Budapest /1978/.
4. Cluskey, J.E., Wu, J. V., Wall, J. S. and Inglett, G.E. Cereal Chem. 50, 475 /1973/.
5. Wu, J. V., and Shingfellow, A.C. Cereal Chem. 50, 489 /1973/.
6. Wu, J. V., Sexson, K.R., Cluskey, J.E. and Inglett, G.E. J. of Food Science 42, 1383 /1977/.
7. Kalbasi - Ashtori, A. and Hammond, E. G. JAOCS. 54, 305 /1977/.
8. Caldwell, E.F. and Pomeranz, J."58th Am. Ann. Ass. of Cereal Chem. /1973/.
9. Kállai, L. and Kralovánszky, V.P. "A takar-mányozás biológiája" /"Biology of Feeding"/ Mezőgazdasági Kiadó, Budapest, /1978/.
10. Clark, W. L. and Potter, G. C. Abstr. of Papers Am. Chem. Soc. 161: AGFD 26 /1971/.
11. Briggle, L. W., and Smith, R.T. Crop Science 15, 547 /1975/.
12. Pomeranz, J. and Robbins, G. S. Proceedings Am. Soc. of Brewing Chem. 15 /1971/.
13. Joungs, V.L., Püskülcü, M. and Smith, R. R. Cereal Chem. 54, 803 /1977/.
14. Joungs, V.L. Cereal Chem. 55, 591 /1978/.
15. Weber, E. J. "58th. Ann. Am. Ass. of Cereal Chem.", Missouri, /1973/.
16. "Agrobotanika" /"Agrobotany"/ /Hortobágyi, T., ed./, Mezőgazdasági Kiadó, Budapest, /1970/.
17. Stuke, H. Z. Pflzucht. 42 /1959/.
18. "Feedstuffs",Reference, Issue 49, /1977/.
19. "Gabonaipari Kézikönyv" /"Textbook of Cereal Industry"/ /Tomay, T.,ed./, Mezőgazdasági Kiadó, Budapest, /1971/.
20. Mitrofanov, A.S. and Mitrofanova, K.S. "A zab" /"The Oat"/, Mezőgazdasági Kiadó, Budapest, /1970/.

21. Smith, W. H. Snack Food 62, 46 /1973/.
22. MSz: 6830 /Hungarian Standard/, Budapest, /1978/.
23. MSz: 6367 /Hungarian Standard/, Budapest, /1976/.
24. Moore, S. and Stein, H. in "Methods in Enzimology" /Colowick, S. P. and Kaplan, N. O. ed./ Vol. VI. 819, Academic Press, New York and London /1963/.
25. MSz: 19928 /Hungarian Standard/, Budapest, /1973/.
26. Tsen, C. C. Anal. Chem. 33, 849 /1961/.
27. Berndorfer-Kraszner, É. Élelmiszervizsgálati Közlemények XVI, 193. /1970/.
28. Berndorfer-Kraszner É., "in Vitamin meghatározási eljárások természetes anyagokban" /"Methods of Vitamin Determinations in Natural Substances"/ Vol. 3. 46, MÉTE, Budapest, /1977/.
29. Dobos, I., Berndorfer-Kraszner, É. "in Vitamin meghatározási eljárások természetes anyagokban" /"Methods of Vitamin Determinations in Natural Substances"/ Vol. 3. 51, MÉTE, Budapest, /1977/.
30. Tamura, G., Takanashi, S.M. in "Methods in Enzimology" /Colowick, S.P. and Kaplan, N.O., ed./ Vol. XVIII/A. 417, Academic Press, New York and London,/1970/.
31. Brown, C.M., Craddock, J.C. Crop Science 12, 514 /1972/.
32. Sahasrabudhe, M.R. J. of the Am. Oil Chemists' Soc. 56, 80 /1979/.

BIOFUNCTIONAL PROPERTIES OF OATS

Max R. Gould

Quaker Oats Company
Barrington, Illinois

James W. Anderson

Veterans Administration Hospital
Lexington, Kentucky

Sean O'Mahony

Nabisco Company
Fairlawn, New Jersey

I. INTRODUCTION

In describing oats it is generally pointed out that the oat grain (groat) has one of the highest protein contents of the cereal grains, the protein quality is good, and oat groats are a good source of thiamin, iron and phosphorus. Some effort has been made to develop oat protein concentrates (1-5). There has been some interest in the gum contained in groats which has been described as a beta-glucan (6-7).

Some previous studies and the work reported here suggests that oats possess some physiological properties which may have implications to man and appear to be unique to oats among the cereal grains. The ingestion of large quantities of rolled oats (140 grams/day) by 21 healthy males was accompanied by a significant reduction in serum cholesterol levels (8). Cholesterol lowering activity of whole ground oats and oat fractions have been studied in chicks fed hypercholesterolemic diets. Cholesterol was significantly reduced by whole oats and oat hulls, and, to a lesser extent, by dehulled oats. Oat hulls were most effective. Oat starch and oat oil had no cholesterol lowering activity (9).

ISBN: 0-12-370960-1

The mechanism for this reduction in serum cholesterol values has not been delineated. The hypercholesterolemic effect of oats may be related to their soluble plant fiber content. Many studies (10) in humans have demonstrated that the ingestion of soluble plant fibers such as pectin, guar gum and bengal gram is associated with a significant reduction in serum cholesterol values.

Some very recent work indicates that a coarse fraction selectively lowers total serum cholesterol in rats while increasing high density lipoprotein cholesterol values. Similar effects were found with guar gum and pectin, but not with cellulose (11).

In this paper we are reporting on two studies which explore the effect of oats on serum cholesterol and in one case on glucose tolerance. No solid scientific conclusions can be drawn from these studies. The numbers involved preclude this, but the observations warrant discussion.

One study was conducted at the John Stuart Research Laboratory of the Quaker Oats Company in Barrington, Illinois. The second was conducted at the Veterans Administration Hospital in Lexington, Kentucky.

The first study indicates that 1½ ounces of Instant Oatmeal consumed daily as part of the diet over a four week period will significantly reduce serum cholesterol among normal healthy people over 40 years of age.

The second study indicates that a coarse fraction of oats consumed at 100 grams per day as part of a high carbohydrate fiber diet (HCF) will effectively reduce serum cholesterol among male adults with hypercholesterolemia. It further showed a beneficial shift in the ratio of low density lipoprotein cholesterol to high density lipoprotein cholesterol.

HCF diets have been shown to be effective in reducing the insulin requirements of individuals suffering from diabetes. When the coarse oat bran fraction was used as the plant fiber source in one individual with this condition, it was more effective than a well established, standard HCF diet.

II. INSTANT OATMEAL STUDY

The study was exploratory in nature. It was designed to give more information on the potential role of oatmeal in regulating serum cholesterol levels in humans.

The study was conducted with fourteen volunteers ranging in age from 23 to 53 and who were, by their own account, in good health. The group consisted of twelve males and two females selected from a total of 24 volunteers. The mean pretest serum cholesterol level was 211 mg/dl. The range was 140 to 270 mg/dl.

Pre-test there was no significant correlation between age and cholesterol values in the test population.

Participants consumed 1.5 ounces of Instant Oatmeal daily at breakfast. They were instructed to follow a normal diet with this one exception. Oatmeal displaced part or all of the normal breakfast at each participant's option.

Fasting blood samples were drawn weekly. One sample was drawn prior to the test. Weekly samples were drawn for a period of four weeks while on the oatmeal containing diet.

Four subjects over forty years of age were sampled two weeks post test. One subject was sampled four weeks post test as well.

Figure 1 is a scatter diagram of the entire test population showing the serum cholesterol levels with time, both pre-test and on test. Due to the relatively normal to low pre-test mean level, a depression in the population overall might be expected to be difficult to observe. This indeed was the case and there was no significant change in the mean value. The mean was relatively constant.

Figure 1.

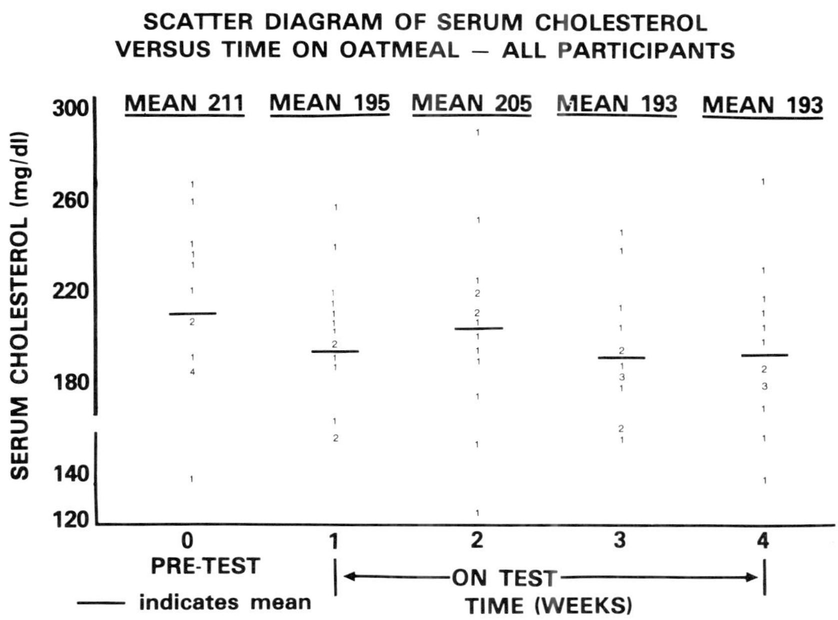

Figure 2 shows the relationship between cholesterol depression and age. There is a correlation significant at the 0.05 level. Depression was greater among the older participants. The correlation is not a strong one. The correlation coefficient is 0.66.

Figure 2.

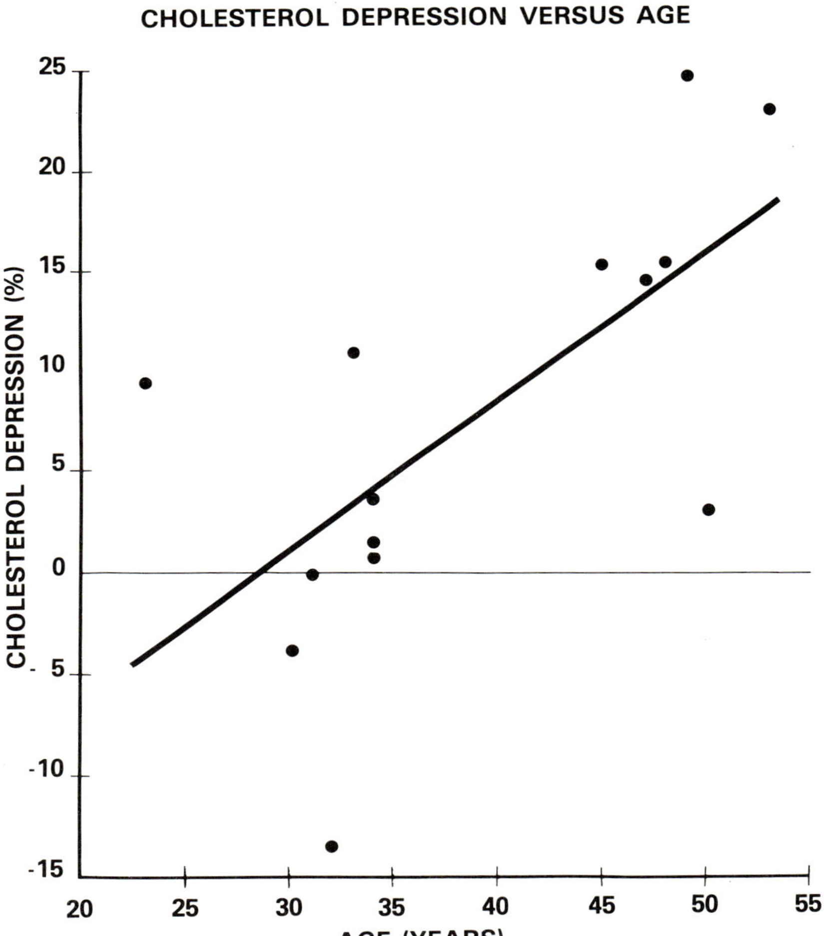

Figure 3 is a scatter plot of four subjects over age 40 showing serum cholesterol versus time. It shows a significant decrease over the test period followed by a "rebound" post test. One individual designated KC in the figure was sampled four weeks post test and appeared to have returned to his normal level after a very dramatic rebound two weeks post test. Again, the correlation is rather weak with an r value of 0.59.

Figure 3.

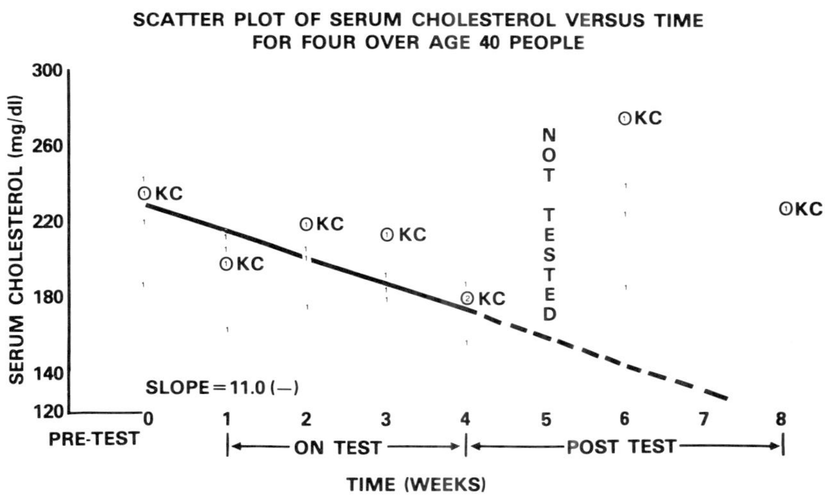

The cholesterol depression is significantly time related for the over age 40 group. That is, the longer on the diet the greater the depression. This might be expected to bottom out at a new equilibrium after some period of time, greater than four weeks, at this level of oatmeal intake.

III. COARSE OAT FRACTION STUDIES

As previously mentioned, the study done at the V.A. Hospital in Lexington, Kentucky involved the use of a coarse oat fraction consumed at the rate of 100 grams per day as porridge, bran muffins or a combination of the two.

Figure 4 presents a schematic flow sheet for production of the coarse fraction used in these studies.

Figure 4

SCHEMATIC FLOW FOR COARSE OAT FRACTION

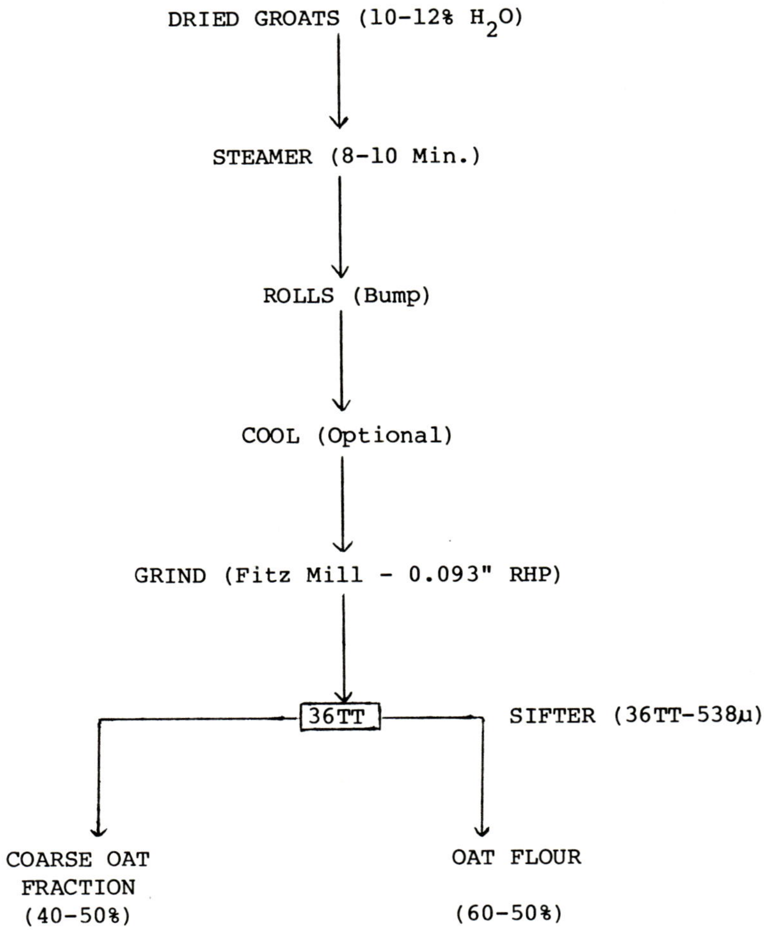

Table I presents a proximate analysis of the coarse fraction compared to the starting oat groats. Notice the shifts in protein, crude fiber, ash, and to a lesser extent fat. Ranges are given for the dietary fiber (Neutral Detergent Residue) simply because we have experienced considerable difficulty with this analysis and particularly with oats. Ranges also are given for the gum due in part to isolation complexities, but also in all probability to varietal differences.

TABLE I. Proximate Analysis Coarse Fraction Versus Starting Groats

	Percent (as is)	
	Groats	Coarse
Protein (Nx6.25)	16.5	21-22
Fat (EE)	6.2	7.5
Moisture	9.3	8.5
Crude Fiber	1.6	2.7
Ash	1.9	2.6
Dietary Fiber (NDR)	6-9	14-30
Beta Glucan (Gum)	4-6	8-12

In the case of dietary fiber, the most representive numbers from our analyses would appear to be about 7% in the whole groat and about 26% in the coarse fraction. Since the fraction is 50% of the groat, this is not possible if we were measuring the same thing in both cases. Obviously we are not. Analyses are by the neutral detergent residue method for dietary fiber which is known to be variable and dependent upon a number of factors.

However, the key point is that, most if not all, of the gum and the dietary fiber are captured in the coarse fraction.

Table II compares the control diet and the HCF diet as to component composition and source of calories. The columns headed g/d refer to grams per day. The columns headed % refer to the percent of total calories coming from that component. The HCF diet applies to a well established one developed by Dr. Anderson in which the cereal plant fiber source is primarily from wheat bran. The values shown also apply to the HCF-Coarse Oat Diet.

TABLE II. Comparison of Diets

	Control		HCF Diet	
	g /d	%	g /d	%
Protein	98	20	96	19
CHO	215	43	351	70
Fat	83	37	24	11
Plant Fiber	27	-	82	-
Cholesterol	450 mg.		50 mg.	

Figure 5 shows the results achieved in reducing serum cholesterol with cholesterol expressed as a percent of the control for both the standard HCF diet and the HCF-oat diet.

Figure 5.

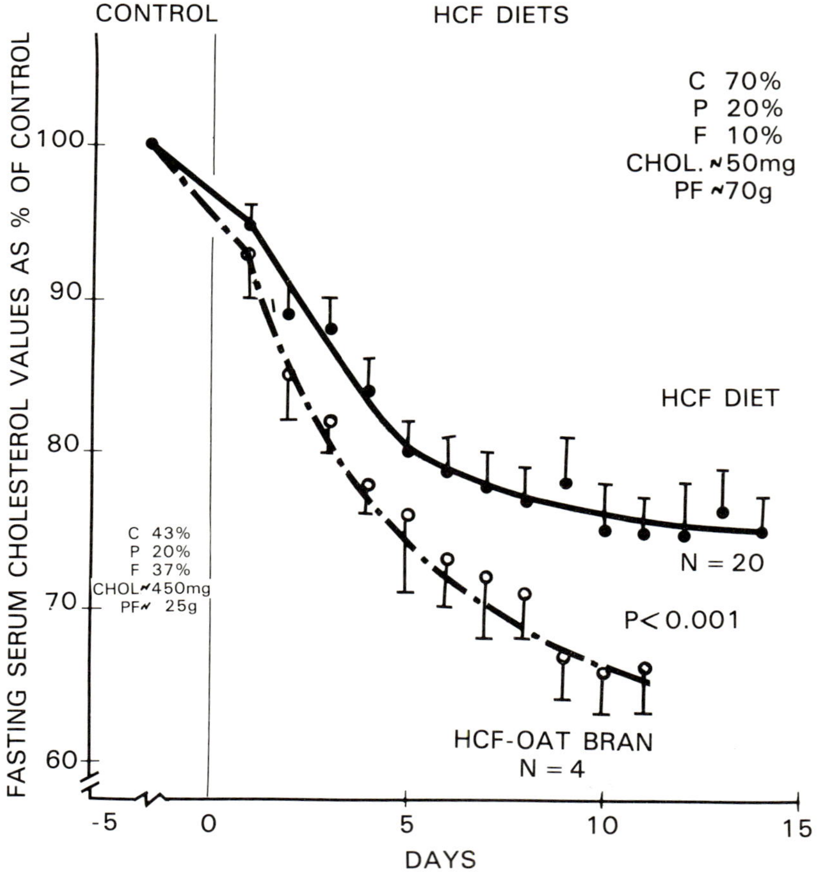

All subjects had hyperlipidemia. The mean total serum cholesterol was 278 ± 41 mg/dl on the control diet. Both the HCF and the HCF-oat diets were very effective in reducing cholesterol when consumed for a period of 10-13 days. The coarse oat HCF diet was significantly more effective.

There is considerable emerging evidence that suggests that high density lipoproteins (HDL) are beneficial in controlling serum cholesterol while low density lipoproteins are detrimental. It has been proposed that HDL serves a carrier function in clearing cholesterol from arterial tissues (12-13). HDL cholesterol is believed to have a protective effect with regard to the development of cardiovascular disease (14). Miller has provided an excellent review of the subject (15).

Figure 6 attempts in a simple way to show graphically how the "good guys", high density lipoproteins, and the "bad guys", low density and very low density lipoproteins, function. It is postulated that the low density lipoproteins carry cholesterol to the artery walls where it may be deposited while high density lipoproteins can pick up cholesterol which is returned to the liver and catabolized.

Figure 6.

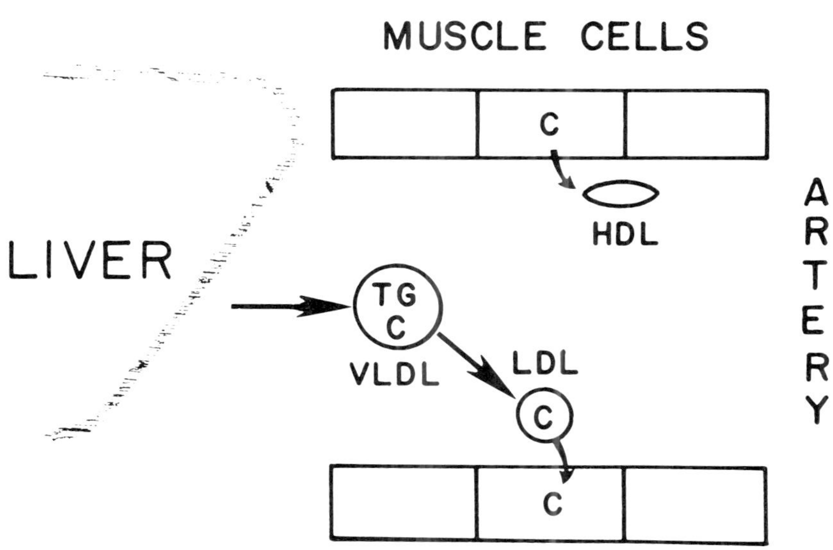

Table III shows the cholesterol disbribution by lipoprotein type comparing the control with the HCF-oat diets. There is not only a reduction in the total serum cholesterol, but a dramatic shift toward the HDL type when the HCF-oat diet is used.

TABLE III. Cholesterol Distribution
Control vs. HCF-Oat

(Means)	Control	HCF (Oat Coarse)	
Total Cholesterol (mg/dl)	278±41	179±23	(P<0.02)
LDL Cholesterol (mg/dl)	174±17	73±11	(P<0.005)
HDL Cholesterol (mg/dl)	27±4.3	49±3.2	(P<0.02)
LDL/HDL Ratio	6.9±1.2	1.5±0.2	(P<0.05)
TC/HDL Ratio	11.0±1.0	3.6±0.1	(P<0.05)

± Standard error of the Mean.

Figure 7 is a bar graph showing the same information.

Figure 7.

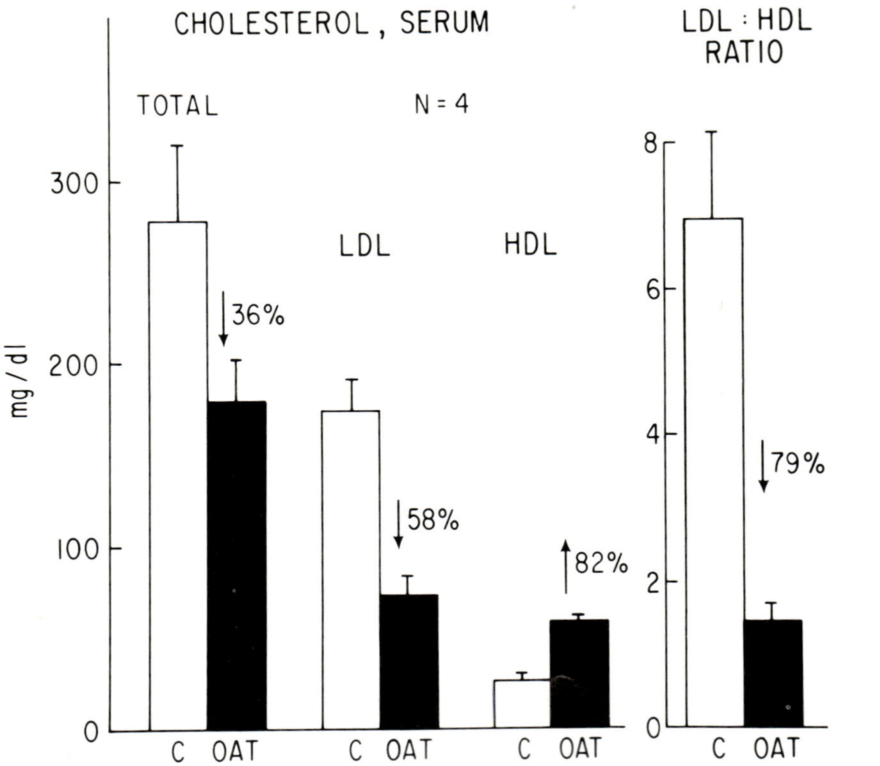

It shows a 36% reduction in total cholesterol, a 58% reduction in LDL cholesterol, an 82% <u>increase</u> in HDL cholesterol for HCF-oat diet, and a dramatic 79% decrease in the LDL/HDL ratio compared to the control.

A similar shift has been observed in animals using guar gum, pectin or coarse oat fraction (11).

The standardized HCF diets have been very effective in reducing total serum cholesterol, VLDL, LDL and HDL cholesterol, but do not demonstrate this shift in type. It is well to remember, however, that the HCF-oat study only involved four subjects. Much additional research is needed.

A number of experienced clinicians have advocated high carbohydrate diets for diabetic patients. This minority has argued with the conventional wisdom that carbohydrate restriction was necessary in the management of diabetes (16-19).

The strongest evidence which supports the concept that high-carbohydrate diets are beneficial in the management of diabetes was provided by Singh (20).

He demonstrated that high carbohydrate (approximately 60% of energy) diets were accompanied by lower insulin doses in his patients with diabetes. It is also noteworthy that these high carbohydrate diets also provided generous quantities of plant fiber from whole wheat chupatties.

The studies of Jenkins and colleagues (21) demonstrated that fiber rich meals were followed by less hyperglycemia than were fiber poor meals. Other workers have confirmed that fiber supplementation is associated with significantly lower post meal hyperglycemia (22-23). Various plant fibers appear to delay the absorption of glucose and result in a slower rise in the blood sugar.

One of the four patients used in the HCF-coarse oat study is a diabetic and was receiving 20 units of insulin per day. Figure 8 shows that this patient was able to gradually reduce his insulin requirements to zero after 10 days on the HCF-oat diet while maintaining control of his fasting plasma glucose within a normal range. For comparative purposes a plot of a group of eight diabetics requiring insulin and given the standardized HCF diet is included.

Figure 8.

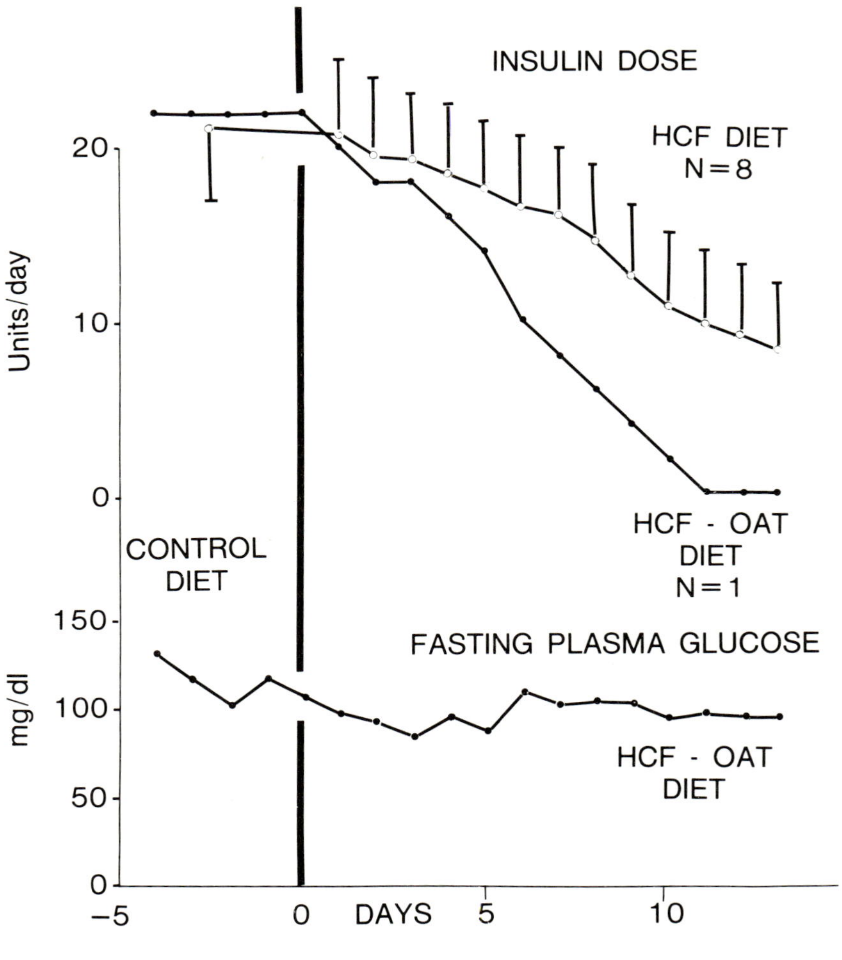
INSULIN DOSE
HCF DIET
N=8
Units/day
20
10
0
HCF - OAT
DIET
N=1
CONTROL
DIET
150
FASTING PLASMA GLUCOSE
mg/dl
100
50
0
HCF - OAT
DIET
−5
0
DAYS
5
10

IV. SUMMARY OF RESULTS

A. Instant Oatmeal Study

Thus, in summary of the instant oatmeal study a significant serum cholesterol depression was observed in a group of people over 40 years of age while on a diet containing 1.5 ounces of instant oatmeal per day consumed at breakfast for a period of four weeks.

There was no significant cholesterol depression among the whole test group ranging in age from 23-53.

B. Coarse Oat Fraction Studies

Four subjects suffering from hypercholesterolemia consumed a high carbohydrate fiber diet containing 100 grams of a coarse oat fraction daily. Significant reductions in serum cholesterol, low density and very low density lipoprotein cholesterol were observed. High density lipoprotein cholesterol was elevated by this diet. This effect is thought to be beneficial. Similar diets using primarily wheat bran as the cereal fiber source, while effective, were not as effective, and did not show the shift toward high density lipoprotein cholesterol.

One of the four subjects had diabetes mellitus and was taking 20 units of insulin per day. By the end of 10 days on the HCF-coarse oat diet, his insulin dosage had been reduced to zero and his serum glucose was in a normal range.

These preliminary studies are difficult to interpret because of the basic differences between the control and the HCF diets. The HCF diets are lower in total fat, saturated fat, and cholesterol, and are higher in starch and plant fiber than the control diets. Thus the specific contribution of coarse oat fraction is confounded by these other factors. These preliminary studies have encouraged further studies to examine the influence of coarse oat ingestion in individuals where the diets are kept as constant as possible except for the addition of oat bran. The protein, carbohydrate, simple carbohydrate, complex carbohydrate, total fat, fatty acids and cholesterol content of the coarse oat diet is virtually identical to that of the control diet. However, the oat bran diet contains approximately 50% more plant fiber than contained in the control diet. This study is being completed now and will be subsequently reported.

REFERENCES

1. Cluskey, J., Wu, Y., Wall, J. and Inglett, G. (1973). Cereal Chemistry 50, 475.
2. Wu, Y., Cluskey, J., Wall, J. and Inglett, G. (1973). Cereal Chemistry 50, 481
3. Wu, Y. and Stringfellow, A. (1973). Cereal Chemistry 50, 489.
4. Youngs, V. (1974). Journal of Food Science 39, 1045.
5. D'Appolonia, B. and Youngs, V. (1978). Cereal Chemistry 55 (5), 736.
6. Clark, W. (1972). Agrologist 1 (6), 8.
7. Wood, P., Siddiqui, I. and Paton, D. (1978). Cereal Chemistry 55 (6), 1038.
8. De Groot, A., Luyken, R., and Pikaar, N. (1963). Lancet 2, 303.
9. Fisher, H. and Griminger, P. (1967). Proceedings of the Society for Experimental Biology and Medicine 126, 108.
10. Truswell, A. (1977). Nutrition Reviews 35, 51.
11. Chen, W. and Anderson, J. (1979). Proceedings of the Society for Experimental Biology and Medicine, In press.
12. Miller, G. and Miller, N. (1975). Lancet 1, 16.
13. Miller, N. and Miller, G. (1975). Lancet 1, 1033.
14. Hulley, S., Cohen, R. and Widdowson, G. (1977). JAMA 238, 2269.
15. Miller, N. in Lipids 13 (12), 914. American Oil Chemists Society, Champaign, Ill. (1978)
16. Kiehm, T., Anderson, J., and Ward, K. (1976). American Journal Clinical Nutrition 29, 895.
17. Anderson, J. (1977). American Journal Clinical Nutrition 30, 402.
18. Rabinowitch, I. (1935). Canadian Medical Association Journal 33, 136.
19. Anderson, J. (1977). Cereal Foods World 22, 12.
20. Singh, I. (1955). Lancet 1, 422.
21. Jenkins, D., Leeds, A., Gassull, M., Cochet, B. and Alberti, K. (1977). Annals Internal Medicine 86, 20.
22. Anderson, J. and Chen, W. (1979). American Journal Clinical Nutrition 32, 346.
23. Anderson, J. and Ward, K. (1978). Diabetes Care 1, 77.

PEARL MILLET: ITS CHEMISTRY AND UTILIZATION[1/]

Hoseney, R. C. and Varriano-Marston, E.

Department of Grain Science and Industry
Kansas State University
Manhattan, Kansas, USA

I. INTRODUCTION

A number of species of millets are grown in various parts of the world. From an economic standpoint the following are the most important:

> Pearl millet (Pennisetum americanum L. Leeke), also called bulrush millet in Africa and bajra in India and often identified in literature by the older name of Pennisetum typhoides.
>
> Proso, common or broom corn millet (Panicum miliaceum), widely grown in the Soviet Union, Manchuria, and China.
>
> Finger millet (Eleusine coracana), also called ragi, widely grown in India and Africa.

In this paper we shall deal only with the chemistry and structure of pearl millet. Pearl millet is one of the most drought tolerant of man's food crops and is grown primarily in the dry sahel zone across Africa and in the semi-arid areas of India. It is the first crop one sees growing as one comes out of the desert and provides the only means of survival for many people

---

[1/]Contribution No. 80-24A Department of Grain Science and Industry, Kansas Agricultural Experiment Station, Manhattan, KS 66506.

ISBN: 0-12-370960-1

throughout its distribution. As water becomes more available, grain sorghum generally replaces pearl millet. Practically all the pearl millet grown is consumed directly as human food, with only minor amounts being fed to animals.

On a worldwide basis there are about equal areas planted to millets (all types) as grain sorghums. For the years 1967 through 1971, annual cultivation of pearl millet averaged 31 million hectares and production averaged 18 million metric tons. This accounted for 4.9% of the world's cultivated lands planted to cereals and 1.8% of total cereal production (Pomeranz, 1975). In a crop such as pearl millet, which is often grown as a crop of last resort, a low yield is not necessarily bad. A poor yield is infinitely better than no yield, for no yield means starvation. Thus a consistently low-yielding line is much preferred to a potentially high-yielding line that may fail completely in a poor year.

Pearl millet is generally planted in the spring after the first substantial rain of the rainy season. The crop is notoriously poor at emerging through a hard crust of soil. Thus, it is not planted in dry soil to await rain, which will provide moisture but will also give a hard crust to the soil and lead to poor stands. Planting after the rain will avoid the problem. Once the millet has emerged it generally is very tolerant of mother nature's whims and almost always will produce some seed.

It has only been in the last 20 years that plant breeders have made serious attempts at developing improved varieties of pearl millet (Brunken et al., 1977). The plant height varies widely and the variability in head type is quite large (Fig 1). The world collection of pearl millet is small compared to the land races available. An attempt to expand and maintain the world collection of pearl millet is under way at the International Crop Research Institute for the Semi-arid Tropics (ICRISAT) in Hyderabad, India. The highly cross-pollinated nature of pearl millet is a great problem with regard to maintenance of identity; genetic contamination at growouts has been considerable (van der Maesen, 1978).

Plant breeders have provided the impetus for studies on the chemistry, nutrition, and food uses of pearl millet. Their success in producing seeds with improved yields heightens concerns about the nutritional and functional quality of new pearl millet cultivars. The following paper provides a summary of pearl millet chemistry and utilization and delineates areas where more research is needed.

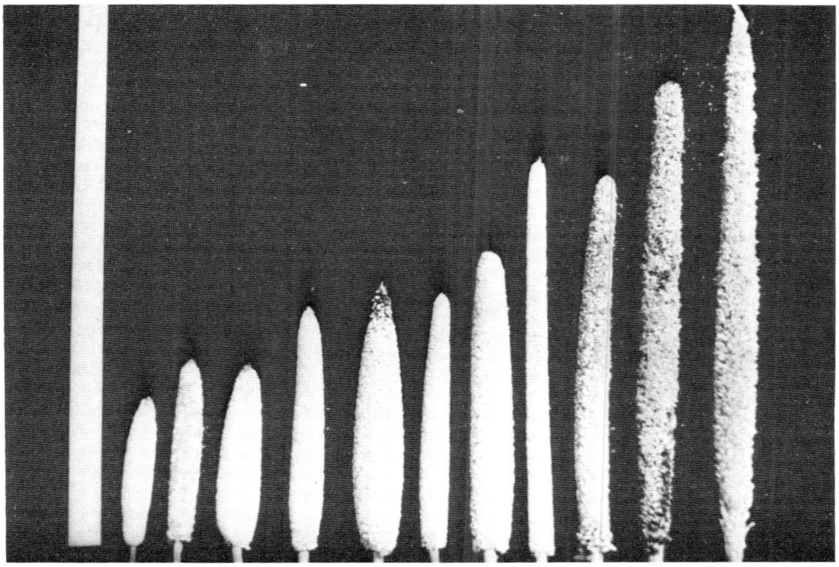

FIGURE 1. A selection of head types found in pearl millet. Courtesy of Dr. Gary Paulsen, K.S.U.

II. SIZE AND STRUCTURE OF THE GRAIN

Kernels of pearl millet are generally tear shaped and about 1/3rd the size of sorghum kernels; an average 100 kernel weight is about 0.89 g compared to about 2.8 g for sorghum (Freemen and Bocan, 1973). Different cultivars of millet vary widely in seed sizes (Fig. 2)

Gross structural features of fractured pearl millet kernels (Fig. 3) indicate that (a) the germ is large in proportion to the rest of the kernel and (b) in any given sample the ratio of hard to soft endosperm varies considerably. A closer view of the pericarp (Fig. 4) shows that it contains three layers: the epicarp, the mesocarp, and the endocarp. Unlike in sorghum, no starch granules have been found in millet mesocarp (Badi et al., 1976a; Sullins and Rooney, 1977). In addition, Sullins and Rooney (1977) found that the testa layer was absent in all 5 pearl millet samples examined.

FIGURE 2. An example of large (lower left), small (lower right), and medium (upper left) pearl millet seed sizes compared with grain sorghum (upper right).

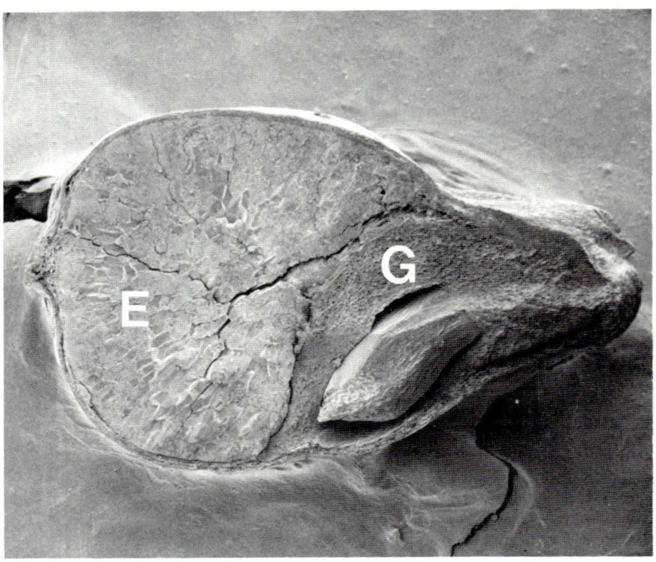

FIGURE 3. A fractured kernel of pearl millet showing the relatively large germ (G) to endosperm (E) ratio. Magnification 25x.

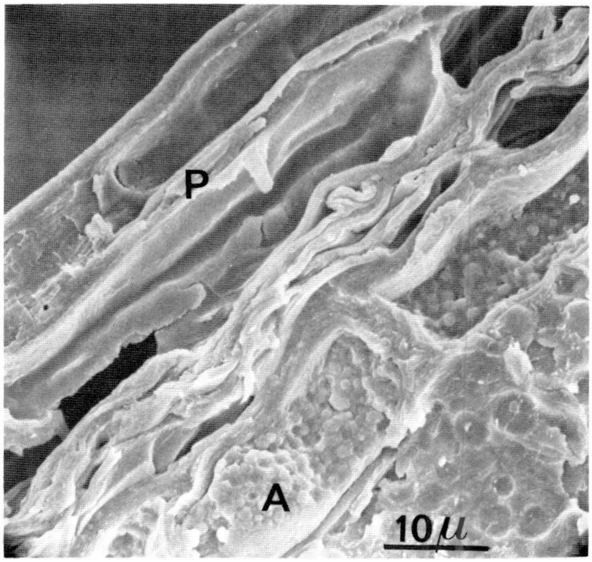

FIGURE 4. A SEM micrograph of the outer edge of pearl millet showing the pericarp (P) and aleurone (A) layer.

A single layer of aleurone cells surrounds the periphery of the kernel and is firmly attached to the pericarp. Those cells, along with the pericarp and germ, contain most of the lipid that is present in the grain (Lai and Varriano-Marston, 1979).

The hard endosperm (Fig. 5) contains polygonal starch granules embedded in a protein matrix which in turn contains protein bodies ranging in size from 0.3 to 4 μm (Badi et al., 1976a; Sullins and Rooney, 1977; Adams et al., 1976). Transmission electron micrographs (Fig. 6) indicate that these protein bodies have a well-defined internal structure as evidenced by the pattern of electron scattering.

The soft or floury endosperm of pearl millet contains loosely packed spherical starch granules (Fig. 7) and no protein bodies. Sullins and Rooney (1977) have suggested that the matrix protein concentration in the floury endosperm cells is higher than that of the same protein in the hard endosperm cells.

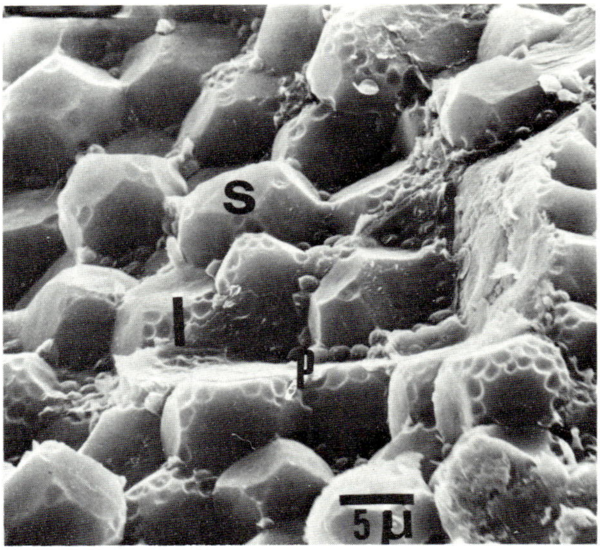

FIGURE 5. A SEM micrograph of the hard endosperm of pearl millet showing the starch (S), indentation in the starch (I), and protein bodies (P).

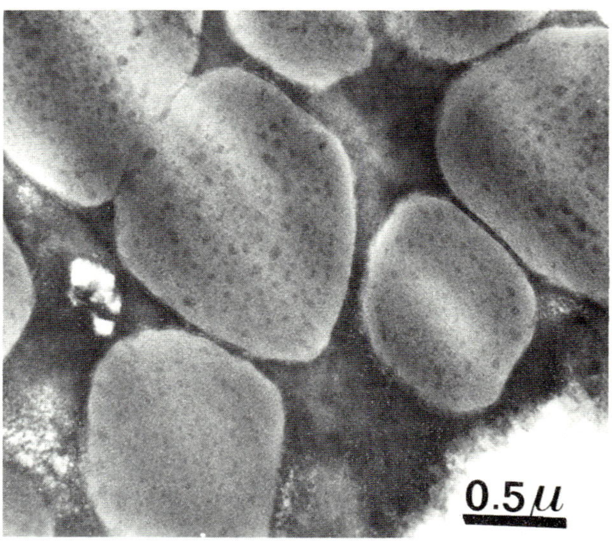

FIGURE 6. A TEM micrograph of the protein bodies from pearl millet.

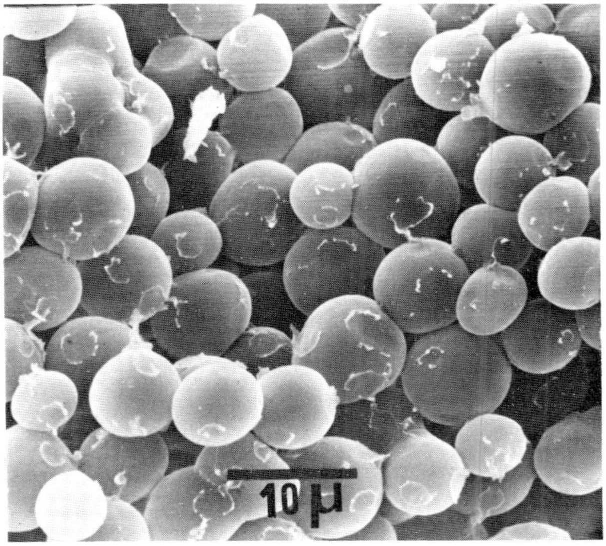

FIGURE 7. A SEM micrograph of the soft endosperm of pearl millet showing the spherical starch granules.

III. CHEMICAL COMPOSITION

A. Protein

It is always difficult, if not impossible, to make valid generalizations about the protein contents of various cereals. Factors that can affect the protein content of an individual sample are too numerous. For example, the environment, availability of nutrients (added or indigenous), the genetics of the sample, and the yield of the sample compared to its yield potential are all factors that affect a sample's protein content. Thus, to compare the protein content of individual samples of different grains is meaningless, and even when large numbers of samples are evaluated it is dangerous to draw firm conclusions. With those disclaimers, we will now say that pearl millet has high protein content when compared with samples of other small grains grown under similar conditions. Whether the higher protein content is inherent in pearl millet or the result of one or more of the above factors is open to question.

1. Effect of Nitrogen Application. Sawhney and Naik (1969) found that N applied at the rate of 88 Kg/ha increased the protein content of pearl millet (from 12.6 to 13.4%). Similar results (Deosthate et al., 1971) showed that nitrogen fertilization up to 120 Kg N/ha increased both yield and protein content. At higher levels of nitrogen, the yield decreased and the protein content remained essentially constant.

2. Amino Acid Composition. Representative amino acid composition data for pearl millet are given in Table 1.

Table 1. Range of Amino Acids in Pearl Millet Meals (g AA/16g N).

Amino Acid	High	Low
Tryptophan	1.9	0.8
Lysine	3.9	1.9
Histidine	3.1	1.6
Arginine	6.2	3.1
Aspartic Acid	11.3	5.0
Threonine	4.3	2.0
Serine	5.9	3.0
Glutamic Acid	25.4	11.6
Proline	11.8	3.9
Glycine	6.4	2.2
Alanine	8.8	6.4
½ Cystine	3.3	1.0
Valine	6.2	4.4
Methionine	2.6	1.2
Isoleucine	6.0	3.1
Leucine	12.7	7.6
Tyrosine	5.0	2.4
Phenylalanine	5.9	3.9

The data illustrates the wide variability found in pearl millet for most amino acids. For example, lysine ranges from 1.8 to 3.9 g/16 g N. The variability may be a reflection of the wide genetic base or the lack of man's effectiveness in breeding the cereal. Generally, pearl millet is comparable to the other cereal grains; it is deficient in lysine, tryptophan, threonine, and the sulfur containing amino acids. However, the data show that at least certain lines of millet are high in lysine and tryptophan when compared to other cereals.

The amino acid composition varies with different anatomical parts of the grain. Flour milled from pearl millet was found to be lower in lysine, arginine, and glycine, and higher in methionine than was the whole meal (Badi et al., 1976a). Conversely, the bran was high in lysine, arginine, glycine, and aspartic acid and low in glutamic acid and the sulfur-containing amino acids.

3. Protein Classes. Cereal proteins are generally classified according to the solubilities in a series of solvents. The different classes of proteins have different amino acid compositions and thus a knowledge of the amount of protein in each solubility class gives a rough estimation of the sample's amino acid composition. The alcohol-soluble prolamines are of special interest because they are generally poor in lysine and other nutritionally critical amino acids (Pomeranz, 1975). Prolamines are relatively rich in glutamine and proline. Because the prolamines are a major fraction of the grain protein, most cereals are deficient in lysine, tryptophan, etc. The development of high-lysine maize, barley, and sorghum has resulted from depressed synthesis of prolamines (Mertz, 1975). It is not known whether the high-lysine millets also have low prolamine.

Relatively little has been reported on the distribution of various protein classes in pearl millet or on the amino acid composition of different protein classes. Narayamurti and Aiyar (1930) extracted proteins from pearl millet meals with successive solvents of water, 6% salt, 70% cold ethanol, 70% hot ethanol, and 0.04% NaOH. Percentages of the total protein extracted with each solvent were 13%, 8.7%, 33.3%, 7%, and 9.1%, respectively. Thus, 71% of the protein in the meal was recovered in the extracts. Abhyander et al. (1939) reported similar results with similar solvents.

Sawhney and Naik (1969) studied the distribution of nitrogen in the protein fractions of several pearl millet varieties as fractionated with a modified Mendel-Osborne procedure (Naik, 1968). In all the samples analyzed, the prolamines and glutelins predominated, accounting for about 60% of the protein present. The albumin fraction averaged 15% and the globulins 9%. The proportions of the different fractions varied from cultivar to cultivar. The prolamine and glutelin fractions ranged from 21 to 38% and from 24 to 37%, respectively, of the total protein present. In the different samples, nitrogen recovery ranged from 72 to 92%.

Abhyander et al. (1939) isolated a prolamine and two globulin fractions from pearl millet and analyzed each fraction for 4 amino acids. One of the globulins coagulated at 43 to 45°C and the other at 82 to 85°C. The analysis for arginine, histidine, cysteine, and lysine showed considerable differences in the amino acid composition of the two globulins. The lysine content of the prolamine fraction (2.5%) was essentially equal to the lysine content of the total protein. Narayamurti and Aiyar (1930) extracted pearl millet with 70% ethanol and analyzed this prolamine fraction for essential amino acids. Surprisingly they found rather high levels of tryptophan (2.8g/16 g N) and moderate amounts of lysine (1.1g/16 g N).

Sawhney and Naik (1969) reported amino acid analyses (Table 2) for 4 protein solubility classes from one pearl millet sample.

Table 2. Amino Acid Composition of Different Protein Fractions of HB-1 Pearl Millet (Sawhney & Naik, 1969) (g AA/16g N).

Amino Acid	Albumin	Globulin	Prolamine	Glutelin
Aspartic Acid	12.46	11.42	7.48	6.36
Threonine	2.12	2.64	3.46	2.92
Serine	4.81	5.62	6.12	5.35
Glutamic Acid	16.36	18.71	22.24	21.68
Proline	8.82	7.46	10.23	9.42
Glycine	7.35	4.12	1.23	2.74
Alanine	8.11	6.02	9.41	10.42
Valine	5.64	7.34	3.24	4.12
Cysteine	2.92	2.62	1.42	1.24
Methionine	1.94	2.12	1.02	1.34
Isoleucine	4.12	2.48	3.32	3.83
Leucine	6.11	5.62	12.06	10.18
Tryosine	3.28	4.31	4.62	4.44
Phenylalanine	3.82	4.14	3.82	4.94
Lysine	3.92	3.74	1.66	2.14
Histidine	2.18	2.22	2.14	1.64
Arginine	4.36	4.82	3.04	4.92
Tryptophan	1.52	0.54	2.84	0.69

As expected, compared to the other two fractions, the albumins and globulins contained higher concentrations of basic amino acids such as lysine, arginine, and histidine as well as the sulfur-containing amino acids. Prolamines were found to be rich in glutamic acid, proline, and leucine as well as in tryptophan. Pearl millet prolamine apparently differs markedly from that of other cereals in being unusually high in tryptophan; although, like that of other cereals, pearl millet prolamine is rich in glutamic acid and proline and deficient in lysine.

The prolamines of proso millet were reportedly soluble in t-butyl alcohol (Jones et al., 1970). Treatment of pearl millet with t-butyl alcohol solubilized protein but did not solubilize the protein bodies presumed to be prolamines (Badi et al., 1976a). The authors concluded that pearl millet may contain two alcohol-soluble proteins.

B. Carbohydrates

1. Starch. As in all cereals, starch composes the major carbohydrate of pearl millet, ranging from 56 to 65% of the grain (Freeman and Bocan, 1973; Sheorain and Wagle, 1973). Starch granules range in size from 4 to 12 μm, and are polygonal and spherical in shape (Fig. 8).

FIGURE 8. A SEM micrograph of starch isolated from pearl millet.

Unlike in other cereal starches, the large hila is clearly visible (Fig. 9) in fractured polygonal granules (Badi et al., 1976b).

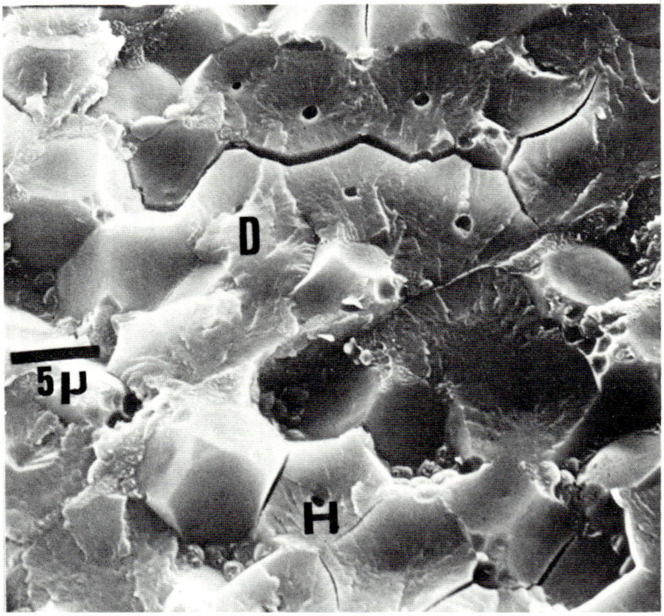

FIGURE 9. A SEM micrograph of the hard endosperm of pearl millet showing damaged starch (D) and the hila (H).

Some physicochemical characteristics of pearl millet starch are shown in Table 3. Iodine affinities of 4.43 to 5.05% suggest that millet starch contains an amylose content equal to that of most other cereals. Beleia et al. (1979) found amylose contents of 20 to 22% for 5 pearl millet cultivars. Badi et al. (1976b) reported 17% amylose in starch from one pearl millet population.

Initiation and end-point gelatinization temperatures vary among cultivars but fall within the ranges reported for sorghum and corn starches. Starch pasting properties also vary among cultivars (Table 3), particularly with respect to the viscosity after the 1-hr holding period at 95°C. At 95°C, Freeman and Bocan (1973) found a decrease in viscosity of the paste while Badi et al. (1976b) observed the opposite effect.

Table 3. Physicochemical characteristics of Pearl Millet Starch

Property	Reference[a] 1	2	3
Iodine affinity (% d.b.)	---	4.84,5.05	4.43-4.75
Amylose content (% d.b.)	17	---	20-22
Intrinsic Viscosity	---	1.74,1.50	---
Gelatinization temp. °C			
initiation	51	59,67	59-63
end-point	69	70,75	68-70
Amylograph pasting curves			
Pasting temp. °C	72	---	76.5
Peak viscosity (BU)	560	600,660	400-500
Viscosity 1 hr at 95°C (BU)	640	400,600	340-460
Viscosity after cooled to 50°C (BU)	780	660,760	500-640
Viscosity after 1 hr at 50°C (BU)	755	600,690	460-620
Cold water binding capacity (% d.b.)	---	---	83.6-99.5

[a]
1. Badi *et al.* (1976) using 50 g starch; 1 sample studied
2. Freeman and Bocan (1973) using 35 g starch; 2 samples studied
3. Beleia *et al.* (1979) using 50 g starch; 5 samples studied

Solubility and swelling power tests provide information on the behavior of pearl millet starch during heating. Figure 10 provides an example of the swelling and solubility patterns obtained for millets. Values for starch swelling power at

85°C range from 10.97 to 13.51; solubility varies from 3.2 to 7.89% (Freeman and Bocan, 1973; Beleia et al., 1979).

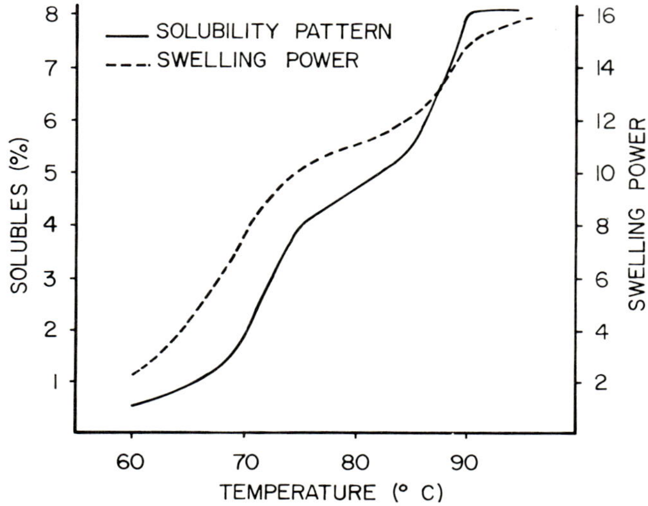

FIGURE 10. Solubility and swelling patterns of pearl millet starch.

Pearl millet flour has been reported to give low amylograph peak viscosities (Badi et al., 1976b; Beleia et al., 1979). Because millet starch gives a peak viscosity similar to that of sorghum and cornstarch, this implies that pearl millet flour has a high amylase activity. Studies on sound millet indicate that amylase activity is 8-15 times greater than that reported for sound wheat (Sheorain and Wagle, 1973; Varriano-Marston et al., 1979). The high activity observed in millet flour is due to a high alpha-amylase activity (Varriano-Marston et al., 1979).

Pearl millet starch has been shown to be more resistant than sorghum starch to attack by pancreatic alpha-amylase or by amyloglucosidase (Sullins and Rooney, 1977). However, the mode of attack of native amylases on pearl millet starch has not been thoroughly investigated. Studies on fermented millet meal indicate that the enzymes preferentially attack the spherical granules (Fig. 11) and that starch hydrolysis

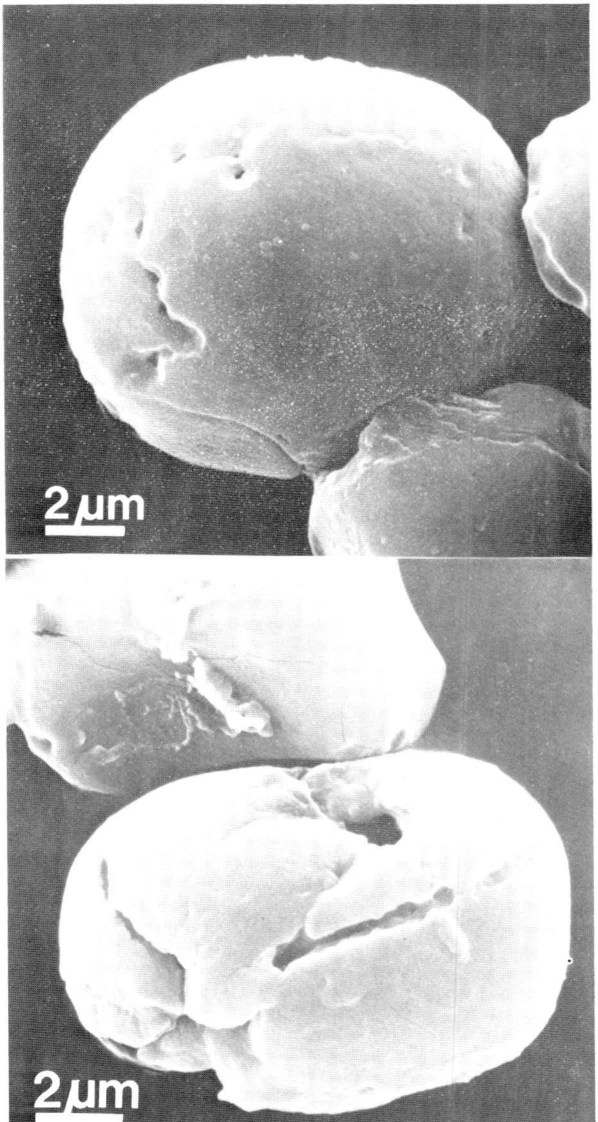

FIGURE 11. A SEM micrograph of an enzyme attack of a spherical starch granule.

is more vigorous at the center of the granule than at the periphery (Varriano-Marston et al., 1979). Others have also suggested that degradation of starch by amylases is more complete in the granule center than the outer portions (Sullins and Rooney, 1977).

2. Pentosans. Bailey et al. (1979) separated pentosans from pearl millet on the basis of their solubility. Pentosans soluble in 80% ethanol had relatively high glucose and galactose contents (28.4% and 18.5%, respectively). Galactose was the principal sugar in pentosans extracted with water at 50°C. Pentosans solubilized in water at room temperature or in 0.5 N alkali consisted predominantly of arabinose and xylose. In general, the water-soluble pentosans from millet contained less glucose than comparable fractions from grain sorghum.

The differences in sugar composition of the various pentosan fractions affects the configuration of the molecules. Sedimentation coefficients determined by Bailey et al. (1979) suggested that the pentosans soluble in water at R.T. or in alkali had branched structures while pentosans solubilized in water at 50°C has a linear configuration.

3. Other Carbohydrates. Although fiber may play a significant role in human nutrition, no information is available on the chemical structure or characteristics of fiber in pearl millet. Crude fiber values of whole pearl millet grain have been reported to range from 1.96% to 3.88% (Carr, 1961; Reichert and Youngs, 1977; Desikachar, 1977). Reichert and Youngs (1976) dehulled pearl millet on a barley pearler, an abrasive mill, an attrition mill, or with a wooden pestle and reported crude fiber values of 0.58, 0.72, 0.96, and 1.12%, respectively.

Data on the types and levels of free sugars present in pearl millet are limited. Lineback and Pompipom (1978) studied one pearl millet population and reported a total free sugar content of 10 mg glucose/g flour.

C. Lipids

1. Composition and Characteristics. The free lipid content of pearl millet cultivars varies from 3.03% to 7.40% (Sharma and Goswami, 1969; Goswami et. al., 1969a & b; 1970a & b; Rooney, 1978; Lai and Varriano-Marston, 1979). American cultivars generally have a higher free lipid content than those from Africa or India.

Acidity and saponification values of free lipids from pearl millet varied from 2.02 to 11.95 mg/100 g and 190 to 333 mg/g, respectively (Sharma and Goswami, 1969; Agawal and Sinka, 1964).

The composition of the free lipid fraction from pearl millet has also been studied using thin-layer chromatography (Pruthi and Bhatia, 1970; Badi et al., 1976a; Lai and Varriano-Marston, 1979). Triglycerides were found to be the major components of the free nonpolar lipids (Fig. 12a).

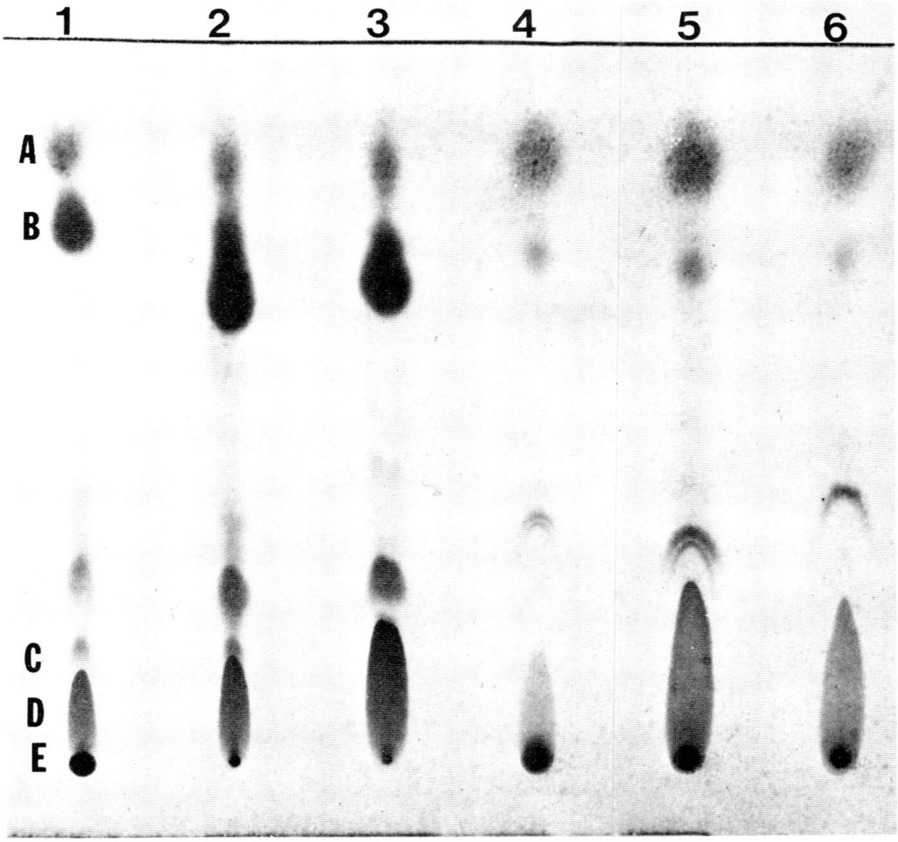

FIGURE 12a. Thin-layer chromatograms of nonpolar free and bound lipids from wheat, sorghum, and millet. Pattern 1, free wheat lipids; pattern 2, free sorghum lipids; pattern 3, free millet lipids; pattern 4, bound wheat lipids; pattern 5, bound sorghum lipids; and pattern 6, bound millet lipids. The spots are tentatively identified as A, hydrocarbons and steryl esters; B, triglycerides; C, digeycerides; D, free fatty acids; and E, unresolved polar lipids.

Monogalactosyl diglyceride was the major component of the free polar lipids (Fig. 12b).

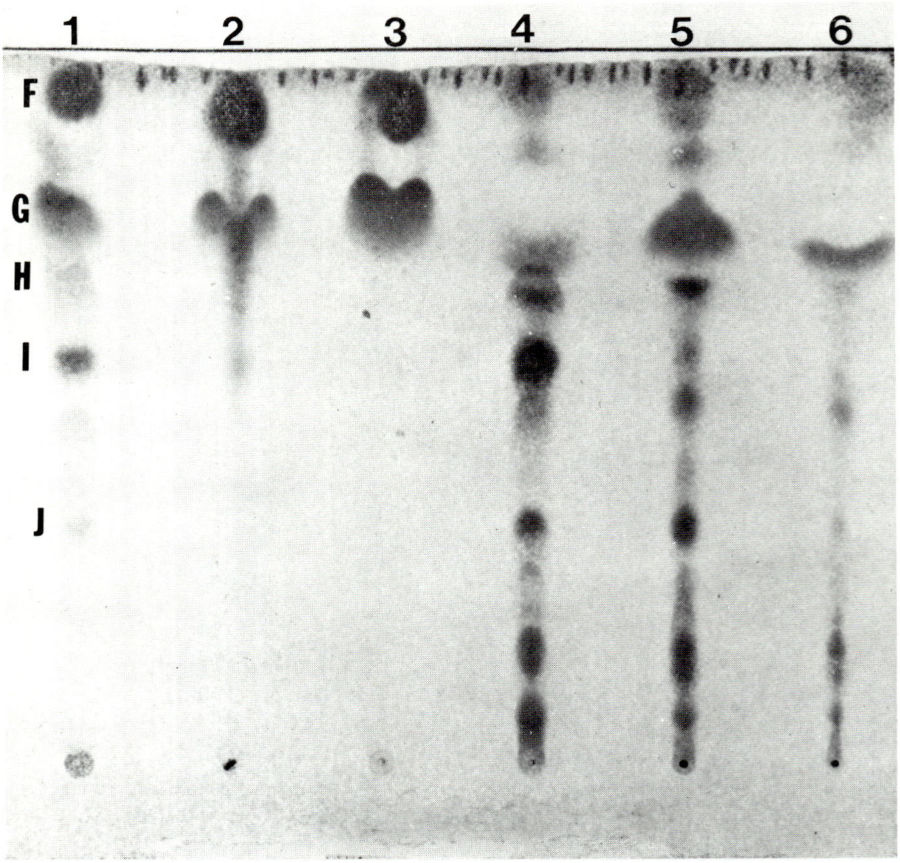

FIGURE 12b. Thin-layer chromatograms of polar free and bound lipids from wheat, sorghum, and millet. Pattern 1, free wheat lipids; pattern 2, free sorghum lipids; pattern 3, free millet lipids; pattern 4, bound wheat lipids; pattern 5, bound sorghum lipids; and pattern 6, bound millet lipids. The spots are tentatively identified as F, unresolved nonpolar lipids; G, monogalactosyl diglyceride; H, phosphatidyl ethanolamine; I, digalactosyl diglyceride; and J, phosphatidyl choline.

The fatty acids present in the free lipids were studied by Agawal and Sinka (1964) using a urea adduct method. They found that oleic acid was the major fatty acid (53.84%), followed by linoleic (34.88%), palmitic (10.8%), stearic (0.28%), and myristic (0.20%). A bromination method indicated that there was no linolenic acid present in the free lipid fraction.

On the other hand, Jellum and Powell (1971), using gas chromatography, studied 65 lines of pearl millet and found that linoleic acid was the major fatty acid present (40.3 to 51.7%) in the free lipid fraction. Oleic acid ranged from 20.2 to 30.6%; palmitic acid ranged from 17.7 to 25.0%. In addition, these authors showed that pearl millet lines contained on the average 3.69% linolenic acid, 3.9% stearic acid, 0.64% arachidic acid, and 0.55% palmitoleic acid. The data of Jellum and Powell (1971) show that differences in fatty acid values reported in the literature for pearl millet not only are due to genetic differences and seed maturity but also are affected by lipid extraction procedures.

Little work has been published on the bound lipids in pearl millet. Pruthi and Bhatia (1970) studied two Indian varieties and found that the lipids extracted with water-saturated n-butanol composed 0.5% of the millet grain. On the other hand, Lai and Varriano-Marston (1979) reported bound lipid contents of 0.58%-0.90% for 18 samples of pearl millet grown in Kansas.

The major lipid components of the polar, bound lipid fraction are phosphatidyl choline, sterol glycosides,and di- and monogalactosyl glycerides. Acylglycerols were the major component of the nonpolar, bound lipids (Pruthi and Bhatia, 1970; Badi et al., 1976a; Lai and Varriano-Marston, 1979).

2. Changes During Storage. Pearl millet quality quickly deteriorates once it has been ground into a meal (Reddy, 1978). Although studies on the storage of millet flour have been conducted, the types of millet under study were not clearly identified (Carovale and Quaglia, 1973; Feron and Bouquet, 1948; Nechaev et al., 1973).

Alterations in the lipid components may be responsible for the deterioration in millet flour quality during storage. Carnovale and Quaglia (1973) found a 3% decrease in the amount of ether extractable lipids after 3 mo. storage of millet flour at $30^{\circ}C$, 95% R.H. However, the authors did not account for the "lost" lipids. In addition, their results contradict the data of Nechaev et al. (1973) who showed essentially no change in total ether extractable lipids in millet during storage.

Carnovale and Quaglia (1973) also noted changes in fatty acid composition in their millet flours during storage. Palmitic acid levels decreased 2.2 to 4.5% while levels of stearic, oleic, and linolenic acids declined to a lesser degree during the storage period. Conversely, linoleic acid levels in the free lipid fraction increased during storage. Infrared studies on the extracted lipids showed increases in the absorption of the -COOH group, indicating an increase in free acidity level. In general, their data suggest that hydrolytic as well as oxidative changes in lipid components occur during storage. These changes were more rapid and pronounced in lower grade flours.

Work by Thiam et al. (1976) suggests that although lipolytic activity occurs in millet flour during storage at 30°C, 90-95% R.H., the major cause of quality deterioration is microbial fermentation. Propagation of microorganisms during storage resulted in the consumption of sugars and the liberation of alcohols. Essentially no lipoxygenase activity was found.

D. Vitamins and Minerals

1. Minerals. Ash content in pearl millet lines varies from 1.46% to 3.88% (Goswami et al., 1969 a & b, 1970 a & b; Deosthale et al., 1971; Burton et al., 1972; Varriano-Marston and Hoseney, 1979). These values are high compared to those for wheat, corn, or sorghum (Casey and Lorenz, 1977).

Some of the mineral constituents of pearl millet have been identified; a summary of those data is presented in Table 4. The values for Na, Ca, Mg, and Cu fall within the range reported for wheat. However, some varieties of millet contain very low levels of Ca. The Mn and Mo contents of millet are considerably lower than those found in wheat while in many cases the P, K, and Fe contents of the grain are considerably higher than values reported for wheat. High Fe values may be due, in part, to the equipment used to mill the grains prior to analysis. For example, Carr (1961) reported an Fe content of 46 mg/100 g of grain for pearl millet ground on a stone containing high Fe levels.

X-ray energy dispersive analysis has been used to study the location of minerals in pearl millet grains (Varriano-Marston and Hoseney, 1979). High levels of Si and K were found in the covering layers while P was concentrated mainly in the germ. About equal amounts of Fe were found in the germ and covering layers. The endosperm portions exhibited low mineral concentrations; the predominant elements detected were S, K, and Fe.

Table 4. Minerals in Pearl Millet (P. americanum)[a]

Element	mg/100g (d.b.)
Na	2.2-4.9
K	366-543
Ca	7-117
Mg	71-156
P	631-1353
Fe	2.1-11.7[b]
Cu	0.42-0.58
Mn	0.66-0.92
Zn	0.10-3.8
Mo	0.014-0.024

[a]Data obtained from: Carr, 1961, Goswami et al., 1969 a,b; Shah and Mehta, 1959; Burton et al., 1972; Varriano-Marston and Hoseney, 1979.

[b]Carr (1961) reported an Fe content of 46 mg/100 g of grain for pearl millet ground on stones containing high levels of Fe.

2. Vitamins. Little work has been reported on the vitamin content of pearl millet. Carr (1961) reported thiamine and riboflavin contents of 0.22 mg/100 g and 0.26 mg/100 g, respectively, for one sample of millet. Current work in the authors' laboratory indicates that there is little variability in riboflavin content among cultivars; 18 cultivars showed riboflavin contents ranging from 0.21 to 0.38 mg/100 g (unpublished data). These values are similar to those published for wheat.

No journal articals have been published on the levels of niacin, vitamin A, or vitamin E in pearl millet. However, Hariharan et al. (1965) reported a vitamin A content of 153 I.U./100 g for their "poor pearl millet diet."

E. Pigments

Pearl millet varies widely in color from an off-white to dark brown. The most common colors are yellow and slate grey. In general the slate grey is preferred.

Color plays a role in consumer acceptance of millet grains. Therefore, characterization of the pigments in pearl millet is an important step in the study of flour quality.

In some parts of Africa, millet grains are soaked overnight in water containing sour milk or tamarind pods (Vogel and Graham, 1979). This step not only alters the flavor of the resultant products but also whitens the grain. Reichert and Youngs (1979) investigated the phenomenon of millet bleaching and found that the discoloration was pH dependent. Soaking in a solution of 0.2N HCl or citric acid was as effective as sour milk or tamarind pods in bleaching the grain pigments. Further studies by Reichert (1979) suggested that C-glycosylflavonoids (glucosylvitexin, glucosylorientin, and vitexin) were responsible for the natural grey color of pearl millet grains, and that these pigments were concentrated in the peripheral regions of the grain.

IV. NUTRITIONAL QUALITY

In some developing nations, a high percentage of the calorie intake of individuals comes from cereal grains. This great dependence on cereal grains in the diet has fostered investigations into their nutritional quality.

Information on the nutritional quality of pearl millet for human food has been extrapolated from studies conducted with weanling albino rats. The use of rats as a test animal is open to question. The rat's much higher protein requirement, as a result if its faster growth rate, is one factor that must be considered when extrapolating rat data to humans. Another factor is the relative inactivity of rats in test cages. The shortcomings of using rats as test animals are dealt with quite well by Vaghefi et al. (1974). Generally, protein efficiency ratios (PER) and weight gains are used as guides to nutritional quality. It must be remembered that PER is only a measure of the quality (amino acid balance) of the protein. It does not measure the usefulness of a sample or type of grain as a food. For PER value to have meaning, other factors such as vitamins or minerals in the diet must not be limiting and the protein content of the diet must be 10% or less. Unfortunately, in some cases, poor methodology may only provide erroneous results, particularly when PER studies are conducted using dietary protein levels above 10%.

The PER's for pearl millet alone or supplemented with amino acids or other sources of plant protein are summarized in Table 5. The PER for unsupplemented pearl millet is higher than that reported for wheat or sorghum (Pushpamma et al.,

Table 5. Protein Efficiency Ratios of Pearl Millet Alone or Supplemented with Other Nutrients.

Diet	Level of protein in the diet (%)	PER	Reference
Pearl millet			
	11	1.67	Hariharan *et al.*, 1965
	5	1.3	Rao *et al.*, 1964
	10	1.4	Rao *et al.*, 1964
	10	1.8	Rao *et al.*, 1964
	11	1.84	Pushpamma *et al.*, 1972
Pearl millet + amino acids			
none	11	1.84	Rao *et al.*, 1964
Lysine	11	2.41	Pushpamma *et al.*, 1972
Tryptophan	11	1.83	Rao *et al.*, 1964
Lys + Trp	11	2.34	Rao *et al.*, 1964
Lysine	10	3.41	Daniel *et al.*, 1965
Lys + threonine	10	3.54	Daniel *et al.*, 1965
Pearl millet + plant protein			
Bengal gram	10	2.16	Rao *et al.*, 1964
Red, black + green gram	10	2.05-2.10	Rao *et al.*, 1964
Coconut cake	10	1.60	Rao *et al.*, 1964
Amaranth leaf	10	1.73	Rao *et al.*, 1964
Grams + amaranth leaf	10	2.04-2.30	Rao *et al.*, 1964

1972; Rao et al., 1964; Oke, 1977). The addition of lysine to a pearl millet diet results in a dramatic increase in the PER. This is commensurate with the fact that, in many millet cultivars, lysine is the limiting amino acid. The amino acid profiles in millet may be complemented by other plant proteins. Including legumes in the diet significantly improves the PER (Table 5).

The effect of vitamin and mineral supplements on the growth rate of rats fed pearl millet is shown by Hariharan et al. (1965). The data of Daniel et al. (1965) show that with diets inadequate in minerals and vitamins the supplementary value of lysine was not realized.

Milling processes and products influence pearl millet PER and digestibility (Goussault and Adrian, 1977). The traditional wooden mortar milling of pearl millet gives a product which has a PER equal to that of the original grain. The SOTRAMIL process decreases the PER from 1.83 to 1.75 going from the whole grain to the flour. Dehusking and decortication by the SEPIAL process progressively lowered the PER.

Most of the PER studies on pearl millet have been done on the raw grain. Recently Dassenko (1979) has shown that cooking millet in the form of chapaties lowers the PER. She found the reduction in PER to be much greater when chapaties were made from the whole meal rather than from 67% extraction flour.

Because energy is a basic need of man, one can measure the nutrient requirements based on calories (Hansen, 1973). Oke (1977) has applied this technique to pearl millet and finds that millet satisfies the requirements for protein, iron, and vitamins B_1 and B_2. Thus, as has been shown by millions of people in India and Africa, pearl millet is an excellent food.

V. MILLING

A number of traditional and mechanical processes are used to mill pearl millet. Those processes not only alter the nutritional quality of the products (as was mentioned above) but also often alter the meal's end-use performance.

A. Traditional Processes

The traditional wooden mortar and pestle method of milling pearl millet (Fig. 13) is still used in the rural areas of many developing countries (Carr, 1961; Reichert and Young, 1977; Goussault and Adrian, 1977; Adrian et al., 1976). The grain plus a small amount of water (10 to 20%) is placed in the mortar and the grain is dehulled by a pounding action. After drying, the bran and fine material are removed by winnowing with a closely woven basket (Fig. 14). Home grinding mills such as the round mill stones shown in Figure 15 may .

then be used to grind the grain into a fine flour (Vogel and Graham, 1979). Other grinding methods such as flat stones and additional pounding with mortar and pestle are also used. The analysis for a flour produced by these traditional processes was 10.3% protein, 3.55% cellulose plus lignin, and 2.7% lysine (Goussault and Adrian, 1977). An excellent description of traditional processing of cereal grain in Southern Rhodesia is given by Carr (1961).

FIGURE 13. A lady using a mortar and pestle to dehull pearl millet. Courtesy of Richard Vanderlip, KSU.

FIGURE 14. Winnowing pearl millet. Courtesy of Richard Vanderlip, KSU.

FIGURE 15. Grinding grain with a hand operated stone mill.

B. Mechanical Milling Processes

The traditional milling processes are laborious. Most villages have grinders, usually diesel powered, that are hammer or burr type mills (Vogel and Graham, 1979). Thus for a fee or percentage of grain a villager can get his grain ground to a whole meal. The fineness of the grind will depend more on the grinder available than on the consumer desires. However, this method is very time saving and much less work than hand pounding.

Several industrial milling processes have been proposed for producing pearl millet flour (Goussalt and Adrian, 1977). The SOTRAMIL process was developed in Zinder, Niger. In this process, washed grain is graded and then dehusked on a horizontal millstone (Bavaria dehusking machine). An attrition apparatus is then used to grind the grain, and the flour is separated by sifting. The extraction rate is 65-75%, and the particle size averages 40 μm. The process reduces the cellulose and lignin content from about 4.50% in the whole grain to 1.00% in the flour, and the protein content drops about 2%. However, the level of lysine in the flour is almost the same as that in the whole grain.

The French SEPIAL process is another method that has been used to dehusk pearl millet (Goussault and Adrian, 1977). Dehusking is achieved by placing dampened grains into an apparatus where the rotation of a vertical arm equipped with paddles results in the removal of the pericarp by friction of the grains against each other. After the husks have been mechanically separated from the grains, the husked grain (approximately 92.5% of the total grain) can be further decorticated to remove a protein fraction (aleurone and parts of germ and scuttelum) representing 8.5% of the grain. Decorticated grain can then be ground to yield a meal that represents 80% of the total grain. The decorticated grain contains about the same protein content as the total grain but less cellulose and lignin and less lysine.

Other types of equipment have been used to mill pearl millet. Perten (1977) studied the applicability of the Buhler laboratory roller mill MLU 202 for processing pearl millet and sorghum, and found that wheat milling technology was not so suitable as were other methods for these grains. He concluded that the millet must be decorticated prior to flour production because millet bran and germ are easily pulverized. The results are high ash and fat values in the flour fractions. In addition, the wheat milling equipment produced coarser flour than did grinding the decorticated grain in an attrition or hammer mill.

The Palyi-Hansen milling system (deMan et al., 1973) consists basically of two parallel plates provided with band saw blades set in a radical fashion around a circular strip on the plates. One of the plates is fixed to the housing of the huller and the other one to a rotating drum. Decortication of millet typically gives about 76% debranned grain, with ash, fiber, protein, and fat all decreased in relation to the original grain (Rasper, 1977).

In comparison of attrition and abrasive mills Reichert and Youngs (1976) concluded that the abrasive mill was generally more suitable for dehulling sorhgum and millet. Their conclusion was based on loss of color per loss of material weight as well as kernel-cracking analysis. The chemical composition of products produced on the above mills was compared with that of grains dehulled traditionally by mortar and pestle (Reichert and Youngs, 1977). In general, mechanically dehulled grains lost more oil, ash, and protein than did traditionally dehulled grain at the same extraction. Crude fiber was removed more efficiently with an abrasive mill than with an attrition or the traditional method.

VI. FOOD USES

Pearl millet used as food can take many forms and names. However, if we look at the foods carefully we find a relatively few general types (Vogel and Graham, 1979).

A. Unleavened Bread

The chapati or roti of India is similar to the well known wheat chapati. However, millet flour does not possess the dough-forming porperties of wheat flour. Thus, considerable skill is required to produce a thin, flexible chapati that will remain soft for one day. The dough is formed and baked on a hot griddle. An ideal product will puff to give a semi-light texture. In some areas hot water is used to partially gelatinize the starch and thus make the chapati easier to work with; however, the final product is not so light or desirable. The ideal product is light, flexible, uniformly round, and firm but not tough to the bite and will retain these properties for one day. Generally, an ideal chapati is not made from millet.

B. Leavened Bread

These products are not leavened with yeast or chemicals but generally with steam. The Sudanase kisra is an example of this product. The Sudanase generally use sorghum flour, but millet can be and is used. The flour-water mixture is often allowed to set overnight to produce a nice sour taste from the lactic acid bacteria. A starter is often used. The mixture should be a very thin batter. It is quickly spread to a thin layer on a hot (150°C) iron plate. Within a few seconds it can be stripped from the plate to give a thin moist sheet. The kisra is generally light, in both color and texture, moist (30 to 50% moisture), quite flexible, and pleasantly acidic in taste. The starch is gelatinized and there is no gritty taste.

C. Rice Substitute Products

Dehulled, whole, or cracked grains are steamed and consumed similarly to rice. Although not a rice substitute, couscous of Senegal is a steamed agglomerated product. Liquid is gradually added to millet flour to form small beads or agglomerates. The agglomerates either are steamed or are dried first and then steamed.

D. Thin Porridge

Thin porridges are called Uji or a host of other names. Texture depends on flour particle size and whether the product is fermented or not. Many types of seasoning can be used. The carboxylic acids produced during fermentation have been identified (Banigo and Muller, 1972).

E. Thick Porridge

Thick porridges are called tuwo and many other names. Products are frequently moulded or shaped. They are much like thin porridges except that less water is used. Therefore, the porridge should be stiff enough to mould smooth without cracks but should not stick to the fingers or the roof of the mouth, or crumble in the hand. Thick porridges act as main dishes in many countries and may be served more than once a day.

F. Nonalcoholic Beverages

Nonalcoholic beverages are popular in Sudan. They are generally made from sorghum, but millet could be used. They are made essentially by the process used for kirsa except that the batter is thinner and the product is dried to a thin flake. The dry flake is then soaked in water to produce the drink. Often malted flour is used as part of the starting material, various spices are added, and the mixture is fermented for 3 to 5 days.

G. Beer

Large amounts of sorghum and millet are consumed as beer. The grains are used both as a source of fermentable carbohydrate and as an enzyme source. The homemade beers of Africa are generally made with no added yeast and thus depend on wild yeast and other organisms to produce alcohol. Along with the alcohol, lactic acid is of course produced, which lowers the pH and helps give the product stability. The beer often has a high solid content and is quite nutritious.

In South Africa the sorghum brewing industry is quite large, with yearly production equal to 10^9 liters (Novellie, 1977), and that value does not include beer made at home. No millet is used in South Africa, but its use increases as one goes north. Pearl millet is used mainly for malting.

In an evaluation of pearl millet, both as a malting grain and as an adjunct, Singh and Tauro (1977) concluded that it was a satisfactory adjunct but lacked sufficient enzyme activity to replace barley as a malting grain. Later work from the same laboratory (Dhamija and Singh, 1978) showed that use of 25% pearl millet as an adjunct at levels up to 25% of the total maize adjunct gave comparable beers, both analytically and organoleptically.

H. Composite Flours for Breadmaking

The use of pearl millet flour to partly replace wheat flour has met with reasonable success. Generally, there is a rather large drop in loaf volume as the percentage of millet is increased; however, all reports indicate that bread containing millet flour has excellent flavor (Dendy et al., 1970; Basse, 1978; Badi et al., 1976b; Casier et al., 1977).

I. Snack Foods

Traditional snack foods are limited; however, roasted grain or popped grain would both certainly qualify. Use of pearl millet as a substitute for all or part of wheat flour in cookies has been reported (Badi and Hoseney, 1976). Work at the Food Research Center in Khartoum, Sudan, has also stressed the production of snacks from sorghum and millet (Badi et al., 1978).

REFERENCES

Abhyander, V. S., Kotastane, W. V., and Naryana, N. (1939). Proc. Indian Acad. Sci. section B, 9, 126.

Adams, C. A., Novellie, L., and Liebenberg, N. V. D. (1976). Cereal Chem. 53, 1.

Adrian, J., Goussault, B., Arnal-Peyrot, F., Samson, M. F., and S.E.P.I.A.L. (1976). Agron. Tropicale 30, 43.

Agawal, P. N., and Sinka, N. S. (1964). Indian J. Agron. 63, 288.

Badi, S. M., Hoseney, R. C., and Casady, A. J. (1976a). Cereal Chem. 53, 478.
Badi, S. M., Hoseney, R. C., and Finney, P. L. (1976b). Cereal Chem. 53, 718.
Badi, S. M., and Hoseney, R. C. (1976). Cereal Chem. 53, 733.
Badi, S. M., Perten, H., Mustafa, A. I., and Abert, P. (1978). Abstract A5-2. Cereal Foods World 23, 485.
Bailey, A. V., Sumrell, G., and Burton, G. W. (1979). Cereal Foods World 24, 23.
Banigo, E. O. I., and Muller, H. G. (1972). J. Sci. Fd. Agric. 23, 101.
Basse, Marie-Therese (1978). Ceres. May-June. p. 38.
Beleia, A., Varriano-Marston, E., and Hoseney, R. C. (1979). Cereal Chem. (submitted).
Brunken, Jere, DeWitt, J. M. J., and Harlan, J. R. (1977). Econ. Bot. 31, 163.
Burton, G. W., Wallace, A. T., and Rachie, K. O. (1972). Crop Sci. 12, 187.
Carovale, E., and Quaglia, G. B. (1973). Ann. Technol. Agric. 22, 371.
Carr, W. R. (1961). Brit. J. Nutr. 15, 339.
Casey, P., and Lorenz, K. (1977). Baker's Digest 51(1), 45.
Casier, J. P. J., de Paepe, G., Willems, H., Goffings, G., and Noppen, H. (1977). In "Proceedings of a Symposium on Sorghum and Millets for Human Food" (D. A. V. Dendy, ed.), p. 127. Tropical Products Institute, London.
Daniel, V. A., Leela, R., Subramanyarajurs, T. S., Rao, S. Venkat, Hariharas, K., Rajalakshimi, M., and Parpia, H. A. B. (1965). J. Nutr. Dietet. 2, 134.
Dassenko, S. (1979). Ph.D Thesis, Kansas State University, Manhattan, Kansas.
de Man, J. M., Banigo, E. O. I., Rasper, V., and Gade, H. (1973). Can. Inst. Food Sci. Technol. 6, 188.
Dendy, D. A. V., Clarke, P. A., and James, A. W. (1970). Tropical Sci. 12, 131.
Deosthale, Y. G., Rao, K. V., Nagarajan, V., and Pant, K. C. (1971). Ind. J. Nutr. Dietet. 8, 301.
Desikachar, H. S. R. (1977). In "Proceedings of a Symposium on Sorghum and Millets for Human Food" (D. A. V. Dendy, ed.) p. 41. Tropical Products Institute, London.
Dhamija, S. S., and Singh, D. P. (1978). J. Food Sci. Tech. 15, 197.
Feron, F., and Bouquet, F. (1948). Anales de la Nutrition et de L'alimentation. 2, 537.
Freeman, J. E., and Bocan, B. J. (1973). Cereal Sci. Today 18, 69.
Goswami, A. K., Sharma, K. L., and Sharma, K. P. (1969a) J. Nutr. and Dietet. 6, 287.

Goswami, A. K., Sharma, K. P., and Gupta, B. K. (1969b). J. Nutr. and Dietet. 6, 291.
Goswami, A. K., Sehgal, K. L., and Gupta, B. K. (1970a). Ind. J. Nutr. and Dietet. 7, 5.
Goswami, A. K., Sehgal, K. P., and Sehgal, K. L. (1970b). Ind. J. Nutr. and Dietet. 7, 67.
Goussault, B., and Adrian, J. (1977). In "Proceedings of a Symposium on Sorghum and Millets for Human Food" (D. A. V. Dendy, ed.). p. 13. Tropical Products Institute, London.
Hansen, R. G. (1973). Nutr. Rev. 31, 1.
Hariharan, K., Subramanyarajurs, T. S., Desai, B. S. M., Rao, S. Venkat, Rajalakshmi, D., Swaminathan, M., and Parpia, H. A. B. (1965). J. Nutr. Dietet. 2, 196.
Jellum, M. D., and Powell, J. B. (1971). Agron. J. 63, 29.
Jones, R. W., Beckwith, A. C., Khoo, U., and Inglett, G. E. (1970). J. Agr. Food Chem. 18, 37.
Lai, C. C., and Varriano-Marston, E. (1979). Cereal Chem. (submitted).
Lineback, D. R., and Ponpipom, S. (1978). Die Stärke. 29, 52.
Mertz, E. T. (1975). In: "Protein Nutritional Quality in Foods and Feeds." (M. Friedman, ed.), p. 1. Marcel Dekker, New York.
Naik, M. S. (1968). Indian J. Genet. 28, 142.
Narayamurti, D., and Aiyar, C. V. R. (1930). J. Indian Chem. Soc. 7, 945.
Nechaev, A. P., Busareva, N. N., Denisenko, Ya. I., and Kuznetsov, D. J. (1973). Applied Bioch. & Micro. 9, 617.
Novellie, L. (1977). In "Proceedings of a Symposium on Sorghum and Millets for Human Food" (D. A. V. Dendy, ed.), p. 73. Tropical Products Institute, London.
Oke, O. L. (1977). In: "Proceedings of a Symposium on Sorghum and Millets for Human Food" (D. A. V. Dendy, ed), p. 121. Tropical Products Institute, London.
Perten, H. (1977). In "Proceedings of a Symposium on Sorghum and Millets for Human Food." (D. A. V. Dendy, ed.), p. 53. Tropical Products Institute, London.
Pomeranz, Y. (1975). In: "Protein Nutritional Quality of Foods and Feeds" (M. Friedman, ed.), p. 13. Marcel Dekker, New York.
Pruthi, T. D., and Bhatia, I. S. (1970). J. Sci. Fd. Agric. 21, 419.
Pushpamma, S., Parrish, D. B., and Deyoe, C. W. (1972). Nutr. Reports Intern. 5, 93.
Rao, S. Venkat, Swaminathan, M., and Parpia, H. A. B. (1964). J. Nutr. Dietet. 1, 128.
Rasper, V. F. (1977). In: "Proceedings of a Symposium on Sorghum and Millets for Human Food" (D. A. V. Dendy, ed.) p. 59. Tropical Products Institute, London.

Reddy, Geervani (1978). Home Science College, Hyderabad, India. Personal communication.
Reichert, R. D., and Youngs, C. G. (1976). Cereal Chem. 53, 829.
Reichert, R. D., and Youngs, C. G. (1977). Cereal Chem. 54, 174.
Reichert, R. D., and Youngs, C. G. (1979). Abstract, Cereal Foods World 24, 23.
Reichert, R. D. (1979). Cereal Chem. (submitted).
Rooney, L. W., (1978). Cereal Chem. 55, 584.
Sawhney, S. K., and Naik, M. S. (1969). J. Genetics and Plant Breeding 29, 395.
Shah, H. C., and Mehta, B. V. (1959). Soil Sci. 87, 320.
Sharma, K. P., and Goswami, A. K. (1969). J. Nutr. and Dietet. 6, 316.
Sheorain, V. S., and Wagle, D. S. (1973). J. Fd. Sci. Tech. 10, 184.
Singh, D. P., and Tauro, P. (1977). J. Food Sci. Tech. 14, 255.
Sullins, L. W., and Rooney, R. D. (1977). In "Proceedings of a Symposium on Sorghum and Millets for Human Food" (D. A. V. Dendy, ed.), p. 91. Tropical Products Institute, London.
Thiam, A. A., Drapron, R., and Richard-Molard, D. (1976). Ann. Technol. Agric. 25, 253.
Vaghefi, S. B., Makdani, D. D., and Mickelsen, O. (1974). Amer. J. Clinical Nutr. 27, 1231.
van der Maesen, L. J. G. (1978). Genetic resources at ICRISAT. International Crop Research Institute for the Semi-Arid Tropics, Hyderabad, India.
Varriano-Marston, E., and Hoseney, R. C. (1979). Cereal Chem. (submitted).
Varriano-Marston, E., Huang, Grace, and Gottneid, D. (1979). Cereal Chem. (submitted).
Vogel, S., and Graham, M. (1979). "Sorghum and Millet: Food Production and Use", International Development Res. Center, Ottawa, Canada.

THE PENTOSAN - STARCH RATIO IN RELATION TO QUALITY OF MILLED RYE PRODUCTS*

D. Weipert and H. Zwingelberg

1. INTRODUCTION

Although rye in world cereal cultivation does not occupy the position like that of wheat, the knowledge of technological properties of rye is not just only of academic value. Yearly about 23.8 Mio.t. of rye are produced in the world (FAO-Year book 1977) - that is only 6% in comparison with the amount of wheat produced. The major cultivation and from all the major consuming sectors are in West and North Europe and U.S.S.R. Alone Europe shares 56.7% and U.S.S.R. a further 35.6% of world rye quantity. Here also the ratio of rye production to wheat is substantially narrow. In West Europe and Federal Republic of Germany the rye crop amounts to about one third of wheat crop and in Poland rye production is even more than that of wheat. The reasons for intensive rye cultivation in these countries are of agricultural (climate, soil and crop rotation) and ethnological (eating habits, taste and tradition) sphere. In these countries rye has a strong place both with respect of crop rotation and also nourishment of people and as such considerable part of rye produce is milled for bread production.

The diagram for rye milling process in comparison to wheat is short and simple (Fleschsig, 1955). As a rule dark and thorough milled flours are produced whereby priority is given for high flour yield from first flour systems. In the last years rye milling industry has brought out a tremendous development with orientation to different quality raw material (protein rich and vitreous in America and Russia, low protein and more starchy type in West Europe) and different rye varieties (Ajzikovic 1954; Drews, Seibel 1976; Rozsa, 1976). In contrast

**Publication number 4565 Federal Institute for Cereal and Potato Processing.*

ISBN: 0-12-370960-1

to American and Russian milling technology where relatively small number of flours are produced, in West Europe especially in Federal Republic of Germany, 8-10 types of rye flours with different ash content are offered. In these flours, the types 997 and 1150 (ash 1 - 1, 25%) with 61% and the type 1800 (whole meal with different granulation) with 20% take major share in the rye flour production. This high offer of more flour types point to extraordinary production of many sorts of bakery products in Federal Republic of Germany (a total of 200 bread and 1200 small bakery items). Consumption of milled rye products shows in comparison to wheat flour about 25% to 75%, though the number of bread types with recipes where rye milled products take a share are very high (Seibel, Brümmer, Stephan 1978). In these recipes of bread types the advantages of milled rye products, like strong typical rye taste especially by sour doughs, the possibility of varying the taste with addition of milled rye products in different proportions, physiological value and also improvement in shelf life are utilized.

The processing value of rye is evaluated from the pentosans (swelling substance) and starch (Drews, Seibel 1976; Bolling, Weipert 1977; Drews 1979). The constitution and the ratio of these two components - swelling substance and starch - are marked as the important characteristics of rye quality in baking as well as in milling technology. Some publications optimum pentosan-starch ratio is reported from the reduction flours (Drews 1965, 1970; Drews and Seibel 1974; Drews, Zwingelberg, 1977). Other authors experimented to change the pentosan to starch ratio with addition of chemically isolated wheat and rye pentosans (Casier *et al.*, 1969; Casier 1973; 1975; Rotsch 1966). This change had shown like the addition of enzyme pentosanase (Rotsch 1966) positive effects on baking results of wheat and rye bakery products.

As industrial separation and purification of rye pentosans through chemical means until now are associated with many problems, our Institute has started work with an aim to influence the pentosan - starch ratio through milling technology. It is to study thereby the possibility to obtain a flour with desired qualities.

This work stands as a continuation of Drews initiated work in our Institute some years ago, supported and carried on by some of us (Weipert, Zwingelberg 1978, 1979) with some published and ready but yet to be published findings.

2. MATERIALS AND METHODS

2.1 Material and Preparation

As experimental samples rye of 1977 and 78 crops are used. The selected samples correspond to an average quality rye grown in East Westfalen (Table I). For carrying out the experiments rye is cleaned, conditioned to 15% and after 3 hours rest, milled in a newly installed experimental mill.

2.2 Milling Diagram

The milling diagram of rye experimental mill is presented in Fig. 1. The rye passes first through smooth crusher rolls before it is milled in 5 break and 8 corrugated grind (reduction) or middling systems. The grinding is supported by bran finishing of the III. and V. break streams and by sifting in centrifugal machines of the last 4 reduction streams. The cover of plain sifter is selected with an aim to separate starch and pentosan rich flours. This separation is possible with break and first four reduction streams, in other words the last 4 reduction type of streams are excluded.

The milling capacity of the experimental mill is 200-300 kg/h.

3. EXPERIMENTAL RESULTS

3.1 Flour Streams

Represented in the milling diagram are 27 individual flours including bran and shorts, of that 10 flours from break systems, 4 further from bran finishers and the rest from reduction systems. All streams are separated to fine and coarse flour fractions through appropriate sieving.

TABLE I. Quality Characteristics of Grain

	Rye	
	Crop 1977	*Crop 1978*
Ash, % d.m.	1.72	1.73
Protein, % d.m.	10.0	8.8
Pentosans, % d.m.	8.2	8.1
Soluble matter, % d.m.	15.8	15.9
Starch, % d.m.	63.4	63.8
Maltose, % d.m.	4.2	3.4
Falling number (7g/25ml) sec	115	85
Amylogram Max., AU	340	300
Temperature in Max., ^{o}C	65	61.5
Swelling curves, AU	380/165/130	185/110/90

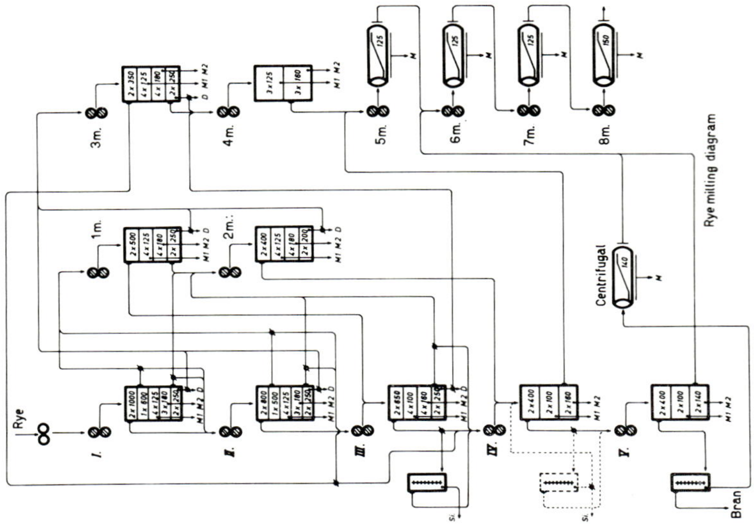

Fig. 1. Milling diagram of rye experimental mill.

The efficiency of such separation is demonstrated in the scanning election microscopic pictures of fine and coarse fractions of second break flour. Both these milled products differ substantially in starch and swelling substance content. The fine fraction is rich in starch (85%) and poor in swelling substance (1.5% pentosans and 5% protein) (Fig. 2). Obviously

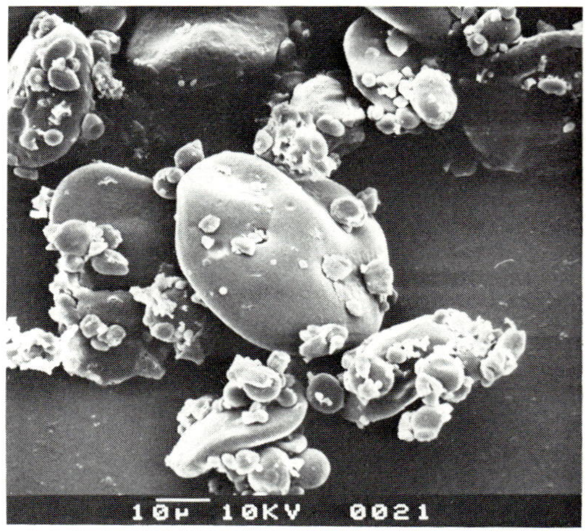

Fig. 2. Fine flour fraction from II.break stream.

here the starch granules are released free from the matrix of swelling substance. Such flour has a degree of fineness almost all below 75 μ.

The swelling substance rich flour (67% starch, 7% pentosans, 10% protein) is essentially coarser and mostly in the range of 75-150 μ (Fig. 3). The picture shows that the starch and swelling matter are present in the form of conglomeration which in turn explains the coarseness of the flour. Simultaneously the starch in this conglomeration is protected by swelling substance.

3.2 Basic Flours

Depending on the degree of fineness and content of swelling substance the obtained flour fractions are blended together into three, named by us as, basic flours.

First flour A - contains all fine fractions
Second flour B - contains all course fractions
Supplementary flour C - contains the flour streams of 5. to 8. reduction and V. break systems.

After applying the corrections with respect to remaining products, the basic flours stand in extraction, composition and quality as shown in the following tables and pictures of scanning electron microscope.

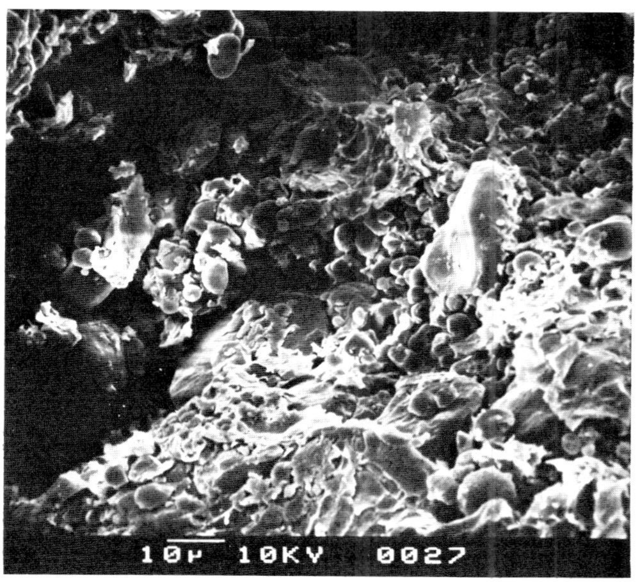

Fig. 3. Coarse flour fraction from II. break stream.

TABLE II. Composition of Basic Flours

	Yield (%)	*Percent d.m.*					*Pentosans to starch ratio*
		Ash	*Starch*	*Pentosan*	*Protein*	*Maltose*	
First flour (A)	60	0.62	80.0	2.9	6.7	2.5	1 : 27.6
Second flour (B)	18	1.23	62.0	8.2	11.6	4.3	1 : 7.6
Supplementary flour (C)	9	3.20	34.0	13.8	17.4	8.2	1 : 2.5

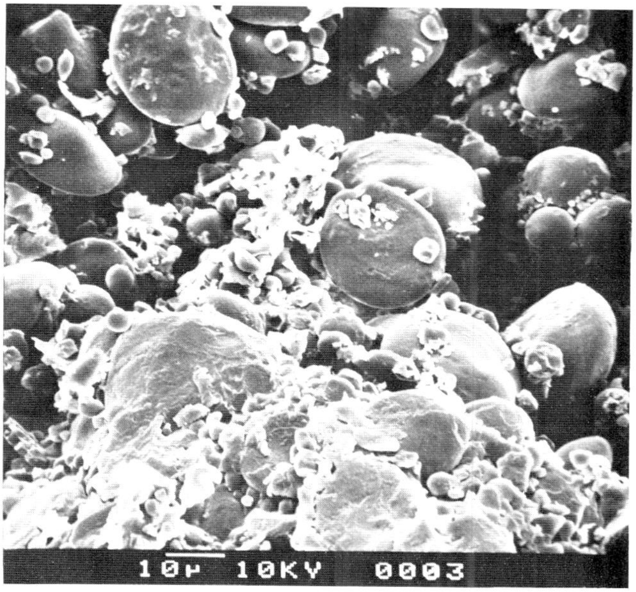

Fig. 4. First flour A - poor in swelling substance.

From the available analytical data (Table II) it is to be inferred that first flour A, is rich in starch and poor in swelling substance; the second flour B, is still rich in starch and simultaneously rich in swelling substance, whereas the supplementary flour C clearly is poor in starch but rich in swelling substance.

The quality differences of flours allow clearly to be seen in the pictures of scanning electron microscope. In first flour A (Fig. 4) which is poor in swelling substance the starch granules are preponderantly free, like in the fractions of II. break systems described already. In second flour B, which is rich in swelling substance (Fig. 5), quantity-wise still richly existing starch is only partly filled by the swelling substance. Very clearly in the picture of supplementary flour C (Fig. 6) only the matrix of swelling substance can be seen while the weakly represented starch is enveloped by the swelling substance.

The swelling matter content and the coarseness of the fractions in the basic flours are reflected in the quality of these flours (Table III).

According to the fact that the first and second flours are mainly derived from the endosperm of rye kernel and differ only in starch and swelling substance content, the difference in total water binding capacity is due to pentosan content of the flours. This water binding influences the values of amylo-

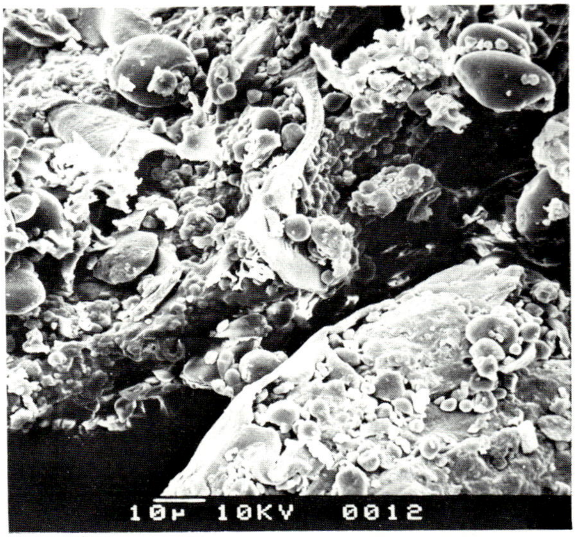

Fig. 5. Second flour B - richer in swelling substance.

grams and swelling curves. The first flour, poor in swelling substance has at the same time low content of enzymatically easily reducable soluble substances and a low α-amylase activity. The high proportion of soluble substances and also high α-amylase activity are directed into the second and the supplementary flours.

The ash and the extraction balance gave a total flour yield of 87% with an ash content of 1.06%. All three basic flours mixed in corresponding proportions gave the rye flour type 1150.

With their composition basic flours are so far in the practical milling not produced because in this form not suitable for preparing bread. This is shown in the pictures of products from baking test (lactic acid method) in comparison with the total flour type 1150 (blended from all flour component) (Fig. 7). Expectedly all products showed series of baking faults. Low water binding in the dough from flour A resulted in a flat form of the product. On the other hand high water content caused an insufficient grain structure and elasticity of crumb by second flour B and especially by supplementary flour C. Likewise in sour dough baking tests also inbalance of flour components have shown the effects (Fig. 8). The baker must vary the water addition to flours extremely so as to obtain dough of required processing quality.

TABLE III. Quality Characteristics of Basic Flours

	First flour (A)	Second flour (B)	Supplementary flour (C)
Pentosans total (% d.m.)	2.9	8.2	13.8
Pentosans soluble (% d.m.)	1.3	3.0	3.3
Soluble from total %	45.7	36.8	23.8
Soluble matter %	9.8	17.6	29.0
Total water binding capacity, g/100 g flour	91.1	173.9	178.1
α-amylase activity, ICC-U	2.5	4.4	15.6
Falling number, sec.	109	102	76
Amylogram, Max., AU	360	555	175
Temperature at Max. °C	62.5	65.0	64.0
Swelling curves,AU			
Starting viskosity	70	1500	1970
Viscosity at 42°C	55	1920	1130
End viscosity	40	1350	680

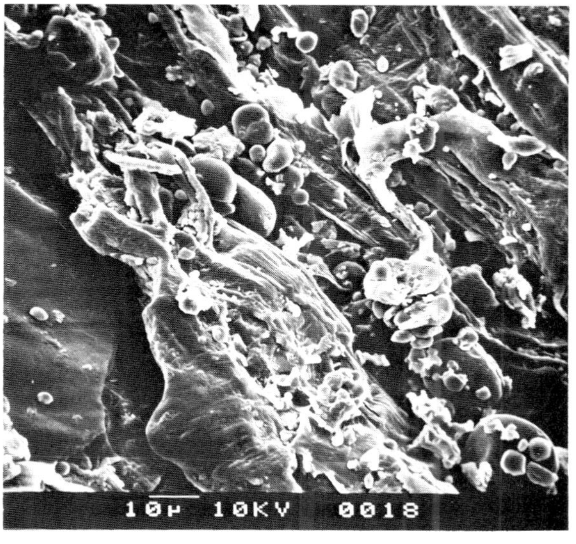

Fig. 6. Supplementary flour C - richest in swelling substance.

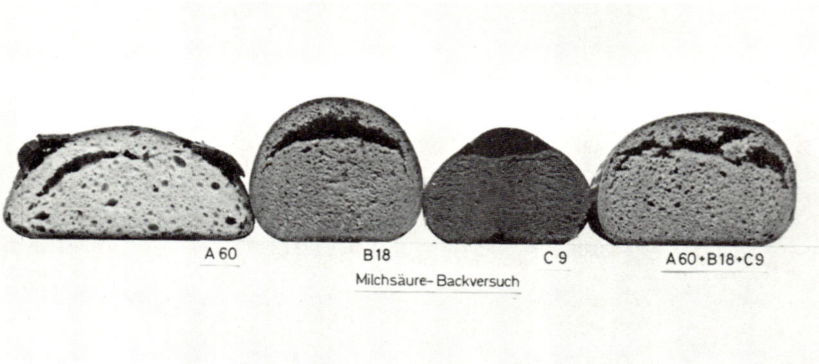

Fig. 7. Breads from basic flours and mixed flour type 1150 (lactic acid method)

Fig. 8. Breads from basic flours and mixed flour type 1150 (sour dough method)

In spite of the corrections applied in water addition the baking results with respect to form and crumb qualities of these products are not satisfactory in comparison with products of the flour type 1150. More over the flours and baking products show a deviation from the otherwise usual colour.

3.3 *Blending of Flours*

In general the flour quality is mainly dependent on raw material, so the milling diagram is of only second importance. The diagram of milling process is developed such that in the course of milling similar quality flours are brought together in basic flours, so as to obtain later with blending the flours of specific processing qualities. In order to examine the possibilities, the three so called basic flours are mixed in different proportions. The idea is that the starch rich first flour A builds the base for baking quality of rye flour and on that foundation the swelling substance rich flours B and C are mixed, so that the pentosan - starch ratio is altered as desired.

Addition of second and supplementary flours to basic flour A is followed out of economical reasons only to the extent of quantity obtained during milling - flour B 3 to 18% and flour C 1.5 to 9%.

The flour resulted with the mixing of first flour A and second flour B in different proportions showed a shift in the pentosan to starch ratio from 1:25 to 1:18.5. With blending of first flour A with supplementary flour C a similar shift in composition is achieved (pentosan - starch ratio 1:24.8 to 1:17.2). Substantial high shift in ratio is attained with simultaneous addition of second and supplementary flours to first flour A. Here the ratio changed from 1:23 to 1:14. In this last step of mixing a flour that corresponds in extraction and quality to the flour type 1150 is obtained.

The processing value of the obtained flours is tested with indirect methods and also with baking tests. The change in composition is reflected in dough characteristics, grain structure and elasticity of crumb - same way in lactic acid method, as also in sour dough baking tests (Fig. 10). At this point without entering into itemwise details of evaluation of individual baking products, it can be ascertained that from blending $A_{60}+B_{12}+C_6$ with a pentosan - starch ratio of 1:16, products are obtained with the best characteristics. The flours with pentosan - starch ratio in the vicinity of 1:16 have given likewise products that are also assessed as good.

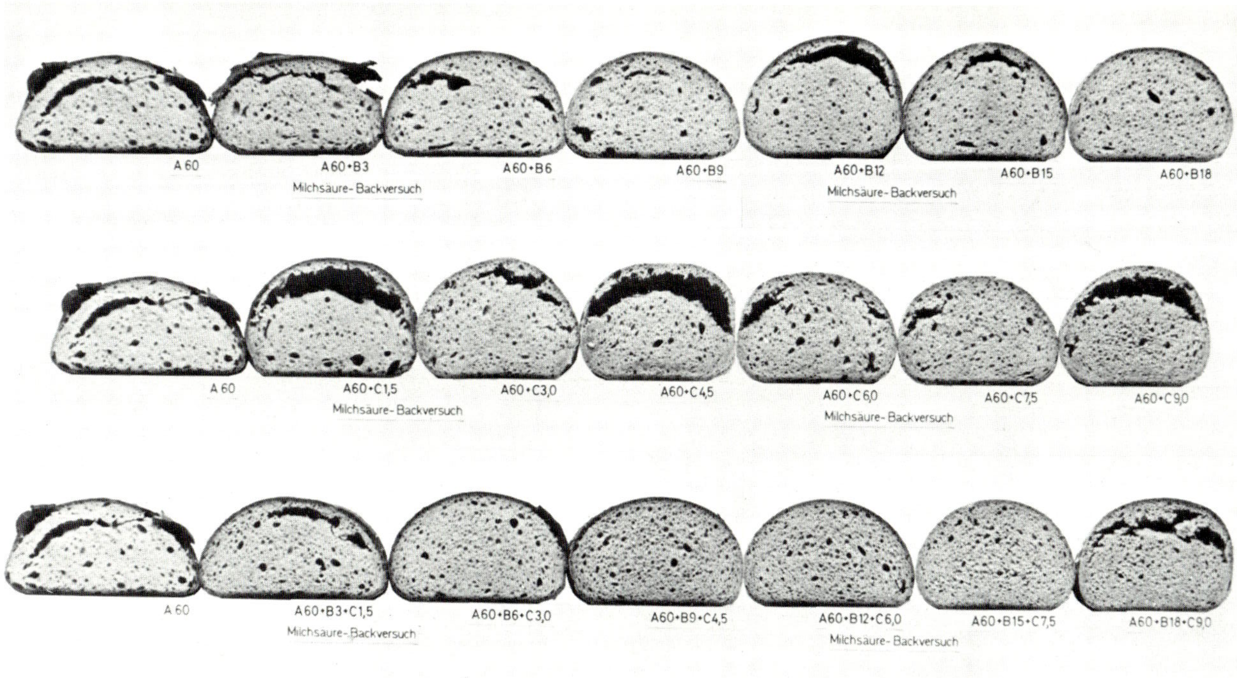

Fig. 9. Breads from blended basic flours (lactic acid method)

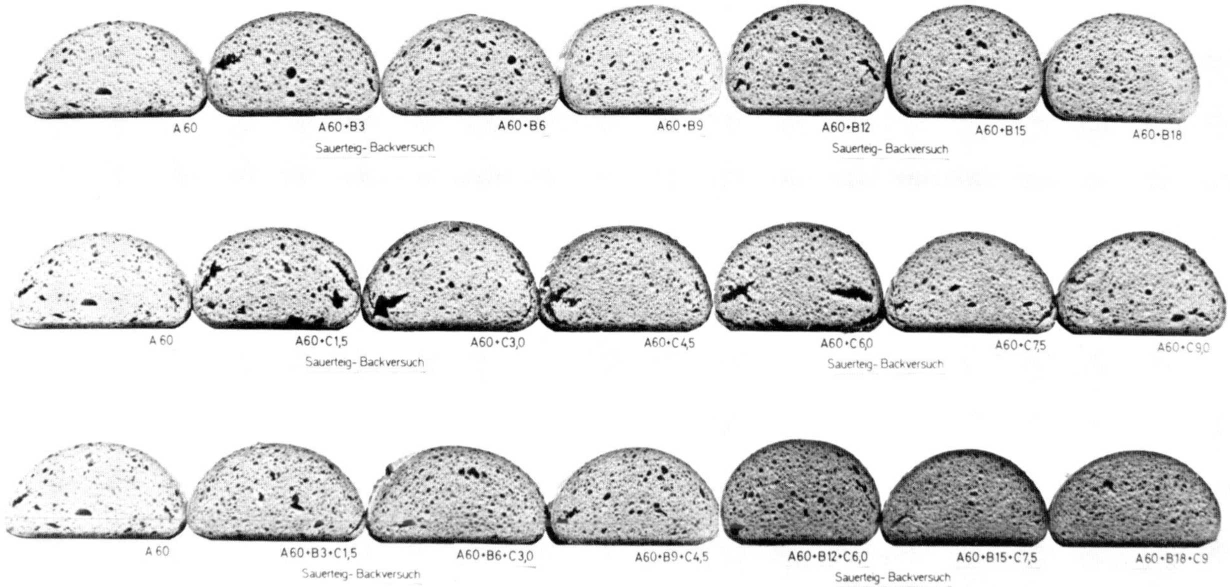

Fig. 10. Bread from blended basic flours (sour dough method)

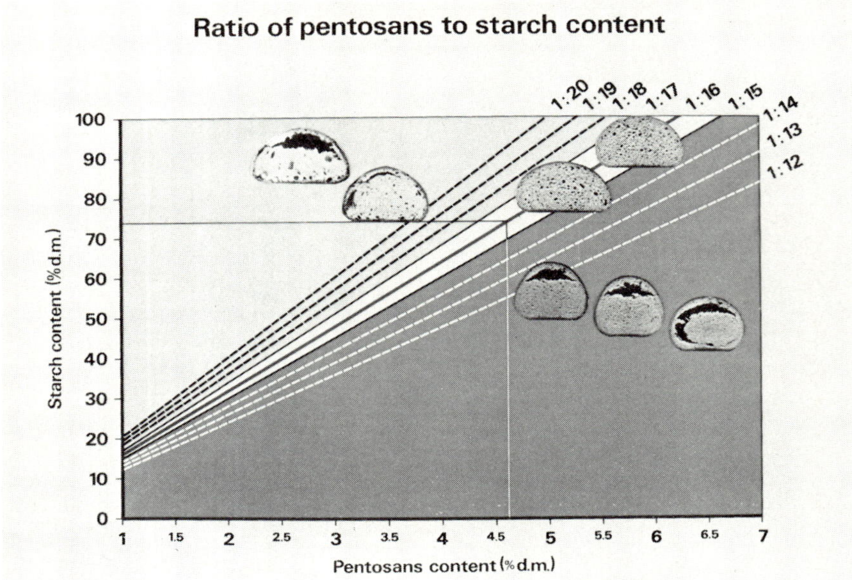

Fig. 11. Schematic representation of influence of pentosan starch ratio on bread quality.

4. DISCUSSION AND CONCLUSIONS

It is the task of mills to offer flours with good processing and baking qualities. The tools available for this purpose are choice of raw material, appropriate steps in milling diagram and blending of flours. In this work it is attempted to bring an improvement in processing characteristics of flours with shift in swelling substance to starch ratio through blending of flour streams.

With consideration of products from the described flour combinations a schematic representation of the influence of pentosan - starch ratio on the processing qualities is prepared (Fig. 11). The cereal processing industry is known to care for the processing value of a flour in combination with flour type. With a flour type not only composition but also quality of components is hinted. Out of normal quality rye flours, the type 997 shows in Federal Republic of Germany as of the best baking quality flour for preparing customarily offered breads (Drews 1965, 1970). Here it appears that the favorable pentosan - starch ratio 1:16 is responsible for the good baking quality. With shift of ratio in favor of starch follows the change in processing value with reduction in dough yield flowy and soft dough. Often the form of bread is flat and affected with split and tight crumb, weak flavor and taste and low shelf life. On the other hand the shift in favor of pentosans results in a higher dough yield and to a certain extent strong dough. The product shows often an insufficient grain, a weak, soft and burdened with much free water crumb. Better shelf life and strong typical taste are additional characteristics. With such shift in ratio a simultaneous change in flour and product color is to be counted.

Comparing the ratio of 1:16 in flour with that of 1:8 in grain it should be concluded that the pentosan content in rye kernel is too high. So the milling means not only just grinding of grain but also bringing a reduction of pentosan content in flour, while at the same time pentosans concentrate in bran and shorts. The pentosan content of these products is between 20 to 30%. The quality of these products is further affected through the increased content of soluble substances. The optimum pentosan - starch ratio is also applicable if in the place of second and supplementary flours, the pentosan rich shorts (in proportion 4 to 8%) and bran (2 to 5%) are added. The products (Fig. 12) show, as long as ratio of 1:16 is maintained, better crumb characteristics than the product of type 1150 with a pentosan - starch ratio of 1:13 or 1:14.

This schematic representation and the optimum pentosan-starch ratio of 1:16 can lose their effect if the quality of the starch or the pentosans deviate substantially from the usual. It is often the case with the grains damaged with sprouting (Weipert, Zwingerlberg 1979). All these pilot scale trials show that with the described milling technological measures (in diagram of milling process) a good separation of starch and pentosans with protein is attainable in rye milling. Thereby certain pentosan rich fractions are left over.

The pentosans, known to influence the water absorption, dough qualities and lastly crumb characteristics of products can be utilized to improve the qualities of other flours including the weak wheat flours.

The results allow to expect that a more intensive separation is possible. Such special flours, on one hand starch rich and on the other hand pentosan rich, will find multiple use in the food industry. Further experiments are conducted with the obtained flours for incorporating in the gingerbread products. Addition of these flour streams to wheat flours have brought with their composition a series of changes in the qualities of the products (Fig. 13). Especially the pentosan rich fractions could be included as "universal baking factor" (Casier 1975, 1976) in bakery products prepared with other and noncereal starches. Experiments were conducted by us in this direction also. With this offer of relatively simple method of getting rye special flours, in our opinion, a series of applications in the food industry are open.

Fig. 12. Breads from flours obtained with blending of shorts and bran to first flour A.

5. SUMMARY

With the use of specially developed milling diagram for the separation of starch and pentosan rich fractions, rye flours with extremely different properties are obtained. These flours named by us as basic flours, different in fineness and in composition make compilation of flours with desired qualities.

From the experimental results of mixed flours, with different pentosan - starch ratios, it can be ascertained that a pentosan starch ratio of 1:16 is optimum for the baking quality of rye flours. In composing such flours with optimum baking quality, the pentosan rich fractions are left over which in turn can be utilized for improving the baking quality of other flours including weak wheat flours.

With the use of described milling diagram, obtained special flours can be utilized for different purposes in the food industry. Especially the pentosan rich fraction offers for incorporation in the bakery products out of noncereal starches.

6. LITERATURE

Ajzikovic, L. E., Horcev, B. N. 1954: Tehnologija proizvodstva psenicnoj i rzanoj muki.- Zagotizdat, Moskva (rus.)

Bolling, H., und Weipert, D. 1977: Roggenqualität, erläutert an der Interdependenz zwischen Sorte, Umwelt, Mahl- und Backeigenschaften.-Getreide, Mehl und Brot 31, 307-312.

Casier, J. P. J., Stephan, H., de Paepe, G., Nicora, L. 1969: Einflus der wasserunlöslichen Endospermpentosane auf die Teig- und Gebäckeigenschaften bei der Verarbeitung von Weichweizen.-Brot und Gebäck 23, 2, 231-236.

Casier, J. P. J., de Paepe, G., Brümmer, J. M. 1973: Einfluß der wasserunlöslichen Weizen- und Roggen-Pentosane auf die Backeigenschaften von Weizenmehl und anderen Rohstoffen.-Getreide, Mehl und Brot 27,1, 36-44.

Casier, J. P. J. 1975: Einfluß von wasserunlöslichen Endosperm-Pentosanen von Weizen und Roggen auf die Teigeigenschaften und Backfähigkeit von Weichweizen und anderen stärkereichen Rohstoffen wie Maniok, Reis Sorghum, Millet usw..-Fermentation 71, 3, 117-134.

Casier, J. P. J., de Paepe, Willems, H., Goffings, G. and Noppen, H. 1976: Bread from starch tropical crops. II. Bread production from pure millet and sorghum flours using cereal endosperm - cell wall-pentosan as a universal baking factor.-Sorghum and millets for human food.-International Association for cereal chemistry, Symposium

Drews, E. 1965: Die Schleimstoffe des Roggenmehles und ihre Bedeutung für seine Qualität.-Getreide und Mehl 15, 4, 37-44.

Drews, E. 1970: Grundlagen der Roggenbackfähighkeit.-Die Mühle und Mischfuttertechnik 107, 6, 74-76: 7, 87-88.

Drews, E., Seibel, W. 1974: Erkenntnisse über die Roggenbackfähigkeit und ihre Auswirkung auf die Diagrammführung.-Getreide, Mehl und Brot, 28, 12, 307-312.

Drews, E., Seibel, W. 1976: in Rye: Production, Chemistry and Technology.-AACC St. Paul, Minnesota.

Drews, E., Zwingelberg, H. 1977: Ergebnisse langjähriger Roggenuntersuchungen über Sorte, Standort und Erntejahr hinsichtlich des Mahlverhaltens.-Getreide, Mehl und Brot, 31, 2, 34-37.

Drews, E. 1979: Mahleigenschaften in Abhängigkeit von der Kornbeschaffenheit und ihrer Auswirkungen auf die Mehlqualität bei Getreide.-Die Mühle und Mischfuttertechnik 116, 23, 311-314.

FAO year book 1977.

Flechsig, J. 1955: Fachkunde für Müller.-Verlag Moritz Schäfer, Detmold.

Rotsch, A., 1966: Uber die Beziehung des Schleimstoffgehaltes des Roggens zur Backqualität.-Brot und Gebäck, 20, 7, 132. 136.

Rosza, T. A. 1976: in Rye: Production, Chemistry and Technology. AACC, St. Paul, Minnesota.

Seibel, W., Brümmer, J. M., Stephan, H. 1978: in Advances in Cereal Science and Technology Vol. II.-AACC St. Paul, Minnesota.

Weipert, D., Zwingelberg, D., 1978: Versuche zur Herstellung von Roggenmehlen mit definiertem Quellstoff-Stärk-Verhältnis.-Getreide, Mehl und Brot, 33, 2, 34-40.

Weipert, D., Zwingelberg, D., 1979: unpublished data.

THE TECHNOLOGY OF SORGHUM PRODUCTS

L. W. Rooney
M. N. Khan
C. F. Earp

Cereal Quality Laboratory
Texas A&M University
College Station, Texas

I. INTRODUCTION

The grain of Sorghum bicolor (L.) Moench is an important world cereal crop ususally ranking fifth in terms of total production. Sorghum will be used to refer to the caryopsis of Sorghum bicolor (L.) Moench plants. Much confusion in terminology exists throughout the world on nomenclature of sorghum. In the U. S., various names have been used for sorghum including: grain sorghum, sorghum grain, milo, milo maize, maize, hegari, kafir corn, kafir, guinea corn, Egyptian wheat, gyp corn, broom corn, Egyptian rice, rice corn, sudan grass, sorgo, and others. In other countries, a variety of terms have been used which include: the great millet, cholam, jowar, pearl durra, kaoliang, feterita, and others. We are not attempting to list all the names under which sorghum is known around the world. Most publications use the scientific name of the species which helps promote communication and eliminates confusion among the workers. Until 1974 the United States grain standards used the term grain sorghum but the official term used presently is sorghum. Grain sorghum is used to refer specifically to short sorghum plants grown for grain production. The grains of tall or short sorghum plants are the same because only four genes are responsible for the height of sorghum plants.

The purpose of this manuscript is to review certain aspects of sorghum utilization and to provide specific information concerning improvements in sorghum that have occurred recently which facilitates its use in various processes and products. No attempt has been made to cover all of the literature available on sorghum product technology.

Dogget (1970), Wall and Ross (1970), Rao (1971) and Hulse (In press) have provided general references on sorghum production and utilization which would be of interest to readers desiring more information. Some of the data presented here

ISBN: 0-12-370960-1

is unpublished data from our research on sorghum quality.

A. World Production and Use of Sorghum

The total production, utilization and yield of sorghum at five year intervals from 1960 through 1978 is presented in Table I (USDA). World production of sorghum has increased more than 50% since 1960. Per hectare yields of grain have also increased about the same. In the United States, sorghum (750 million bushels per year) is the third most important cereal crop following corn and wheat.

Sorghum production in terms of the percentage of total world production for the 1975 crop was: North America 37%, Central America 7%, South America 11%, Africa 20%, Western Asia less than 1%, South Asia 21%, East Asia 1%, Oceania 2%, and Europe less than 1%. Undoubtedly, these figures on production and utilization of sorghum are too low because of the problems of obtaining reliable information on sorghum production in Asia and Africa where it is produced mainly by small farmers for use on their farms and never enters grain marketing channels. A good estimate is that at least 75% of total world sorghum production is consumed directly as human food. Production estimates for 1977-78 for the major sorghum producing countries were: the United States (20.1 tons), India (9.8), Argentina (6.5), Mexico (3.4), and Australia (0.6), which accounted for more than 75% of world production of sorghum. Sorghum production in the western hemisphere has increased from 31% in 1950 to 54% of the total world production in 1975. However, sorghum production has decreased from 34% to 20% in Africa and from 34% to 22% in Asia. Sorghum production in Argentina and Mexico has increased until it is 17% of the world's sorghum. Sorghum production in Mexico continues to increase rapidly because of high yields in certain areas relative to corn yields.

The major use of sorghum in the western hemisphere has traditionally been as a livestock feed with 1-2% of total production being used for food and industrial products. It is used extensively for hay production and grazing. However, in Africa and India nearly all sorghum grain is utilized directly as human foods. The leaves, stalks and roots of sorghum are used as livestock feed, building material and firewood.

The export of sorghum is also primarily for livestock feed. Japan, Israel, Venezuela, Belgium, Luxumbourg, the Netherlands, and the United Kingdom are major importing countries. Japan imports the most sorghum. At times, India has imported sorghum which was used for food. Sorghum and soy fortified sorghum grits were imported into the sub Sahelan areas of Africa. Major exporting countries are the United

TABLE I - WORLD SORGHUM SUPPLY-DISTRIBUTION, 1960-1978

Time period	Area harvested	Grain Yield	Production	Consumption: Feed	Consumption: Other
	Million hectares	Tonnes/hectare	Million Tonnes	Million Tonnes	
1960	45.4	.94	42.5	14.6	24.3
1965	47.2	1.04	49.2	21.5	33.2
1970	49.7	1.23	61.2	29.8	34.9
1975	48.1	1.37	65.7	29.3	36.4
1978	47.5	1.44	68.6	30.9	35.7

Source: Foreign Agriculture Service, USDA.

States, Argentina, Australia and South Africa. At times, France and the Sudan are also exporters.

B. Sorghum Plant

The sorghum plant is a coarse annual with culms or stems 2-15 ft. or higher. The vegetative structure of the sorghum plant is similar to corn. The culms are made up of 7-18 or more nodes and internodes. The diameter of the stalk at the base varies from a fraction of an inch to 2 in. The stalks may be juicy or dry, sweet or nonsweet. A leaf occurs at each node; and nodes are alternate and on opposite sides of the culm. The surface of the culms, sheaths, and leaves are glaucous. The leaf sheaths are long and overlap in dwarf cultivars grown for grain.

The inflorescence or "head" of sorghum is a panicle ranging from 3-20 inches in length and 1.5-8 inches in width. Shape of the sorghum panicle varies from a loose, grassy to a compact type. Each flower has both female and male organs and is self-pollinated. Each panicle contains from 800 to 3,000 kernels. The kernels are enclosed in floral tracts (lemma and palea) which usually thresh free of the kernel during harvesting. Sometimes, the lemma and palea remain attached to the kernel as in many sorgos, grassy types, and broomcorn. The sorghum kernel and floral tracts are inside the glumes. Color of the glumes range from light brown to red to purple. Sometimes, the phenolic pigments responsible for the color migrate from the glume and discolor the kernel.

1. Agronomy and Genetics of Sorghum. Sorghum is mainly cultivated in the hot, arid areas of the world because it has drought tolerance and can withstand the hot temperatures and dry conditions. The sorghum plant has many diverse types. The world collection of sorghum germplasm contains about 17 to 18 thousand entries and more are being currently added. A great deal of variation in plant and grain characteristics, with genetic variations in insect and disease resistance, grain yield and quality characteristics, exist in the world collection of sorghums. New mutant genes of sorghums have been found or induced which alter the composition of the crop and may make it possible to develop sorghum with improved quality characteristics. Progress toward a sorghum with a higher lysine content has been made, but it is slow, and suitable types with acceptable agronomic and quality characteristics have not been found.

A dramatic increase in sorghum production resulted from development of hybrids in the mid 1950's in the United States. The utilization of hybrid vigor in combination with improved fertility and cultural management practices has led to substantial increases in grain yields. For instance, sorghum production and yields in the United States have increased 364% since 1950. This was accomplished in spite of increasing insect and disease hazards.

C. Kernel Structure

Sorghum is a naked caryopsis which varies from 3 to 80 mg per kernel. The size and shape of the caryopsis varies widely among sorghums in the world collection. For U.S. commercial hybrid sorghums, the kernels are flattened spheres 4.0 mm long by 3.5 mm wide by 2.5 mm thick with a weight of 25-35 mg per kernel. The sorghum caryopsis is composed of three major portions; the outer covering (pericarp), the storage tissue (endosperm), and the growing point (germ). The pericarp consists of four distinct layers. The outside layer is the epicarp which contains pigments and wax. The middle layer, or mesocarp, contains small starch granules. The mesocarp of some sorghums (i.e. pearly types) is nearly absent and does not contain starch granules. The inner portion of the pericarp (endocarp) is composed of cross and tube cells. The presence of a starchy mesocarp and the testa is controlled by several genes. The trend in sorghum varieties and hybrids is to select those with a thin pericarp (no starch in the mesocarp) and without a testa. The testa, sometimes referred to as the seedcoat, undercoat, subcoat, nucellar layer or integument, contains pigments and is not present in most sorghum cultivars. The testa is present in sorghums that are relatively resistant to birds. These are the high tannin sorghums which have reduced nutritional value.

The endosperm is composed of an aleurone layer and a starchy endosperm. The aleurone layer is one cell layer thick, composed of small, block-like cells that are high in protein, fat, ash, vitamins and enzymes. The starchy endosperm consists of peripheral (subaleurone), corneous, and floury portions (Watson-1967, Sanders-1955). The peripheral endosperm area is characterized by the prevalence of small block-like cells filled with large quantities of protein interspersed with small starch granules. The endosperm cells beneath the pericarp area, are larger and more elongated with the cells containing less protein and larger, more numerous starch granules. This area is the corneous (hard, flinty or horny) endosperm area and is translucent. The floury (soft) endosperm structure resembles a honeycomb in which starch granules are encased in a protein matrix. The protein matrix decreases in concentration from the outside to near the center of the kernel. This is readily apparent by observation of endosperm sections of sorghum kernels. In general, size of starch granules increase from the outside to the inside of the kernel. The cells of the corneous endosperm appear to be completely filled with starch and protein, whereas in the floury portion, the starch granules are not as tightly encased in the protein matrix. There are small intracelluar voids which probably cause diffusion of light which accounts for the floury or opaque appearance. The floury endosperm is more easily broken into small particles during processing than the corneous endosperm. The starch granules of the floury endosperm cells are more spherical while those of the corneous endosperm cells are more likely to be polyhedral. Sorghum genotypes that produce nearly all-floury or all-corneous endosperm textures exists. Environment as well as genetics influence the proportion of corneous to floury endosperm in the sorghum kernels. The structure of the endosperm of floury, corneous, intermediate and waxy sorghum kernels differ initially and respond differently to attack by enzymes (Sullins and Rooney, 1974).

The sorghum kernel is composed of 80 to 82% starchy endosperm, 6-8% bran (pericarp plus associated layers) and 10-12% germ. In our experience, sorghums with a thin pericarp without a testa and with a small germ can contain as high as 86% or more of endosperm. Yields of pearled sorghums have exceeded 86% but the pearled materials contained some germ although the pericarp was completely removed.

1. Composition of Sorghum Grain. The proximate composition of sorghum is listed in Table II. Sorghum is similar to corn (*Zea mays*) in composition. Sorghum composition varies significantly due to genotype and environment. Sorghum composition is probably more variable than most other cereals because it is produced under more variable conditions. In general, sorghum contains one percentage point more protein than corn and one percentage point less oil than corn. Generally, grain yield is inversely related to protein content of sorghum. Fiber and ash composition are similar to other cereals with a naked caryopsis.

a. Carbohydrates. Starch, cellulose, simple sugars and pentosans comprise approximately 80% of the dry weight of the kernel with starch usually 70-75%. Sorghum starch contains 20-30% amylose and 70-80% amyopection. In general, the properties of sorghum starch are similar to corn starch. Waxy sorghum starches that are 100% amylopection exist and have been used in food products. The gelatinization temperature range of sorghum starch is 67-77°C which is slightly higher than that of corn (Watson-1960). Glucose, fructose, sucrose, and traces of raffinose and stachyose are found in sorghum. The pentosan content of sorghum ranges from 2.6-5.2% of the

TABLE II: PROXIMATE ANALYSIS OF SORGHUM GRAIN

Number of Samples	Protein		Ether Extract		Crude Fiber		Ash		Nitrogen Free Extract		References
	Mean	Range	Mean	Range	Mean	Range	Mean	Range	Mean	Range	
25	10.9	8.8-14.5	3.9	3.6-5.2	3.0	2.3-3.5	3.0	1.6-4.5			Bressani & Rios 1962
31	10.1	8.1-13.7	2.8	1.8-5.2	2.5	1.4-4.9	2.6	1.3-4.5	71.8 Starch	68.4-74.5	Bredon 1961
8	10.7	8.5-11.9	3.6	1.6-4.6	4.7	3.4-7.3	1.7	1.4-2.2			Aucamp 1961
1160	12.5	8.7-16.8									Miller 1958
1159			3.4	1.4-6.1							"
1085					2.7	0.4-13.4					"
1133							2.2	1.2-7.1	79.2	65.3-85.3	"
5	13.6	11.8-16.6	2.8	2.5-3.4			1.6	1.3-1.95	71.5	67.0-75.2	Lamar et al 1972
234	11.0		3.4	2.0-5.3	2.4		1.8			81.0	Rooney 1973

dry grain and is located mainly in the pericarp and germ (Karim and Rooney-1972 a, 1972 b).

b. Lipids. Sorghum oil is concentrated in the germ, pericarp and aleurone layer at a level of about 3.5%. The germ contains more than 79% of the total oil. Sorghum contains a few tenths of a percentage of carnauba-like wax. The fatty acid composition consists primarily of linoleic, oleic, palmitic, stearic, and linolenic acids. The free lipid fraction, extracted with petroleum ether, consists of hydrocarbons, tri- di- and mono glycerides, free fatty acids and some polar lipids. The bound lipid fraction, extracted with water-saturated n-butanol at room temperature contains lecithin, lysolecithin and other components (Rooney-1978).

c. Proteins and amino acids. Lysine is the first limiting amino acid, followed by threonine and tryptophan (Table III). Lysine is negatively correlated to protein content. Like other cereals, the percentage of lysine in the protein decreases as protein content increases, even though the total lysine as a percentage of dry weight increases. Sorghum proteins contain slightly more tryptophan than corn proteins.

Sorghum proteins are difficult to extract (Wall and Blessin-1969). The albumins (water soluble), globulins (dilute salt-soluble), prolamines (alcohol soluble) and glutelins (dilute base soluble) fractions are summarized in Table IV. The albumins and globulins, located primarily in the germ, aleurone layer, and pericarp have the highest level of lysine and other essential amino acids as listed in Table V. Kafirin (prolamine fraction) is high in glutamic acid and the non-polar amino acids. It is low in the essential amino acids lysine and methionine. The kafirin fraction is positively correlated with sorghum protein content. Usually, when protein content increases, due to nitrogen fertilization or other reasons, it is the prolamine fraction which increases the most, thus further diluting the percent of essential amino acids. The kafirin proteins are located in the endosperm in spherical protein bodies which are associated with starch granules in a protein matrix. The glutelins are insoluble in neutral solvents and are the major constituent of the matrix protein in the sorghum endosperm.

D. Sorghum Pigments

Two major types of pigments in sorghum are carotenoids and polyphenolic compounds. Common sorghum does not contain significant quantities of carotenoids. However, yellow endosperm sorghums have a significantly higher carotenoid content; but, more than 50% of the carotenoids may be lost in the field prior to harvest. Thus, the carotenoid content of even yellow

TABLE III: ESSENTIAL AMINO ACID COMPOSITION OF SORGHUM AS % PROTEIN

Number of Samples	Protein %	Lys	Leu	Phe	Val	Trp	Met	Thr	His	Ileu	References
58 Samples	12.4	2.0				1.1					Maxson Dissertation, 1973
522 lines, weighted average	12.6	2.1	14.3	5.2	5.4		1.0	3.3	2.0	4.1	
9 lines						1.3					Srinivason, Axtell, Jambunathan 1972
5 Samples	13.6	1.9	13.1	4.8	4.6		1.5	2.9	2.0	3.6	Lamar, 1972
30 Samples	10.4	2.0	13.1	4.8	4.9		1.3	3.0	2.1	3.8	Deyoe and Shellenberger 1965
Range		1.6 to 2.6	10.2 to 15.4	3.8 to 5.5	4.0 to 5.8		.81 to 2.0	2.4 to 3.7	1.7 to 2.3	2.9 to 4.8	
25 Samples		2.9	14.3	4.3	6.0	.704	1.6	3.8	3.3	4.7	Bressani and Rios, 1962
FAO/WHO Suggested Pattern		2.2	2.5	2.5	1.8	.65	2.4	1.3		1.8	FAO/WHO 1971

TABLE IV: PROTEIN FRACTIONS FROM OSBORNE EXTRACTION PROCEDURE

Samples #	Albumins	Globulins	Prolamin	Glutelins	Residue	Extraction Efficiency	References
5	5.7 1.6-9.2	7.1 1.9-10.3	52.7 39.3-72.9	34.4 23.5-45.0		88.4	Virupaksha and Sastry 1968
5	23.4 4.3-30.4	11.6 1.0-17.9	51.8 40.6-60.3	13.7 4.5-34.3	5.2 3.6-7.0		Neucere and Sumrell 1979
5	3.8 2.2-4.9	3.9 2.5-5.8	5.8 3.9-8.5	17.7 14.9-20.8	59.1 52.8-62.6	31.2	Skoch et al. 1970
1	20.0	8.0	30.0	12.0	30.0	70.0	Haikerwal and Mathieson 1971
PROTEIN FRACTION FROM LANDRY-MOUREAUX EXTRACTION PROCEDURE							
	I	II	III	IV	V		
4 Whole Grain	10.5 4.1-16.8	9.4 2.5-18.4	15.6 11.5-18.9	11.2 6.1-17.3	42.0 29.9-54.8	88.7	Jambunathan and Mertz 1972
4 Endosperm	5.1 2.8-8.0	14.0 9.2-19.9	30.0 27.0-35.1	9.4 6.4-11.2	32.8 24.0-35.7	91.2	"
21 Whole Grain	17.5 13.8-22.4	8.0 4.3-13.3	20.5 1.0-30.6	3.2 1.9-4.9	31.1 20.3-44.7	80.2	Meckenstock 1979
3 Whole Grain	17.1 16.4-18.5	5.9 5.3-6.2	19.9 18.1-19.4	4.6 4.5-4.8	32.3 29.9-34.6	79.8 77.0-84.0	Walker 1974

TABLE V: AMINO ACID COMPOSITION OF SORGHUM PROTEIN FRACTIONS AS EXPRESSED AS % OF PROTEIN

Fraction		Samples	Lys	Leu	Phe	Val	Trp	Met	Thr	His	Ileu	References
Gloubulin	H_2O+1%NaCl	1	3.36	6.72	4.77	6.46		2.24	4.87	1.45	3.45	Virupaksha
Prolamin	60% ethanol	1	0.14	15.33	5.84	4.88		1.33	-	0.67	5.04	and
Glutelin	.4% NaOH	1	3.12	12.49	4.9	5.5		-	4.88	3.12	4.07	Sastry 1968
Whole Grain		1	2.02	16.54	5.37	5.37		1.39	3.35	1.95	4.72	
Endosperm Meal		1	1.70	16.58	6.4	7.25		1.51	3.81	2.16	4.91	
Whole Grain (WG)		4	2.6	13.1	5.1	5.1	1.3	1.7	3.3	2.2	3.8	Jambunathan
Endosperm Meal (EM)		4	2.1	16.0	5.8	5.5		1.4	3.45	2.4	4.5	and
WG1	Saline	4	4.5	4.3	2.6	3.9		1.2	3.0	1.8	2.4	Mertz 1972
WG2	Isopropanol	4	.33	17.6	5.4	4.6		1.45	2.65	1.3	3.8	
WG3	Isopropanol+ME	4	.15	20.6	6.88	5.0		1.45	2.78	1.7	4.7	
WG4	Buffer+ME	4	2.53	8.03	4.25	5.1		1.38	5.4	6.0	3.05	
WG5	Buffer+ME+SDS	4	4.55	14.88	6.35	5.53		1.73	4.13	2.7	4.5	
EM1	Saline	4	4.0	5.2	2.5	3.9		.8	3.2	1.5	2.7	
EM2	Isopropanol	4	.25	20.65	6.35	5.1		1.4	3.0	1.35	4.3	
EM3	Isopropanol+ME	4	.23	20.53	6.95	4.98		1.18	2.8	1.56	4.75	
EM4	Buffer+ME	4	1.8	7.73	2.68	4.95		1.35	6.1	6.68	2.88	
EM5	Buffer+ME+SDS	4	3.3	14.63	6.4	6.18		2.58	4.75	3.0	4.78	
WG	Residue	4	4.93	8.23	4.63	6.0		1.7	4.9	3.4	4.1	
EM	Residue	4	4.28	7.6	3.85	5.33		1.43	3.73	3.0	4.53	
1 Albumins	H_2O	5	6.5	6.6	4.9	5.7		1.5	4.6	3.8	3.4	Neucere and
2 Globulins	.5 M NaCl	5	4.5	6.3	5.2	4.8		1.3	4.3	3.9	3.1	Sumrell 1979
3 Prolamins	Butanol	5	.9	15.3	5.3	4.7		1.2	2.9	1.4	3.9	
4 Glutelins	Buffer+SDS	5	3.4	9.2	5.1	5.4		1.6	4.0	3.2	3.6	
Residue Meal		5	1.4	13.7	4.8	4.8		1.3	2.9	2.8	4.0	

ME=Mercaptoethanol; Buffer=Borate Buffer; SDS=Sodium Dodecyl Sulfate endosperm sorghum is not generally of practical significance. However, in certain areas of India where sorghum is a major part of the diet, the yellow endosperm cultivars apparently provide sufficient additional vitamin A in the diet to help prevent blindness.

Sorghum contains a number of polyphenolic pigments. The kind and amount of polyphenols present in the sorghum kernel is determined by genetics and to a lesser extent by environmental conditions during maturation of the grain. The kinds of polyphenols include: flavanoids, anthocyanidins, leucoanthocyanidins, and condensed tannins. The condensed tannins are present in the sorghums that have a testa layer, especially those that are bird resistant. The bird resistant sorghums are grown becuase the birds generally prefer other available food sources. However, birds do consume the bird resistant sorghums when other food sources are scarce. The major disadvantage with the bird resistant sorghums in that they do cause a significant decrease in nutritional value of the sorghums. Apparently, the condensed tannins combine with the proteins making them less available to the animal which in turn depresses the carbohydrate availability.

Bird proof (resistant) sorghums are referred to as brown sorghums because they usually have a brown appearing kernel resulting from the combination of the pericarp and testa color. In U. S. grain markets, bird proof sorghums are classified as brown and usually sell for significantly less in the market. U. S. grain standards for sorghum does not permit more than 10% of brown grain in a yellow sorghum grain sample. However, in the U. S., brown sorghums are grown only in the Southeastern part of the country, and little if any of these find their way into the usual market channels. Other sorghum exporting countries do not keep the brown sorghum segregated from the yellow sorghums. This can cause a misunderstanding on the part of importers of sorghum who feel all sorghums are nutritionally inferior. At best, it causes more variations in nutritional value of sorghum.

Considerable progress has been made in refining analytical methods for determination of "tannins" in sorghum and the biochemical characterization of them. Even so, current methods for analysis of sorghum "tannins" are difficult to interpret because of the various substances that are measured by the different tests. In grain marketing tests, the best way to determine whether a sorghum sample has potentially high levels of condensed tannin polyphenols is to determine the relative percent of brown kernels (those kernels with a testa) in the mixture. This can be done most easily by scratching the surface of the sorghum kernel and looking for the presence of the pigmented testa or by using a bleach test in which potassium hydroxide and sodium hypochlorite (NaOCl) are used to remove the pericarp of the grain. The testa appears as a dark brown or black layer after bleaching. Then, the percentage of kernels containing the testa can be determined. Other methods with potential for use in marketing systems are the Prussian Blue Method and the Vanillin Hydrochloric acid method. However, none of these methods have been evaluated for their effectiveness in analysis and prediction of nutritional value of sorghum in marketing situations. They are all highly correlated when groups of samples containing high and low tannin sorghums are used in making the correlations. However, in further tests, we have also seen a very high correlation in polyphenol contents of brown sorghums as measured by several methods.

The term "tannin" has been used to refer to polyphenolic compounds by many authors. Other authors, when speaking of tannins in sorghums, are talking about the whole complex of polyphenols. In the so called "low tannin" sorghums (without a pigmented testa) there are no condensed tannins or hydrolyzable tannins present. Instead, these "tannins" appear to be polyphenols of the monomeric or dimeric type since they do not precipitate proteins.

The polyphenols of the non-tannin type can cause problems of acceptability of sorghums in various food products by producing off colored products. This is especially pertinent in the production of teau and tortillas in which the sorghum is cooked in alkali. The polyphenols of sorghum are found primarily in the epicarp, the endocarp and the testa of the sorghum kernel. However, sometimes they are found in endosperm due to leaching or actual synthesis. In preliminary studies, the pigmented areas of some sorghum kernels have been removed with a dissecting knife to yield pure endosperm. When exposed to acid or alkali, the pure endosperms did not change color, indicating absence of the polyphenolic compounds.

The polyphenols can be removed partially by proper milling procedures; however, some of them cannot be eliminated without excessive removal of the starchy endosperm. Sorghum kernels that have been attacked by insects and diseases are often heavily pigmented, even the endosperm, which complicates the processing of the grain to achieve an all white product. Sorghum grown under good conditions, however, can be processed to produce a fine, white milled product.

E. Wet Milling

Sorghum has been wet milled on a commercial basis from 1948 to 1970 by a major wet milling firm in Corpus Christi, Texas, in the United States. Approximately 224 thousand tonnes were milled annually. The major products from sorghum wet milling were the same as those obtained from corn wet milling, namely starch, starch derivatives, glucose, oil, germ, gluten, feed meal, and solubles. Technology for wet milling sorghum was worked out and acceptable products were obtained. However, in 1970 sorghum wet milling was replaced by corn wet milling. The exact reasons for the switch are many and varied but relate to an increase in the relative price of sorghum versus corn that occurred when a major feed lot industry was developed in Texas. The price increase along with various other small differences in the milling properties of sorghum caused the industry to wet mill corn rather than sorghum. In general, important differences between the wet milling properties of corn and sorghum are as follows: 1) The recovery of starch from sorghum is slightly lower and it is more difficult to separate the starch from the gluten. 2) The oil yields are lower because sorghum contains one percent less oil than corn. 3) The gluten, a brownish-red color, does not contain high levels of carotenoid pigments, causing it to have limited value for use as a pigment in commercial broiler finishing. 4) Starch is stained by pigments and must be bleached. 5) Sorghum starch requires a little higher temperature to gelatinize, hence requires a longer cooking time. 6) Sorghum starch, because it was bleached, sometimes had slightly different cooking properties and appearances. For instance, the starch might have a scum on the surface resulting from the interaction of the starch with the bleach pigments therein. The overall effect resulted in a poorer quality starch than the original product. Most of these problems are related to the major difference in size between corn and sorghum and the changes in the relative amount of peripheral endosperm in the sorghum kernel. Although sorghum usually contains a slightly higher percentage of starch than corn, some cultivars contain starch in the pericarp and the small starch granules located in the peripheral endosperm are difficult to recover and separate cleanly from the protein surrounding them. In addition, the germ is more tenaciously embedded in the sorghum kernel than in the corn kernel.

The potential of using sorghum in wet milling is greater than it appears because development of new sorghum cultivars and hybrids with improved wet milling properties is possible. For instance, Watson (1970), and Norris and Rooney (1970), have shown that there are types of sorghums that have improved starch recovery in laboratory wet milling experiments. In some pilot plant scale experiments, these same hybrids had improved starch separation capabilities. Therefore, it is possible that sorghum could be useful as a wet milling raw ingredient where it is less costly than corn and especially if the new sorghum hybrids and cultivars with improved wet milling properties can be used. Norris (1971), determined the wet milling properties of several sorghum hybrids, cultivars and lines and found that considerable differences exist in the relative recovery and separability of the component parts of the sorghum kernel among these lines. He found that certain yellow endosperm sorghums and the soft, floury endosperm sorghums had the best overall wet milling properties. Furthermore, Norris and Rooney (1970), reported that the wet milling properties of sorghum grain appeared to be heritable although the number of samples in the study were small. These data suggest strongly that there is a distinct possiblity of developing sorghums with better potential for use in wet milling; however, some of the sorghums that appear to have the best wet milling capabilities such as the soft or floury endosperm types do not have good grain yield potential. Therefore, it would be necessary for the wet milling industry to contract grow the sorghum which might detract from the economics of the situation. However, in many countries of the world, for instance Mexico, sorghum commonly sells for 70-80% of the value of corn. In these cases development of sorghum hybrids specifically for certain industries appears to be practical.

F. Processing of Sorghum for Livestock Feeds

Of all the cereals, sorghum has to be more highly processed to produce maximum weight gains and to improve feed efficiency. Processing causes disruption of the subcellular structure, especially that of the peripheral endosperm of the sorghum kernel, and facilitates movement of digestive enzymes into contact with the starch, protein and other constituents. Now, more emphasis is on methods of wet processing sorghum to reduce the use of energy. In the United States, numerous methods have been devised to process sorghum for use in livestock rations. They include grinding, rolling, steam rolling, pressure cooking and flaking, popping, extruding, micronizing, reconstituting early harvesting and exploding (general). A method that has been quite popular in the large feed lots is to pressure cook the sorghum kernels and flake them in a way similar to the manufacturing of corn flakes, where the corn grits are cooked and flaked. The grain is treated with steam under pressure and put through large rolls which are adjusted to produce a thin flake. This method has improved efficiency.

However, the high cost of energy has caused it to fall into disfavor.

Micronizing has been used extensively to process sorghum and other grains. Micronizing consists of heating the grain with gas fired infra-red burners almost to the point of eversion, which is then followed by dropping the heated kernels through a roller mill. The roller mill is adjusted to produce a relatively thin flake. One advantage of this method is that it requires less energy than steam flaking to obtain the same feed efficiency. Sorghum varieties with a waxy endosperm processed more efficiently than nonwaxy varieties indicating that some sorghums may have less processing requirements (Rusnak et al. unpublished data). Micronized sorghums have been made into acceptable food products such as cookies or candy, where a substitute for oatmeal might be incorporated.

G. Dry Milling of Sorghum

The dry milling processes for sorghum grain vary from cracking, to produce a crude product, to debranning and degermination which produce highly refined fractions of bran, germ, meal and grits. The general methods for dry milling sorghum products have been summarized by Hahn (1969, 1970). The information on dry milling processes for sorghum is generally based on laboratory procedures because the limited number of sorghum dry mills precludes access to industrial methods, production data and markets.

H. Roller Milling

Most approaches to milling of sorghum have been to modify existing wheat milling equipment (Anderson, 1969; Anderson, 1969;Shoup, et al., 1970). Generally, the quality and yields of flour and grits milled from sorghum by roller milling are undesirable if refined products are required. Anderson and co-workers (1969 a, 1969 b) modified the roller milling operation by preprocessing the sorghum to remove bran and/or germ with some success in producing improved yields of good quality grits and flour. The yields of major fractions from roller milling are: bran and germ - 19.3%, break flour - 12.0%, flour less than 100 mesh per inch - 29.1%, endosperm retained on 100 mesh screen - 34.2%, and overs of 70 mesh grits gauze - 5.1%. If the grain is not completely debranned before roller milling, off color, specky endosperm products of limited utility are produced. Modified wheat milling equipment for dry milling sorghum does not achieve sufficient separation of bran and germ from the starchy endosperm when highly refined products are desired.

I. Decortication and Degermination Processes

The best way to obtain optimum separation of sorghum kernel is to use abrasive milling to remove the bran and impaction to degerm the decorticated kernels. Then, the fractions are separated with gravity tables and screening. Additional grinding may be used to release additional germ. Anderson and Burbridge (1971) removed 16-18% of the initial weight of the grain with an abrasive rice mill. Decorticated material was tempered and degerminated by impaction. Composition of products obtained by laboratory dry milling of sorghum with their procedure is presented in Table VI.

Table VI. Typical Yield and Composition of Products Obtained by Milling Sorghum with an Integrated Process.

Fraction	Yield %	Moisture %	Crude Fat %	Crude Fiber %
+14 Grits	50	5.9	0.8	0.2
+20	23	5.9	0.5	0.2
Flour	4	7.3	3.0	0.9
Germ	7	5.2	15.0	---
Feed	16	8.1	9.0	7.8

Anderson and Burbridge(1971)

This process uses conventional, readily available equipment and produces good quality grits and flour in high yields. It can produce highly refined sorghum products, but can easily be adjusted to produce higher yields of less refined products.

Most sorghum mills producing highly refined products use a pearling or decortication procedure followed by degermination and then separation of the particles into various fractions with a combination of screening and gravity tables. The yield of fractions are: grits, 67% bran 12%, germ 11% and fines 10%. The grits had 0.6% oil, 0.8% fiber, 0.5% ash and 92.5% brewers extract. Specifications for dry milled sorghum products have been definitely established. However, as for other cereal flours, particle size, moisture, and fat content are important criteria. The products range from decorticated whole grain to highly refined sorghum flours as listed in Table VII. by R. R. Hahn (1969). The yield of endosperm products varies from 60 to 90% of initial grain depending on degree of refinement and process. The hominy feed is a mixture of bran, germ and must contain more than 5% fat and less than 6% fiber.

Table VII
Analysis of Typical Dry Milled Commercial Sorghum Products

Product	Protein%	Oil%	Fiber%	Ash%
Whole Sorghum	9.6	3.4	2.2	1.5
Pearled Sorghum	9.5	3.0	1.3	1.2
Sorghum Flour - Crude	9.5	2.5	1.2	1.0
Sorghum Flour - Refined	9.5	1.0	1.0	0.8
Sorghum Brewers' Grits	9.5	0.7	0.8	0.4
Sorghum Bran	8.9	5.5	8.6	2.4
Sorghum Germ	15.1	20.0	2.6	8.2
Sorghum Hominy Feed	11.2	6.5	3.8	2.7

Hahn, 1969

J. Abrasive Milling Processes

The sorghum kernel is ideally suited to dehulling by pearling procedures using barley pearler equipment or rice milling equipment (Anderson and Burbridge, 1971). The abrasive action removes the pericarp of kernels with a high proportion of corneous endosperm with a minimum of kernel breakage. The composition of the abraded material depends on the quantity of material removed from the kernel. The protein content of the fines removed by abrasive milling has a broad peak which corresponds to removal of the aleurone layer, the peripheral area and perhaps some of the germ.

Rice milling equipment and other abrasive action mills have been used successfully in various parts of the world to mill sorghum by removal of the bran followed by crushing or grinding to produce flour, meal or grits for consumption. Desikachar (1976) has developed useful procedures for production and use of decorticated kernels in India. Small mills for sorghum have been developed in Nigeria and Botswana (Maiduguri Mill Project, 1976) and a major effort to mill sorghum exists at the Food Research Institute in the Sudan at Khartoum (Perten,

1976). Sorghum grits produced by abrasive milling without degermination are used in significant and increasing quantities as a brewing adjunct in Mexico.

K. Air Classification

Sorghum flours exposed to air classification can yield products with considerable range in composition (Stringfellow and Peplinski - 1966). Protein values for the separated fractions varied from 1.6 to 19% and a total protein shift of 20-40% was obtained. Flour from cultivars of sorghum with soft or floury endosperm were most effectively separated by air classification.

L. Sorghum Grains with Improved Milling Properties

A laboratory method for assessing the performance of small sorghum samples is based on combined degermination and decortication by the abrasive tearing action of a modified barley pearler (Rooney and Sullins - 1969; Maxson et al - 1970). The pearler was fitted with a wire brush instead of the carborundum wheel. The brush removes the pericarp and germ in one operation. The stocks are seperated by air flotation and by use of sieves. The fractions obtained are bran, germ, endosperm chunks, and a fines fraction that consists of pericarp, aleurone layer, germ and starchy endosperm. This method should be highly correlated with commercial milling processes which use abrasive milling followed by degermination. The grit yield ranged from 30.9 to 80.7% for 56 samples of grain from the world collection of sorghum. Sorghum kernels with thin, white pericarps, no testa, and a high proportion of corneous endosperm had the highest yields of endosperm fractions (grits). Several workers (Kapasi-Kakama - 1976; Desikachar - 1976) have demonstrated that sorghum cultivars have significantly different milling properties. In general, the more corneous endosperm sorghums with thin pericarp (no testa) have the best dry milling properties for abrasive milling procedures.

Sorghum hybrids with improved milling properties have been developed in our sorghum improvement program. TAM 670 and TAM 680 were the first USA hybrids developed using exotic germ plasm from the world sorghum collection. The milling properties of seven sorghum hybrids and proximate analyses of their grits and flour were evaluated (Table VIII). The effect of environment on the milling properties of two of these hybrids was tested by growing them in seven locations. The range and mean for several milling properties are listed in Table IX. Grain from these hybrids has consistently improved milling

TABLE VIII

MILLING PROPERTIES OF GRAIN AND PROXIMATE ANALYSES OF GRITS AND FLOUR FROM SEVEN SORGHUM HYBRIDS GROWN AT LUBBOCK, TEXAS (1972)

Hybrid	MILLING PROPERTIES					PROXIMATE ANALYSES			
	Grain	Wt.	Grit	Grit	Flour	Grits		Flour	
	Hardness	lbs./Bu	Density	Yield %	Yield %	Ash%	Fat%	Ash%	Fat%
TAM680	22.0	60.4	1.42	80.7	68.1	0.50	0.47	0.51	1.17
TAM670	24.0	57.4	1.42	80.8	65.2	0.77	0.51	0.47	1.13
ATx399x TAM428	22.3	59.8	1.42	79.7	69.2	0.55	0.76	0.67	1.26
ATx399x Tx2536	20.1	57.3	1.42	77.0	68.9	0.53	0.93	0.75	1.48
RS671	16.4	57.3	1.44	75.7	69.7	0.56	0.91	0.77	1.56
ATx378 Tx2536	19.0	57.5	1.41	76.5	67.3	0.44	0.62	0.75	2.00
RS610	18.0	58.2	1.43	76.4	72.4	0.54	1.22	1.76	1.73

TABLE IX

PHYSICAL PROPERTIES AND DRY MILLING CHARACTERISTICS OF RS671, TAM680, and TAM670 GROWN AT SEVEN LOCATIONS IN 1970

	RS671	TAM680	TAM670
Test weight (lbs./bu)			
Mean	57	60	60
Range	55.7-58.4	59.5-61.0	59.6-67.1
1000 kernel weight(mg)			
Mean	25	28	26
Range	22.9-26.9	26.9-30.6	22.7-28.4
Density			
Mean	1.334	1.368	1.364
Range	1.324-1.349	1.356-1.377	1.341-1.386
Hardness			
Mean	24	43	38
Range	15.4-38.4	37.3-49.4	23.3-50.5
Grit Yield(%)			
Mean	58	70	71
Range	49.5-61.7	63.7-77.4	65.8-74.7
Protein(%) (Nx6.25)			
Mean	12	12	12
Range	9.8-13.3	8.9-13.6	10.5-13.4

properties over standard commercial hybrids such as RS 671. Milling performance with the abrasive procedure is highly, positively related to hardness and density of the grains (Table X). The grain with a red pericarp produced "grits" with the same color and appearance as those from the white pericarp grain sample.

Great potential exists to develop sorghum varieties that produce grain with improved milling properties. A corneous bird resistant sorghum kernel could be developed to be abrasively milled to more completely remove the condensed tannins while still achieving optimum yields of starchy endosperm free of condensed tannins.

TABLE X

Table 4: Correlation coefficients between physical properties of grain from RS671, TAM680 and TAM670 grown at seven locations in Texas (1970)-(n=21).

	Test Wt.	1000 Kernel Wt.	Density	Grit Yield	Hardness
Test Wt.	-	0.68**	0.95**	0.93**	0.68**
1000 Kernel Wt.		-	0.50*	0.37 ns	0.60*
Density			-	0.96**	0.80**
Grit Yield				-	0.74**
Hardness					-

** Significant at 1% level.

* Significant at 5% level.

ns Non-significant

M. Sweet Sorghum for Molasses and Sugar

Sweet sorghum or "sorgo" has been used to produce syrup for many years. Syrup yields depend on the quantity of juice extracted and its solids content. Coleman (1970) described good quality sorghum syrup as mild and sweet with a distinctive sorghum flavor. A light amber color is preferred. Dark syrup is often strong and may have a burnt flavor. Sorghum syrup should have a high viscosity with little or no crystallization. The viscosity of the syrup occurs because of small amounts of starch in the juice.

Syrup is produced in a four step process. First, the juice is pressed from the stalks and stored in shallow pans to allow starch and trash to settle. Next, the juice is clarified. Clarification is usually a heat process, but filtering aids may be used. Open evaporators are used to clarify, evaporate, and finish the syrup. The construction is such that cold juice enters one end and syrup is poured off at the other. In the batch system clarification and concentration takes place in one evaporator. After clarification, the process of concentrating the syrup is done at a low temperature in a vacuum pan to reduce scorching and color development. Steam heating is utilized, thus increasing cost. The syrup is finally cooked at 106°C-110°C for flavor development and to meet specific gravity specifications. Often, invertase or malt diastase is used to break down starch in the semi-syrup to prevent future crystallization or jelling.

A pilot plant in Weslaco has been used to produce sugar from sweet sorghum (Smith - 1969, 1975). Sweet sorghum has the potential to extend the sugar processing season because it matures at a different time than sugar cane. Sorghum juice used for sugar production must have a purity of at least 75% and a low rate of sucrose inversion. Starch is an undesirable component because it interferes with juice concentration. Aconitic acid content should be low to prevent interference with sucrose crystallization. Sucrose production is similar to that of the sorghum syrup, except the sucrose is finally crystallized from the syrup.

The biggest potential use for producing sugar from sweet sorghum is that of a supplementary sugar source to sugar cane and sugar beet. Cost of production can be lowered considerably by using the same production equipment with small adapatations. The sweet sorghum is mature and ready for sugar production during the off-season months of other sugar processing.

N. Sorghum Beer

Two major kinds of beer are made with sorghum. True sorghum beer is produced in most African countries. It is produced in various ways but usually includes the use of sprouted sorghum (mah) with or without additional sorghum flour. An innoculum from a previous batch of beer may be utilized to enhance the fermentation. In other instances, the sorghum flour is allowed to ferment. The fermented material is consumed as a thick beverage which has a high solids content. Sorghum beer production in the Republic of South Africa has been mechanized into large modern brew facilities with a total of a billion liters of beer produced annually. Over the years, Novelli (1959 - 1962) and colleagues have determined the optimum procedures for sorghum beer production and have extensively studied the biochemistry of sorghum malting and brewing.

The use of sorghum in making European beer varies considerably. Significant quantities of sorghum brewers grits were once used as an adjunct in brewing in the United States (Hahn and Stewart - 1965; Stewart and Hahn -1965; Hahn -1966). The beer produced by using sorghum brewers grits has comparable qualities with those produced by using corn grits as the brewers adjunct (Hahn - 1966). Sorghum grits are being used in significant amounts to produce some of the better quality beers available in Mexico. The use of sorghum brewers grits appears to be determined by economic factors and other considerations. The sorghum grits used in the brewing of beer reported by Hahn and co-workers (1965) were made from sorghum with a red pericarp. Present development of sorghum cultivars specifically designed for better milling properties should mean that sorghum grits with superior brewing properties could be available. In summary, the use of sorghum as a brewers adjunct is entirely feasible but it is not being used extensively in the United States at the present time.

O. Pasta Products

The use of sorghum for pasta, alone and mixed with wheat, various starches and nonwheat flours, has been studied extensively by French workers (Miche et al. - 1978). The major problems associated with production of pasta from sorghum are color, texture and modifications in technology to achieve adequate gelatinization. Sorghum pasta made with semolina have a gritty, sandy texture. Grittiness can be overcome by use of flour and by choosing a softer endosperm variety. Up to 75% sorghum flour can be used to make pasta with good technological and cooking quality on a pilot scale. However, the pasta is of lower protein content than desirable. In some sorghum cultivars, brown color due to oxidation of phenolic pigments may occur during pasta formation.

Desikachar (1976) has developed relatively simple procedures for production of vermicelli from sorghum which can be used at the village level.

P. Fermentation Industry

Approximately 2.5 million bushels of sorghum grain are used for production of alcohol annually. The whole grain is ground to produce a 12-60 mesh meal. The grain is cooked and the mash is hydrolyzed by enzymes or acid. The liquid is cooled and inoculated with yeast. After yeast fermentation, the alcohol is recovered by distillation. Beverage alcohol follows a similar procedure except malt enzymes are preferred for hydrolysis. The distillate which is approximately 95% alcohol by volume is filtered through charcoal, blended and aged. Sorghum has been used to produce latic acid, riboflavin, microbial polysaccharides, antibiotics, and citric acid. Use of sorghum as a substrate depends on its price relative to other grains. The use of sweet sorghum and sorghum grain for alcohol for energy production is being seriously considered. Alcohol yield of sorghum is approximately six gallons per bushel (56 lbs).

Q. Nutritional Value of Grain Sorghum

The nutritional value of sorghum has been improved in recent years with the production of U.S. yellow sorghums. Tanksley (1975) summarized information on swine feeding trials, reporting that sorghum has 95% the feeding value of corn (Table XI). His group has found consistently that growing and finishing swine fed a properly supplemented sorghum diet gained weight at about the same rate as those fed a properly supplemented corn diet. But, the swine on the sorghum diet consumed more feed than those consuming corn, thus having a decreased feed efficiency. Sorghum used in these trials were typical hybrids producing U.S. yellow sorghum. Brown sorghums (with a testa) had lower feed efficiency than the yellow sorghums.

TABLE XI. PERFORMANCE OF GROWING-FINISHING SWINE FED CORN AND THREE SORGHUMS WITH DIFFERENT ENDOSPERM TYPES[1]

Criteria	Yellow Corn	Sorghum endosperm type			Avg. of Sorghum Diets
		N-Y[2]	H-Y[2]	Y[2]	
COMBINED SOUTH PLAINS AND BLACKLAND GRAINS					
No. of animals	48	48	47	47	---
Avg. initial wt., lb.	48.1	48.3	48.7	47.7	48.3
Avg. final wt., lb.	216.2	216.4	214.6	215.4	215.5
Avg. feed intake/day, lb.	5.48^{a}	$5.66^{a,b}$	5.73^{b}	5.71^{b}	5.70^{*}
Avg. daily gain, lb.	1.71	1.73	1.68	1.70	1.70
Feed:gain ration	3.21^{a}	$3.27^{a,b}$	3.41^{c}	$3.35^{b,c}$	3.34^{*}
Relative feed efficiency, %	100.0	98.13	93.77	95.64	95.95

a,b,c - Means on the same line with different superscripts differ significantly ($P < .05$).

* - Corn significantly different from the average of the sorghum ($P < .05$).

1 - T. D. Tanksley, Jr. (1975)

2 - N-Y-non-yellow; H-Y-heteroyellow; Y-yellow. These are typical modern U.S. sorghum hybrids.

R. Food Products

Sorghum is one of the important cereals used for food purposes in Asia, Africa and Central America. Various types of foods made from sorghum whole grain, flour or starch are chapati, boiled sorghum, babri, and various kinds of porridges such as ugali, akamu, eko, kamu, koko, ogi etc.(Vogel and Graham - 1978). Sorghum is also used in beer and beverage making in South Africa (Novelli - 1959 - 1962). In Central America, sorghum alone or in combination with corn is used in tortilla making (Nolaso - 1978).

In the United States, sorghum is not used for food purposes. However, various academic institutions have developed baked products in which wheat flour was partially replaced with sorghum flour such as in bread, cookies, cakes and brownies. These experiments were done to determine the feasibility of including sorghum flour in baked products and to help the various tropical countries to reduce the import of costly wheat from U.S., Canada and Australia. Extensive work in this regard has also been done by Tropical Products Institute (TPI) of England (Dendy et al - 1971). Scientists of TPI have developed various bread formulations where they use large quantities of flours from locally grown cereals such as sorghum and corn. In this section we will devote our attention to the characteristics of sorghums for use in foods such as ogi, ugali, tortillas and chapati and summarize the results of the several baking experiments in which wheat flour was replaced with different levels of sorghum flour.

1. Ogi. Ogi is the most important food use of sorghum in Nigeria. It is made by soaking sorghum grain in water at room temperature for 2-4 days. This softens the kernels and permits adequate fermentation. The fermented sorghum is then washed and poured into a mill with large quantities of water added. The slurry of milled grain is then passed through a sieve with U.S. mesh size of 60 to 100. The filtrate is collected in an earthenware pot and allowed to settle. The sediment is ogi.

Sorghums with various grain characteristics are used for ogi making in Nigeria. However, sorghum cultivars with hard, corneous, non-waxy endosperm, plump kernels, white-colored pericarp and low tannin content have acceptable ogi making properties. Cultivars with waxy endosperm are not desirable because the consistency of ogi made from waxy grain is thin (Table XII).

2. Ugali. Ugali is a dumpling made in Kenya from coarse sorghum flour. It is either made from whole sorghum flour or is mixed with maize. It is made by adding flour to boiling

TABLE XII. LABORATORY OGI MAKING PROPERTIES OF SEVEN SORGHUM CULTIVARS

Hybrid	Ogi Yield %	Tannin Content of Ogi (mg/g)	Cooking Properties 2/		Dry Ogi Color 1/			Organoleptic Properties 3/	
			Pasting Tempt.	Pasting Viscosity	L	a	b	Taste	Consistency
GA615	68	11.8	76.5	72	62.9	7.2	11.2	2.2	2.6
NK300	71	11.9	78.3	64	72.2	4.8	10.4	1.4	4.6
TAM680	73	3.1	87.2	45	78.3	3.3	11.8	3.0	3.8
FUNK-G-766W	72	1.3	80.0	43	82.8	1.1	10.3	4.0	3.8
Texicoa 64	72	1.8	74.3	25	79.9	1.9	10.9	4.4	2.4
CSH-5	75	1.2	85.5	48	78.5	2.0	11.6	1.4	3.8
ATx413x NSA954	79	6.2	72.2	37	79.7	4.9	12.1	4.4	2.8

1/ 'L', 'a' and 'b' are values from Hunter Lab Color Difference Meter.

'L' is the intensity of color.

+'b' = yellowness

-'b' = blueness

+'a' = red, -'a' = green

2/ Amylograph data

3/ 1 = Poor

5 = Excellent

water, while constantly stirring to avoid lumps. Ugali is generally served with meat, a vegetable sauce or stew. Cooking quality of different cultivars varies significantly. Pale yellow or white colored ugali is preferred. White corneous grains produce a better ugali than the one made from softer endosperm grain. Texture and keeping qualities of ugali from floury grain are not acceptable. Several sorghums from world collection evaluated at ICRISAT (Murty - 1979) indicated that most of the cultivars had acceptable characteristics for making ugali (Table XIII).

3. Tortilla. Tortilla is a non-leavened staple food made from corn, consumed in large quantities in Mexico and Central America. In Central American countries, such as Guatemala and Honduras, tortillas are also made from sorghum alone or in combination with corn (Nolasco - 1978). This is predominant where corn yields are low due to the prevailing hot and dry climates and where sorghum consistently out yields corn. Sorghum however, is not a preferred cereal for tortilla making because it produces tortillas with light green to dark brown color, when compared to the yellow color of corn tortillas. However, it is feasible to develop a sorghum which could produce fairly acceptable tortillas. The sorghum should have a white or colorless pericarp with corneous to intermediate and non-waxy endosperm. It should also be free of pigments that are leached from glumes to the grain. Sorghum with floury endosperm produces sticky masa. The waxy endosperm sorghum has undesirable cooking and processing properties. Preliminary experiments indicated that highly acceptable tortillas can be made from sorghums with 25% of the pericarp removed by pearling with a barley pearler (Palacios et al. unpublished data). Brown or red pericarp sorghums produce unacceptable dark colored tortillas. The dark color is partially associated with condensed tannins and non-tannin polyphenols.

White pericarp sorghums produce light-greenish yellow tortillas. However, this depends on the environment that exists during maturation of sorghum grain (Table XIV). Weathering of white pericarp grain makes it unacceptable for tortilla making; while hot and dry climate during maturation of white sorghums produces grain with fairly acceptable tortilla making properties. Tortillas with the lightest and most intense yellow color were made with the sorghums from Central America. Maldondi 35-1, a sorghum with excellent quality grown for chapatis in India produced tortillas with light color. A simple color test was developed to evaluate the tortilla making properties of white sorghums in early generations of breeding programs. The test consisted of heating 5 kernels of clean white sorghum with 10N NaOH in a test tube at 60°C for

Table XIII
UGALI AND UGI MAKING PROPERTIES
OF
90 DIFFERENT SORGHUM CULTIVARS 3/

Characteristics 1/	Mean 2/	Range
Color Score	2.55	1.08-5.00
Taste	2.38	1.16-4.66
Texture	2.14	1.11-4.66
Keeping Quality	2.35	1.00-5.00
Ugi gel consistency (dia.mm)	68.47	58-112

1/ Subjective evaluation 1 = Best, 5 = Poor

2/ Average of grain samples from 90 sorghum varieties.

3/ Murty (1979)

2 hours. The color developed at the end of heating ranges from light yellowish green (rated #1) to dark brown (rated #4). Sorghum producing a light yellow color will make light colored tortillas (Khan and Rooney - unpublished data).

4. Chapati. Chapati is non-leavened Indian bread made from 100% sorghum flour. It is consumed predominantly in South-Central India. Significant differences exist among sorghum hybrids or cultivars for chapati making properties such as kneading and rolling qualities of the dough; and percent of water absorbed by flour during dough making. Generally, chapati color, aroma, and taste are affected by the pigmentation of the pericarp. (It was noted that the range of variation within the pearly white group for all characteristics was as broad as that observed in the whole set of samples). Organoleptic properties of chapatis made from corneous endosperm sorghums are acceptable. Sorghums with the floury endosperm absorb more water and produce poor quality chapatis. A good chapati should taste good, be soft in texture, remain palatable for 10 hours at room temperature and should have creamy white color. Tests such as water absorption of the grain (Murty - 1979) and flour (Waniska - 1976); peak viscosity and dough stability (Waniska - 1976), kneading quality score (Murty - 1979), keeping quality of chapati (Murty - 1979 and Waniska - 1976), were developed at Texas A&M University and ICRISAT, to evaluate the chapati making properties of sorghum hybrids in the early stages of breeding programs.

(Tables XV, XVIand XVII). Selections of sorghum for chapati quality is best undertaken from F5 generation. Selection for visual tests such as colorless thin pericarp and corneous endosperm may be carried out until F4 generation followed by laboratory tests for chapati quality from the F5 generation.

S. Baked Products

Sorghum flour does not contain any gluten in its protein, therefore acceptable leavened products cannot be made with 100% sorghum flour. Presence of 70-80% wheat flour is necessary in a baking formula. Several acceptable baked products such as bread, cookies and brownies were made by including 15-20% of sorghum flour in the formula at Texas A&M University (Cereal Quality Lab unpublished data). In this section we will summarize the results of several baking experiments that were conducted at Texas A&M University (Cereal Quality Lab and elsewhere (Dendy et al., 1971).

1. Bread. Several bread making experiments were conducted to evaluate the feasibility of including maximum level of sorghum flour in a bread formula and obtain an acceptable finished product. Wheat flour was replaced with various levels of sorghum flour (5-100%) in a regular straight-dough (Rooney et al. - 1972) and short-time dough (Khan et al. - 1975) baking procedures. In another experiment, bread with various levels of sorghum flour was made, including an emulsifier, sodium-stearoyl-lactylate (SSL)(unpublished). A strong wheat flour (12-14% protein) was used in all these bread making experiments. Results of these experiments indicated that high levels of sorghum flour can be more successfully included in a short-time dough system than in a straight dough formula. A significant finding of the bread making experiments was that water absorption of the dough decreased with the increase in the level of sorghum flour in the formula. This was due to the decrease in the gluten content of the dough and sorghum protein absorbed less water than wheat gluten. Water absorption of the dough containing 100% sorghum flour was about 19-20% less than one mixed with 100% wheat flour (Table XVIII).

Dilution in gluten content also reduced the loaf volume, lightened the crust color and darkened the crumb color of the bread. Light crust color resulted from the decrease in availibility of basic amino acids (Table XVIII). Acceptable bread was obtained when 20-30% of the wheat flour was replaced with sorghum flour. No significant differences were observed between breads with 100% wheat flour and the one made with 5% sorghum and 95% wheat flour. These results were similar to the one reported by Avanzi and Gruff (1939), Borasio (1937),

TABLE XIV: THE EFFECT OF DIFFERENT SORGHUM CULTIVARS AND ENVIRONMENT ON TORTILLA PROPERTIES

Sorghums from:	Diameter			COLOR 1/ L		a		b	
	n	$\bar{x}$	Range	$\bar{x}$	Range	$\bar{x}$	Range	$\bar{x}$	Range
Central America	7	10.4	10.3-1.05	51.9 (58.5)	49.2-54.9 (52.8-61.8)	-1.05 (2.8)	-1.6 to -0.6 (2.2 to 3.6)	16.7 (16.7)	14.4-18.6 (14.4-18.1) 2/
Lubbock Red	3	11.4	10.4-12.7	35.5 (37.7)	34.4-36.8 (36.6-37.9)	1.4 (9.9)	0.8-1.8 (9.4-10.5)	10.0 (12.6)	9.7-10.5 (12.3-12.9)
White	7	10.9	10.3-12.6	40.3 (53.6	39.6-43.7 (52.1-56.0)	0.82 (3.7)	-0.8 to 2.5 (3.0-4.3)	11.7 (16.3)	7.5-13.3 (14.3-17.3)
Temple Red	2	10.4		32.5 (39.8)		3.2 (9.6)		11.1 (14.6)	
White	10	10.8	10.3-12.7	46.1 (55.8)	41.9-50.0 (51.0-61.0)	0.66 (3.8)	-0.7 to 1.6 (3.1-5.1)	15.2 (18.5)	14.0-16.2 (17.2-20.1)
Corn	12	11.2	10.3-12.1	64.9 (62.9)	60.6-68.2	-1.6 (0.7)	1.17 to -0.7	22.2 (20.6)	20.1-23.8

1/ Determined with Hunter Lab Color Meter: L- 100=white, 0=black; -a=red, +a=green; -b=blue, +b=yellow

2/ Values within parenthesis are for grains used in making tortillas.

TABLE XV: VARIABILITY FOR SOME GRAIN AND CHAPATI QUALITY ATTRIBUTES AMONG 367 SORGHUM GENOTYPES

Attribute	Mean	±	s.e.	Range Minimum	Range Maximum
Corneousness score 1/	2.531	±	0.036	1.0	5.0
Grain weight gm/100 seeds	3.622	±	0.045	1.464	7.630
Breaking strength Kg. 3/	8.557	±	0.098	0.770-5.09	14.58
Grain density	1.235	±	0.002	1.002	1.375
% Water absorption of the grain	25.567	±	0.245	12.44	42.10
ml of water for dough	27.646	±	0.125	20.93	38.86
Kneading quality - score 2/	1.157	±	0.018	0.50	3.00
Diameter of chapati	22.059	±	0.067	15.25	27.75
Chapati taste score 2/	2.308	±	0.036	1.00	5.00
texture score	2.231	±	0.027	1.00	4.50
flavor score	1.462	±	0.024	1.00	3.00
keeping quality score	2.833	±	0.033	1.00	5.00

1/ 1=corneous, 5=floury

2/ 1=good, 5=bad

3/ Avg. pressure needed to break 10 grains individually on KM Grain Hardness Tester

TABLE XVI: PROPERTIES MEASURED WITH A FARINOGRAPH OF CHAPATI DOUGHS FROM SORGHUM FLOURS MILLED FROM NON-WAXY SAMPLES WITH DIFFERENT ENDOSPERM TEXTURE 1/

Variables	Endosperm Texture				
	Floury	Inter. Floury	Inter- mediate	Inter. Corn.	Corneous
No. of Samples	1	2	11	1	2
Water Absorption (% of flour)					
Range	-----	72.0-76.0	72.0-77.0	-----	80.0-83.0
Mean	65.0c	74.0b	73.7b	72.0b	81.5a
Mixing Time (min)					
Range	-----	2.7-3.0	2.7-4.5	-----	3.5-5.0
Mean	2.5c	2.8c	3.8b	3.0c	4.3a
Peak Viscosity (B.U.) 2/					
Range	-----	300-340	2.7-4.5	-----	240-260
Mean	240bc	305a	277a	210c	250b
Peak Stability (min)					
Range	-----	4.5-6.5	3.05-5.3	-----	9.5-5.2
Mean	9.0a	5.5b	4.0c	4.2c	4.4c
Factor 3/					
Range	----	8.1-10.2	6.5-12.1	-----	9.1-12.0
Mean	6.0b	9.0ab	10.4a	6.3b	10.6a

1/ Values with the same letters are not significantly different at the 5% level according to DRMT. (Duncan's Multiple Range Test)

2/ B.U. = Brabender Unit. 3/ Factor = (Mixing time)(Peak viscostiy)/(100).

TABLE XVII: RHEOLOGICAL PROPERTIES OF SORGHUM CHAPATIES AFTER 4 HOUR STORAGE PREPARED FROM NON-WAXY SAMPLES WITH DIFFERENT ENDOSPERM TEXTURE 1/ MEASURED WITH AN INSTRON

Variables	Endosperm Texture				
	Floury	Inter. Floury	Inter- mediate	Inter. Corn.	Corneous
No. of Samples	1	2	11	1	2
Stress (N/cm^2)					
Range	-----	22.4-26.9	13.1-25.8	-----	17.0-21.6
Mean	25.1a	24.7a	20.5ab	20.6ab	19.3b
Strain (cm/cm)					
Range	-----	0.14-0.15	0.07-0.14	-----	0.12-0.13
Mean	0.08c	0.14a	0.11b	0.07c	0.12b
Modulus of Elasticity ((N/cm)/Ccm/cm))					
Range	-----	159.3-183.2	137.4-252.6	-----	129.0-185.0
Mean	294.2	171.2b	184.8b	307.8a	157.0b
Resilience ($(N\text{-}cm)/cm^3$)					
Range	-----	0.74-0.96	0.20-0.74	-----	0.53-0.58
Mean	0.51b	0.85a	0.54b	0.34b	0.55b
Dowel Test 2/					
Range	-----	1.2-2.0	1.0-2.4	-----	1.2-2.2
Mean	3.0a	1.6b	1.9b	2.4ab	1.7b

1/ Values with the same letters are not significantly different at the 5% level according to DMRT.

2/ Keeping quality of chapaties was evaluated with the dowel test according to the following scale: 1=no breaking or cracking, 2=slow or partial cracking and 3=rapid breaking.

TABLE XVIII: BREAD BAKING EXPERIMENTS IN WHICH SORGHUM FLOUR WAS SUBSTITUTED FOR WHEAT FLOUR USING DIFFERENT BAKING SYSTEMS.

	Baking Properties									
	Straight Dough Method					Short-time Dough Method				
% of Sorghum Flour	Absorption %	Loaf Volume 0% SSL	Loaf Volume 1% SSL	Crumb Grain 1/	Score Color 1/	Absorption %	Loaf vol. C.C.	Crumb Grain 1/	Score Color 2/	Crust Color 2/
0	65	3050	3200	90	91	69	3225	90	81	28
5	65	3075	3250	90	91	-	-	-	-	-
10	64	2950	3150	90	91	67	3200	80	80	31
15	63½	2825	2950	89	91	-	-	-	-	-
20	61½	2600	2725	87	90	67	2900	80	76	28
25	62	2400	2550	85	89	-	-	-	-	-
30		2350	2425	84	87	65	2800	80	75	32
40						62	2400	50	74	35
50						60	2250	40	70	42
60						58	2025	30	65	44
100						50	1700	Poor	Dark	-

1/ Subjective Evaluation

2/ 'L' value of Hunter Lab Color meter.

Dendy et al. (1972) and Mustafa (1973). SSL improved the baking properties of the bread made with various levels of sorghum flour in a straight dough procedure. Hart et. al (1970) developed a formulation to make bread with 100% sorghum flour and additives such as methyl cellulose, glycerol mono stearate cereal starches and emulsified shortenings.

There were no significant differences in the baking performance of sorghum flour milled from grain of each of 8 sorghum hybrids grown at Texas Agricultural Experiment Stations. Flour from each hybrid produced bread with acceptable volume and crumb grain score. However, flour from red pericarp sorghum imparted slightly dark color to the bread crumb.

2. Cakes. In another baking experiment, cake flour was replaced with 0, 15, 25 and 50% of sorghum flour in a standard AACC (1970) formula. Cakes made with 50% sorghum flour had lower volume, coarser grain and darker crumb than the one made with 15 and 25% of sorghum flour (Table XIX). Sorghum flour had no effect on the tenderness and crust color of the cake. Less air was incorporated into the batter with the increased sorghum flour.

TABLE XIX: BAKING PROPERTIES OF VARIOUS CONCENTRATIONS OF SORGHUM FLOUR IN A HIGH RATIO WHITE CAKE

Concentration of Sorghum Flour	Sp. gr. of Batter	Weight (g)	Volume (c.c.)	Sp. Volume (c.c./g)	Subjective Analysis				
					Crust Color	Crumb Color	Symmetry	Tender-ness	Grain
0	0.93	376	1025	2.73	15	30	5	10	35
15	0.95	380	1025	2.70	15	28	5	10	32
25	0.85	375	1025	2.73	15	25	4	10	28
50	0.83	380	825	2.17	15	20	3	10	25

II. CONCLUSION

In this paper we have attempted to summarize the past work in sorghum processing and utilization. This is by no means a complete summary, but rather a review of the work done in our laboratory and others around the world. Much work remains to characterize sorghum for use in food products. Future research must include development of quick and easy methods for use in determining food quality of sorghums which can be used by plant breeders for selection of sorghum cultivars for food use.

REFERENCES

Akingbala, J.O. (1979) M.Sc. Thesis, Texas A&M University, College Station, TX

Anderson, R. A. (1969) The Northwestern Miller. 276(10):10.

Anderson, R. A., and Burbridge, L. H. (1971). The Northwestern Miller. 6:278.

Anderson, R. A., Montgomery, R. R., and Burbridge, L. H. (1969) Sorghum Cereal Sci. Today. 14:366.

Aucamp, M. C., Grieff, J. T., Novellie, L., Papendiok, B., Schwartz, H. M., and Steer, A. G. (1961). J. Sci. Food Agric. 12:449.

Avanzi, E. and Graff, G. (1939). Industria Saccar. Ital. 32(5):273.

Badi, S. M. and Hosney, R. C. (1976). Intl. Assoc. of Cereal Chem. 37.

Bidwell, G.L., Bopst, L. E., and Bowling, J. D. (1922). U.S. Dept. Agr. Bull. 1129:1-8.

Borasio, L. (1937). G. Risc. 27(1):7.

Bredon, R. M. (1961). Uganda Govt. Dept. Vet. Sci. and Animal Industries (Cyclostyled).

Bressani, R., and Rios, B. J. (1962). Cereal Chem. 39:50-58.

Coleman, O. H. (1970). "Sorghum Productions and Utilization" Avi Pub. Co. Inc, Westport, Conn.

Dendy, D. A. V., James, A. W., and Clarke, P. A. (1971). Sorghum Newsletter. 14:22.

Desikachar, H. S. R. (1976). Int'l. Assoc. of Cereal Chem. 41.

Deyoe, C. W., and Shellenberger, J. A. (1965). J. Agric. Food Chem. 13(5):446-450.

Doggett, H., (1970). "Sorghum". Longmans Green and Co. Ltd. 250.

FAO & WHO. (1971). "Energy and Protein Requirements." FAO Nutr. Meetings, Rept. Series. 52:55

Grain Sorghum Research in Texas. (1971). Texas Agr. Exp. Sta. Consol. PR:2938-2949.
Hahn, R. R. (1970). "Sorghum Production and Utilization." Avi Publishing Co. Conn. 573.
Hahn, R. R. (1969). Cereal Sci. Today. 14:234-237.
Hahn, R. R. (1970). "Sorghum Production and Utilization." Avi Publishing Co. Conn. Ch:16.
Haikerwal, M. and Mathieson, A. R. (1971). J. Sci. Food Agric. 22:142-145.
Hart, M. R. (1970). J. Food Sci. 35:661.
Hirata, Y. and Watson, S. A. (1967). "Starch Chem. and Tech. II." Acad. Press. N.Y. 29.
Hubbard, J. E., Hall, H. H. and Earle, F. R. (1950). Cereal Chem. 27:415.
Hulse, J. (1979). "Sorghum and The Millets: Their Composition and Nutritive Value". Acad. Press (In Press).
Jambunathan, R. and Mertz, E. T. (1972). Purdue University Research Progress Report. 9:43-56.
Kapasi-Kakama, J. (1976). Intl. Assn. for Cereal Chem. Vienna. 21.
Karim, A. and Rooney, L. W. (1972). J. of Food Sci. 37:369.
Karim, A. and Rooney, L. W. (1972). J. of Food Sci. 37:365-368.
Karim, A. and Rooney, L. W. (1972). J. of Food Sci. 37:365.
Khan, M. N. and Rooney, L. W. (Unpublished data).
Khan, M. N. and Rooney, L. W. (1979). 39th Ann. IFT Meeting. St. Louis.
Khan, M. N. and Rhee, K. C., Rooney, L. W. and Cater, C. M. (1975). J. of.Food Sci. 40:580.
Kummerow, F. A. (1946). Oil Soap. 23:273.
Lamar, et al. (1972). Cereal Quality Lab of the Sorghum Improvement Research Program of Texas Agric. Exp. Sta.
Lime, B. J. (1979). "Tropical Foods:Vol. 1". Acad. Press. New York.
Maiduguri Mill Project. (1976). Agricultural Food and Nutrition Sci. Division. I.D.R.C. Canada.
Maxson, E. D., Fryar, W. B., Rooney, L. W. and Krisheaprasad, M. N. (1970). Cereal Chem. 48:478-490.
Miche, Jean-Claude. (1978). Institut de Recherches Agronomigiues Tropicales (IRAT). France.
Murty, D. S. (1979). Cimmyt-ICRISAT Undp. Policy Advisory Committee Meeting. Mexico.
Mustafa, A.M.I. (1973). J. of Food Sci. 5:41.
Neucere, N.J. and Sumrell, G. (1979). J. Agric. Food Chem. 27(4):809.
Nolasco, P. (1979). Personal Communications.
Norman, F. L., Hogan, J. T. and Deobald, H. J. (1965). Cereal Chem. 42:359-367.
Norris, J. R. (1971). Ph.D. Dissertation, Texas A&M Univ.

Norris, J. R. and Rooney, L. W. (1970). Cereal Chem. 47:64.
Novellie, L. (1959). J. of Sci. Food Agric. 10:441-449.
Novellie, L. (1962). J. of Sci. Food Agric. 13:124-126.
Perten, H. (1976). Intl. Assn. of Cereal Chem. 47.
Rao, S. N. R. and Desikachar, H. S. R. (1964). J. of Food Sci. Tech. 1:40-42.
Rao, N. G. P. and House, L. R. (1971). "Sorghum in Seventies". Oxford and IBH Pub. House.
Rasper, V. F. (1973). Can. Inst. Food Sci. Technol. J. 6: 188-192.
Rooney, L. W. (1973). Industrial Uses of Cereal.
Rooney, L. W. (1978). Cereal Chem. 55(5):584-590.
Rooney, L. W. and Clark, L. E. (1968). Cereal Sci. Today. 13:258..
Rooney, L. W., Fryar, W. B. and Cater, C. M. (1972). Cereal Chem. 49(4):399-406.
Rooney, L. W., Gustafson, C. B., Clark, S. P. and Cater, C. M. (1972). J. of Food Sci. 37:14.
Rooney, L. W. and Sullins, R. D. (1969). Cereal Chem: 46: 446.
Rooney, L.W. and Sullins, R.D. (1969). Cereal Chem. 46: 486-490.
Sanders. E.H. (1955). Cereal Chem. 32:12.
Shoup, F.K., Deyoe, C.W., Farrell, E.P., Hammond, D.L. and Miller, G.D. (1970). Food Tech. 24(9):88.
Smith, B.A. (1969). U.S. Patent No. 3,442,704.
Smith, B.A. (1969). U.S. Patent No. 3,464,856.
Smith, B.A., Rombo, R.V., Smith, R.C., DelaCruz, R.A., Griffithe, F.P., and Cowley, W.R. (1972). Proceedings of 14th Congress ISSCT. 1637-1644.
Srinivasan, G., Axtell, J.D. and Jambunathan, R. (1972). Purdue Univ. Research Progress Report (9). 71-74.
Stringfellow, A.C. and Peplinski, A.J. (1966). Cereal Sci. Today. 11:438-440.
Sullins, R.D. and Rooney, L.W. (1974). Cereal Chem. 51: 134-142.
Tanksley, T.D. (1975). Ninth Biennial Grain Sorghum Research and Utilization Conference, Lubbock, Texas.
Virupaksha, T.K. and Sastry, L.V.S. (1968). J. Agric. Food Chem. 15(2):199-203.
Vogel, Sally and Graham, M. (1978). IDRC. Report of a Workshop held in Nairobi, Kenya.
Walker, R.D. (1974). Ph.D. Dissertation, Texas A&M Univ., College Station, Texas.
Wall, J.S. and Blessin, C.W. (1969). Cereal Sci. Today. 14(8):264-271.
Wall, J.S. and Ross, W.M. (1970). "Sorghum Production and Utilization". Avi Publ. Co. Conn.

Waniska, R.D. (1976). M.Sc. Thesis. Texas A&M Univ., College Station, Texas.

Watson, S.A. (1967). "Starch: Chemistry and Technology" Vol. II. Acad. Press, N.Y.

Watson, S.A. (1970). "Sorghum Production and Utilization". Avi Publ. Co. Conn. 602.

Weinges, K., Bahr, W., Ebert, W., Goutz, K. and Marx, H.D. (1969). Fortschr. Chem. Org. Naturstoffe. 27:158.

Index

F

G

H

L

M

N

O

P

Q

R